SYNTHESIS OF FUSED HETEROCYCLES

This is the Forty-seventh Volume in the Series

THE CHEMISTRY OF HETEROCYCLIC COMPOUNDS

THE CHEMISTRY OF HETEROCYCLIC COMPOUNDS

A SERIES OF MONOGRAPHS

EDWARD C. TAYLOR

Editor

SYNTHESIS OF FUSED HETEROCYCLES

G. P. ELLIS

Department of Applied Chemistry
University of Wales Institute of Science and Technology
Cardiff, Wales, UK

A Wiley–Interscience Publication

JOHN WILEY & SONS

Chichester . New York . Brisbane . Toronto . Singapore

Library of Congress Cataloging-in-Publication Data:

Ellis, G. P. (Gwynn Pennant)
  Synthesis of fused heterocycles.

  (General heterocyclic chemistry series)
  'A Wiley–Interscience publication.'
  Includes index.
  1. Heterocyclic chemistry. 2. Chemistry,
Organic—Synthesis. 3. Heterocyclic compounds.
I. Title. II. Series.
QD400.E45   1987   547'.59   86-28944

ISBN   0 471 91431 2

British Library Cataloguing in Publication Data:

Ellis, G. P.
  Synthesis of fused heterocycles.—
  (General heterocyclic chemistry series,
  ISSN 0363-8626)
  1. Chemistry, Organic—Synthesis
  2. Heterocyclic compounds
  I. Title II. Series
  547'.59   QD400.3

ISBN   0 471 91431 2

Phototypesetting by Thomson Press (India) Ltd., New Delhi, and printed and bound in Great Britain at The Bath Press, Avon

Introduction to the Series

The series *The Chemistry of Heterocyclic Compounds*, published since 1950 under the initial editorship of Arnold Weissberger, and later, until Dr. Weissberger's death in 1984, under our joint editorship, was organized according to compound classes. Each volume dealt with syntheses, reactions, properties, structure, physical chemistry, and utility of compounds belonging to a specific ring system or class (e.g. pyridines, thiophenes, pyrimidines, three-membered ring systems). This series, which has attempted to make the extraordinarily complex and diverse field of heterocyclic chemistry as readily accessible and organized as possible, has become the basic reference collection for information on heterocyclic compounds.

However, many broader aspects of heterocyclic chemistry are now recognized as disciplines of general significance which impinge on almost all aspects of modern organic and medicinal chemistry. For this reason we initiated several years ago a parallel series entitled *General Heterocyclic Chemistry* which treated such topics as nuclear mangetic resonance of heterocyclic compounds, mass spectra of heterocyclic compounds, photochemistry of heterocyclic compounds, the utility of heterocyclic compounds in organic synthesis, and the synthesis of heterocyclic compounds by means of 1, 3-dipolar cycloaddition reactions. These volumes were intended to be of interest to all organic chemists, as well as to those whose particular concern is heterocyclic chemistry.

It has become increasingly clear that this rather arbitrary distinction between the two series creates more problems than it solves. We have therefore elected to discontinue the more recently initiated series *General Heterocyclic Chemistry*, and to publish all forthcoming volumes in the general area of heterocyclic chemistry in *The Chemistry of Heterocyclic Compounds* series.

Edward C. Taylor

Department of Chemistry
Princeton University
Princeton, New Jersey 08544

Contents

Preface

Over the last couple of decades, there has been a remarkable expansion in heterocyclic chemistry. This has been accompanied by the publication of new journals and books in the field. Abstracting journals efficiently alert the reader about new syntheses and reactions, but the retrieval of methods of forming a new heterocyclic ring fused to a carbocyclic or heterocyclic ring is difficult. In our own research programmes, we felt the need for such cyclizations to be classified according to the functional groups present in the precursor. We were unable to find any source of such information, and this book was developed from a collection of such references.

It is not possible in one volume to include examples of all types of ring closures, and I apologize to any reader who fails to find his favourite cyclization. Most of the two thousand or so references cited were published since about 1970 and each of these provides further references to earlier work. This enables a collection of relevant references to be obtained with minimum of time and effort. Few if any books are entirely free of errors, and I shall appreciate being informed of any that have escaped my notice.

Without the help of a large computer and a well-written program, I doubt whether it would have been possible to complete the work of writing this book. I am therefore greatly indebted to my colleague Dr. Peter M. May (now of Murdoch University, Australia) who wrote the complex program and patiently showed me how to use it. I am grateful also to Dr. Gillian L. Christie and Mrs. A. E. Bisgood who gave me valuable computing assistance towards the end of the task; Dr.Kevin M. Quinlan too gave valuable help with the program. I would like to thank Professor Hans Suschitzky for advice and encouragement and Professor Edward C. Taylor for a continuing dialogue and for advice. Dr. R. W. White of Chemical Abstracts Service kindly enlightened me on some of the finer points of nomenclature.

My interest in heterocyclization was stimulated over several years by my research colleagues. I am grateful to Dr. T. M. Romney-Alexander for his willing help over a few weeks during which he acted as honorary postdoctoral assistant! I thank the staff of John Wiley for their encouragement and assistance at every stage.

I owe many thanks to Mrs. P. M. Bevan, who patiently transformed the text of my manuscript into a typescript which was a delight to publisher and printer. Finally, I thank my wife for surviving many hours of isolation while the book was being written—we met only at meal times! As always, she provided invaluable and willing help as literary adviser and proof reader.

October 1986   G. P. Ellis

CHAPTER 1

Introduction

I. AIMS

This book is intended to provide a convenient way of locating papers (and reviews) on methods for the synthesis of a heterocyclic ring which is fused to another ring. Figure 1.1 shows diagrammatically the types of cyclizations covered. The α-ring may be either homocyclic or heterocyclic and be of any size; the newly formed β-ring is heterocyclic and contains five to eight atoms at least one of which is nitrogen, oxygen or sulphur (Z). Atoms A and B may or may not be identical and may be carbon or nitrogen (or rarely, sulphur in which case it would not have a substituent, X or Y, attached to it). X and Y are usually functional groups but one of them may be hydrogen. It is rare for both to be hydrogen. When A is a double-bonded or pyridine-type nitrogen, then X is not present or is an N-oxide function.

The book is divided into chapters according to the identities of X and Y, the arrangement being alphabetical as far as possible. For example, reactants in which X and Y are amine and carboxamide, respectively, are considered in Chapter 46 while those where X and Y are hydroxy and methylene are contained in Chapter 99. Where only a few examples of a particular X and Y combination were found in the literature, these are integrated with closely related groups, for

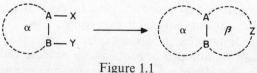

Figure 1.1

example, Chapter 20 contains ring closures of acylamine or amine and thiocyanate. It is sometimes scientifically logical also to discuss two related groups together. The contents list at the beginning should therefore be scanned for any particular combination of X and Y; the index may also be useful.

Another aim of the book is to show the IUPAC-approved way of drawing each of the fused heterocycles whose synthesis is described. From this, it follows that the overall peripheral numbering of each ring system can be deduced by applying IUPAC rules [B-1]. Although there is much to be said for drawing *all* ring systems according to IUPAC recommendations (as in this book), a few of the simpler ones (for example, quinoline) and others (for example, β-lactam antibiotics) are often drawn differently in the literature. Formulae of reactants are drawn so as to require the minimum of alteration during their conversion into the product and are therefore not necessarily drawn in accordance with IUPAC rules.

Thirdly, the approved name of the product ring system is given under the formula but this omits dihydro-, tetrahydro-, etc. prefixes and functional groups except endocyclic carbonyl or thiocarbonyl. The basis of this distinction is that a heterocyclic ring containing one or more endocyclic carbonyl groups usually requires a radically different method of synthesis from the non-carbonyl parent, and compounds of this kind are sometimes of considerable biological or industrial importance; the two types of ring systems are also separated in the index so that they may be more easily found. During the preparation of this book, it was noted that many ring systems were erroneously named in papers. These have been corrected and some notes on the application of IUPAC rules to polycyclic heterocycles are included in an appendix.

II. SCOPE

In recent years, heterocyclic chemistry has greatly expanded its boundaries and complexity. The formation of a new fused heterocyclic ring is an important type of reaction which is often difficult to search for in abstracting journals. The book contains a classified selection of papers, published since about 1970, which describe a wide variety of heterocyclizations. Each of these is likely to contain citations of earlier relevant papers and the reviews cited in the book should provide additional references. In this way, the reader may often locate suitable references more easily than by using abstracting journals. Although this book cannot cover all heterocyclic ring systems, the methods described can be applied to the synthesis of many more by varying the reactant and/or reagent.

Spiro and mesoionic products are not included, neither are reactions which have been proved to proceed by an ANRORC mechanism [551] or other courses in which a heterocyclic ring is opened and closed again into a different one; these types of reactions are already covered by monographs [B-2, B-3]. Exceptions to this rule are cyclic anhydrides, imides, epoxides or diketene. The newly formed fused ring is usually unsaturated or contains one or more endocyclic carbonyl groups but a few examples of saturated heterocyclic rings are included.

III. USE OF THE BOOK

The aims of this book have been outlined in Section I; examples of how these may be achieved and other explanatory comments are now given.

1. One- and Two-step Reactions

Let us assume that the nitro-ester (**1.1**) is to be converted into the novel tricyclic compound (**1.2**) which is assumed for illustrative purposes to be a novel ring system also. No useful references are to be found in Chapter 72 (carboxylic acid or its derivative and nitro) because the nitro group has to be reduced before this kind of cyclization occurs. Chapter 48 which includes cyclization of amino-esters is more promising and Section II.1 offers at least one example of the direct conversion of a nitro-ester into a pyridin-2-one ring and the conditions given [562] are likely to be applicable to the synthesis of triazine (**1.2**) also.

(**1.1**)   (**1.2**)

There are two points worthy of comment regarding this example. First, two-step conversions such as this where the intermediate is not isolated but its identity is known beyond doubt are classified according to the functional groups of that intermediate. It is then to be found amongst similar cyclizations in which the intermediate had been isolated. Ring closures where the structure of an intermediate may be in some doubt are classified according to the functional groups of the compound actually placed in the reaction vessel.

Secondly, potentially relevant information may be gleaned by examining other cyclizations to be found in Chapter 48. For example, a glance at Section I.1 shows that variations in the reductive cyclization of nitro-esters are possible; should the ring closure of (**1.1**) give an unexpected product, knowledge of the alternative course taken by an apparently similar reaction [345] may be useful. Minor changes in reaction conditions may affect yields and Sections II.4 and III.3 of the same chapter show potentially useful variations.

2. Naming of Functional Groups

For most reactants, identification of the functions which participate in the ring closure reaction is easy (as in the example in Section III.1) but several others presented difficulties. Brief comments on how the problems have been resolved may help the reader to appreciate the basis of classification. Several of the difficulties are collected in the (mythical?) compound (**1.3**) which has four functions, each of which has to have an unambiguous name which is to appear in the chapter heading.

(1.3)

The *primary amine group* may also be regarded as a disubstituted hydrazine but since its reactions resemble those of primary amines and can only occur at one of the nitrogen atoms, it is classified as an amine group. For example, the cyclization of (1.4) appears under 1, 2-diamine (Chapter 76).

(1.4)

Pyrazolo[1,5-*b*][1,2,4]−
triazine

[1874]

The *ethoxycarbonyl group*, being attached to a nitrogen, differs somewhat in reactivity from an ester and is classified as a carbamate.

The *carbonyl group* is part of a cyclic urea and is also capable of tautomerism; it cannot therefore be regarded as a typical ketone or amide. Similarly, the thiol can tautomerize to a thione and its properties are likely to differ from those of a

$$ \underset{\displaystyle \overset{\|}{S}}{C-C-C} \quad \text{or a} \quad \underset{\displaystyle \overset{|}{SH}}{C-C-C.} $$

Cyclic —NH—CO— and their sulphur analogues are therefore classified as lactam carbonyl and lactam thiocarbonyl respectively, whatever the size of the ring. Cyclization of lactam (1.5) is classified as between amine and lactam carbonyl (Chapter 11). The difficulties of naming cyclization products which contain an endocyclic lactam carbonyl or thiocarbonyl group are discussed in Section III.3 of this chapter.

(1.5)

xylene, TiCl$_4$, Δ
52–94%

Pyridazino[1,6-*a*]−
benzimidazole

[1959]

Formulae (1.6) and (1.7) show other functions which present problems of classification. The former has a group which may be regarded as either a ureide or a carboxamide and is classified as a carboxamide in this book. Since it is the ester group of (1.7) that reacts in the second cyclization, the conversion is classified under carbamate and ring-nitrogen (Chapter 61).

(1.6)

$HC(OEt)_3, \Delta$
62%

[58]

[1,2,4]Triazolo[1,5-a]-
[1,3,5]triazin-7-one

(1.7)

aq. NaOH
~94%

[1785]

Pyrazolo[1,5-a]-1,3,5-
triazin-4-one,2-thioxo-

3. Approved Names and Formulae

Recommendations have been made by the International Union of Pure and Applied Chemistry (IUPAC) [B-1] regarding the drawing and naming of fused heterocycles; these are followed carefully by *Chemical Abstracts* in their indexes but some authors neglect some or all of the recommendations, especially those relating to the correct alignment of the formula. This is regrettable because a formula which is properly drawn is more likely to have its correct peripheral numbering. The IUPAC report [B-1] and a review [2014] give the basic rules; the appendix to this book contains some additional hints.

The confusion that can arise when the formula of a relatively simple fused heterocycle is wrongly drawn may be illustrated with a tricyclic compound which may be drawn in several ways, some of which are shown in (1.8) to (1.12). None of these is in accord with IUPAC recommendations but (1.13) is the correct version. Peripheral numbering of this is shown and its name is pyrazino[2', 3':4, 5]thieno-[3, 2-d]pyrimidin-4(3H)-one.

(1.8)

(1.9)

(1.10)

(1.11)

(1.12)

(1.13)

Names of the hetrocyclic ring systems synthesized are given under the formulae and efforts have been made to ensure that, as far as possible, these comply with IUPAC recommendations. As mentioned in Section I of this chapter, functional groups do not appear in these names except to distinguish those which have an endocyclic carbonyl (or thiocarbonyl) group from those which do not. Some of these compounds are tautomeric but firm evidence for either keto or enol form is rarely available. For consistency, those tautomeric compounds which have ring-nitrogen atoms are shown as their oxo (or thioxo) form (as indexed in *Chemical Abstracts*) even though some authors have assumed (without any published supporting evidence) that they exist as the enols. Examples are compounds which are mono- or di-carbonyl (or their thiocarbonyl) derivatives of azoles, diazoles, diazines or triazines. For example, compound (**1.14**) is named as its diketo tautomer, (**1.15**), namely, pteridine-2, 4-dione (lumazine), the designation of the added hydrogens being omitted since, in general, the names refer to the ring system and not to individual compounds.

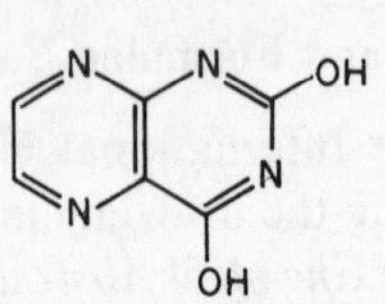

(1.14) 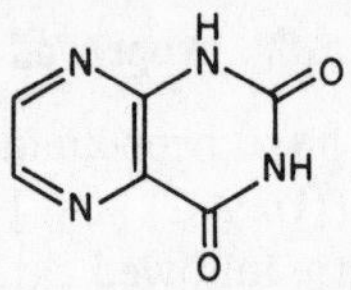   (1.15)

4. Format

Each equation which shows a cyclization is intended to provide as much information as space allows. Figure 1.2 is a typical example and shows the reactants on the left of the arrow. Sometimes, when the reagent is a small molecule or its name is easily abbreviated, it is placed above the arrow. Solvent (s), catalyst and promoter (acid, base or other compound) are shown above the arrow and when the reaction temperature is higher than that of the surroundings, a delta (Δ) is then added above the arrow. Its absence means that the reaction proceeds at or below ambient temperature. When the name of a solvent or reactant is abbreviated, reference should be made to Section IV of this chapter for the full name.

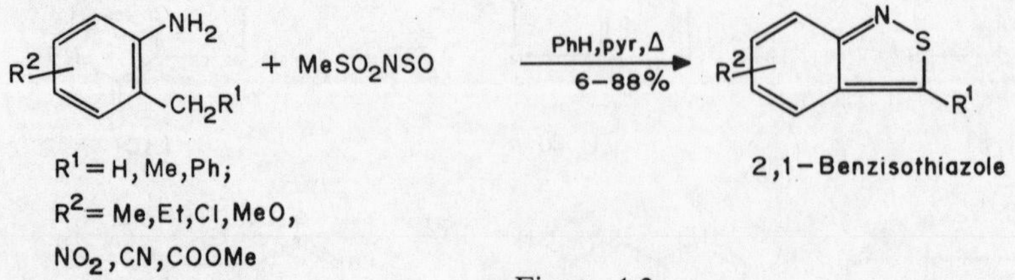

R^1 = H, Me, Ph;   2,1–Benzisothiazole
R^2 = Me, Et, Cl, MeO,
NO$_2$, CN, COOMe

[808]

Figure 1.2

The figures below the arrow indicate the yield quoted in the paper (s) cited; this consists of one value when the equation shows the synthesis of one particular compound, or an approximate value (for example, ≈ 65 per cent) when the yields for several related products fall within a range of ± 5 per cent of the value shown, or as in Figure 1.2, shows the lowest and highest values for variations of the substituents R^1 and R^2 shown under the reactants. As mentioned earlier, the IUPAC-approved name and formula of each product are shown on the right of the arrow and are followed by the reference number(s) (assigned by a computer on which the references are stored). A list of the references is to be found towards the end of the book. References to relevant reviews or monographs are cited either in the text or below the name of the product whichever is the more suitable. Books and monographs are listed separately just in front of the list of references.

With the compressed information supplied, it is hoped that the reader can judge the suitability of a particular reaction to his or her own work and then need only consult the most promising papers.

IV. ABBREVIATIONS

The following abbreviations are used in equations and occasionally in the text.

Ac	acetyl	DMFDMA	N,N-dimethyl-formamide dimethyl acetal
CDI	carbonyldi-imidazole		
DBO	1,4-diazabicyclo[2.2.2]-octane		
		HMPT	hexamethylphosphoric triamide
DBU	1,5-diazabicyclo[5.4.0]-undec-5-ene		
		LDA	lithium di-isopropylamide
DCC	dicyclohexylcarbodi-imide		
		LTA	lead tetra-acetate
DDQ	2,3-dichloro-5,6-dicyano-1,4-benzoquinone	MCPBA	3-chloroperbenzoic acid
		Mes	mesitylene sulphonyl
DEG	diethyleneglycol	mor	morpholine
diox	1,4-dioxan	NBS	N-bromosuccinimide
DMA	N,N-dimethyl-acetamide	NCS	N-chlorosuccinimide
		NIS	N-iodosuccinimide
DMAD	dimethyl acetylenedicarboxylate	PEG	polyethyleneglycol
		pip	piperidine
DMAP	N,N-dimethyl-4-pyridinamine	Phth	phthaloyl
		PPA	polyphosphoric acid
DME	dimethoxyethane	PPE	polyphosphoric acid ethyl ester
DMF	N,N-dimethyl-formamide		
		pyr	pyridine
DMFDEA	N,N-dimethyl-formamide diethyl acetal	RaNi	Raney nickel
		tet	5-tetrazolyl
		Ts	4-tolylsulphonyl

Acetal or Aldehyde and Amine

Cyclization of *o*-aminobenzaldehydes and similar heterocyclic compounds was reviewed in 1980 [1018]. Acetal-amines are included in this chapter because they are often converted into the aldehyde-amines under the reaction conditions, for example, in mineral acids. Cyclization of one or two examples of the oxime or semicarbazone of an amino-aldehyde are included in this chapter.

I.   FORMATION OF A FIVE-MEMBERED RING

1.   Pyrrole

o-Aminophenylacetaldehyde acetals and similar compounds cyclize to indoles on warming with mineral acid. The products are usually 2,3-unsubstituted indoles. When the semicarbazones of such aldehydes are hydrogenolysed, ring closure occurs.

$R = H, Me$ EtOH, HCl, Δ ~69%   X = CH, N

Indole

Pyrrolo[2,3-*c*]pyridine

[1510]

Another ring system synthesized similarly:

Pyrrolo[2,3-d]pyrimidin-4-one

[1739]

Indole

[747]

Reaction of an acetal-amine with an α-bromoketone alkylates the amine which then cyclizes on deacetalation with hot mineral acid.

[20]

2. Imidazole

In a variation on the cyclization of arylacetaldehyde acetals (Section I. 1), this side-chain is attached to an endocyclic nitrogen atom and an imidazole ring is formed.

R^1=H,Me,MeS

Imidazo[1,2-b]pyrazole

[1172, 1991]

3. 1, 2, 3-Triazole

A modification of the aldehyde-amine groups enables this ring to be formed. The oxime of the aldehyde adjacent to an N-amino group is converted by hot PPA to the triazole in good yield but only when the E-form of the oxime is present.

[1329]

[1,2,3]Triazolo[1,5-a]pyridine

(review [1694])

4. Isoxazole

When an *o*-nitrobenzaldehyde is reduced with tin-acetic acid, cyclization to the 2,1-benzisoxazole occurs; reviews of the chemistry of these heterocycles have been published [1674, 1900].

[546]

2,1-Benzisoxazole

II. FORMATION OF A SIX-MEMBERED RING

1. Pyridine

The conversion of an *o*-aminoaldehyde into a pyridine ring by reaction with a ketone is a well-known cyclization called the Friedländer synthesis (review [1093]). The ketone has a —CH_2CO— group which may be part of an open chain or a ring (as in the case of the tautomeric 4-hydroxycoumarin or 4-oxobutyrolactone) and cyclizes under basic conditions, for example, in the presence of piperidine, alkoxide or aqueous alkali. In the Borsch modification of the Friedländer synthesis, the imine (Schiff's base) derived from the aldehyde reacts more readily with the ketone [31].

R^1=alkyl,Ph,heteroaryl;
R^2= H,Me,Ph,COOH

X=CH,N   [36, 1198]

Quinoline

1,7-Naphthyridine

(review [1648])

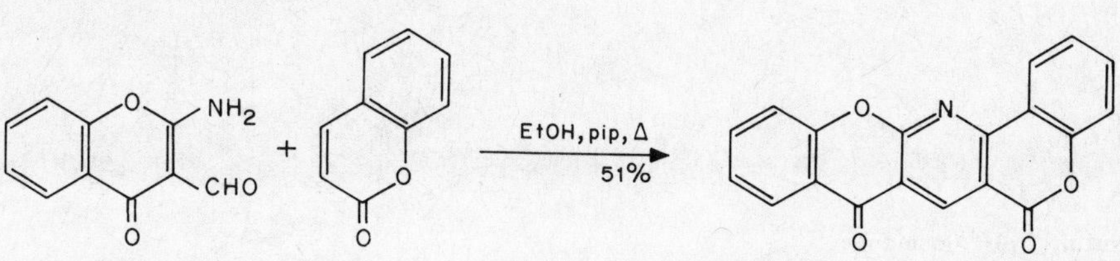

Bis[1]benzopyrano[2,3-b:3',4'-e]-
pyridine-6,8-dione

[1040]

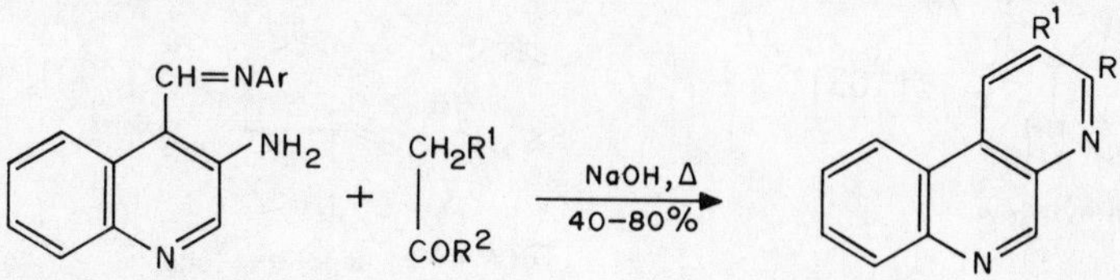

$R^1 = H, Ph; R^2 = Ph, 4-BrC_6H_4$ or
$R^1R^2 = (CH_2)_4; Ar = 4-MeC_6H_4$

Pyrido[2,3-c]quinoline

[31]

Other ring systems synthesized similarly:

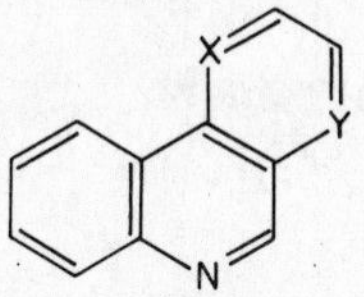

[92]

X = CH, Y = N
Benzo[f][1,7]naphthyridine
X = N, Y = CH
Benzo[h][1,6]naphthyridine

[113]

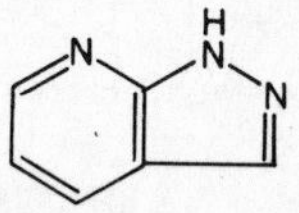

Pyrazolo[3,4-b]pyridine

[453]

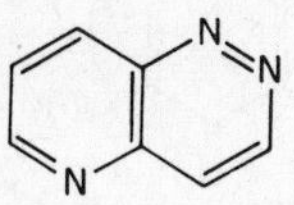

Pyrido[3,2-c]pyridazine

[827]

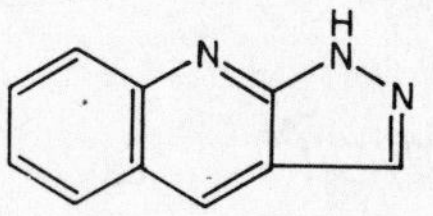

Pyrazolo[3,4-b]quinoline

[1629]

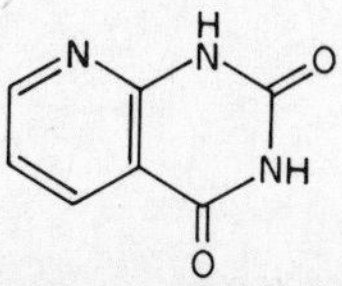

Pyrido[2,3-d]pyrimidine-
2,4-dione

[1759]

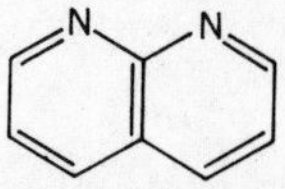

1,8-Naphthyridine
(review [1648])

[35]

Isoxazolo[5,4-b]quinoline

[40,91]

X=CH,N

[1]Benzopyrano[4,3-b]quinolin-6-one
[1]Benzopyrano[4,3-b][1,8]naphthyridin-6-one

[103]

Quino[3,2-c][1,8]naphthyridine

[490]

X=CH,N

Furo[3,4-b]quinolin-1-one
Furo[3,4-b][1,8]naphthyridin-1-one

Benzyl cyanides and malonic acid derivatives, under basic conditions, condense with aminoaldehydes to form a new pyridine ring but weakly acidic or neutral conditions suffice with the more reactive malonic derivatives.

$$\text{+ ArCH}_2\text{CN} \xrightarrow[\text{19-81\%}]{\text{EtONa,EtOH,}\Delta}$$

[437]

R=H,alkyl; Ar=aryl,heteroaryl

Pyrido[2,3-d]pyrimidine

R=H,alkyl,Cl,MeO

[322]

[1]Benzopyrano[2,3-b]pyridin-5-one

Other ring systems synthesized similarly:

[37, 38, 1903]

X=CH,N

Quinoline
1,8-Naphthyridine

Alkynoic esters or nitriles annulate aminoaldehydes on heating, sometimes with a tertiary amine.

R^1=H,alkyl,Cl,MeO;
R^2=COOEt,CN

[1]Benzopyrano[2,3-b]pyridin-5-one

[322, 1384]

An appropriately positioned aldehyde (or protected aldehyde) group in a side-chain reacts with the o-amino group to give a pyridine ring (cf. the formation of a pyrrole ring, Section I.1).

Pyrido[2,3-d]pyrimidine

[746]

2. Pyridin-2- or -4-one

o-Aminobenzaldehyde is cyclized by reaction with ethyl cyanoacetate and a base to give a new pyridine ring (see Section II.1) but a 5-aminopyrazole-4-carbaldehyde in boiling acetic acid gives a pyridin-2-one ring in high yield.

Pyrazolo[3,4-b]pyridin-6-one

[113]

A synthesis of the interesting 3-hydroxyquinolin-4-ones is by condensation of o-aminobenzaldehydes with glyoxal bisulphite in the presence of potassium cyanide.

R=H,Me

4-Quinolinone

[19]

3. Pyrimidine and Pyrimidin-2-one

Formamide or formamidine provides the required C-N fragment to form a pyrimidine ring. A doubly fused pyrimidine is formed when the amino and acetal groups are attached to different rings.

Pyrimido[4,5-*d*]pyrimidine   [57]

Pyrazolo[3,4-*d*]pyrimidine   [113]

R = Ph,COOEt

Thieno[2,3-*e*][1,2,3]triazolo-
[1,5-*a*]pyrimidine   [676]

Reaction of an aminoaldehyde with an isocyanate gives moderate yields of a fused pyrimidin-2-one.

R = Me,Ph

[1]Benzopyrano[2,3-*d*]pyrimidine-
2,5-dione   [376]

4. Pyrazine

Suitably placed amino and aldehyde groups react to form a pyrazine ring; the former may sometimes be prepared *in situ* by a Curtius rearrangement of an acyl azide.

$$H_2O, \Delta \quad 74\%$$

[452]

Pyrrolo[1,2−a]thieno[3,2−e]−
pyrazine

5.   1,4-Oxazine

This ring is formed by heating an o-acylamino-formylmethoxybenzene in an acidic medium.

$$PhH, TsOH, \Delta \quad 66\%$$

[1795]

1,4−Benzoxazine

III.   FORMATION OF A SEVEN-MEMBERED RING

1.   1,4-Diazepin-2- or -5-one

An unusual reduction-cyclization occurs to a nitroaldehyde with the formation of a saturated —NH—CH$_2$— bond which is part of a diazepine ring [1218], presumably by further reduction of a —N = C bond but when a less active catalyst (Pd-BaSO$_4$) is employed in a similar reaction, the imine linkage is retained [1934, 1936].

$$H_2, Pd-C \quad 67\%$$

[1218]

Pyrazolo[5,1−c][1,4]benzodiazepine

The following ring systems were synthesized similarly:

[1934]

[1936]

Pyrido[2,1−c][1,4]benzodiazepin-12-one

Pyrrolo[2,1−c][1,4]benzodiazepin-5-one

2. 1,4-Oxazepine

This ring is formed by a similar reduction of a nitro group and spontaneous condensation of the amino and aldehyde groups to give the expected —N = CH—bond which is a part of the new oxazepine ring.

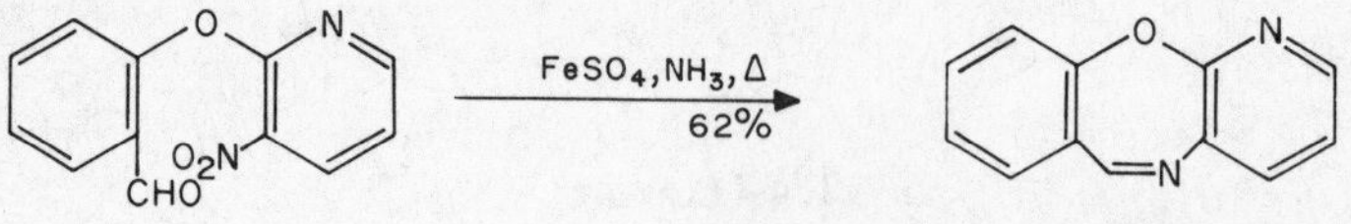

Pyrido[2,3–b][1,4]benzoxazepine

Other ring systems synthesized similarly:

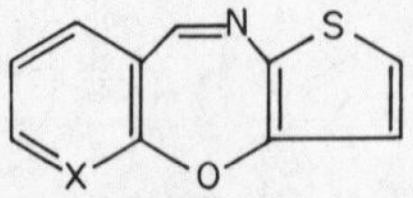

[1227]

X=CH,N
Thieno[3,2–b][1,4]benzoxazepine
Pyrido[3,2–f]thieno[3,2–b][1,4]-
oxazepine

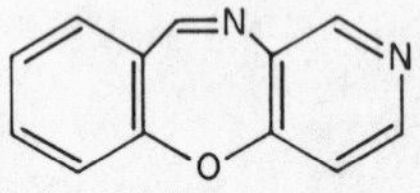

[1037]

Pyrido[4,3–b][1,4]benzoxazepine

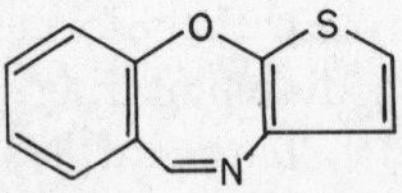

[1227]

Thieno[2,3–b][1,4]benzoxazepine

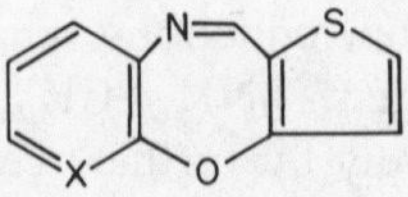

[1227]

X=CH,N
Thieno[3,2–b][1,5]benzoxazepine
Pyrido[2,3–b]thieno[2,3–f][1,4]-
oxazepine

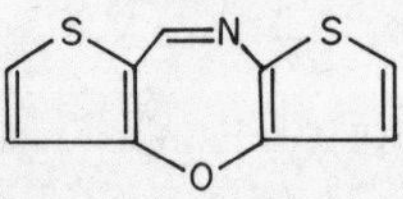

[1227]

Dithieno[3,2–b:2′,3′–f][1,4]oxazepine

CHAPTER 3

Acetal and Ring-carbon or Ring-nitrogen

In this chapter, cyclizations of the following two types are listed, X containing one or more carbon and/or nitrogen, oxygen or sulphur atoms and R is an alkyl group or $(OR)_2$ may be an ethylenedioxy ring.

Although aldehydes are usually more readily available than their acetals, halogeno- and amino-acetaldehyde dialkylacetals are often used as reagents and give acetals as products. The latter can be directly cyclized to a heterocyclic ring.

I. FORMATION OF A FIVE-MEMBERED RING

1. Pyrrole

Under Friedel–Crafts conditions, acetal groups in a side-chain attack an endocyclic carbon or nitrogen with the formation of a pyrrole ring. The indole sulphonamide is formed only when the *ortho* position is unsubstituted [892].

Pyrrolo[3,2,1-*kl*]phenothiazine   [759]

R = H,4-Me,4-halogeno,4-MeO   Indole   [892]

R = H,CH₂COOEt   Pyrrolo[2,1-*b*][1,3]oxazine   [1712]

2. Imidazole

When the side-chain (see preceding section) contains a nitrogen atom, an imidazole ring is formed under Friedel–Crafts conditions. This reaction appears to be regioselective towards nitrogen as in the following example.

[1374]

Imidazo[2,1-*c*]pyrido[2,3-*e*][1,4]–
oxazine

II. FORMATION OF A SIX-MEMBERED RING

1. Pyridine

In a modified Pomeranz–Fritsch cyclization under very mild conditions, an acetal in acid solution reacts with an adjacent position of the benzene ring to give a reduced pyridine ring in good yield but when a similar reaction is done in boiling dioxan with chlorosulphonic acid, or in orthophosphoric acid, a high yield of the annulated pyridine or dihydropyridine is usually obtained.

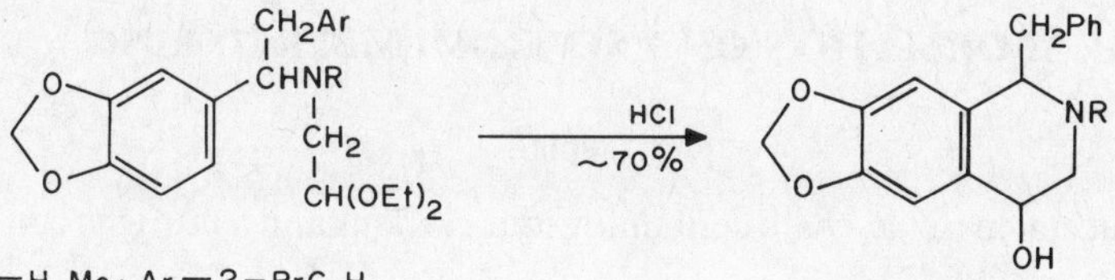

[311]

R=H,Me; Ar=2-BrC$_6$H$_4$

Isoquinoline

1,3-Dioxolo[4,5-*g*]isoquinoline

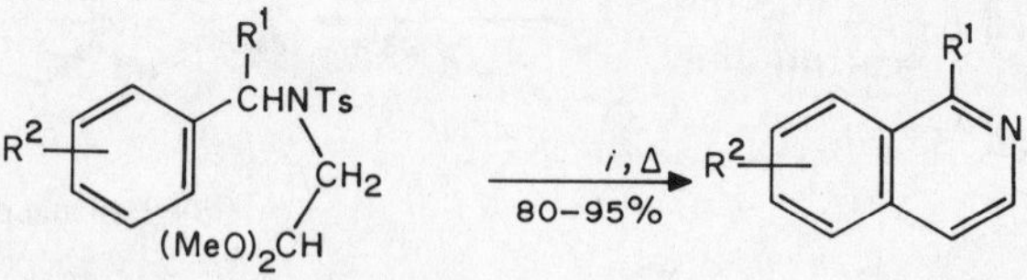

[263, 588, 760]

R^1=H, alkyl; R^2=H, alkoxy.

i, HCl–diox. or ClSO$_3$H

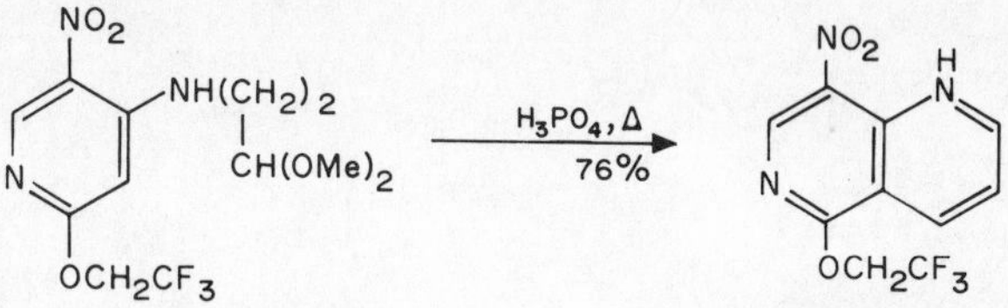

[272]

1,6–Naphthyridine

(review [1648])

2. Pyrazine

An acetal group attacks an endocyclic imine under the influence of phosphorus oxychloride and PPA to form a six-membered ring—a pyrazine in this example.

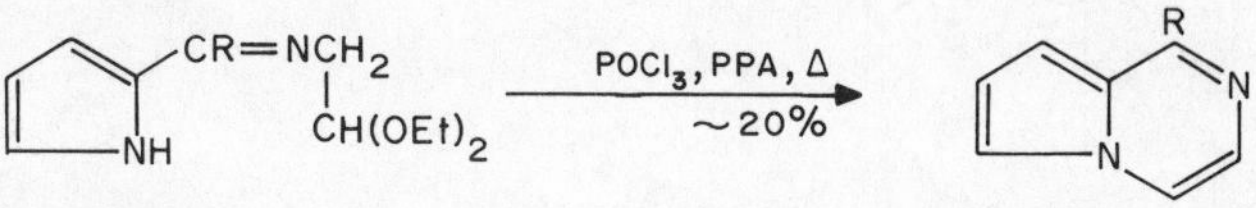

[1740]

R=H, Me

Pyrrolo[1,2–*a*]pyrazine

(review [1659])

3. Pyran

2*H*-Chromenes (2*H*-benzopyrans) may be prepared by the regioselective cyclization of an acetal-ether under mildly acidic conditions.

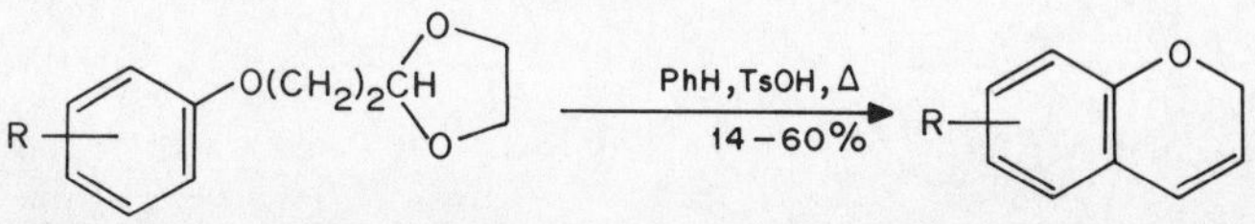

[1210]

R=H, Me, PhCH$_2$O

2*H*-1-Benzopyran

III.   FORMATION OF A SEVEN-MEMBERED RING

1.   Azepine

A terminal acetal in a side-chain containing four carbons and a nitrogen can form a fused azepine ring.

3–Benzazepine

CHAPTER 4

Acylamine and Aldehyde or Ketone

I. FORMATION OF A FIVE-MEMBERED RING

1. Pyrrole

The best of several methods of cyclizing N-acyl 2-benzoylanilines uses the dichloroacetanilide as precursor; potassium cyanide is believed to form a cyanohydrin at the benzoyl carbonyl before ring closure under basic conditions.

3H−indole   [2]

2. Oxazole

Under the influence of PPA and phosphorus oxychloride, an oxazole ring is readily formed from an acylamine-ketone.

[1]Benzopyrano[3,4-d]oxazol-4-one   [12]

II. FORMATION OF A SIX-MEMBERED RING

1. Pyridin-2-one

Hot ethanolic alkali converts a 2-acylaminoketone into a fused pyridin-2-one ring.

[886]

Thieno[3,4-*b*]pyridin-2-one

2. Pyrimidine or Pyrimidin-2-one

An aldehyde or ketone containing an adjacent acylamino group reacts with ammonia to form a pyrimidine ring but a 2-ethoxalylamino-aldehyde gives a fused pyrimidin-2-one ring [9].

[1372]

Quinazoline
(reviews [874, 1671])

Other ring systems synthesized similarly:

[9]

Pteridine

[9]

2-Pteridinone

3. 1,4-Oxazine

When the alkyl ketone is joined to the ring through an ether oxygen, a 1,4-oxazine ring is formed, the position of the double bond depending on whether the *N*-acyl group is hydrolysed or not. The stability of 1,4-benzoxazines vary but a 3-t-butyl group stabilizes this molecule.

[1211, 1795]

R = tBu, Ph

1,4-Benzoxazine

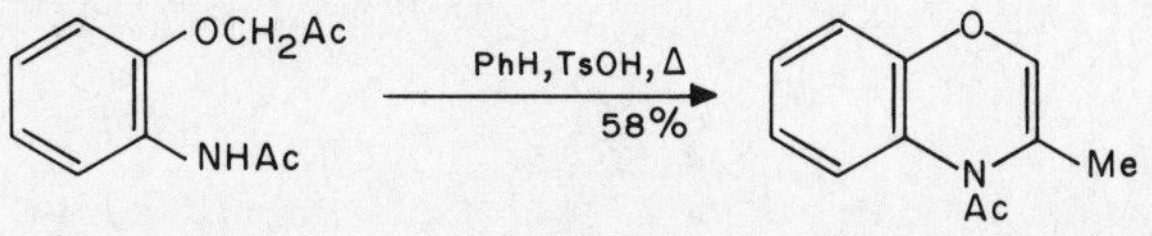

[1795]

III. FORMATION OF A SEVEN-MEMBERED RING

1. 1,4-Diazepine or 1,4-Diazepin-2-one

When the side-chain contains a suitable number of atoms, the diazepine is formed in good yield.

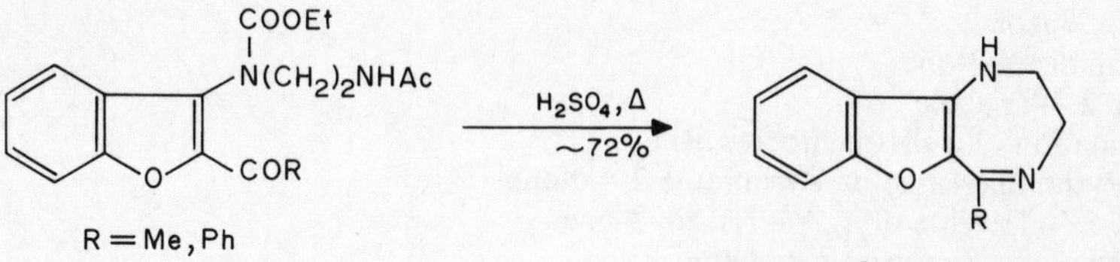

[856]

Benzofuro[3,2−e]−1,4−diazepine

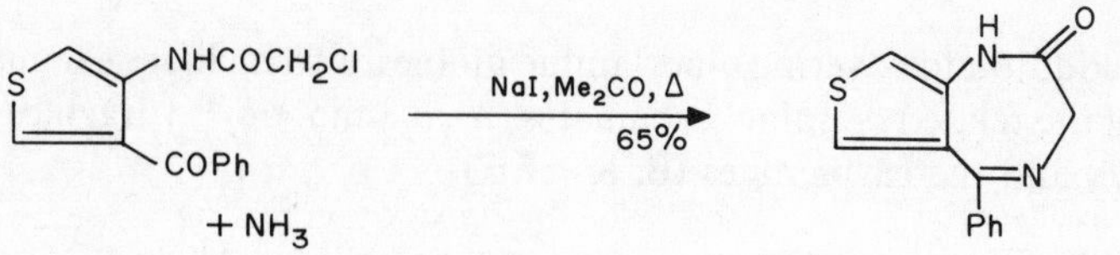

[885]

Thieno[3,4−e]−1,4−diazepin−2−one

CHAPTER 5

Acylamine and Amine

This chapter, in addition to covering o-acylamino amines, also includes examples of cyclizations of the related o-aminocarbamates (A), o-aminoacylhydrazines (B, R = alkyl or aryl) and the carbamates (B, R = EtO).

This chapter, in addition to covering *o*-acylamino amines...

(A)   (B)

I. FORMATION OF A FIVE-MEMBERED RING

1. Imidazole

Cyclization of a 2-acylaminoaniline into a benzimidazole is often a convenient method because it avoids the need to prepare pure, but readily oxidized diamine. Heating the stable 2-acylaminoanilines with mineral acid is the most frequently used method of cyclization (for example, for the synthesis of 5-ethoxy-2-methylbenzimidazole [1896]) but phosphorus oxychloride in hot diethylaniline or an arylamine with TEA-P_2O_5 also give good yields. Prolonged reaction with hot water is sometimes preferable where acidic conditions are to be avoided.

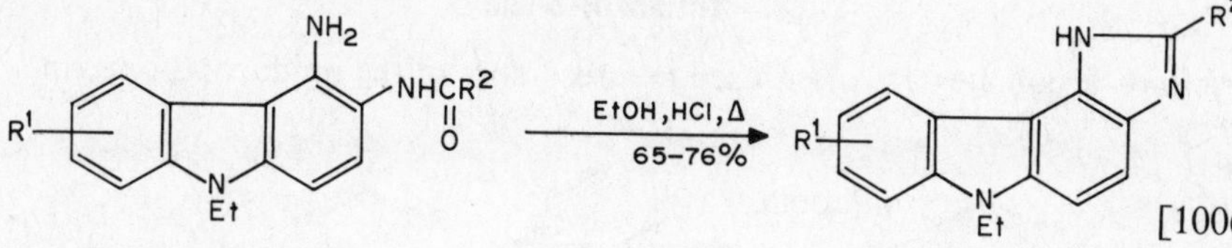

$R^1 = H, NH_2; R^2 = H, Me$

EtOH, HCl, Δ
65–76%

Imidazo[4,5-c]carbazole [1006]

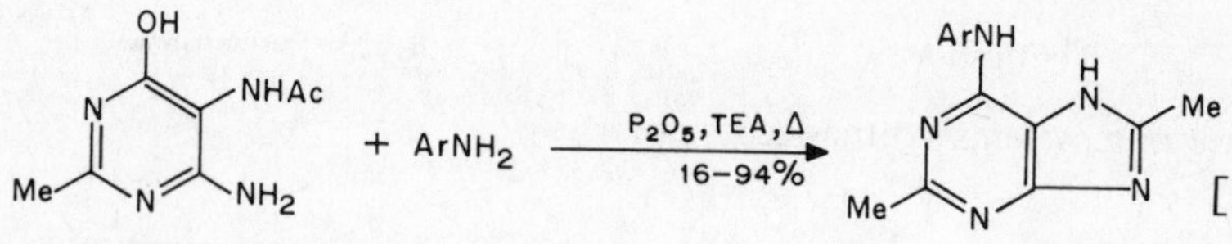

+ $ArNH_2$

Ar = Me–, Cl–, F–C_6H_4

P_2O_5, TEA, Δ
16–94%

Purine [1827]

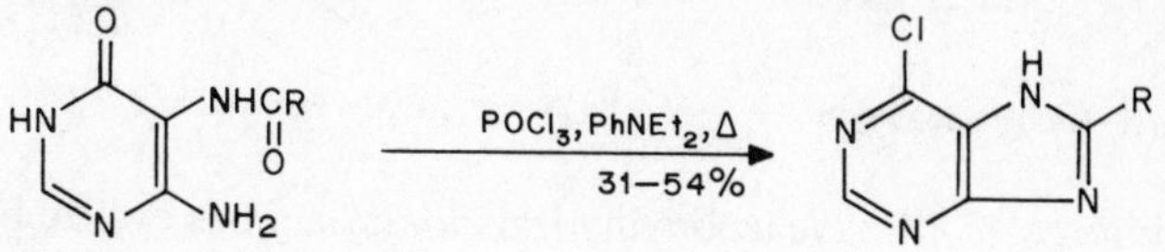

R = Me, Et, CH_2OH

$POCl_3$, $PhNEt_2$, Δ
31–54%

[1009]

H_2O, Δ
47%

[1199]

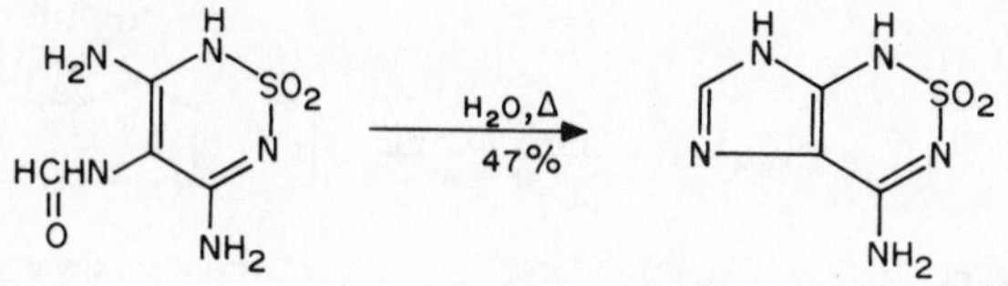

Imidazo[4,5-c][1,2,6]thiadiazine
2,2–dioxide

Two reactions effected under mild conditions and in which the acyl group is retained on the nitrogen are of interest but a mixture of positional isomers is obtained when R ≠ H and varying amounts of the N-1 deacylated thiol are also formed.

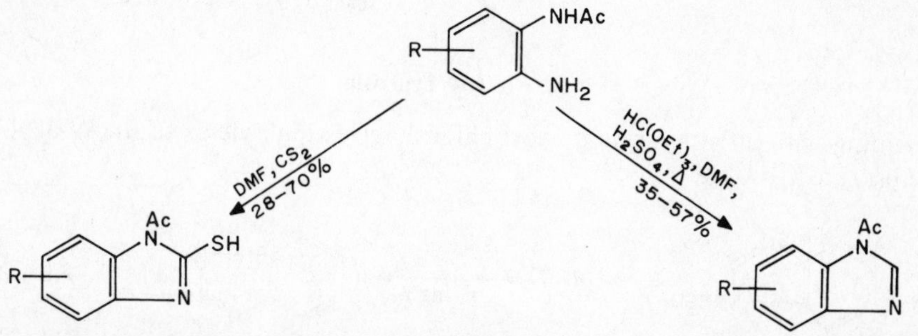

DMF, CS_2
28–70%

HC(OEt)$_3$, DMF,
H$_2$SO$_4$, Δ
35–57%

R = Me, MeO, Cl, AcNH, COOMe

Benzimidazole
[312]

2. Imidazol-2-one

o-Amino-carbamates are often used as precursors of the imidazol-2-one ring.

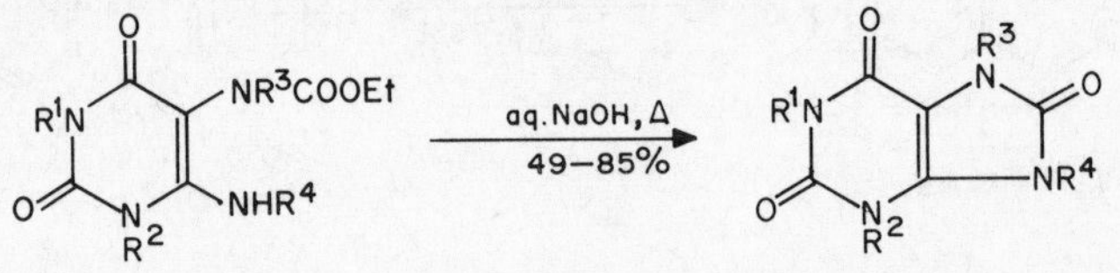

$$R^1-R^4=H,Me \qquad\qquad 2,6,8-Purinetrione$$

[1702]

Other ring systems synthesized similarly:

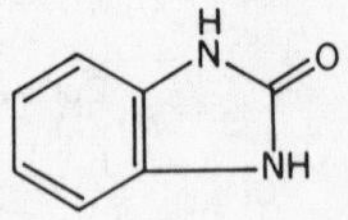

2-Benzimidazolone [1810]

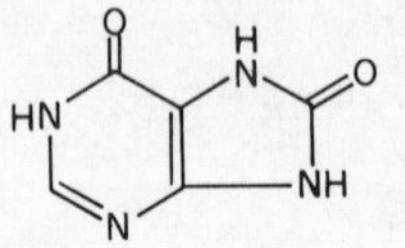

6,8-Purinedione [1702]

o-Acylhydrazino- or *o*-ethoxycarbonylhydrazino-arylamines cyclize by treat-ment with ethyl chloroformate-pyridine or (for the carbamates) methanolic hydrogen chloride.

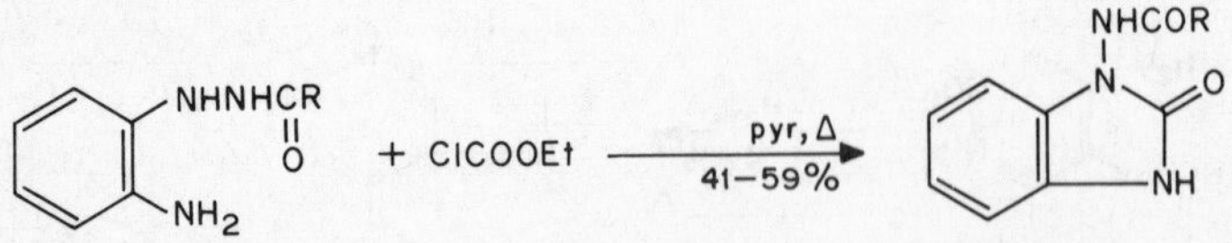

R=Me,Et,Ph,EtO   2-Benzimidazolone [1796]

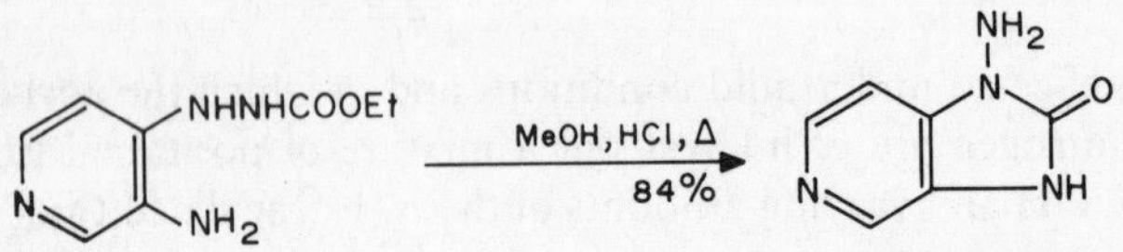

Imidazo[4,5-*c*]pyridin-2-one [1205]

3. 1, 2, 4-Triazole

Heating with an anhydride or acyl chloride gives high yields of the *N*-acyl-2-substituted triazole.

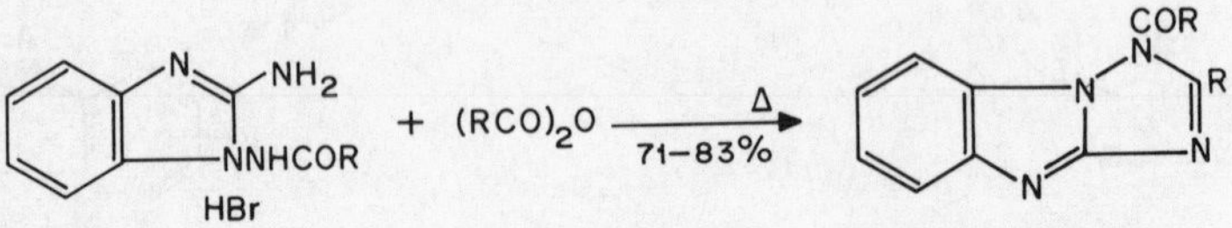

R=Me,Et,Ph

[1,2,4]Triazolo[1,5-*a*]-
benzimidazole [1554]

II. FORMATION OF A SIX-MEMBERED RING

1. Pyrimidin-4-one or Pyrimidine-2, 4-dione

An amine and an adjacent carbamate (or an ethoxycarbonylhydrazinyl) group cyclize thermally or in hot quinoline to give pyrimidine-2, 4-dione. When the precursor is an acylhydrazide, acid-induced cyclization yields a pyrimidin-4-one.

[956]

$R^1, R^2 = H, Me$

Pyrimido[4,5−d]pyrimidine−
2,4,5,7−tetraone

[994]

2,4−Quinazolinedione

[1357]

$R = Ph, 4-NO_2C_6H_4$

4−Quinazolinone

An alternative method of cyclizing 2-amino-N^2-acylhydrazides is to treat them with an orthoester in hot DMF.

[271]

Pyrazolo[3,4−d]pyrimidin−4−one

2. 1, 2, 4-Triazine or 1, 2, 4-Triazin-3-one

Cyclization of a 2-amino-acylhydrazine occurs spontaneously when the 2-nitro-precursor is reduced either catalytically or chemically; a triazine is formed except when an ethoxycarbonylhydrazine is present. This leads to the formation of a triazinone ring.

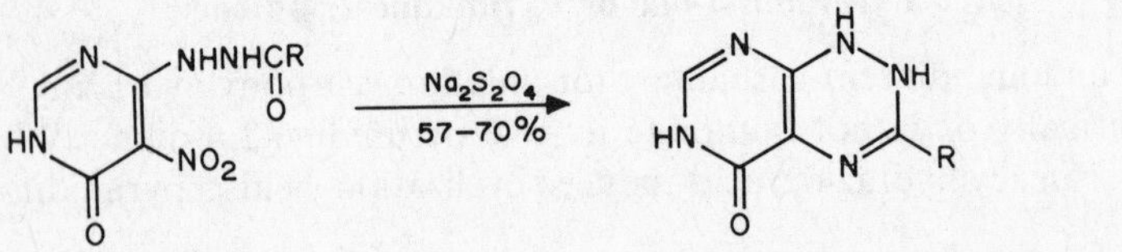

R = Me, PhCH₂, HOCH₂

Pyrimido[5,4-e]-1,2,4-triazin-5-one

[844]

Other ring systems synthesized by one of the methods mentioned:

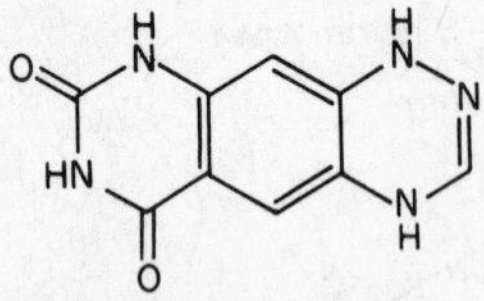

[1215]

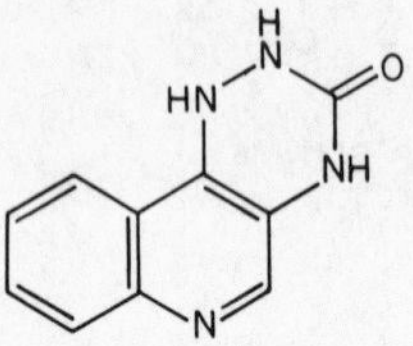

Pyrimido[5,4-g]-1,2,4-
benzotriazine-6,8-dione

[656]

1,2,4-Triazino[5,6-c]quinolin-3-one

III. FORMATION OF A SEVEN-MEMBERED RING

1. 1, 3-Diazepine

When the amino and acylamino groups are separated by four carbon atoms, treatment of the compound with phosphorus pentachloride in boiling chloroform or with thionyl chloride-pyridine gives a moderate yield of the benzodiazepine.

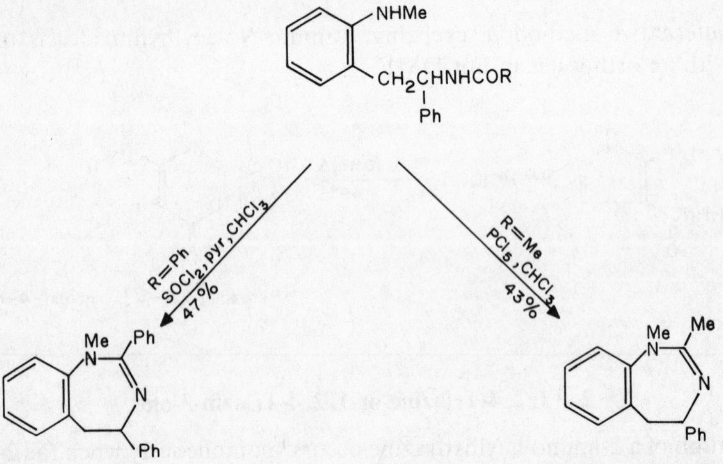

[8]

1,3-Benzodiazepine

CHAPTER 6

Acylamine or Carbamate and Carboxamide or Nitrile

I. FORMATION OF A SIX-MEMBERED RING

1. Pyrimidin-4-one or Pyrimidine 1-Oxide

A 2-acylamino-carboxamide is converted into a pyrimidin-4-one ring by heating under nitrogen [50, 1074, 1328] or by treatment with a base (alkali or ammonia) [11, 216, 1002] or with hydrogen chloride-ethanol [1206].

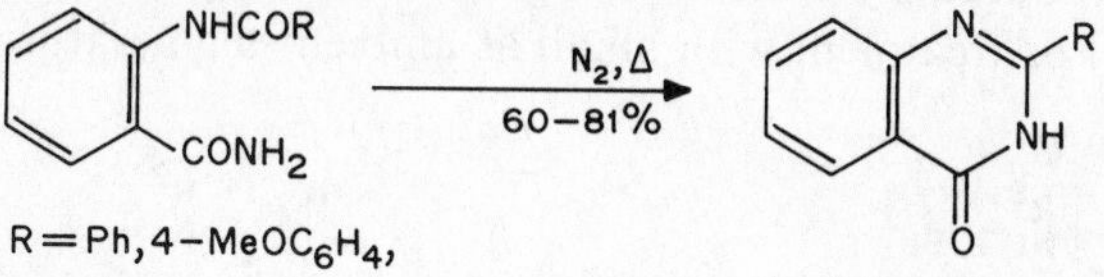

[50, 1328]

R=Ph, 4-MeOC$_6$H$_4$, PhCH=CH

4-Quinazolinone

Other ring systems synthesized similarly:

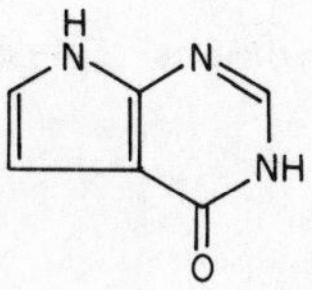

[1483]

Pyrrolo[2,3-*d*]pyrimidin-4-one

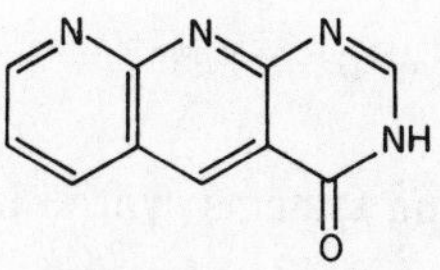

[11]

Pyrimido[4,5-*b*][1,8]-naphthyridin-4-one

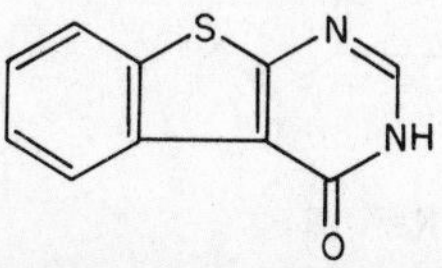

[1002, 1206]

[1]Benzothieno[2,3-*d*]-pyrimidin-4-one

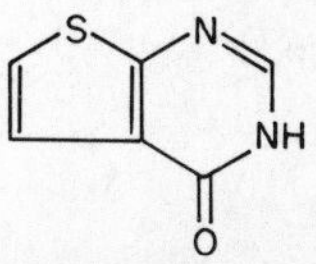

[1074]

Thieno[2,3-*d*]pyrimidin-4-one

Triethyloxonium fluoroborate in boiling methylene dichloride converts 2-acylamino-carboxamides into pyrimidin-4-ones.

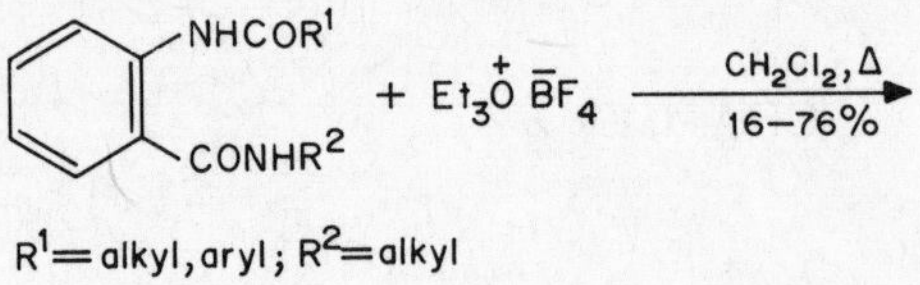

4−Quinazolinone

[1412]

When the carboxamide group is replaced by its more reactive oxime (prepared from the nitrile and hydroxylamine), a fused pyrimidine *N*-oxide is formed in high yield.

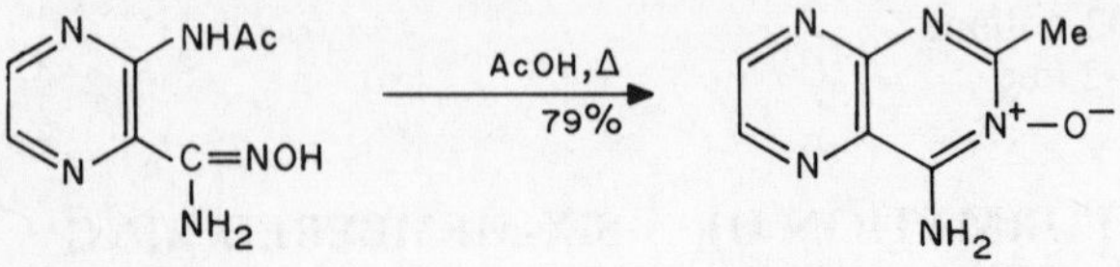

Pteridine 3−oxide

[244]

2. Pyrimidine-2, 4-dione

This ring is produced by either heating a 2-carbamoyl-carbamate in a high-boiling alcohol or treatment with alkali at ambient temperature [784, 1859].

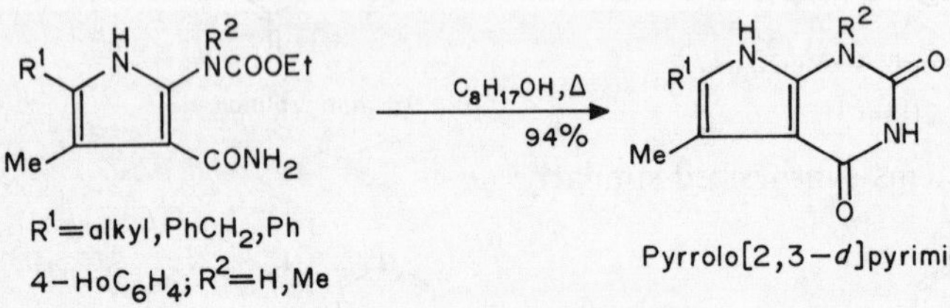

Pyrrolo[2,3−*d*]pyrimidine−2,4−dione

[54]

Other ring systems synthesized using alkali:

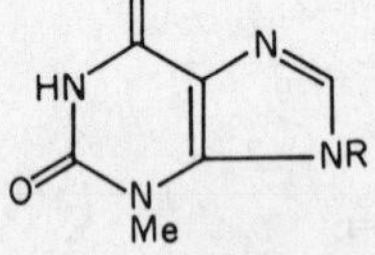

2,6−Purinedione

[1859]

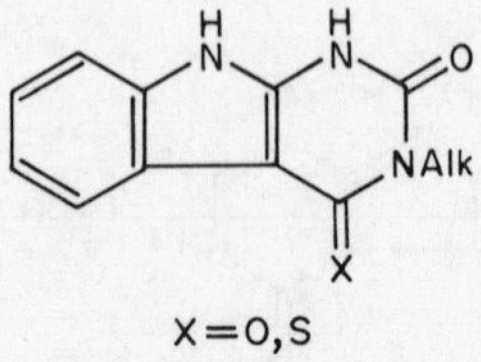

X=O,S

Pyrimido[4,5−*b*]indole−2,4−dione
Pyrimido[4,5−*b*]indol−2−one,4−thioxo−

[784]

3. 1,3-Oxazin-2-one

A reversed carbamate (oxygen is attached to the principal ring) which has a neighbouring nitrile group is cyclized by heating with an aryl isocyanate and a tertiary amine.

1,3-Benzoxazin-2-one

CHAPTER 7

Acylamine and Carboxylic Acid or Ester

A few reactants containing the thioacylamine (-NHCSR), carbamate (-NHCOOEt) or thiocarbamate (-NHCSOEt) group are included.

I. FORMATION OF A FIVE-MEMBERED RING

1. Pyrrole

Dimethyl acetylenedicarboxylate (DMAD) reacts with an *o*-acylamino carboxylic acid in acetic anhydride with the formation of a pyrrole ring. The reactions of DMAD with *N*-heterocycles have been reviewed [1725].

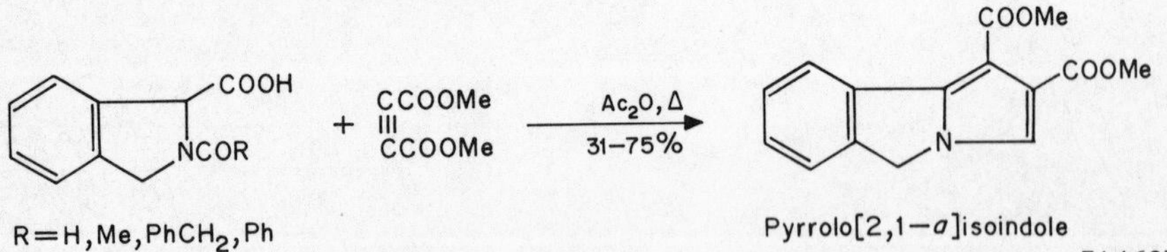

R=H,Me,PhCH$_2$,Ph

Pyrrolo[2,1−*a*]isoindole

[1463]

II. FORMATION OF A SIX-MEMBERED RING

1. Pyridin-2-one

When the ester group is separated from the ring by two carbon atoms, cyclization is induced by hot mineral acid; otherwise, lithium borohydride-THF is effective.

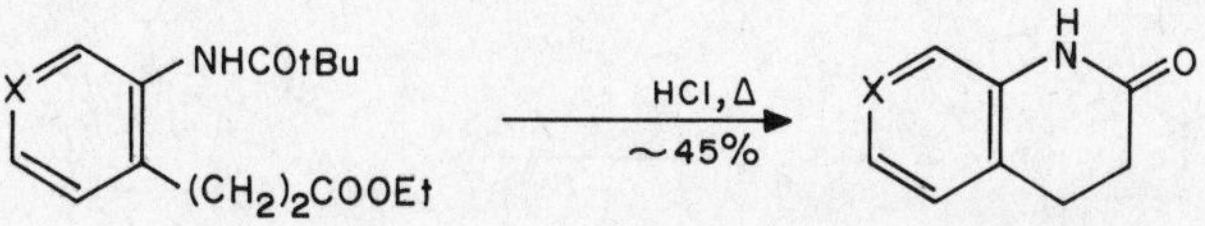

X=CH,N

2−Quinolinone

1,7−Naphthyridin−2−one

[313]

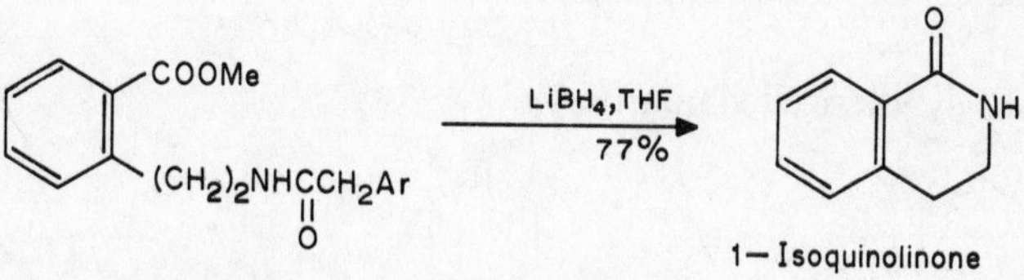

Ar = 4−MeOC$_6$H$_4$

1− Isoquinolinone

[339]

2. Pyrimidin-4-one

A wide variety of reagents and conditions have produced this fused ring from acylaminocarboxylic acids or their esters [874]. The reagent has to supply the 3-nitrogen atom which may or may not be substituted in the product. For the latter, ethyl carbamate or formamide is chosen but ammonium chloride may be used provided the cyclization is effected in the presence of hot phosphorus pentoxide and a t-amine.

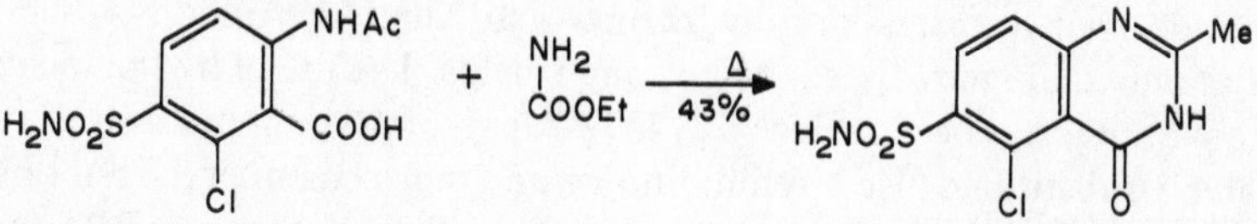

4 − Quinolinone

[925]

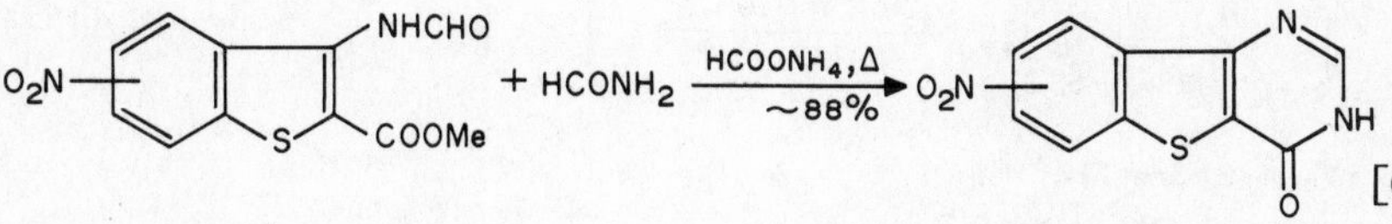

[1]Benzothieno[3,2−d]pyrimidin−4−one

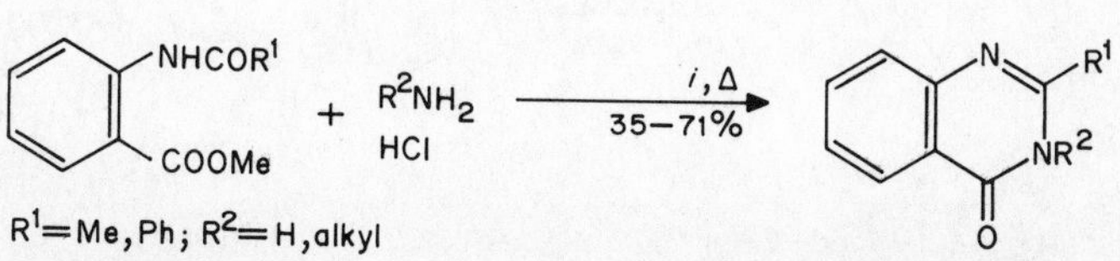

[675]

R^1=Me, Ph; R^2=H, alkyl

i, C$_6$H$_{11}$NMe$_2$− P$_2$O$_5$

4 − Quinazolinone

[1628]

1,2,3−Triazolo[4,5−d]−
pyrimidin−4−one
(review[2023])

[1625]

Other ring systems synthesized similarly:

[1]Benzothieno[2,3−d]−
pyrimidin−4−one

[69, 375]

Pyrido[2,3−d]pyrimidin−4−one

[1627]

Pyrrolo[2,3−d]pyrimidin−4−one

[778]

2-Acylaminocarboxylic esters and hydrazine yield 3-aminopyrimidin-4-ones; the thioacylamino analogues appear to give high yields. The use of hydrazine in heterocyclic synthesis has been reviewed [1437]. Alkyl- or aryl-amines instead of hydrazine give 3-substituted fused pyrimidinones and replacement of the ester by a carboxylic acid group needs the presence of triethyl phosphite, PPA or phosphorus pentachloride and phosphorus oxychloride.

R^1 = alkenyl; $PhCH=CH(CH_2)_2$,
4−MeC_6H_4; R^2=Me,Et; X = O,S

4−Quinazolinone

[14, 1157]

R^1 = alkyl, $ClCH_2$, Ph;
R^2 = H, Cl, SO_2NH_2;
R^3 = Bu, Ph, 2−MeC_6H_4, 2−Cl−pyrid−2−yl.
i, $P(OEt)_3$, PCl_5−$POCl_3$, PPA or PCl_3

[82, 476, 855, 925]

[14]

4-Quinazolinone

3. 1,3-Oxazin-6-one

When the source of N^3 (such as an amine) is omitted from the reactions of carboxylic acids or esters mentioned in the preceding section, A 1,3-oxazin-6-one ring is obtained. The cyclization is effected in boiling pyridine or acetic anhydride (for carboxylic acids), or cold sulphuric acid (for esters).

R^1 = Me, Ph; R^2 = H, Me; R^3 = H, Cl, SO_2NH_2
i, P(OPh)$_3$—pyr, or Ac$_2$O or H$_2$SO$_4$ at 18 °C

3,1-Benzoxazin-4-one
[32, 476, 925, 1955]

4. 1,3-Thiazine-6-thione

When ethyl 2-acylaminobenzoates are heated with Lawesson's reagent (**7.1**) [1826] cyclization to a 3,1-benzothiazine-4-thione occurs.

R = alkyl, Ph

[353]

3,1-Benzothiazine-4-thione

$4-MeOC_6H_4P$ ... $PC_6H_4-4-MeO$

(**7.1**)

III. FORMATION OF A SEVEN-MEMBERED RING

1. 1,2-Diazepin-3-one

Carbamate and ester groups in side-chains interact at room temperature when stirred with hydrogen chloride-ether to give a high yield of this fused ring compound.

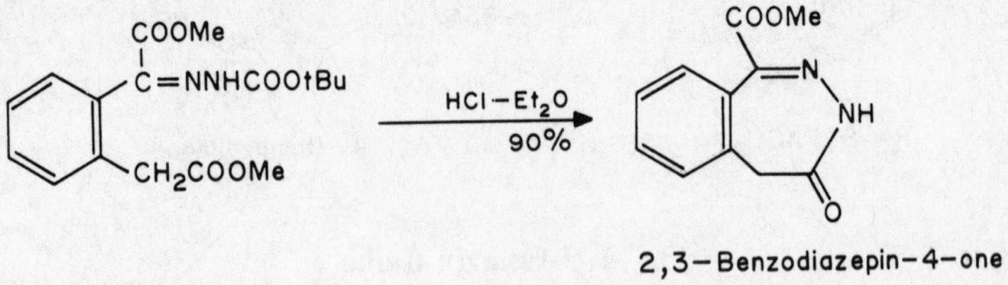

[587]

2,3-Benzodiazepin-4-one

CHAPTER 8

Acylamine or Amine and Ether or Thioether

I. FORMATION OF A FIVE-MEMBERED RING

1. Pyrazole

In this cyclization, an anion from a —COCH$_2$CO— or a —CH$_2$COOH displaces a methylthio group and the amine attacks the carbonyl to form a pyrazole ring.

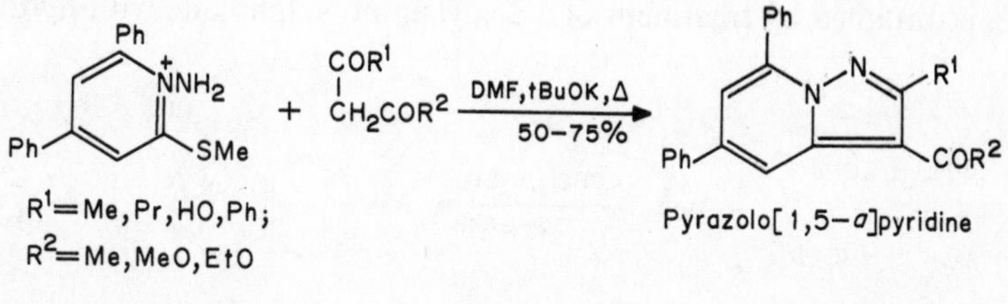

R^1=Me,Pr,HO,Ph;
R^2=Me,MeO,EtO

Pyrazolo[1,5-a]pyridine

[820]

2. 1,2,4-Triazole

A methylthio group attached to π-deficient ring is usually readily displaced and in this base-catalysed reaction one molar proportion each of base and aryl nitrile react with the methylthio-amine to form a triazole ring.

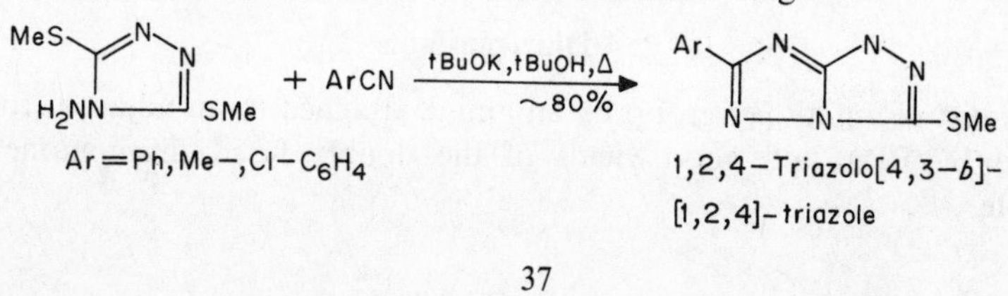

Ar=Ph,Me—,Cl—C$_6$H$_4$

1,2,4-Triazolo[4,3-b]-
[1,2,4]-triazole

[1535]

3.   1, 3, 4-Oxadiazole

A thermal cyclization of a 2-methylthio-acylamine yields a fused oxadiazole.

$$\xrightarrow[10-94\%]{\Delta}$$

[121]

R = alkyl, PhCH$_2$, Ph,
MeO-, Cl-, NO$_2$-C$_6$H$_4$

1,3,4-Oxadiazolo[3,2-*a*]-
pyrimidine

II.   FORMATION OF A SIX-MEMBERED RING

1.   Pyrimidine

Displacement of a nuclear methylthio group by an amine attached to another ring gives a new *N*-containing ring.

$$\xrightarrow[56-70\%]{\substack{1.\text{SnCl}_2,\text{HCl} \\ 2.\text{NaOH}}}$$

R^1=H,Me,Cl; R^2=H,Br

Pyrimido[2,1-*b*]quinazolin-4-one

[1137]

2.   1, 2, 4-Thiadiazine 1-oxide

This ring is obtained by treatment of a 2-acylamino-sulphoxide with hydrazoic acid.

$$+ \; \text{NaN}_3 \quad \xrightarrow[36-96\%]{\text{CHCl}_3, \text{H}_2\text{SO}_4}$$

[1954]

R^1=H,Cl;
R^2=Me,ClCH$_2$,Cl$_2$CH

1,2,4-Benzothiadiazine 1-oxide

III.   FORMATION OF A SEVEN-MEMBERED RING

1.   1, 4-Diazepin-5-one

Displacement of an ethoxy group by an amine attached to an adjacent ring is promoted by PPA and good yields of the doubly fused diazepinone are obtainable.

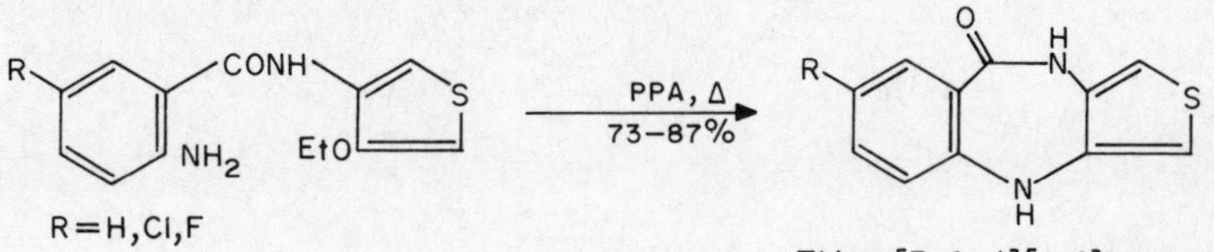

R=H,Cl,F

PPA, Δ
73–87%

[493]

Thieno[3,4–*b*][1,4]benzodiazepin–9–one

Another ring system synthesized similarly:

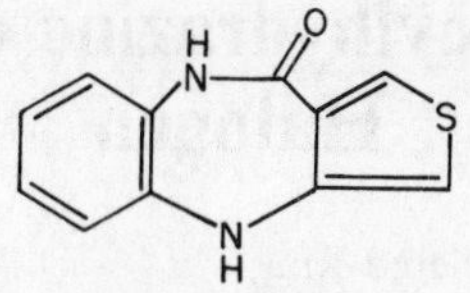

[493]

Thieno[3,4–*b*][1,5]–
benzodiazepin–10–one

CHAPTER 9

Acylamine, Acylhydrazine or Amine and Halogen

I. FORMATION OF A FIVE-MEMBERED RING

1. Pyrrole

2-Iodoanilines, when treated with an aldehyde or ketone and t-butoxide-liquid ammonia and irradiated in liquid ammonia, give good yields of indoles. Alternatively, when copper (I) iodide-mediated arylation of sodium enolates is applied to 2-iodoaniline in this way, ethyl acetoacetate yields ethyl 2-methyl-indole-3-carboxylate in 60 per cent yield.

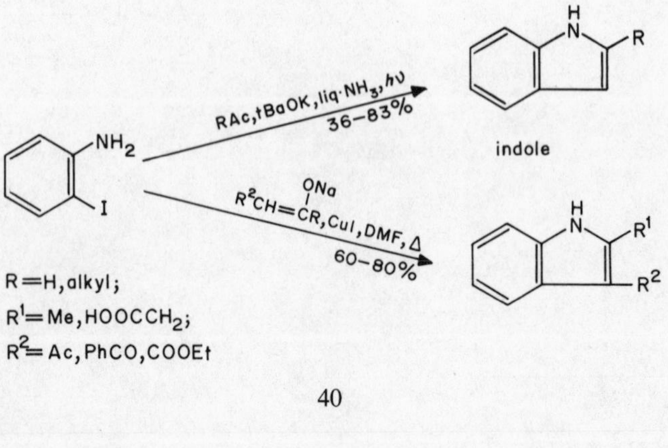

$R = H,\, alkyl;$

$R^1 = Me,\, HOOCCH_2;$

$R^2 = Ac,\, PhCO,\, COOEt$

When 1, 2-di(chloroaryl)ethylenediamines are heated, the product depends largely on the positions of the chlorine atoms in the rings. The 2, 6-dichlorophenyl derivatives (both racemates and meso-forms) produce high yields of a 4-chloro-2-(2, 6-dichlorophenyl) indole but the racemates also give a smaller yield ($\approx$ 23 per cent) of an indolo[2, 3-b]indole (9.1). 2, 4-Dichlorophenyl compounds do not cyclize to indoles.

[505]

R = Me, Et

[1880]

(9.1) Indolo[2,3–b]indole

When the amine and halogen are on different rings, a doubly fused pyrrole is formed by heating the substrate either in acetic acid or with palladium tetrakistriphenylphosphine in THF.

[1229]

X = Br, Cl

Pyrrolo[3,2–b]indole

[1566]

Pyrido[3,4–b]indole

2. Oxazole, Thiazole or Thiazole-2-thione

Cyclization of a halogenoacylamine can yield either an oxazole or a thiazole depending on the reagent chosen; PPA-phosphorus oxychloride gives the

oxazole while phosphorus pentasulphide-pyridine produces a thiazole. The
chemistry of thiazolopyridines has been reviewed [1865].

R = Me, Ph, 3-ClC$_6$H$_4$, 2-furyl,

2-thienyl;

i, PPA, POCl$_3$, X = O

ii, P$_2$S$_5$-pyr, X = S

X = O, S

Oxazolo[5,4-b]pyridine

Thiazolo[5,4-b]pyridine

[1499]

Several reagents convert 2-halogenoamines into a thiazole ring, for example, a
thioxoester and a Grignard reagent, the copper salt of a carbothioic acid, or an
isothiocyanate.

R^1 = H, Cl; R^2 = Ph, H-ClC$_6$H$_4$,

2-furyl, 2-thienyl

Thiazolo[5,4-b]pyridine

(review [1865])

[1707]

R^1 = H, Me; R^2 = H, Me, 3,4-benzo

Benzothiazole

[1013]

R = alkyl, aryl, PhCH$_2$, PhCO,

Ph(CH$_2$)$_2$, COOEt

Thiazolo[5,4-b]pyridine

[1288]

R^1 = H, Me, Cl, CF$_3$; R^2 = H, Me, Bu

i, (Ph$_3$P)$_2$NiCl, NaBH$_3$CN

[1952]

Other ring systems synthesized similarly:

[402]

Pyrazolo[3,4-*d*]thiazole

[1228]

Thiazolo[4,5-*c*]pyridine

When carbon disulphide (review of its reactions [2017]) is heated with the halogenoamine in a basic medium, a thiazole-2-thione is formed in moderate yield.

[86]

Naphtho[2,3-*d*]thiazole-2-thione

3. Isothiazole-5-thione

The relatively stable trifluoromethyl group is hydrolysed on heating with sodium sulphide and DMSO; an isothiazole-5-thione ring is formed.

R = Me, Ph

Na_2S, DMSO, Δ
33—55%

[564]

2,1-Benzisothiazole-3-thione

II. FORMATION OF A SIX-MEMBERED RING

1. Pyridin-2-one or Pyridin-4-one

2-Alkenoic esters react with halogenoamines under the influence of palladium chloride and TEA and under pressure with the formation of a new pyridin-2-one ring.

MeCN, $PdCl_2$, TEA, Δ
40—70%

[306]

R^1 = H, alkyl, Ph; R^2, R^3 = H, Me

Pyrido[2,3-*d*]pyrimidin-7-one

Strongly basic conditions induce cyclization of *o*-chloroacylhydrazines to a pyridin-4-one in a regiospecific synthesis which yields the antibacterial compound amifloxacin in high yield.

4-Quinolinone   [1864]

2. Pyrazine

Amine and reactive chlorine functions on different rings (joined by a nitrogen atom) interact to form a doubly fused pyrazine ring on heating in DMSO.

Pyridazino[3,4-*b*]quinoxaline   [1927]

3. 1,4-Oxazine

Benzoins (α-hydroxyketones) condense with halogenoamines to give a 1,4-oxazine ring. Base-induced ring closure of the suitably placed functions is an efficient method of synthesizing phenoxazines (review [1923]); a Smiles rearrangement (review [1883]) may occur under these conditions. Diphenyl ethers (R^3 = H) which failed to cyclize under basic conditions gave acceptable to high yields when heated for long periods with dimethyl methylphosphonate which also cyclized an ether in which NHR^3 was replaced by NMe$_2$. Phenoxypyridines cyclized in a shorter time.

Pyrazolo[4,3-*b*][1,4]oxazine   [1305]

R^1=H,Me,Cl; R^2=H,Cl;   Phenoxazine   [920,934]
R^3= Me, Me$_2$N(CH$_2$)$_3$ (review[1923])
i , K$_2$CO$_3$–DMF, X = CH   Pyrido[3,2-*b*][1,4]benzoxazine
ii, (MeO)$_2$POMe, X = CH,N

4. 1, 3-Thiazine-2, 4-dithione

The *o*-trifluoromethylaniline mentioned in Section I.4 reacts with carbon disulphide and sodium sulphide to give this six-membered ring in good yield.

[564]

3,1−Benzothiazine−2,4−dithione

5. 1, 4-Thiazine

As in the formation of the 1, 4-oxazine ring, basic conditions may be used for the formation of this ring, but a Smiles rearrangement is likely to occur (see Section II.3). A strongly acidic medium (hydrochloric acid) avoids this complication but

[1581]

Dipyridazino[4,5−*b*:4′,5′−*e*][1,4]−
thiazine−1,9−dione

[1581]

Dipyridazino[4,5−*b*:4′,5′−*e*][1,4]−
thiazine−1,6−dione

[1581]

Dipyridazino[4,5−*b*:4′,5′−*e*][1,4]−
thiazine−1,6−dione

[1581]

Dipyridazino[4,5−*b*:4′,5′−*e*][1,4]−
thiazine−4,6−dione

cyclization is accompanied by debenzylation. A thorough study of this reaction [1581] shows that four products, namely, 10-benzyl-1,6-dione, 10-benzyl-4,6-dione, and the debenzylated 1,6-dione and 1,9-dione, may be synthesized from the two isomeric sulphides shown in the equations. In a study of another related cyclization [506], a Smiles rearrangement was observed to occur in acetic or dilute mineral acid.

Other ring systems synthesized similarly:

Pyridazino[4,5-*b*]pyrido-
[3,2-*e*][1,4]thiazin-9-one

Pyridazino[4,5-*b*]pyrido-
[3,2-*e*][1,4]thiazin-6-one

[506]

Reactions of this kind may be catalysed by copper–copper(I) iodide to give high yields of phenothiazines, provided they are conducted under nitrogen.

R = Me,Cl

Phenothiazine

[88]

6. 1,3,4-Oxadiazine

Pyrolysis of a 2-bromomethyl-*NN*-diacylamine yields a fused oxadiazole ring.

Ar^1,Ar^2= Ph,Cl–,MeO–,
NO$_2$–C$_6$H$_4$

[1,2,3]Triazolo[1,5-*d*]–
[1,3,4]oxadiazine

[1732]

III. FORMATION OF A SEVEN-MEMBERED RING

1. 1,4-Diazepin-5-one

Heating a 2-chloropyridine *N*-oxide containing an amino group on another ring in a high-boiling solvent with mineral acid gives a high yield of the cyclized product but, in the absence of the *N*-oxide function, PPA (review [B-21]) is a more efficient medium.

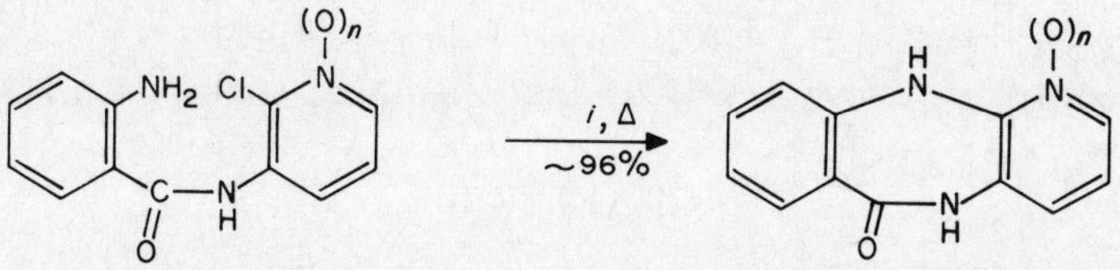

$i, n = 1$, DEG monomethyl ether,
$n = 0$, PPA

Pyrido[2,3-b][1,4]-
benzodiazepin-6-one
(1-oxide)

[855]

CHAPTER 10

Acylamine or Amine and Hydroxy or Thiol

I. FORMATION OF A FIVE-MEMBERED RING

1. Pyrrole or Pyrrol-2-one

Cyclocondensation under dehydrating conditions converts 4-aminoalcohols into fused pyrroles in moderate yields; since phosphorus pentachloride is also an effective reagent, the reaction may then be dehydrohalogenation.

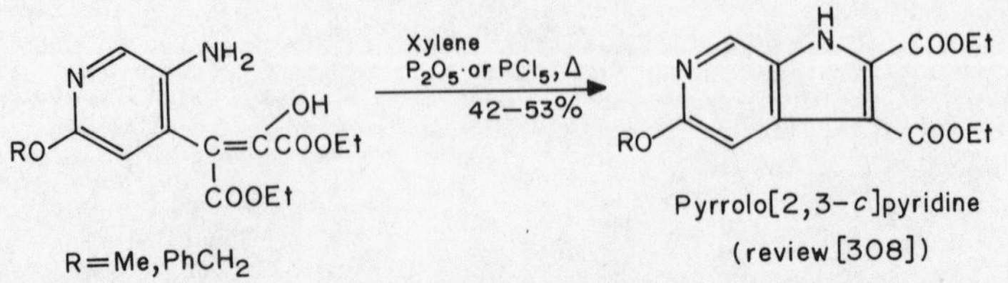

[565]

Pyrrolo[2,3-c]pyridine
(review [308])

Carbon monoxide (generated *in situ*) reacts with a carbonium ion (formed by loss of hydroxyl ion) and cyclizes with the amine group to produce a fused pyrrol-2-one ring.

48

[1912]

Indol−2−one

2.   Oxazole

2-Acylaminophenols cyclize readily into benzoxazoles under the influence of hot mineral acid, all the atoms required being present in the substrate. Dean and Stark conditions are often effective. On the other hand, a wide variety of reagents can be used to cyclize 2-aminophenols. It is likely that the acylamine may be present as an intermediate when an acyl chloride, a carboxylic acid or an imidate is used. Formamide has also been used [96, 325].

[785]

R=H,NO$_2$; Ar=4−Me−,4−NO$_2$−C$_6$H$_4$,
PhCH=CH

Benzoxazole

[98, 1296]

R^1= H,NO$_2$;
R^2= aryl,heteroaryl
i,TEA,H$_2$SO$_4$,xylene,
Dean and Stark method or pyr.

X=CH,N
Benzoxazole
Oxazolo[4,5−b]pyridine

[98, 1485, 1559]

R =alkyl,aryl,heteroaryl
i,RCOOH, PPA or Me$_3$Si ester of PPA;
ii,RC=NH,EtOH
      |
      OEt

Methyl cyanodithioic ester reacts at both functional groups to give a complex product but the selenium analogues of both ester and amide give good yields at moderate temperatures.

$R^1 = H, NO_2$ [1630]

$R^1, R^2 = alkyl$ [1231]

$R^1 = H; R^2 = aryl, heteroaryl$ [403]

Condensation with an aromatic aldehyde gives the imine (Schiff's base) which, on oxidation, cyclizes to the oxazole. 3,3-Dimercapto-1-phenyl-2-propen-1-one on heating under Dean and Stark conditions behaves similarly.

$R^1 = H, NO_2; R^2 = Ph, Me_2N-,$
$MeO-, HO-C_6H_4, 2-thienyl$

Benzoxazole [98]

[1373]

2-Aminobenzoxazoles may be prepared from the aminophenol by reaction with dichloromethylenesulphonamide, the pyrrolidinecarbimidoyl chloride or diphenyl cyanocarbimidate, all of which appear to give good yields under relatively mild conditions.

Benzoxazole [99, 305, 1754]

i, $R^2SO_2N=CCl_2$, PhMe, $R^2 = Me, aryl$;
   $R^1 = NHSO_2R^2$

ii, $R^1N=C-N$⟩ (pyrrolidine), DME, TEA; $R^1 = 4-NO_2C_6H_4$
        $|$
        Cl

iii, $(PhO)_2C=NCN$, $iPrOH$; $R^1 = CN$

3. Oxazol-2-one or Oxazole-2-thione

The C-2 atom of these two rings may be provided by phosgene, carbonyldiimidazole or thiophosgene.

[354]

X = O, S
Pyrano[2,3−e]benzoxazole−2,6−dione
Pyrano[2,3−e]benzoxazol−6−one,
2−thioxo−

Other ring systems synthesized similarly:

[1698]

Oxazolo[4,5−b]pyridin−2−one

[354]

X = O, S
Pyrano[3,2−f]benzoxazole−2,8−dione
Pyrano[3,2−f]benzoxazol−8−one,2−thioxo−

[104]

R = H, Me, MeO, NO₂

2−Benzoxazolone

Carbon disulphide can be the source of a 2-thione group of N-substituted benzoxazole-2-thiones.

[1167]

R = PhCH₂, 4−Me₂NC₆H₄CH₂,
2,4−(MeO)₂C₆H₃CH₂

2−Benzoxazolethione

4. Thiazole

Treatment of a 2-mercapto-N-acylamine with hot mineral acid annulates 2-acylamino-thiols into a thiazole ring. A similar ring is formed by heating with an α-bromoketone.

Thieno[3,2-d]thiazole   [737]

Imidazo[2,1-b]thiazole   [930]

Malononitriles, imidates or isothiocyanates react with amino-thiols under relatively mild conditions to form a fused thiazole in high yield.

$R^1 = H_2, PhCH=$;
$R^2 = Ph, PhCH_2$

Benzothiazole   [848]

                                                                         [1419]

                                                                         [959]

Compounds containing an activated methyl or methylene are capable of reacting in one several ways. 2-Methylbenzimidazole yields 2-(benzimidazol-2-yl)–benzothiazole while the very reactive benzoylacetonitrile reacts only at the carbonyl group. The ester group of ethyl benzoylacetate, however, is the most reactive function when it is heated in xylene with o-aminothiophenol. Aroyl chlorides give a rather low yield. A more promising method is to treat the amino-thiol with a carboxylic acid in the presence of the trimethylsilyl ester (PPSE) of polyphosphoric acid. Selenium homologues of carboxylic derivatives, DMFDMA and chloroimines are also sources of the C-2 atom of a fused thiazole.

                                                                         [1683]

X = CH, N

Benzothiazole
Thiazolo[5,4-b]pyridine
(review [1865])

i or ii or iii or iv
40-93%

[334, 403, 1231, 1296, 1559]

$$\underset{\text{43\%}}{\overset{\overset{\text{O}}{\parallel}}{PhCCH_2CN, TsOH, PhMe, \Delta}}$$

[150]

$$\underset{\text{64\%}}{\overset{\overset{\text{O}}{\parallel}}{PhCCH_2COOEt, xylene, \Delta}}$$

[1373]

$$\underset{\sim 66\%}{ArN=CN, DME, TEA, \Delta}$$

[1754]

$$\underset{\text{92\%}}{R^4SO_2N=CCl_2, PhMe, \Delta}$$

[305]

Benzothiazole

i, RMe, S, Δ,
R = benzinidazol-2-yl;
ii, RCOCl, AcONa, AcOH, Δ,
  R = 3,5-tBu$_2$-4-HOC$_6$H$_2$;
iii, MeCOOH, PPSE; R = Me;
iv, RCX, EtOH or pyr, Δ; R = alkyl, heteroaryl, X = EtO, NH$_2$
    $\overset{\parallel}{Se}$
Ar = 4-NO$_2$C$_6$H$_4$

Benzothiazoles carrying a 2-alk-2-enyl group are difficult to prepare because of the inaccessibility of the corresponding carboxylic acid precursors but the problem has been eased by condensing the amino-thiol with alk-2-en-1-ynamines.

$$\underset{\sim 68\%}{\overset{Et_2O}{\longrightarrow}}$$

[1813]

R^1 = Et, Pr; R^2 = H, Me

5. 1,2,3-Oxathiazole 2-oxide

2-Tosylaminophenol reacts with thionyl chloride to give the benzoxathiazole 2-oxide in high yield.

$$+ \quad SOCl_2 \quad \underset{\text{81\%}}{\overset{PhH, \Delta}{\longrightarrow}}$$

[1099]

1,2,3-Benzoxathiazole 2-oxide

6. 1, 2, 3-Thiadiazole

Acylation of both amino and thiol groups followed by reaction with nitrosyl chloride gives a good yield of this thiadiazole.

[1130]

1,2,3-Benzothiadiazole

II. FORMATION OF A SIX-MEMBERED RING

1. Pyridine

A dimethylaminomethyl group can form a reactive methide ($=CH_2$). When this is adjacent to a carbonyl, a primary arylamine condenses thermally (through the Schiff's base) to form a new pyridine ring.

[732]

[1]Benzopyrano[4,3-*b*]quinolin-6-one

2. Pyrazine

Annulation of a 2-hydroxyethylideneimine to a neighbouring amino group proceeds at ambient temperature in DMSO to give a fused pyrazine ring.

[33, 1209]

2,4-Pteridinedione

3. Pyran

A doubly fused pyran is formed by heating a compound containing amino and phenolic groups attached to different rings.

[881]

[1]Benzopyrano[3,2-*c*]pyridin-1-one

4. 1,4-Oxazine

The additional pair of carbon atoms required for oxazine formation may be provided by a 1,2-dibromoethane, α-halogenoketone, α-halogenoquinone or 2-chloroacrylonitrile. A rather unusual source of two carbon atoms is cyanogen di-N-oxide (obtained from dichloroglyoxime and alkali); it reacts at low temperature with aminophenols to yield the di(hydroxyimino)benzoxazine.

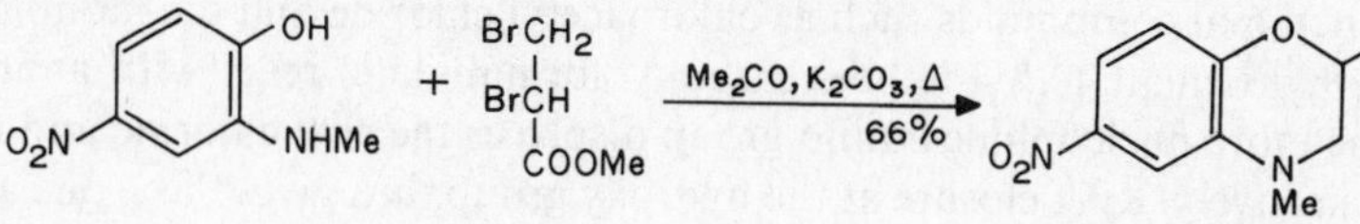

[326]

1,4−Benzoxazine

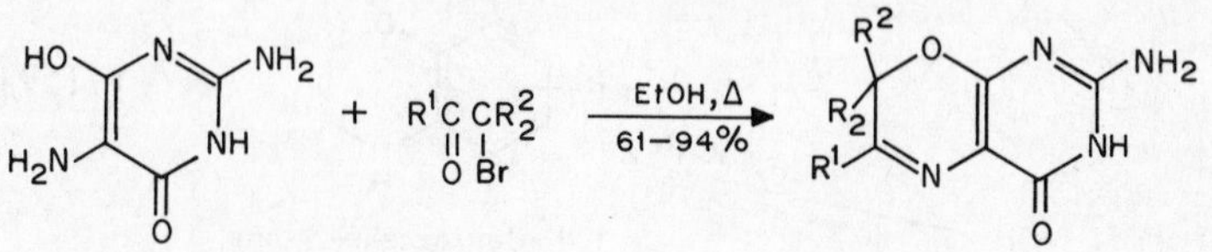

[651, 1077]

$R^1=Me, C_5H_{11}, Ph(CH_2)_n (n=1-4)$, or $4-MeOOC C_6H_4$; $R^2=H, Me$

Pyrimido[4,5-b][1,4]oxazine

Another ring system synthesized similarly:

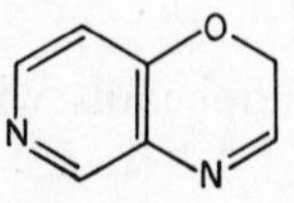

[824]

Pyrido[4,3-b]-1,4-oxazine

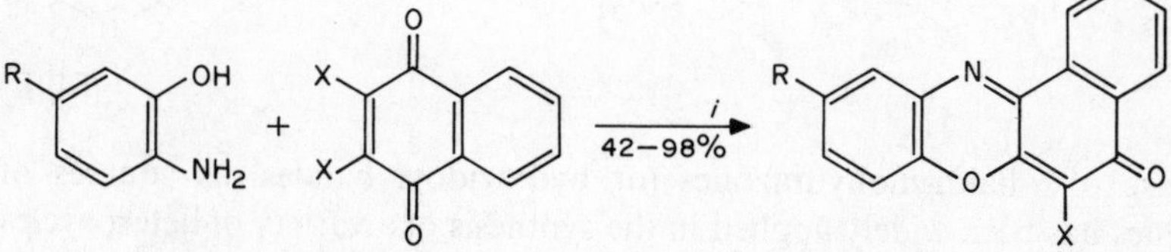

[834]

$R=H, Cl, Me$; $X=Br, Cl$
i, PhH, AcOK or MeOH, KOH

Benzo[a]phenoxazin-5-one

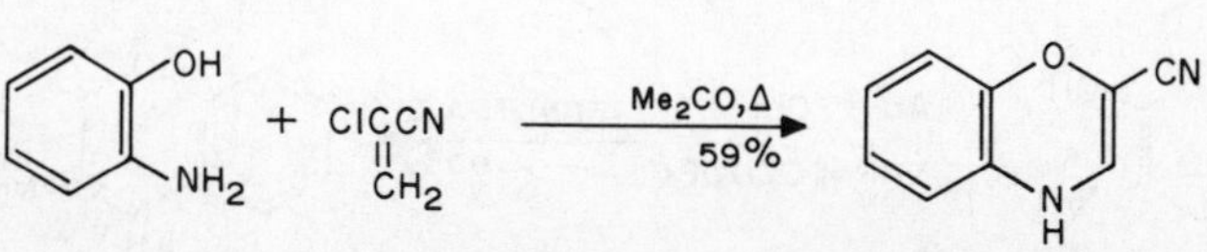

[450]

1,4−Benzoxazine

[1843]

5. 1,4-Oxazin-2- or -3-one (or -thione)

When difunctional compounds such as chloroacetyl chloride and the monoimi-
date of diethyl oxalate (ethyl (ethoxycarbonyl)formimidate) react with aminophenols, the more nucleophilic amino group displaces the acyl chlorine and the
ethoxy, respectively. Ring closure at the hydroxy group thus gives the 3-one and
2-one, respectively. The use of a phase transfer reagent (benzyltriethylammonium
chloride) gives high yields of the pure oxazin-3-one [1159].

ClCH₂COCl,Δ
~87%

[1159]

1,4-Benzoxazin-3-one

COOEt
EtOC=NH,EtOH,Δ
R=H   83%

R=H,Me,Cl,NO₂

[1419]

1,4-Benzoxazin-2-one

1,4-Benzoxazin-2-ones are obtained by reaction of aminophenols with α-
oxocarboxylic acids or esters.

EtOH,H₂,
RaNi
41–82%

COR¹
COOR²,Δ

[498]

The use of α-halogenohydrazones (or hydrazidoyl halides) as sources of
nitrilimines has been widely applied in the synthesis of a variety of heterocycles
and has been the subject of a review [1753]. Their reaction with o-hydroxyamines
is capable of yielding more than one product but the following example gives a
single oxazinone.

Ar   Cl
NHN=CCOOEt

EtOH,TEA,Δ
~90%

[1085]

Ar=Ph,4-MeC₆H₄

6. 1,3- or 1,4-Thiazine

An aminobenzyl alcohol when heated with thiourea in an acidic medium gives a 3,1-benzothiazine.

R = H, Cl

3,1-Benzothiazine

[1374]

In situ oxidation of a 3-(aminoethylthio)catechol to a quinone may occur during the synthesis of 1,4-benzothiazine. The chemistry of this system has been reviewed [1772].

1,4-Benzothiazine

[961]

α-Halogeno-hydrazones have a wide applicability in the synthesis of heterocycles; this is demonstrated in a review [1753]. Another review is confined to reactions of these hydrazones with sulphur compounds [1065]. Several examples illustrate the base-catalysed synthesis of a fused 1,4-thiazine ring in high yields from different halogeno-hydrazones. Phenacyl halides in acid solution give similar products [824].

Ar = 3-CF$_3$-, 2-COOMe-C$_6$H$_4$

[825]

Ar = Ph, 4-MeOC$_6$H$_4$

[1064]

Ar = Ph, MeOC$_6$H$_4$

Pyrido[4,3-*b*]-1,4-thiazine

[824]

Readily available enols, such as those of 1,3-diones, undergo thermal cyclization with amino-thiols to give 2-acyl-1,4-benzothiazines. Alkynones readily react with di(2-aminophenyl)disulphides to yield the 2-acyl derivatives. A reduced benzothiazine is obtained by base-induced reaction with 1,2-dibromoethane [151].

R^1=H,Me,MeO,Cl;
R^2=Me,Ph,MeO;
R^3=Me,Ph

[761, 1048, 1438]

R^1=H,Ph,COOMe;
R^2=Ac,COOMe,COOEt

1,4-Benzothiazine

[806]

A 1,4-thiazine ring fused on both sides to a benzene ring is the parent compound of the medicinally important phenothiazines. These may be obtained from amino-thiols by reaction with activated o-bromonitrobenzenes, p-benzoquinones or a 2-chlorophenol (see also Chapter 81, Section II.3). Thermal cyclization of an aminothiol is achieved by treatment with a cyclic 1,3-diketone [149].

R=H,Br,Cl,NO_2

Phenothiazine   [148]

3-Phenothiazinone   [583]

i,Zn,NaOH,EtOH

[581]

7. 1,4-Thiazin-2- or -3-one

This ring is readily formed by heating the amino-thiol with chloroacetic acid in aqueous alkali, or with a quaternary ammonium salt or with diketene (the reactions of this versatile reagent have been reviewed [2018]).

1,4-Benzothiazin-2-one   [1085]

[1159,1749]

[1114]

R = H, Me, Cl, NO₂

i, CHCl₃, PhCH₂ṄEt₃C̄l, NaHCO₃
or aq. NaOH

1,4-Benzothiazin-3-one

III. FORMATION OF A SEVEN-MEMBERED RING

1.   1,4-Oxazepine-5,7-dione

Carbon suboxide (1,2-propadiene-1,3-dione) (review of reactions: [100]) reacts with aminophenols under mild conditions to give moderate yields of the oxazepine.

[208]

1,5-Benzoxazepine-2,4-dione

2.   1,4-Thiazepine or 1,4-Thiazepin-7-one

An αβ-unsaturated aryl ketone reacts with an amino-thiol to give a 1,5-benzothiazepine in good yield. Diketene (review [2018]) in benzene gives a mixture of isomeric benzothiazepinones in contrast to the benzothiazin-3-one formed by the same reagent in DMSO (Section II.7).

1,5-Benzothiazepine   [1498]

Ar = Ph, HO–, MeO–,
Br–C₆H₄

[1114]

1,5-Benzothiazepin-2-one

CHAPTER 11

Acylamine, Amine or Diazonium Salt and Lactam Carbonyl

Heterocycles which contain a carbonyl (or thiocarbonyl) adjacent to an unsubstituted nitrogen atom in the ring are tautomeric and can react in either oxo or enol form depending on the reagents and conditions. The term 'lactam carbonyl' (or thiocarbonyl) covers both of these possibilities. In this chapter, isolated examples are given of an amidino or hydrazino group reacting with a lactam carbonyl.

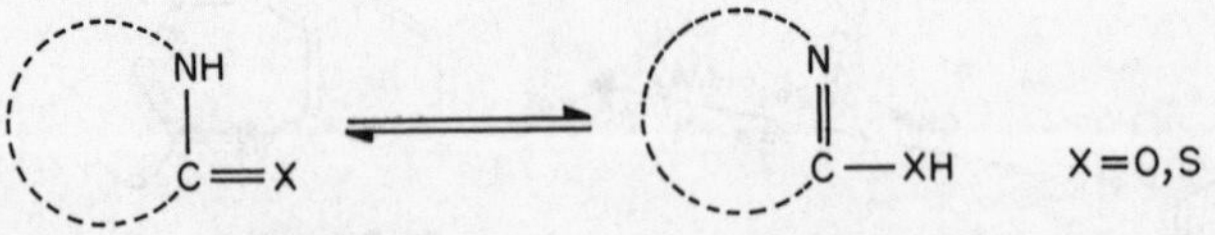

I. FORMATION OF A FIVE-MEMBERED RING

1. Pyrrole

Cyclodehydration in high yield occurs between amino and lactam carbonyl groups by heating the compound in DMF.

60

1,2,4−Triazino[5,6−*b*]−
indole−3−thione

[758]

2. Imidazole

The cyclization of a 2-acylamino-lactam carbonyl by reaction with a primary arylamine takes different courses according to the degree of steric hindrance at the ortho positions of the aryl ring. In this example, Ar^1 has substituents at both 2 and 6 positions while Ar^2 has only one or no ortho substituent.

Purine

[1794]

$Ar^1 = 2,6-R_2C_6H_3, R = Me,Cl;$
$Ar^2 = 2-Me-,2-Cl-,2-F,$
$2-Ph-C_6H_4$

An amine and lactam carbonyl in different rings may react in hot acetic acid to form a doubly fused imidazole ring.

R = CN,COOH

[1,2,4]Triazino[2,3−*a*]−
benzimidazol−3−one

[1561]

Another ring system synthesized similarly:

[1959]

Pyridazino[1,6−*a*]−
benzimidazole

3. 1, 2, 4-Triazole

An *N*-amidino group adjacent to a lactam thiocarbonyl cyclizes when treated with methyl iodide at ambient temperature but an electron-attracting aromatic ring (for example, 2-pyridyl or 4-nitrophenyl) at Ar^2 inhibits the reaction.

$$H_2NCAr^2 \quad \xrightarrow[\sim 65\%]{EtOH, MeI} \quad [1861]$$

$Ar^1, Ar^2 = Ph, 4-MeC_6H_4$

[1,2,4]Triazolo[1,5-c]-
quinazoline

Refluxing a 2-thioxo-hydrazinomethyl heterocycle in benzene causes ring closure to a triazole and the same ring may be formed from an N-amino-thiolactam, an aryl nitrile and a strong base.

$$\xrightarrow[55\%]{PhH, \Delta} \quad [911]$$

1,2,4-Triazolo[3,4-b]-
[1,3,4]oxadiazole

$$+ \; ArCN \quad \xrightarrow[50-84\%]{tBuOK, tBuOH, \Delta} \quad [1844]$$

$Ar = Ph, Me-, Cl-, MeO-C_6H_4$
3- or 4-pyridinyl

1,2,4-Triazolo[4,3-b]-
[1,2,4]triazole

4. Oxazole

An oxazole ring may be obtained by dehydrative cyclization of a hydroxy-amine with hot acetic anhydride or an acyl chloride.

$$\xrightarrow[68\%]{Ac_2O, \Delta} \quad [1989]$$

Oxazolo[5,4-b][1,8]-
naphthyridin-8-one

Another ring system synthesized similarly:

[1296]

Oxazolo[4,5-b]pyridine

5. Thiazole

Several reagents can convert an amino-lactam thione (or its tautomer) into a fused thiazole ring, for example, an orthoester, ethyl xanthate salt, DMFDMA or a tetra-alkoxymethane.

Thiazolo[4,5-*b*]pyrazine   [1070]

Thiazolo[4,5-*b*]pyrazine   [1070]

X = CH,N
Benzothiazole
Thiazolo[5,4-*b*]pyridine   [1683]

Thiazolo[5,4-*d*]pyrimidine   [698]

R = Et, Pr

6.   1, 2, 3-Thiadiazole or 1, 2, 3, 4-Thiatriazole

Diazotization of a 2-thioxo-amine converts these functions into a fused thiadiazole ring and when the amino group is attached to a ring nitrogen, a thiatriazole ring is obtained.

R = H, alkyl

1,2,3-Thiadiazolo[5,4-*d*]-
pyrimidin-4-one   [1224]

[1,2,4]Triazolo[4,3-*d*]-
1,2,3,4-thiatriazole   [911]

Another ring system synthesized similarly:

[1989]

[1,2,3]Thiadiazolo[5,4-*b*][1,8]−
naphthyridin−8−one

7. 1,3,4-Thiadiazole

This ring is formed when an *N*-amino-thiolactam is heated with a carboxylic acid and phosphorus oxychloride (or mineral acid) or an aryl isothiocyanate. A fused 1,3,4-thiadiazole-2-thione is obtained in low yield when a carbon disulphide-pyridine mixture is warmed with the amino-thiolactam [917].

$$+ R^2COOH \xrightarrow[20-92\%]{POCl_3, \Delta}$$ [1985]

$R^1 = Me, Ph; R^2 = MeCH=CH,$
$Ph, PhCH=CH$

[1,3,4]Thiadiazolo[2,3-*c*]−
[1,2,4]triazin-4-one

$$+ RNCS \xrightarrow[\sim56\%]{DMF, \Delta}$$ [1534]

$R = Ph, Me-, Cl-, Br-,$
$MeO-C_6H_4, PhCO$

1,2,4−Triazolo[3,4-*b*]−
[1,3,4]thiadiazole

Other ring systems synthesized similarly:

[1985]

1,2,4−Triazolo[3,4-*b*]−
[1,3,4]thiadiazole

[917]

1,2,4−Triazolo[3,4-*b*]−
[1,3,4]thiadiazole-6-thione

[138, 1540]

[1,3,4]Thiadiazolo[2,3-*c*]−
[1,2,4]triazin-4-one

II. FORMATION OF A SIX-MEMBERED RING

1. Pyridine

When amino and lactam carbonyl groups are in different rings linked by a methylene group, heating in acetic acid results in the formation of a pyridine ring.

R^1 = H, Me; R^2 = H, NH_2; R^3 = H, O

i, R^3 = H: EtOH, AcOH;

ii, R^3 = O: $FeSO_4$, $Ba(OH)_2$, NH_3

1,2,4-Triazino[5,6-*b*]-
quinolin-3-one

[303, 2016]

2. Pyridazine

Inter-ring annulation may be effected by reaction of a diazonium salt and a lactam carbonyl, the two nitrogen atoms bridging the two rings. The chemistry of heterocyclic diazonium compounds has been reviewed [1326].

$NaNO_2$, HCl, Na_2SO_3, Δ
57%

[1512]

1,2,4-Triazino[5,6-*c*]-
cinnolin-3-one

3. 1,4-Oxazine

3-Bromo-3-methylbutan-2-one cyclizes the (tautomeric) aminohydroxypyrimidinone (**11.1**) to form a new 1,4-oxazine ring.

+ AcCMe₂ (Br)

EtOH, $NaHCO_3$, Δ
76%

[1790]

Pyrimido[4,5-*b*][1,4]oxazin-4-one

4. 1,4-Thiazine or 1,4-Thiazin-3-one

A fused thiazine ring is formed in high yield by heating a 2-amino-thiolactam (or its tautomer) with phenacyl bromide or an α-halogeno-ester. In a strongly basic medium, the latter reaction leads to a thiazin-3-one.

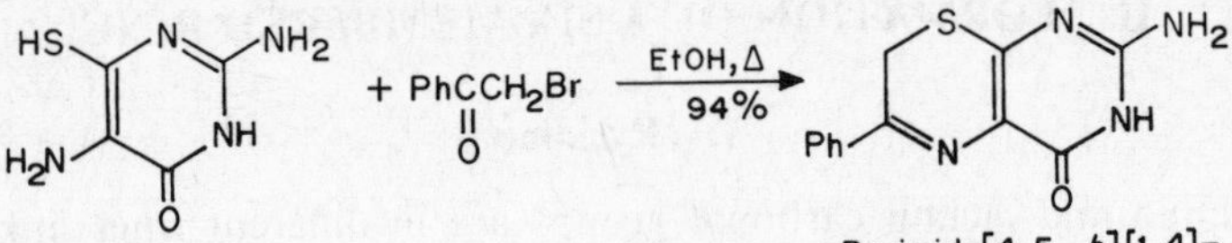

[1047]

Pyrimido[4,5-b][1,4]-
thiazin-4-one

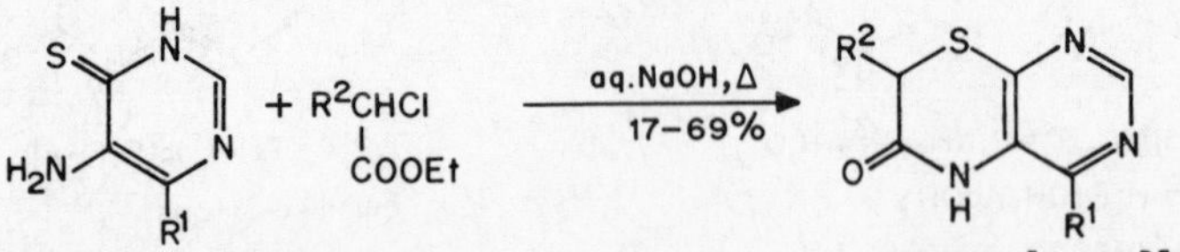

[1841]

R^1 = MeO, EtO, Cl, HS;
R^2 = H, Me, CH$_2$COOH

Pyrimido[4,5-b][1,4]-
thiazin-6-one

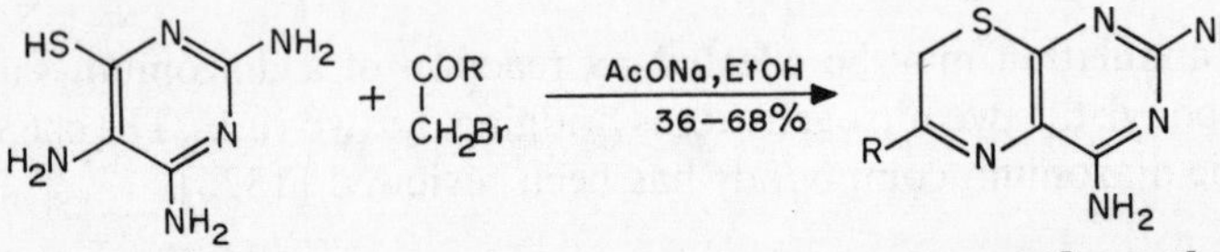

[1610]

R = Ph, COOEt

Pyrimido[4,5-b]-
[1,4]thiazine

5. 1,3,4-Thiadiazine

When the α-bromo-ketone (see previous section) is cyclic, a doubly fused thiadiazine is obtained on heating the bromoketone with an N-amino-thiolactam in ethanol or in ethanol containing carbonate.

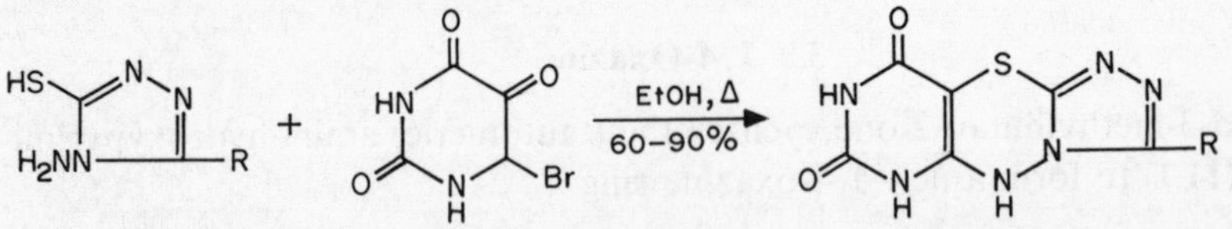

R = H, alkyl

Pyrimido[4,5-e]-1,2,4-triazolo-
[3,4-b][1,3,4]thiadiazine-7,9-dione

[1811]

Another ring system synthesized similarly:

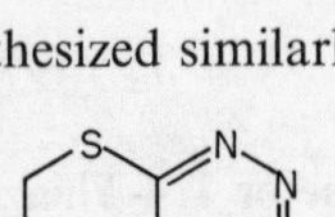

[917]

1,2,4-Triazolo[3,4-b]-
[1,3,4]thiadiazine

III. FORMATION OF A SEVEN-MEMBERED RING

1. 1,4-Diazepin-5-one

In this inter-ring cyclodehydration between an amino and lactam carbonyl, heating with phosphorus pentoxide gives good results.

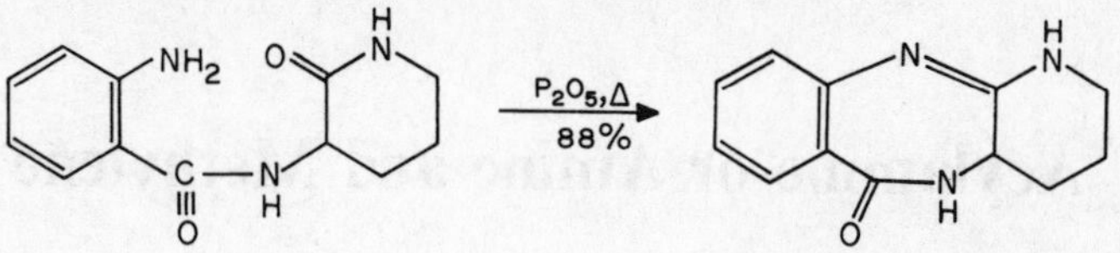

Pyrido[2,3-*b*][1,4]-
benzodiazepin-6-one

[858]

2. 1,3,4-Thiadiazepine

The versatile *N*-amino-thiolactam can be converted into this seven-membered ring by heating with a 3-arylpropynal.

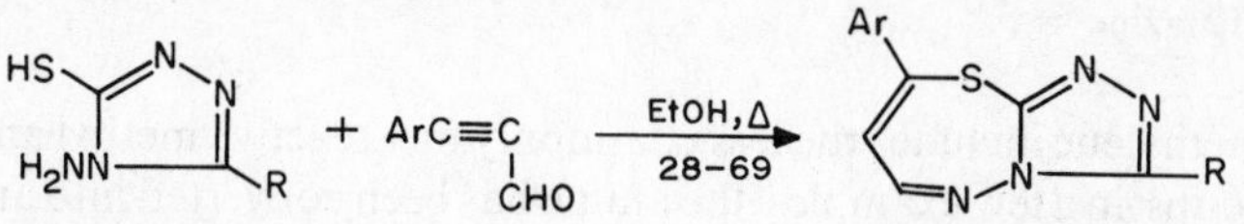

R=PhCH₂,Ph,Br-,F-,
Me₂N-C₆H₄; Ar=Ph,4-BrC₆H₄

1,2,4-Triazolo[3,4-*b*]-
[1,3,4]thiadiazepine

[479]

Acylamine or Amine and Methylene

The term methylene includes the less commonly met reactive methyl and reactive methine groups; in a few examples, the amine has been converted into an imine. In this chapter, a carbamate is regarded as an acylated amine.

I. FORMATION OF A FIVE-MEMBERED RING

1. Pyrrole

Basic conditions generate an anion which attacks the neighbouring acylamine group to form a pyrrole ring—a reaction known as the Madelung indole synthesis [B-8]. Lithiation and treatment with an ester converts an *N*-trimethylsilyl-*o*-toluidine into a 3- or 2,3-substituted indole in moderate yield [2007].

R = alkyl, aryl   tBuOK, tBuOH, Δ / 60–92%   Indole   [550]

+ CH₂=CHCN   tBuOK or NaH / 25%   Pyrrolo[1,2-*a*]quinoline   [906]

Another ring system synthesized similarly:

[906]

Pyrrolo[2,1-*a*]isoquinoline

An imine (prepared from the amine and an aldehyde) cyclizes readily (even at ambient temperature) with a reactive methylene to yield a fused pyrrole ring. A reactive methyl condenses with oxalate ester and the product cyclizes at ambient temperature in acid solution by reaction with a neighbouring ethoxymethylen-imine.

[1505]

Ar=Ph,2-MeOC₆H₄, 2-pyridyl

Indole

Pyrrolo[3,2-*c*]pyridazine 2-oxide [1987]

2. Pyrazole

The formation of an anion is a step in this cyclization too and the strength of the base that is needed varies inversely with the acidity of the methylene hydrogen atoms. Trichloroacetonitrile supplies a carbon atom in a base-catalysed synthesis of pyrrolopyrazoles (review [1600]).

[1279]

Pyrazolo[3,4-*c*]pyridine

[95]

R=H,Ph

Pyrrolo[2,3-*c*]pyrazole

Another ring system synthesized similarly:

[1599]

Pyrazolo[4,3-b] pyridine

3. Imidazole

Oxidative cyclization occurs when a carbamate containing a neighbouring methylene group is treated with a mild oxidant. The uses of lead tetra-acetate in organic chemistry have been reviewed [B-36].

$$\xrightarrow[\sim 78\%]{PhNO_2 \text{ or } LTA}$$

[1122]

R = PhCH$_2$, Ph

2,6-Purinedione

4. Isothiazole

This ring is formed by treatment of an *o*-toluidine with *N*-sulphinylmethanesulphonamide. The chemistry of the resulting ring system, 2,1-benzisothiazole, has been reviewed [1775].

+ MeSO$_2$NSO $\xrightarrow[6-88\%]{PhH,pyr,\Delta}$

[808]

R^1=H,Me,Ph; R^2=Me,Et,Cl,MeO,
NO$_2$,CN,COOMe

2,1-Benzisothiazole

Other ring systems synthesized similarly:

[809]

Benzo[1,2-*c* : 3,4-*c'* : 5,6-*c''*]-
triisothiazole

[809, 810]

X = CH,N
Benzo[1,2-*c* : 3,4-*c'*]-
diisothiazole
Isothiazolo-
[4,3-*e*]-2,1,3-benzo-
thiadiazole

5. Furazan

Oxidative cyclization of the 2-aminoisonitroso derivative (prepared from the CH_2 and alkyl nitrite) yields a fused furazan.

[1199]

[1,2,5]Oxadiazolo[3,4-c]-
[1,2,6]thiadiazine 5,5-dioxide

II. FORMATION OF A SIX-MEMBERED RING

1. Pyridine or Pyridin-2-one

Pyridine ring formation occurs under basic conditions (as for pyrazole, see Section I.2). A carbon atom is supplied by DMFDMA, triethyl orthoformate, LDA-carbon dioxide or a reactive carboxylate group. A secondary amine often gives a higher yield than a primary amine. Thiazolopyridines have been reviewed [1865].

R^1=H,Cl,F,CF$_3$;R^2=H,alkyl,Cl CH$_2$,
Cl$_2$CH,CF$_3$,CH$_2$=CH; R^3= 4-MeC$_6$H$_4$,
2-thiazolyl

Quinoline

[1462]

X=CH,N

X=CH,N
Thiazolo[4,5-b]pyridine
Thieno[3,2-b]pyridine

[1423]

R=H,Me

4-Quinolinone

[1134, 1135]

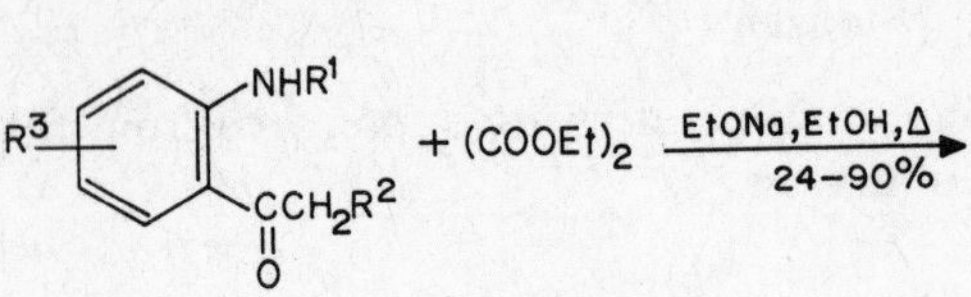

$R^1 - R^3 = H, alkyl, Ho, MeO, Cl, Ph$

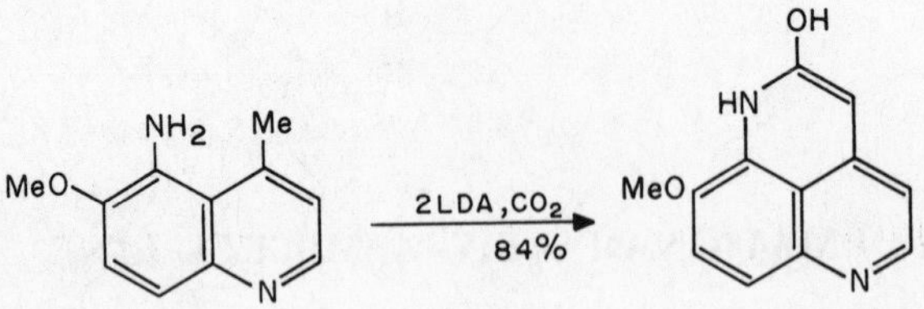

[1014]

Benzo[ij][2,7]naphthyridine

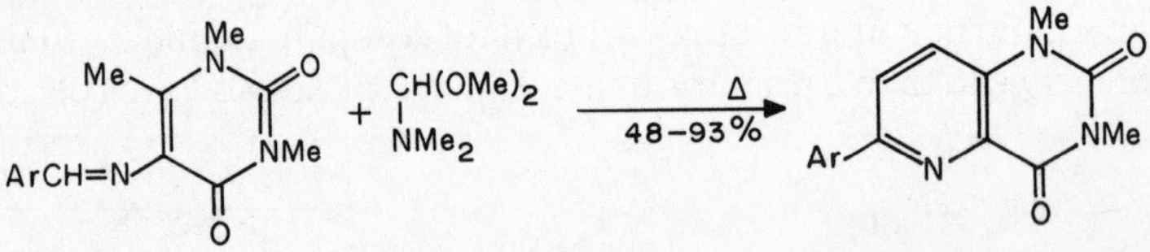

[1906]

Pyrido[3,2-d]pyrimidine-
2,4-dione

[586]

2. 1, 2, 4-Triazin-3-one

An aryldiazonium salt (review [B-28]) couples with a reactive methylene and reaction with an adjacent carbamate ester group leads to cyclization in high yields.

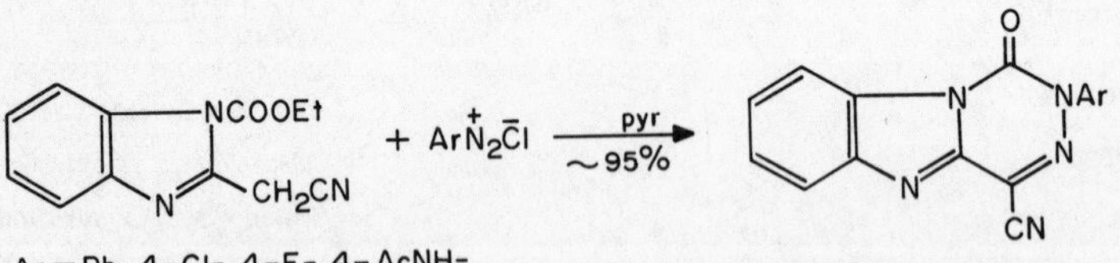

Ar = Ph, 4-Cl-, 4-F-, 4-AcNH-,
4-COOEt-C_6H_4

[1,2,4]Triazino[4,5-a]-
benzimidazol-1-one

[1000]

3. 1, 4-Thiazine

In the presence of a strong base such as lithium diisopropylamide, a proton is abstracted from a methylene adjacent to sulphur, sulphoxide or sulphone and good yields of benzothiazines ([1772] for a review) are obtained.

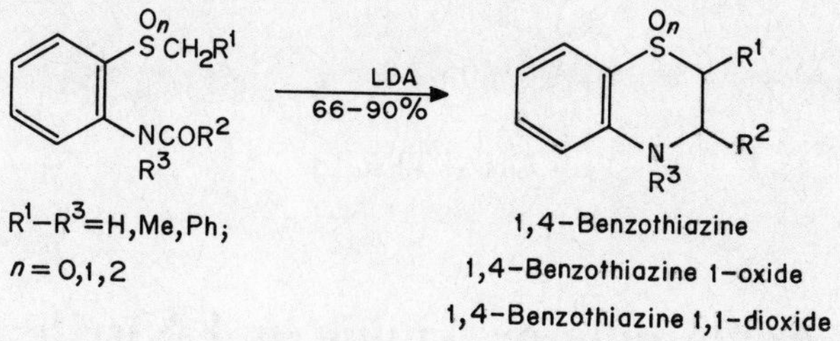

LDA
66–90%
R1–R3=H,Me,Ph;
n=0,1,2
1,4-Benzothiazine
1,4-Benzothiazine 1-oxide
1,4-Benzothiazine 1,1-dioxide

[1156]

Acylamine or Amine and Nitrile

The cyclization of o-amino-nitriles was comprehensively reviewed in 1970 [525].

I. FORMATION OF A FIVE-MEMBERED RING

1. Pyrrole

Catalytic reduction of a 2-aminobenzyl nitrile followed by heating the resulting diamine is a convenient method (called the Pschorr–Hoppe reaction) of preparing some indoles. It is also possible to reduce the nitro-nitrile but the overall yield is lower.

$NR^1R^2 = NMe_2, NHPh, N(CH_2)_4, N(CH_2)_5;$
$R^3 = H, Me, MeO, F, CF_3, NH_2, SO_2NMe_2$

Another ring system synthesized similarly:

[17]

Pyrrolo[3,2-*d*]pyrimidine

2. Pyrazole

When hydrogen chloride is passed into a solution of an *N*-amino-nitrile, a pyrazole ring is obtained in high yield.

[1631]

R^1=Me,MeS; R^2=CN,COOMe   Pyrazolo[5,1-*c*]-1,2,4-triazole

3. Imidazole

Reduction of a 2-nitrophenylcyanamide in a hot acidic solution yields a 2-aminobenzimidazole.

[374]

Benzimidazole

II. FORMATION OF A SIX-MEMBERED RING

1. Pyridine or Pyridin-2-one

Malononitriles with piperidine as catalyst react with many amino-nitriles to form a fused pyridin-2-one ring in good yield but failed to condense with the pyrazole (13.1) although ethyl cyanoacetate and a strong base converted the pyrazole into a pyrazolopyridine in rather low yield.

[126]

R^1=2-HOC$_6$H$_4$CH$_2$;   1,6-Naphthyridine-2,7-dione
R^2= CN,PhCO

[1138]

(13.1)

Pyrazolo[3,4-*b*]pyridine

The reaction with malonic acid derivatives may be in two steps, the intermediate 2-aminobenzylidenemalonate being isolated and then cyclized. In this way, the synthetically useful 2-aminoquinoline-3-carboxamide may be prepared.

[38]

R=Br,Cl,MeO,MeS,PhCH₂O

Quinoline

2. Pyrimidine

When an *o*-acylamino-nitrile is heated with a fourfold excess of a secondary amine, a fused pyrimidine ring is formed in high yield (review [874, B-5]) but a twofold excess of dimethylammonium chloride yields a thieno[2, 3-d]pyrimidin-4-one [434].

[434]

R^1=H,Me,Ph; R_2^2N=(CH₂)₅N,
c-C₆H₁₁N,O[(CH₂)₂]₂N

Thieno[2,3-*d*]pyrimidine

A fused pyrimidine ring is obtained from amino-nitriles by reaction with formamidine (or its salt) or chloroformamidine in refluxing ethanol, t-butanol, ethoxyethanol or diglyme.

[1804]

R=PhCH₂O(CH₂)₂OCH₂

Pyrazolo[4,3-*d*]pyrimidine

[1471]

R=Cl,PhS

Pyrido[3,2-*d*]pyrimidine

Other ring systems synthesized similarly:

[111]

Purine

[108, 110, 127–130, 405, 1750]

X = CH, N
Quinazoline
Pyrido[2,3-d]pyrimidine

[1376]

Thieno[3,2-g]pteridine

[404]

Pteridine

An orthoester reacts first with the amino group to give an —N=CHOEt function; ammonia, an amine or hydrazine displaces the labile ethoxy and the —N=CHNHR group reacts with the neighbouring nitrile group to form a pyrimidine ring. Another cyclization which proceeds under mild conditions is that with an imidate or thioimidate. The nitrile itself may also be converted into an imidate which is cyclized at ambient temperature by methoxycarbonyl isothiocyanate (review [860]).

$\xrightarrow[97\%]{HC(OEt)_3, MeNH_2, Ac_2O}$ [115]

$\xrightarrow[25\%]{EtOC=NH, EtOH, \Delta}$ COOEt [1419]

Quinazoline

R = Sugar residue

$\xrightarrow[60\%]{DMF}$ [2030]

Purine

R = Me, pentosyl

$\xrightarrow[\sim 60\%]{CHCl_3, pyr, \Delta}$ [700]

Pyrazolo[3,4-d]pyrimidine

Other ring systems synthesized similarly:

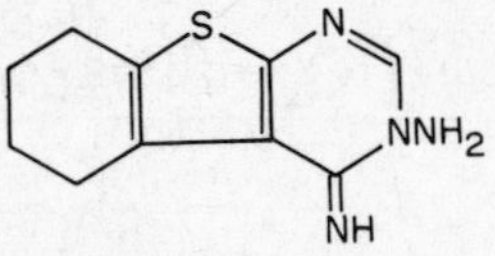

[557]

[1]Benzothieno[2,3-*d*]-
pyrimidine

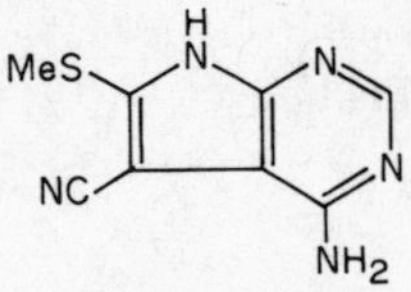

[56]

Pyrrolo[2,3-*d*]pyrimidine

Amidines (which are freed *in situ* from their salts by a base) or guanidines usually give high yields of the 2,4-diaminopyrimidine derivative.

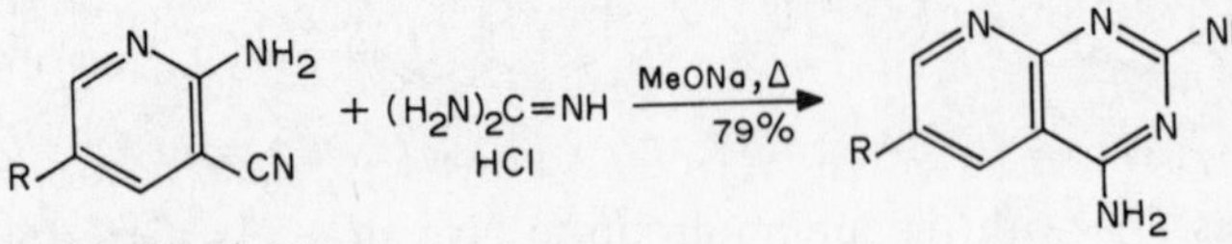

[107, 746, 1750]

R = Me, (MeO)$_2$CH,
4-tBuOOCC$_6$H$_4$

Pyrido[2,3-*d*]pyrimidine
(review [1647])

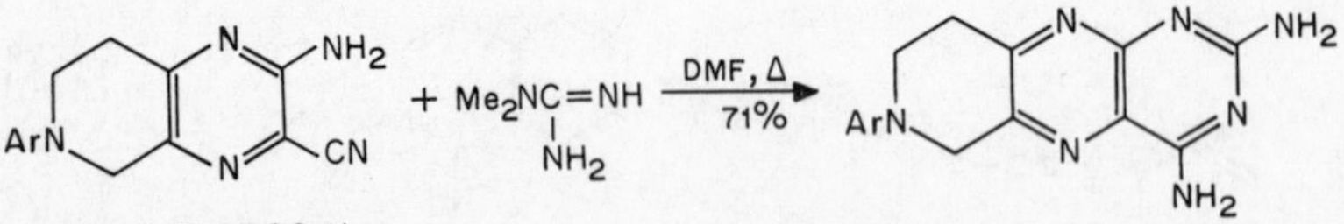

[1778]

Ar = 4-EtOOCC$_6$H$_4$

Pyrido[3,4-*g*]pteridine

Other ring systems synthesized similarly:

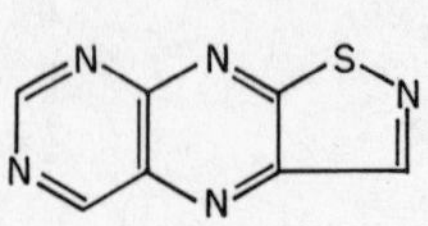

[503]

Isothiazolo[4,5-*g*]pteridine

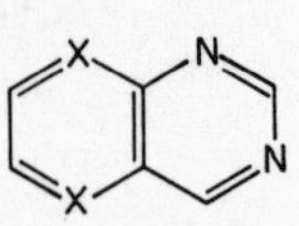

[125, 404, 405, 569]

X = CH,N
Quinazoline
Pteridine

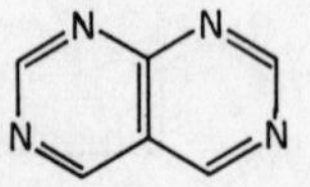

[57]

Pyrimido[4,5-*d*]-
pyrimidine

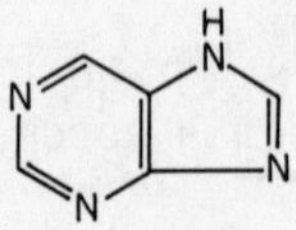

[111]

Purine

Thieno[3,2-g]pteridine

[1376]

Thieno[2,3-d]-
pyrimidine

[129]

Pyrido[4',3':4,5]thieno
[2,3-d]pyrimidine

[130]

o-Aminonitriles react with other nitriles in the presence of hydrogen chloride to give a 2-substituted 4-aminopyrimidine ring while N,N-dimethyl-phosgeniminium salt with anthranilonitrile gives a high yield of 4-chloro-quinazoline under mild conditions.

$R^1 = Me, Ph;$
$R^2 = Me, PhCH=CH$

[63,1789]

X = NPh,O,S
Pyrrolo[2,3-d]pyrimidine
Furo[2,3-d]pyrimidine
Thieno[2,3-d]pyrimidine

R = H,Me,Cl

[816]

Quinazoline

3. Pyrimidin-2- or -4-one or Pyrimidine-2,4-dione or their Sulphur Analogues

Treatment of N-acyl (or thioacyl) anthranilamide-type substrates with either acid or base converts them into pyrimidine-4-ones (or -thione).

$R^1, R^2 = Me, Ph;$
$R^3 = H, Me$

[18, 1416, 1660]

X = CH,N
4-Quinazolinone
Pyrido[2,3-d]
pyrimidin-4-one

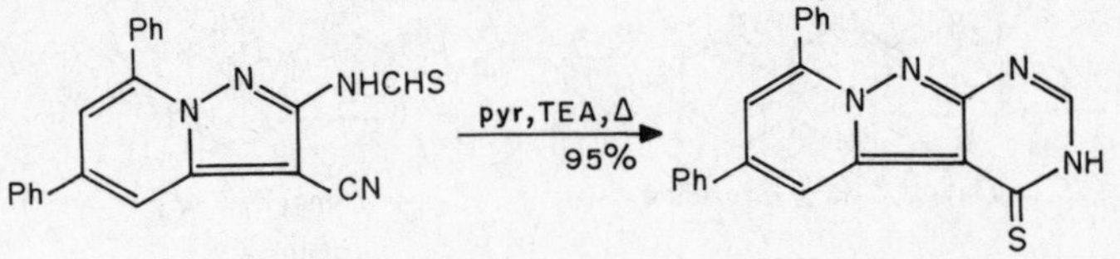

[1212]

Pyrido[1',2':1,5]pyrazolo-
[3,4-d]pyrimidine-4-thione

Another ring system synthesized similarly:

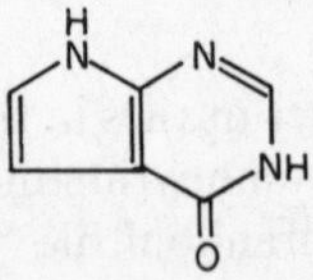

[1483]

Pyrrolo[2,3-d]pyrimidin-4-one

Several reagents which contain a C=O or C=S convert aminonitriles into a new pyrimidine-2-thione or -2,4-dithione. Carbonyl sulphide can produce a 4-thioxo-2-one derivative in alkaline medium.

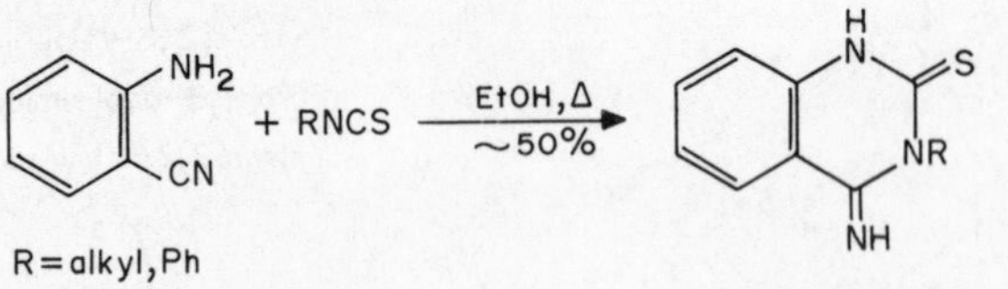

[114]

R = alkyl, Ph

2-Quinazolinethione

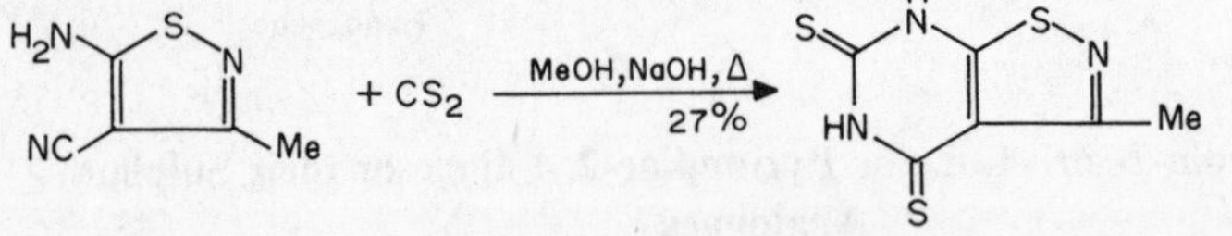

[1355]

Isothiazolo[5,4-d]pyrimidine-4,6-dithione

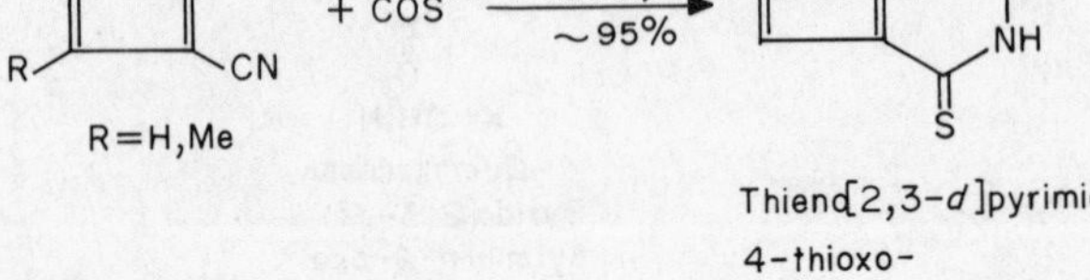

[254]

R = H, Me

Thieno[2,3-d]pyrimidin-2-one,
4-thioxo-

Other ring systems synthesized using carbonyl sulphide:

Pyrazolo[3,4-*d*]pyrimidin-
6-one, 4-thioxo-

[254]

Pyrazolo[4,3-*d*]pyrimidin-
5-one, 7-thioxo-

[254]

4. 1,3-Thiazine-2-thione or 1,4-Thiazine

Annulation of an amino-nitrile by reaction with carbon disulphide (review [2017]) in pyridine produces a 1,3-thiazinethione ring in good yield.

$$+ CS_2 \xrightarrow[76\%]{pyr, \Delta}$$

[1834]

Pyrrolo[2,3-*d*][1,3]-
thiazine-2-thione

When 2-aminophenylthioacetonitrile is treated with ethanolic hydrogen chloride, 3-imino-1,4-benzothiazine is obtained.

$$\xrightarrow[62\%]{HCl-EtOH, \Delta}$$

[131]

1,4-Benzothiazine

5. 1,2,6-Thiadiazine 1,1-Dioxide

Hexamethylsilazane facilitates acylation of an amino group by formation of —NHSiMe$_3$; treatment of this successively with sulphamoyl chloride and alkali leads to the formation of the thiadiazine ring.

$$+ (Me_3Si)_2NH \xrightarrow[28\%]{i, ii}$$

i, H$_2$NSO$_2$Cl
ii, NaOH

[109]

Pyrazolo[3,4-*c*][1,2,6]-
thiadiazine 2,2-dioxide

III.   FORMATION OF A SEVEN-MEMBERED RING

1.   Azepin-2-one

When both acylamino and nitrile groups are in side-chains, a seven-membered ring is produced when the compound is heated with aqueous alkali.

$R^1 = MeO, OCH_2O$; $R^2 = H, ArCH_2$;
$Ar = MeOC_6H_4$

3-Benzazepin-2-one   [807]

2.   1,4-Diazepin-5-one

An extension of the reaction in the preceding section to amino-nitriles which have another nitrogen in the side-chain leads to a diazepinone being formed in good yield.

Thieno[3,2-f]-1,4-diazepin-5-one   [652]

CHAPTER 14

Acylamine, Acylhydrazine, Amine or Carbamate and Nitro

Reactions in which the nitro-amine is reduced *in situ* to a diamine are discussed in Chapter 76.

I. FORMATION OF A FIVE-MEMBERED RING

1. Imidazole 3-Oxide

Mild partial catalytic reduction of a 2-nitro-*N*-acylaniline is an improved method of synthesizing benzimidazole 3-oxides but a (phenyl) substituent on the amino nitrogen is essential for this cyclization.

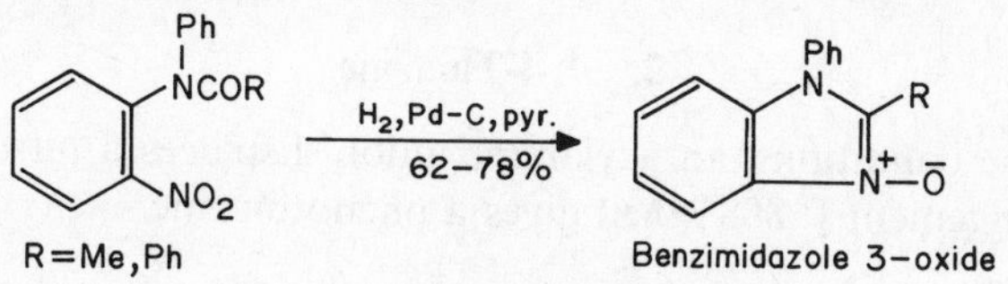

[393]

Benzimidazole 3-oxide

2. Furazan or its 1-Oxide

Benzofurazan 1-oxides (also called benzofuroxans) are of considerable interest and have been reviewed [1690, 1726]. These compounds are obtained by treatment of a nitroamine with alkaline hypochlorite.

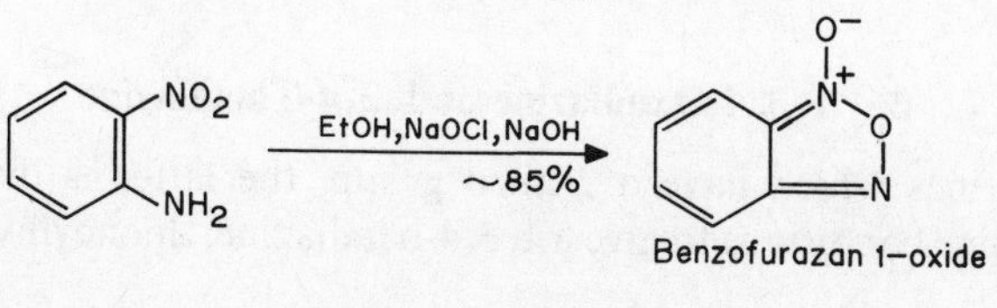

[1675]

Benzofurazan 1-oxide

Other ring systems synthesized similarly:

[1236]

[1609]

[1,4]Dioxino[2,3-*f*]-2,1,3-
benzoxadiazole 1-oxide

[1,2,5]Thiadiazolo[3,4-*e*]-
2,1,3-benzoxadiazole 1-oxide

Thermolysis of a 2-nitrocarbamate is a convenient method of synthesizing a benzofuran.

[1435]

R=H,Me,MeO,Cl,NO$_2$

Benzofurazan

II.   FORMATION OF A SIX-MEMBERED RING

1.   1,2,4-Triazine 1-Oxide

o-Nitroanilines may be converted into 1,2,4-benzotriazine 1-oxide by reaction with cyanamide either thermally or in an acidic medium.

[328,1374]

R=H,Me,Cl,MeO
i, Δ or HCl,AcOH,Δ

1,2,4-Benzotriazine 1-oxide

2.   1,4-Thiazine

Under alkaline conditions, an acylamine anion displaces a nitro group and a Smiles rearrangement [1883] then gives a phenothiazine.

[1050]

R=H,Cl,Br,CF$_3$,COOH

Phenothiazine

3.   1,3,4-Oxadiazine or 1,3,4-Thiadiazine

In acylhydrazines which have a 2-nitro group, the latter is displaced under strongly alkaline conditions to give a 1,3,4-oxadiazine; thioacylhydrazines yield

1, 3, 4-thiadiazines.

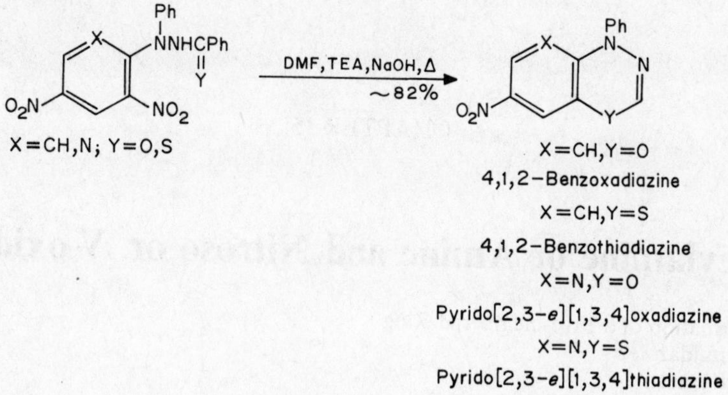

$$\text{X=CH,N; Y=O,S}$$

DMF, TEA, NaOH, Δ
~82%

[361]

X=CH,Y=O
4,1,2-Benzoxadiazine
X=CH,Y=S
4,1,2-Benzothiadiazine
X=N,Y=O
Pyrido[2,3-e][1,3,4]oxadiazine
X=N,Y=S
Pyrido[2,3-e][1,3,4]thiadiazine

CHAPTER 15

Acylamine or Amine and Nitroso or *N*-oxide

I. FORMATION OF A FIVE-MEMBERED RING

1. Imidazole

A 2-nitroso-amine may be cyclized by heating with an aldehyde to give a fused imidazole ring. Some 7-hydroxyxanthine (7-hydroxypurine-2, 6-dione) was also formed.

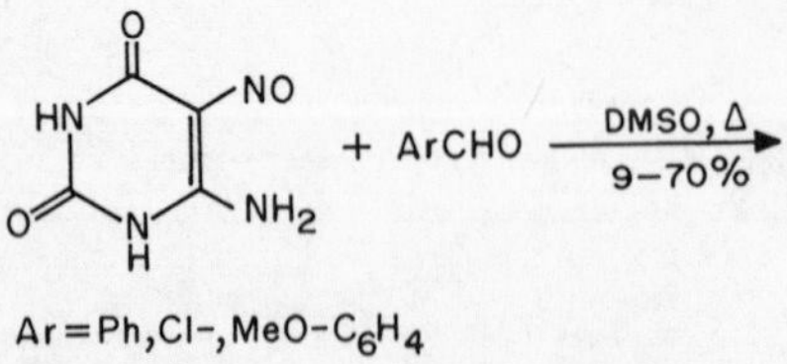

Ar = Ph, Cl-, MeO-C$_6$H$_4$ 2,6-Purinedione

[838]

An α-amino *N*-oxide may be cyclized by heating with a phenacyl halide.

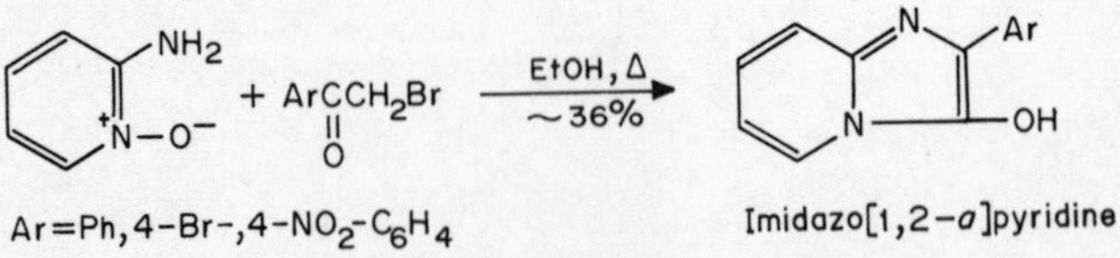

Ar = Ph, 4-Br-, 4-NO$_2$-C$_6$H$_4$ Imidazo[1,2-*a*]pyridine

[406]

Another ring system synthesized similarly:

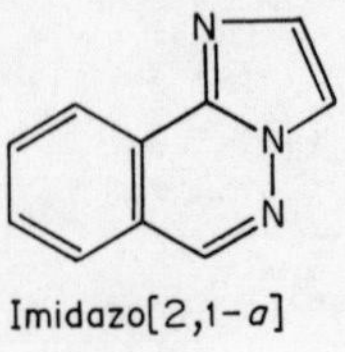

Imidazo[2,1-*a*]
phthalazine

[1704]

86

2. Furazan

Oxidation of a 2-nitroso-amine with lead tetra-acetate gives a high yield of a furazan ring at ambient temperature.

R=H,Me,NH₂,Me₂N,MeS,Ph

[1,2,5]Oxadiazolo[3,4-d]
pyrimidine

[1184]

II. FORMATION OF A SIX-MEMBERED RING

1. Pyrazine or Pyrazin-2-one

Two additional carbon atoms to form a pyrazine ring may be supplied by an activated acetylene, malononitrile, or pyrimidinetriamine. The potassium salt of an enolized 1, 3-diketone also gives a pyrazine but a 2-ketoaldehyde yields an *N*-hydroxypyrazinone. The chemistry of quinoxalines has been reviewed [1669].

Phosphorus-containing anions have also supplied two carbon atoms to complete a pyrazine ring in reactions which proceed at ambient temperature. When R^2 is COOEt, pteridin-7-ones are obtained in high yield.

2,4-Pteridinedione

[323]

R=Me,MeS

Isothiazolo[4,5-b]pyrazine

[503]

[503]

Isothiazolo[4,5-g]pteridine

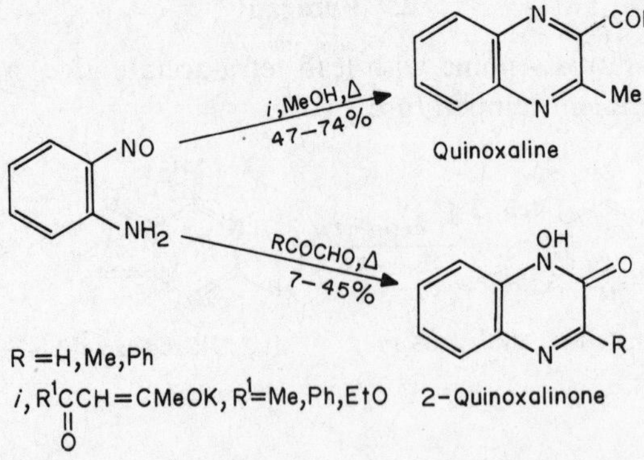

[1208]

R = H, Me, Ph

i, R^{1}CCH=CMeOK, R^1=Me, Ph, EtO   2-Quinoxalinone
‖
O

[1506]

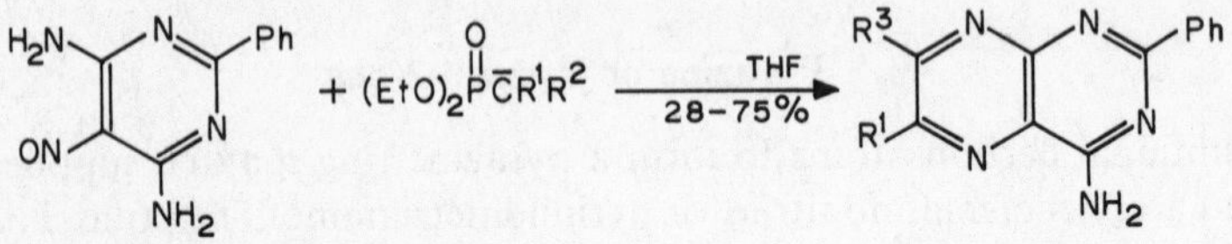

R^1=H, Me; R^2=CN, R^3=NH$_2$;
R^2=PhCO, R^3=Ph

Pteridine

2. 1,2,4-Triazine

With hydrazine hydrate and a benzaldehyde, a 1,2,4-triazine ring is formed.

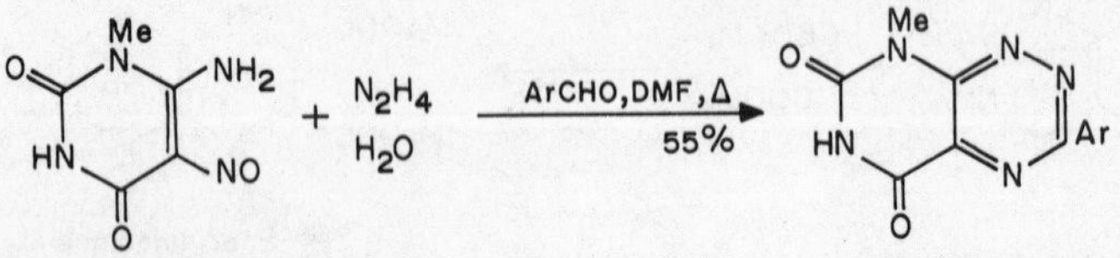

[462]

Ar = Ph, 3,4-OCH$_2$OC$_6$H$_4$

Pyrimido[4,5-*e*]-1,2,4-
triazine-5,7-dione

CHAPTER 16

Acylamine, Acyloxy, Amine or Hydroxy and Phosphorane

I. FORMATION OF A FIVE-MEMBERED RING

1. Pyrrole

A C-triphenylphosphinylmethyl adjacent to an acylamino group cyclizes under basic conditions to give high yields of the fused pyrrole. The reactions of phosphorus-containing compounds used in heterocyclic synthesis have been reviewed [1724].

$$R^1 = H, Ph; \quad R^2 = Me, CH_2 = CMe,$$
$$Ph, 4-MeO-, 4-NO_2-C_6H_4$$

Indole   [411]

2. 1,2,4-Triazole

Iminophosphoranes which have a neighbouring amino group cyclize to a triazole in high yields at ambient temperature on reaction with an arylaldehyde.

$$Ar^1, Ar^2 = Ph, 4-Cl-, 4-MeO-C_6H_4$$

[1,2,4]Triazolo[5,1-c]–
[1,2,4]triazin-4-one   [1282]

3. Furan

A 2-acyloxybenzyl alcohol or bromide reacts with triphenylphosphine hydro-
bromide to give the benzylphosphonium salt which is converted by triethylamine
into a benzofuran.

R^1 = Me, Ar; R^2 = HO, Br;
R^3 = 3,4-OCH$_2$O, MeO, AcO

Benzofuran   [742, 1493]

2-Hydroxybenzylphosphonium bromide is cyclized to a benzofuran by
treating it successively with DMAP, a benzoic acid and DCC.

$$Ar =$$

$$R = Ac, MeOCH_2$$

[1700]

II. FORMATION OF A SIX-MEMBERED RING

1. 2H- or 4H-Pyran

A 2-hydroxybenzyltriphenylphosphonium bromide and a ketone in the presence
of a strong base cyclizes to a 2H-benzopyran on heating in an inert solvent but the
corresponding 2-acyloxy salt (prepared in situ from the bromide) cyclizes to a 4H-
benzopyran in good yield on heating in an inert solvent (without base).

R^1 = H, Ac; R^2 = H, MeO, Me, Ar;
R^3 = H, Me, Ph; R^4 = H, Me

2H-1-Benzopyran   [628]

R^1 = alkyl, Ph; R^2 = H, PhCOO;
R^3 = Ac, COOMe, Ph

4H-1-Benzopyran   [627]

2. Pyran-4-one

Chromones are formed when a 2-acylphenol containing a triphenylphosphine group is heated with an acylating agent or its *O*-acyl derivative is heated in ethanol-toluene.

$$R^1=Me, Ph; R^2=MeO, PhNH \qquad 1-Benzopyran-4-one \qquad [1401, 1755]$$

$$R = Me, CH_2=CMe, Ph, 4-MeOC_6H_4 \qquad [1392]$$

3. 1,3-Oxazin-4-one

An iminophosphorane derived from a 2-acyloxycarboxamide cyclizes at ambient temperature to give a 1,3-benzoxazin-4-one.

$$R^1=Me, Ph; R^2=H, Me \qquad 1,3-Benzoxazin-4-one \qquad [1823]$$

CHAPTER 17

Acylamine or Acylhydrazine and Ring-carbon or Ring-sulphur

I.   FORMATION OF A FIVE-MEMBERED RING

1.   Pyrazole

An *N*-acylhydrazine on treatment with Vilsmeier reagents (review [1676]) produces a pyrazole ring but the yield is depressed by a side-reaction in which a 2-$CCl=CMe_2$-pyrazolopyridazin-4-one is produced.

[1722]

Pyrazolo[3,4-*d*]-
pyridazin-4-one

2.   Oxazole

Thermal cyclization of a 1-acylaminonaphthalene-8-azide gave rather low yields of the napth[1,2-*d*]oxazole by an acid-catalysed decomposition which also involves the azide group. A similar but intermolecular mechanism probably explains the cyclization of 3-azidobenzanilide.

R = Ph, 4-NO$_2$-, 4-MeO-C$_6$H$_4$ Naphth[1,2-d]oxazole

[1596]

Benzoxazole

[1956]

3. Thiazole

The diacyl derivatives of benzoquinone imine are cyclized in acid by reaction with thiolacetic acid; this is a convenient method of obtaining 2,6-disubstituted benzothiazoles. Plumbophosphates have been used to convert thiobenzanilides into benzothiazoles under mild conditions.

R = Me, Ph   Benzothiazole

[16]

R = H, Me, Cl
i, [Pb(H$_2$PO$_4$)$_2$(HPO$_4$)$_2$]H$_2$

[1849]

4. 1, 2, 4-Dithiazole

The comparatively rare annulation to an endocyclic sulphur simultaneously converts it into a tetravalent atom.

R^1 = Et, PhCH$_2$;   [1,2,4]Dithiazolo[1,5-b]-
R^2 = H, Me   [1,2,4]dithiazole

[1743]

II. FORMATION OF A SIX-MEMBERED RING

1. Pyridine or Pyridin-4-one

Cyclization of a 2-acylaminoethylbenzene with phosphorus oxychloride, phosphorus pentachloride or phosphorus pentoxide is known as the Bishler–Napieralski reaction—a much used method discovered in 1893 for the synthesis of reduced isoquinolines. Amongst recent variations is the replacement of the phosphorus compounds mentioned by polyphosphoric ester [264] or PPA trimethylsilyl ester [1559] and the use of nitriles (such as acetonitrile, propionitrile or benzonitrile) as solvents; it appears that the latter have a role other than as solvent. Such reaction conditions have enabled acylated 1,2-diarylethylamines containing electron-releasing substituents in the aryl rings to be cyclized. The newly formed ring may be fused to two other rings.

[61, 264, 489, 1057, 1169]

Isoquinoline

$R^1 = CMe_2COOEt, 4-NO_2C_6H_4$;
$R^2 = 3,4-(MeO)_2C_6H_3$; $R^3 = H, MeO$.
i, $PhMe-POCl_3$, $MeCN-POCl_3$, Et_2O-PPE
or $MeCN-PCl_5$

[1324]

Pyrido[3,4-b]indole

[435]

Ar = 3-MeOC_6H_4

Phenanthridine
(review [1616])

Other ring systems synthesized similarly:

[1982]

Pyrido[4,3-b]indole

[355]

Pyrrolo[3,2-c]pyridine

Acylated aromatic (including heterocyclic) primary amines undergo a very useful cyclization when subjected to the Vilsmeier reagents and the newly formed pyridine ring may be simultaneously chlorinated when an excess of phosphorus oxychloride is used (review [1676]). Sensitive groups such as CH_2Cl are unaffected. Another variation is to increase the proportion of reagents even further and thus synthesize a 2-chloro-3-formylpyridine ring.

$R^1 = Me, MeO; R^2 = H, alkyl, ClCH_2, Cl_2CH, CN, CH_2COOMe, Ph$

Other ring systems synthesized similarly:

Thieno[2,3-*b*]pyridine
(review [1669])   [1478]

Thieno[3,2-*b*]pyridine   [1478]

Thieno[3,4-*b*]pyridine   [1478]

N-Formyl-*N*-methylaniline may be cyclized in high yield to a quinolinone by reaction with malonic monoester chloride. This is an efficient method for the preparation of quinolinone-3-carboxylic acids.

4-Quinolinone

2. Pyrimidin-4-one

This ring may be formed by cyclodehydration of an *N*-acylarylamine and ethyl glycinate mixture.

$R^1 = H, Me; R^2 = alkyl, PhCH_2$

4-Quinazolinone

[1069]

3. Pyrazine

Phosphorus oxychloride brings about the cyclization of an acylamino group to a reactive ring like that of pyrrole.

$R = Me, tBuCH_2$

Pyrido[2,3-*h*]pyrrolo-[1,2-*a*]quinoxaline

[711]

III. FORMATION OF A SEVEN-MEMBERED RING

1. 1,4-Diazepine

A diazepine ring fused at three of its sides is produced by the action of phosphorus oxychloride on a suitable substrate containing an acylamino group.

$R^1 = H, Cl; R^2 = H, alkyl, Ph;$
$R^3 = H, CF_3$

Indolo[1,7-*ab*][1,5]benzodiazepine

[928]

2. 1,4-Oxazepine

Acylated 2-aminoethoxybenzenes cyclize to the oxazepines by the action of phosphorus oxychloride with or without phosphorus pentoxide.

$R = H, MeO, OCH_2O, Ph$
$Ar = Ph, 3-ClC_6H_4$

1,4-Benzoxazepine

[340, 1574]

3. 1, 4-Thiazepine

PPA is an effective reagent for cyclizing acylamino groups on to reactive benzene rings.

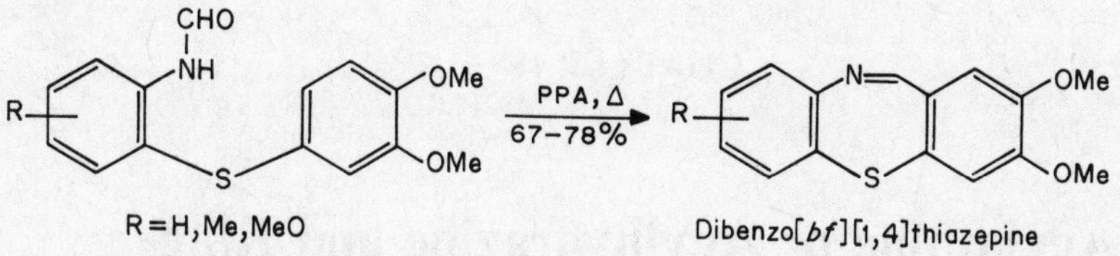

R = H, Me, MeO

Dibenzo[*bf*][1,4]thiazepine

[952]

CHAPTER 18

Acylamine or Acylhydrazine and Ring-nitrogen

I. FORMATION OF A FIVE-MEMBERED RING

1. Imidazole

Annulation of an acylamino group on to an adjacent endocyclic nitrogen atom to form an imidazole ring is a frequently used method. The amide carbon atom becomes C–2 of the new ring and a variety of substituents can thus be introduced at this position by choice of the acyl group. It is not necessary to isolate the pure amide as it can be formed from the amine and treated *in situ* with the cyclizing agent—usually phosphorus oxychloride or PPA, but phosphorus trichloride-TEA in THF reduces tar formation in some cyclizations [929].

R^1=H,Ph; R^2=H,Me,PhSCH$_2$,
N$_3$CH$_2$,COOEt,MeOCH$_2$,ClC$_6$H$_4$,
3-CF$_3$C$_6$H$_4$, 4-ClC$_6$H$_4$(CH$_2$)$_2$;
i, POCl$_3$–Cl(CH$_2$)$_2$;
ii, PCl$_3$–TEA, THF

[553, 792, 929, 1327]

X=CH,N

Imidazo[1,5-*a*]pyridine
Imidazo[1,5-*a*]pyrazine

R=H,Ph

Imidazo[1,5-*b*]pyrazole

[269]

Other ring systems synthesized similarly:

Imidazo[1,5-b]
[1,2,4]triazole [652]

Imidazo[5,1-f][1,2,4]-
triazin-4-one [741, 792, 1214]

Imidazo[5,1-f][1,2,4]-
triazin-4-one [1214]

[1]Benzothieno[2,3-d]imidazo-
[1,5-a]pyrimidin-10-one [1002]

Reaction of 2-acylaminopyrimidin-4-one with PPA varies with temperature: at about 110 °C, cyclization to the imidazo[1, 5-a]pyrimidin-4-one occurs normally but at about 190 °C, the pyrimidine ring is opened and a good yield of imidazo[4, 5-b]pyridine is obtained.

~190 °C

~110 °C

[695]

Imidazo[4,5-b]pyridine
R = alkyl

Imidazo[1,5-a]-
pyrimidin-4-one

2. 1, 2, 4-Triazole

Cyclization of a 2-acylhydrazinopyridine gives a 1, 2, 4-triazole fused at its 3, 4-bond. This widely used method may be effected in acetic acid, formic acid, PPA, phosphorus oxychloride or methoxyethanol.

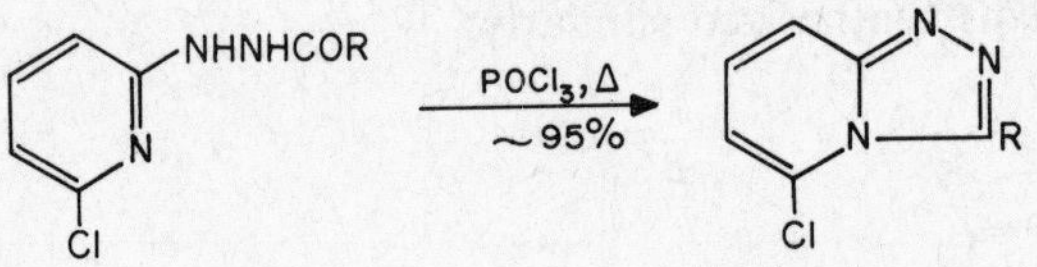

$$\xrightarrow[\sim 95\%]{POCl_3, \Delta}$$

[1148]

R=H, alkyl, ClCH₂

1,2,4-Triazolo[4,3-a]-
pyridine

Other ring systems synthesized similarly:

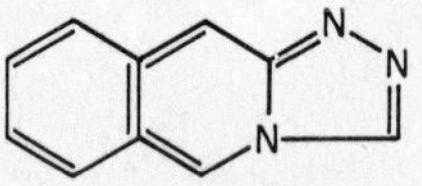

[1163]

1,2,4-Triazolo[4,3-b]-
isoquinoline

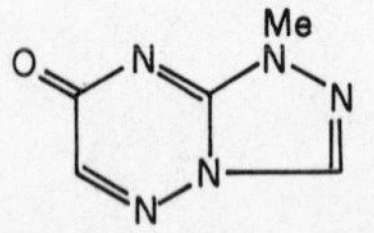

[748]

1,2,4-Triazolo[4,3-b]-
[1,2,4]triazin-7-one

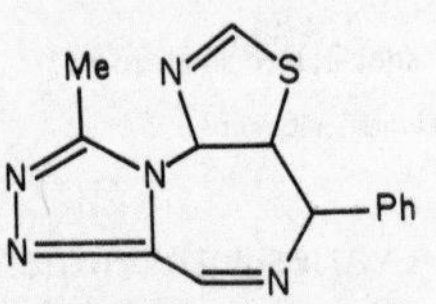

[983]

Thiazolo[5,4-f][1,2,4]-
triazolo[4,3-a][1,4]-
diazepine

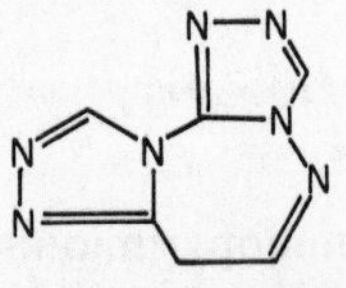

[1213]

Bis[1,2,4]triazolo[4,3-b:4',3'-d]-
[1,2,4]triazepine

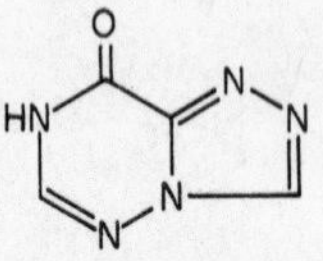

[1214]

1,2,4-Triazolo[3,4-f]-
[1,2,4]triazin-8-one

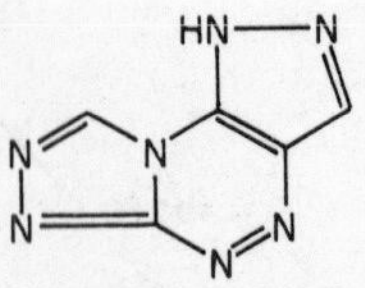

[1474]

Pyrazolo[3,4-e][1,2,4]triazolo-
[3,4-c][1,2,4]triazine

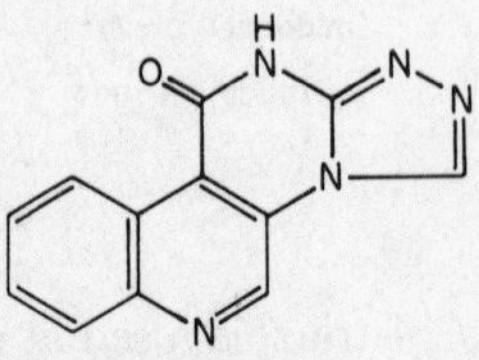

[394]

1,2,4-Triazolo[3',4':2,3]pyrimido-
[4,5-c]quinolin-11-one

II. FORMATION OF A SIX-MEMBERED RING

1. Pyrimidine

A doubly fused pyrimidine ring is formed when the acylamino group is suitably placed with respect to an endocyclic nitrogen of another ring.

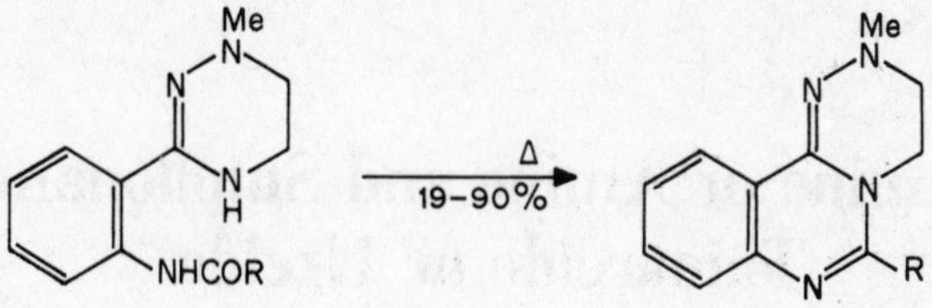

R=H,Ph,Me-,Br-,MeO-,CN-C$_6$H$_4$ [1,2,4]Triazino[4,3-c]--
quinazoline

[1328]

Another ring system synthesized similarly:

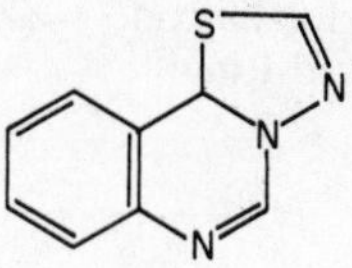

1,3,4-Thiadiazolo[3,2-c]quinazoline

[433]

CHAPTER 19

Acylamine or Amine and Sulphonamide, Thioureide or Ureide

I. FORMATION OF A FIVE-MEMBERED RING

1. Imidazole or 2-Imidazolone

When a 2-amino-thiourea is heated with mercury(II) oxide (review [B-22]), hydrogen sulphide is eliminated and a 2-aminobenzimidazole is formed but the corresponding urea in an acidic medium gives the benzimidazolone.

II. FORMATION OF A SIX-MEMBERED RING

1. 1,2,4-Thiadiazine 1,1-Dioxide

This ring is formed when 2-acylamino-sulphonamide is treated with aqueous alkali [971], ammonium hydroxide [183] or alkoxide [1345]. The aminosulphonamide requires a carboxylic acid [183, 1543], an aldehyde [93] or an orthoester [459] to convert it into the same ring and yields are usually high. The imidazole (19.1, R = H) resisted cyclization under some conditions but successive treatment with benzoyl thiocyanate and alkali gave the thiadiazole [117].

R^1 = H, alkyl, Cl, SO$_2$NH$_2$;
R^2 = Me, COOEt, AcCH$_2$, Ph
i, aq.NaOH, Δ; ii, NH$_4$OH, EtOH
iii, MeONa, MeOH

[183, 971, 1345]

1,2,4–Benzothiadiazine
1,1–dioxide

(19.1)
R = H, Me, PhCH$_2$

[459]

Imidazo[4,5–e]–1,2,4–
thiadiazine 1,1–dioxide

The imine (prepared from 2-aminobenzenesulphonamide and ethyl acetoacetate) cyclizes in high yield when heated with alkoxide but disproportionation of the side-chain occurs under the basic conditions to give 3-methyl-1,2,4-benzothiadiazine 1,1-dioxide in 96 per cent yield [1334].

[1334]

2.   1,2,4-Thiadiazine-3-thione 1,1-Dioxide

Amino-sulphonamides are cyclized by warming with thiophosgene; the 3-thione or its tautomeric 3-thiol is formed.

[1333]

1,2,4–Benzothiadiazine–
3–thione 1,1–dioxide

III. FORMATION OF A SEVEN-MEMBERED RING

1. 1, 2, 4-Triazepin-7-one

An N^1-aroylthiosemicarbazide reacts with an amino group on the benzene ring under the influence of DCC to give high yields of the triazepinone ring.

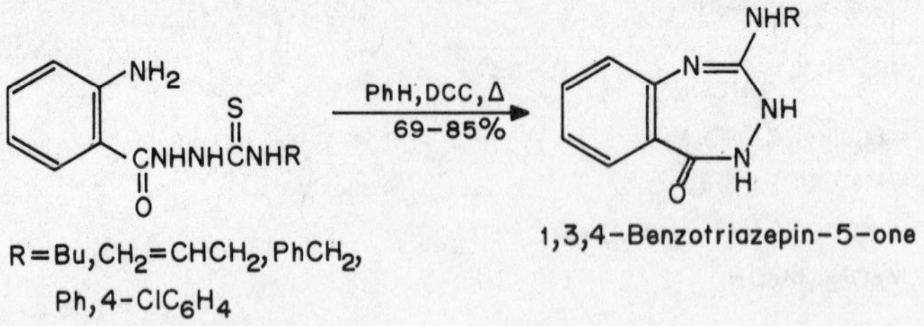
NH2
S
CNHNHCNHR
O
PhH,DCC,Δ
69–85%
NHR
N
NH
N
H
O
1,3,4-Benzotriazepin-5-one
R=Bu,CH2=CHCH2,PhCH2,
Ph,4-ClC6H4

[936]

CHAPTER 20

Acylamine or Amine and Thiocyanate

I.   FORMATION OF A FIVE-MEMBERED RING

1.   Thiazole

2-Amino- or 2-acylamino-thiocyanates readily rearrange and cyclize on heating in ethanol, mineral acid or ethyl benzoate to give a 2-amino- or 2-acylamino-thiazole ring.

Pyrazolo[3,4-*d*]thiazole   [402]

Other ring systems synthesized similarly:

Thieno[2,3-*d*]thiazole   [455]

Benzothiazole   [374]

Thieno[3,2-*d*]thiazole   [782]

Thiazolo[5,4-*d*]thiazole   [1641]

Thieno[3,2-*d*:4,5-*d*']bisthiazole   [805]

105

The amino-thiocyanate may more conveniently be formed *in situ* by treatment of the arylamine with sodium thiocyanate-bromine.

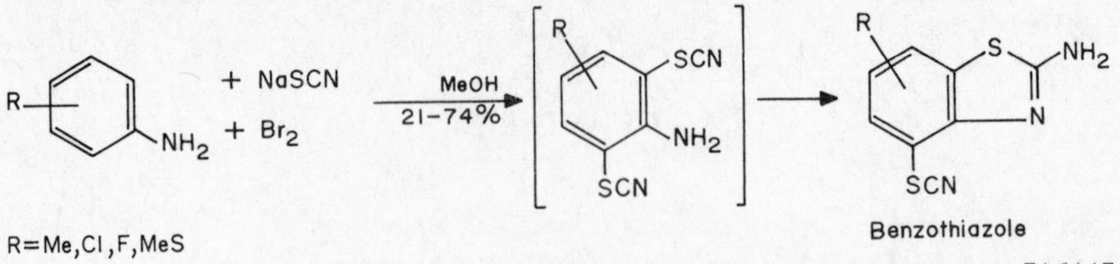

R=Me,Cl,F,MeS

Benzothiazole

[1611]

2. 1, 2, 3-Thiadiazole

Diazotization of an amino-thiocyanate gives good yields of the fused thiadiazole.

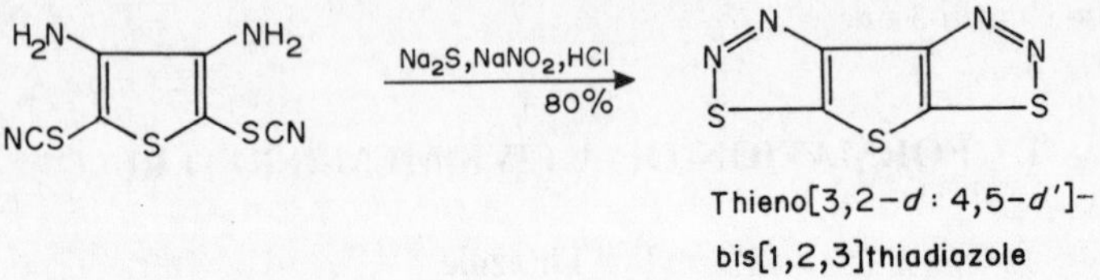

Thieno[3,2–*d* : 4,5–*d'*]–
bis[1,2,3]thiadiazole

[805]

II. FORMATION OF A SIX-MEMBERED RING

1. 1, 4-Thiazin-3-one

The thiocyanate group behaves quite differently when treated with bromoacetic acid and alkali; the C—N moiety is lost and the product is a fused thiazinone ring.

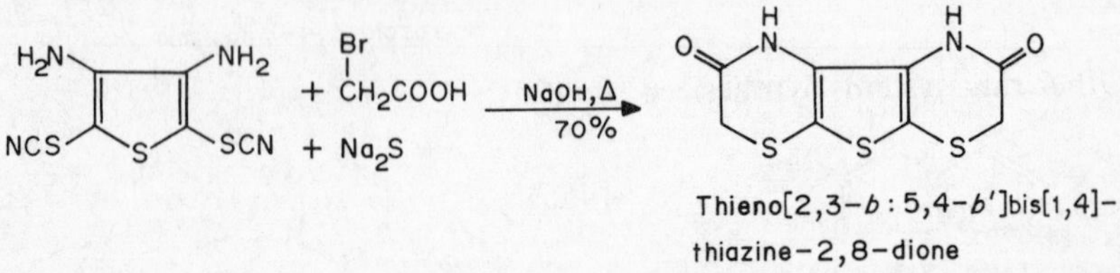

Thieno[2,3–*b* : 5,4–*b'*]bis[1,4]–
thiazine–2,8–dione

[805]

Acyl Halide and Ring-carbon or Ring-nitrogen

Cyclizations of compounds containing an acyl halide function are usually discussed with those of the parent acid but in view of the relatively large number of cyclizations involving reaction between an acyl halide and a ring-carbon or ring-nitrogen atom, these are considered separately in this chapter.

I. FORMATION OF A FIVE-MEMBERED RING

1. Pyrrol-2-one

In spite of its low basicity, a pyrrole nitrogen is acylated intramolecularly by an acyl halide and a pyrrolizin-3-one is formed.

[818]

3–Pyrrolizinone

II. FORMATION OF A SIX-MEMBERED RING

1. Pyridazin-4-one

Intramolecular cyclization of a hydrazonomalonyl chloride leads to the formation of a fused pyridazin-4-one ring.

107

R = H, MeO, NO$_2$

4-Cinnolinone

[352]

2. Pyrazine-2, 5-dione

This ring is formed by reaction of an indole-2-carbonyl chloride with a glycine ester.

+ PhCH$_2$NHCHCOOEt
        |
        R^2

R^1 = H, PhCH$_2$O; R^2 = H, Me

Pyrazino[1,2-a]indole-1,4-dione

[1216]

3. Pyran-4-one

A Friedel–Crafts intramolecular acylation of a phenoxyheterocyclic acid chloride forms a new pyran-4-one ring—a reaction which is a variation on the more common attack on a benzene ring by a carboxyl group (see Chapter 73).

[1]Benzopyrano[2,3-b]
quinoxalin-12-one

[7]

III.   FORMATION OF A SEVEN-MEMBERED RING

1.   Azepin-3- or -4-one

Intramolecular Friedel–Crafts acylation is used to prepare these rings, the catalyst being aluminium chloride. The parent dicarboxylic acid (**21.1**, R = OH) failed to cyclize with PPA but one of the acid chloride groups reacted preferentially under very mild conditions (0 °C) in the presence of aluminium chloride.

R^1 = H, Br, Cl, NO$_2$, CN;
R^2 = H, Cl

Pyrrolo[2,1-b][3]benzazepin-11-one

[547,548]

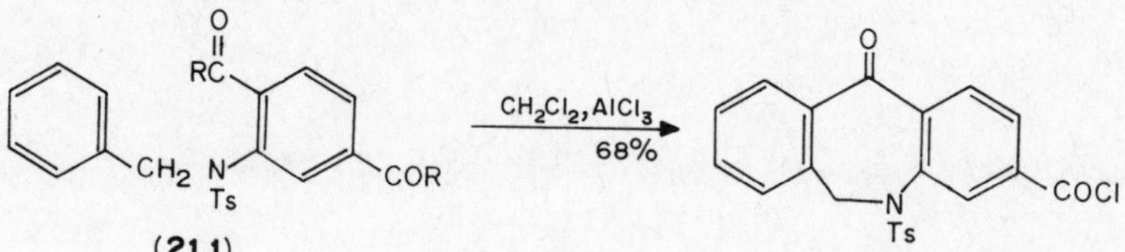

(21.1)

R=Cl

Dibenz[b,e]azepin–11–one

[436]

2. Oxepin-4-one or Thiepin-4-one

A doubly fused oxepin-4-one or thiepin-4-one ring is formed by an intra-molecular Friedel–Crafts reaction using a mild catalyst and dichloromethane as solvent.

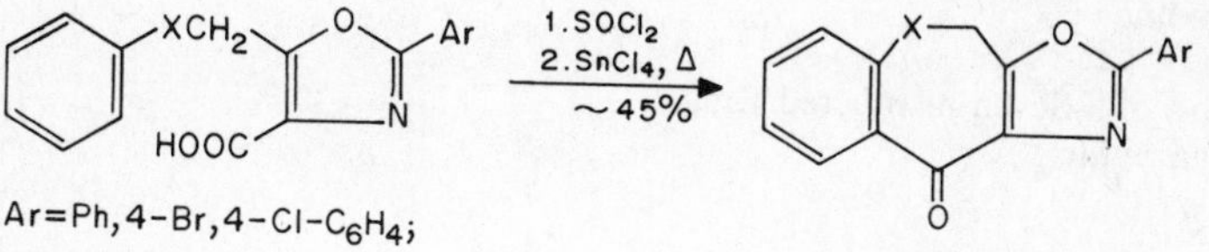

[270]

Ar=Ph,4-Br,4-Cl-C$_6$H$_4$;
X=O,S

X=O,S

[1]Benzoxepino[4,3-d]oxazol-4-one
[1]Benzothiepino[4,3-d]oxazol-4-one

Other ring systems synthesized similarly:

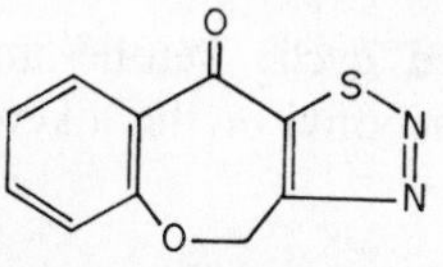

[270]

[1]Benzoxepino[3,4-d]–
1,2,3-thiodiazol-10-one

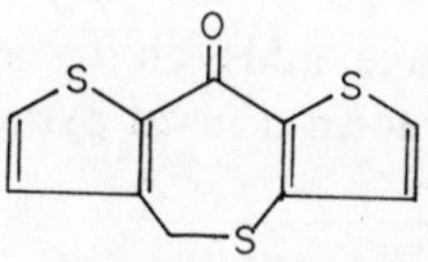

[549]

Dithieno[3,2-b:2',3'-e]–
thiepin-9-one

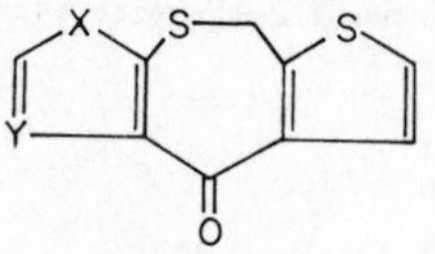

[549]

X=S,Y=CH
Dithieno[2,3-b:3',2'-e]–
thiepin-4-one

X=CH,Y=S
Dithieno[3,2-b:3',2'-e]–
thiepin-4-one

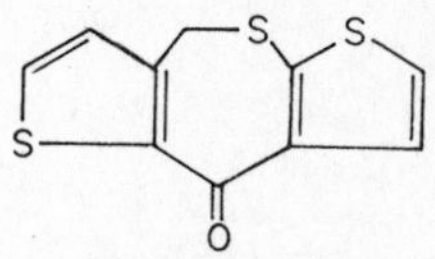

[549]

Dithieno[2,3-b:2,3-e]–
thiepin-4-one

CHAPTER 22

Aldehyde or Ketone and Alkene or Alkyne

I. FORMATION OF A FIVE-MEMBERED RING

1. Pyrazole

Phenylhydrazine adds on to an $\alpha\beta$-unsaturated cyclic ketone under mild conditions to form a fused pyrazole ring. The carbonyl or the alkene may be endocyclic.

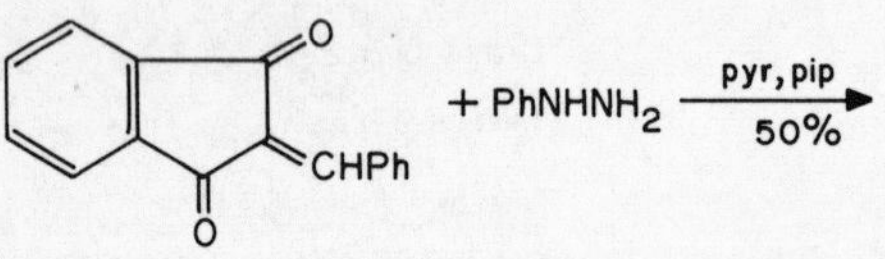

[440]

Indeno[1,2-*c*]pyrazol-4-one

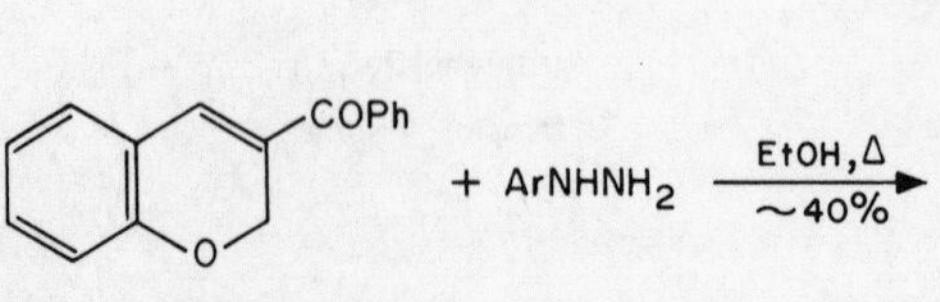

[634]

Ar = Ph, 4-$NO_2C_6H_4$

[1]Benzopyrano[4,3-*c*]-pyrazole

II. FORMATION OF A SIX-MEMBERED RING

1. Pyridine or Pyridin-2-one

Cyclic $\alpha\beta$-unsaturated ketones may be converted into a pyridine ring by prolonged heating with either ammonium formate or ethyl cyanoacetate-ammonium acetate. In the former reaction, two molecules of reactant combine to form a doubly fused pyridine. The indanone (**22.1**) yields a mixture of products.

R=alkyl,PhCH$_2$,4-Br-,
4-COOEt-C$_6$H$_4$
Ar=Ph,4-MeO$_6$H$_4$

Dipyrrolo[3,4-*b*: 3′,4′-*e*]−
pyridine−3,5−dione

[319]

(**22.1**)

R=H,55%,R=CN,35%
Indeno[1,2-*b*]pyridin-2-one

[318]

Thermal cyclization of a 2-vinyl-aldehyde oxime occurs in decalin in good yield.

R=H,Me,Ph

Thieno[3,2-*c*]pyridine

[1405]

Another ring system synthesized similarly:

[405]

Pyrido[4,3-*b*]indole

2. Pyridazine

This ring is formed when an oxoalkyne is treated with hydrazine and an acid in boiling toluene.

Furo[3,4-c]pyridazine

[442]

3. Pyrimidine

A fused 2-aminopyrimidine ring may be built from a 2-enone and guanidine.

$R=PhCH_2, 4-BrC_6H_4$

Pyrrolo[3,4-d]pyrimidin-7-one

[474]

4. Pyran

Under acid catalysis at ambient temperature, an o-alkenyl-ketone is converted into a pyran ring in high yield.

R = Me, Et, Ph, PhCH=CH

Pyrano[4,3-c]pyrazol-3-one

[1061]

III. FORMATION OF A SEVEN-MEMBERED RING

1. 1,2-Diazepine

Conversion of the substrate aldehyde or ketone to its tosylhydrazone and heating this with an alkoxide causes ring closure to a fused 1,2-diazepine in good yield.

$R^1 = H$, alkyl; $R^2 = H$, Me, Ph;
$R^3 = H$, MeO

2,3-Benzodiazepine

[1583, 1604]

Other ring systems synthesized similarly:

Thieno[2,3-d][1,2]-
diazepine

[1603]

Thieno[3,2-d][1,2]-
diazepine

[1603]

CHAPTER 23

Aldehyde or Ketone and Azide

I. FORMATION OF A FIVE-MEMBERED RING

1. Pyrazole

Reaction of the carbonyl group with hydrazine hydrate followed by heating (sometimes with acetic acid) yields a new pyrazole ring in high yield. An *o*-azidoketoxime behaves similarly on heating in toluene (only the *E*-oxime of a diphenyl ketone cyclizes). This shows that one nitrogen atom of the azido group remains to form the new ring; similarly, an *o*-azidoaldehyde phenylhydrazone cyclizes thermally to give a 2-phenylaminoindazole. A nitrene is probably an intermediate in several of these cyclizations [B-23].

R=H,Me,PhCH₂,Ph,
4-MeOC₆H₄CH₂;
1-naphthylmethyl

Indazole   [843, 1697]

R=H,Me,Ph

Pyrazolo[3,4-*d*]pyrimidine-
4,6-dione   [81]

$R^1 = H, Cl; R^2 = H, Me, Ph$
$R^3 = HO, PhNH$

[584, 728]

Indazole

2. Isoxazole

When an o-azido-aldehyde or -ketone is heated, an isoxazole ring is formed in good yield. The ketone may be either cyclic or acyclic.

$Ar = Ph, 4-alkyl,$
$4-Br-, 4-MeO-,$
$4-NO_2-C_6H_4$

2,1-Benzisoxazole

[1644]

Other ring systems synthesized similarly:

[139]

a

Anthra[1,9-cd]isoxazol-6-one

[81]

Isoxazolo[3,4-d]pyrimidine-
4,6-dione

II. FORMATION OF A SIX-MEMBERED RING

1. Pyridine

When an o-azidoaldehyde is heated with aniline, the nitrene initially generated attacks the phenyl ring and a pyridine is formed. Thermolysis of a side-chain azide containing a nuclear acetal also produces a pyridine ring.

[81]

Pyrimido[4,5-b]quinoline-2,4-dione

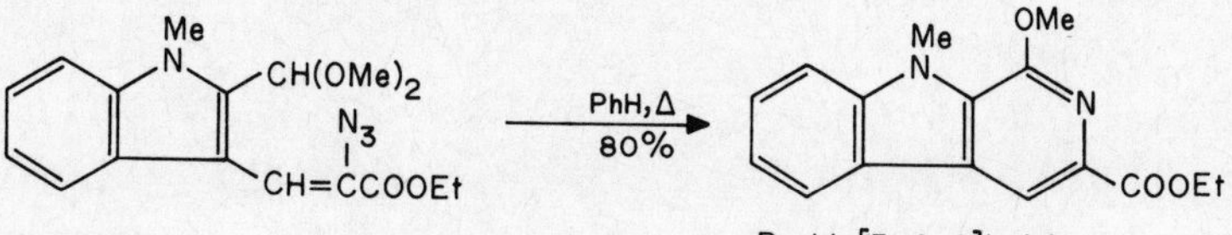

Pyrido[3,4-*b*]indole

[866]

Isoquinolines are formed thermally framazido-ketones [379].

2. 1, 2, 3-Triazine

When an azidoaldehyde is heated with triethyl phosphite, the three nitrogen atoms are retained and a fused 1, 2, 3-triazine ring is formed.

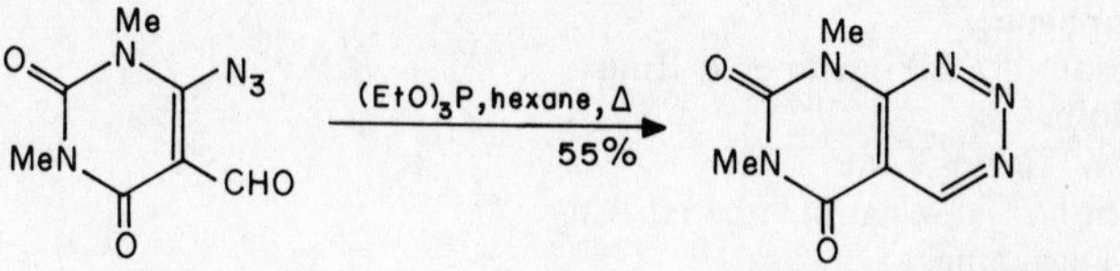

Pyrimido[4,5-*d*]-1,2,3-triazine-5,7-dione

[81]

CHAPTER 24

Aldehyde or Ketone and Carbamate

I. FORMATION OF A FIVE-MEMBERED RING

1. Thiophene

A reversed thiocarbamate placed ortho to an aldehyde function reacts with sodium chloroacetate to give a fused thiophene ring.

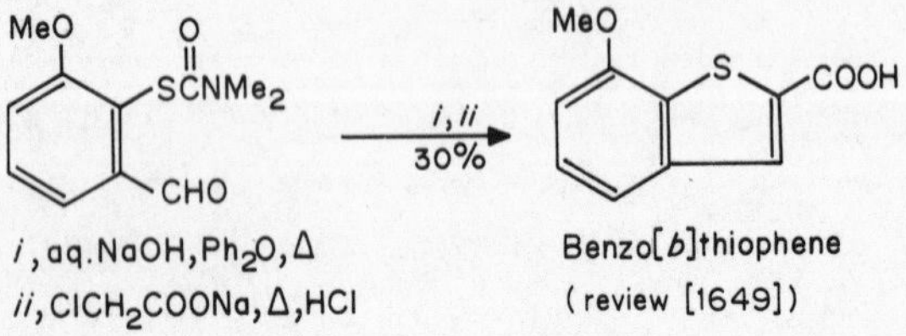

i, aq.NaOH, Ph$_2$O, Δ
ii, ClCH$_2$COONa, Δ, HCl

Benzo[b]thiophene
(review [1649])

[726]

II. FORMATION OF A SIX-MEMBERED RING

1. Pyridine

Base-catalysed cyclization of a 2-acyl-carbamate with acrylonitrile gives moderate yields of a dihydropyridine ring.

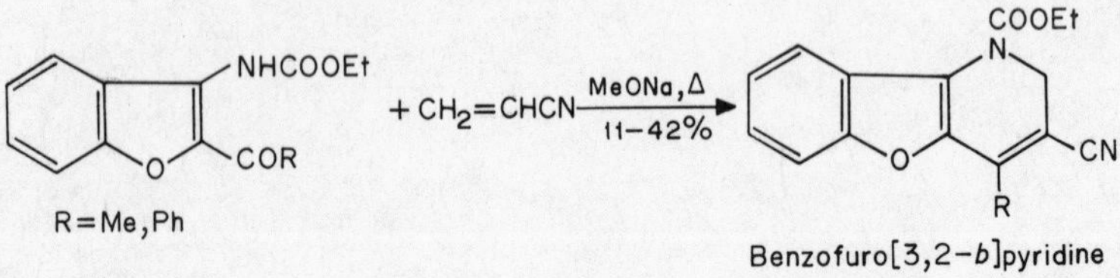

R = Me, Ph

Benzofuro[3,2–b]pyridine

[856]

116

2. 1, 2, 4-Triazin-3-one

Heating a phenylhydrazone of an aldehyde or ketone which also has a carbamate function results in ring closure in good yield.

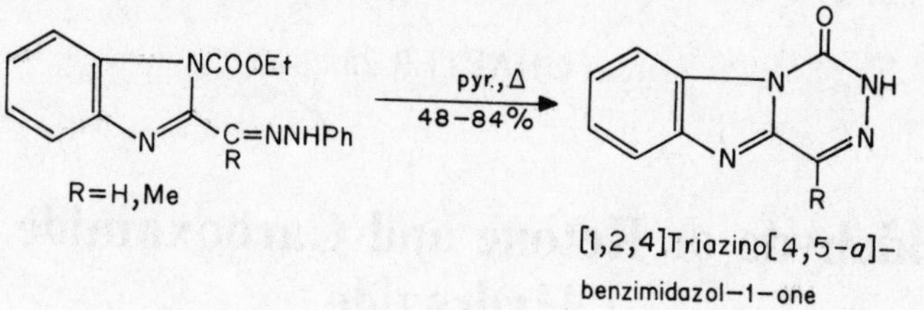

$$\text{R} = \text{H, Me}$$

[1171]

[1,2,4]Triazino[4,5-a]-
benzimidazol-1-one

III. FORMATION OF A SEVEN-MEMBERED RING

1. 1, 4-Diazepine

A benzyl carbamate is a useful protecting group for an amine as the benzyl can be removed by acid or hydrogenolysis leaving the N-carboxyl which decomposes on warming. This route is used in a ring-closure which is the final step in a synthesis of a diazepine ring. The medicinally important benzodiazepines have been reviewed [B-24, 1895].

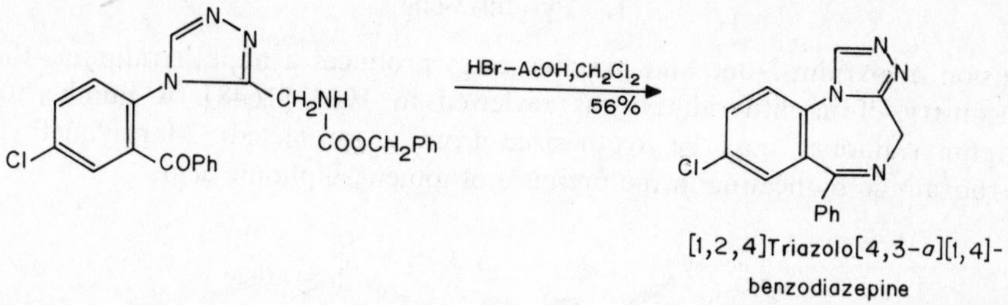

[1,2,4]Triazolo[4,3-a][1,4]-
benzodiazepine

[1348]

CHAPTER 25

Aldehyde or Ketone and Carboxamide or Hydrazide

I. FORMATION OF A SIX-MEMBERED RING

1. Pyridin-2-one

Fusion of pyridin-2-one and pyridine rings produces a naphthyridinone; the chemistry of naphthyridines was reviewed in 1983 [1648]. A number of naphthyridinones may be synthesized from a (protected) 2-formylmethyl-carboxamide by heating in the presence of toluenesulphonic acid.

1,7-Naphthyridin-8-one   [1708]

Other ring systems synthesized similarly:

1,6-Naphthyridin-5-one   [1708]

X=CH,Y=N
2,6-Naphthyridin-1-one

X=N,Y=CH
2,7-Naphthyridin-1-one   [1708]

In the presence of a strong base, the N-anion of a keto-carboxamide is produced in preference to that from the methyl group and a fused pyridin-2-one is formed.

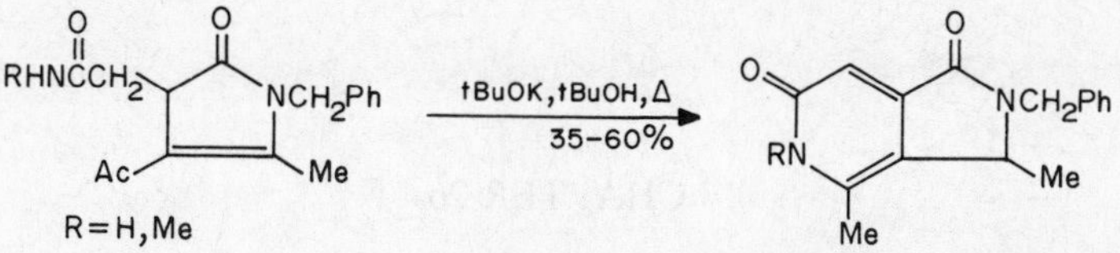

Pyrrolo[3,4-*c*]pyridine-1,6-dione

[143]

2. 1,3-Oxazin-4-one

On heating with acetic anhydride, 4-oxoquinoline-3-carboxamides are converted into, 1,3-oxazin-4-one analogues in high yield.

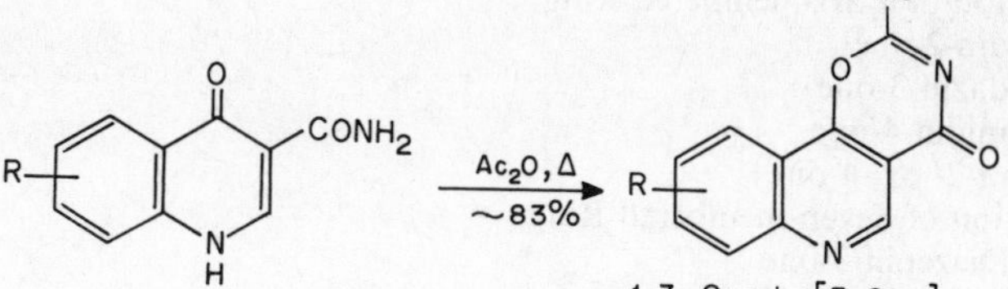

R=Me, Me_2

1,3-Oxazino[5,6-*c*]-quinolin-4-one

[1003]

II. FORMATION OF A SEVEN-MEMBERED RING

1. 1,2-Diazepin-3-one

The hydrazone of a 2-phenacyl-hydrazide cyclizes to a diazepinone on heating with acetic acid.

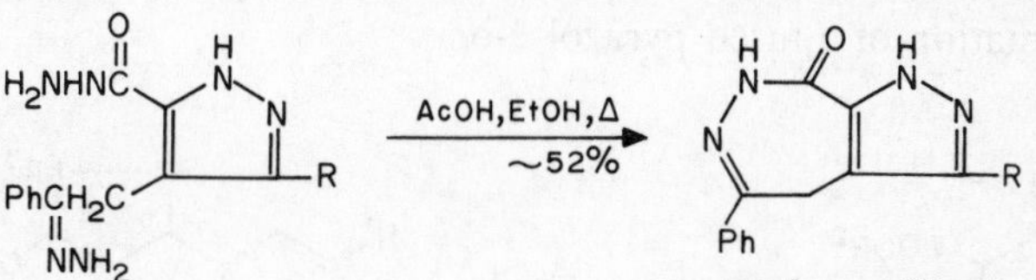

R=Me, Ph

Pyrazolo[3,4-*d*][1,2]-diazepin-8-one

[162]

CHAPTER 26

Aldehyde or Ketone and Carboxylic Acid or Ester

I. FORMATION OF A FIVE-MEMBERED RING

1. Pyrazol-3-one

A cyclic 2-oxocarboxylic acid or ester condenses with hydrazine or arylhydrazine with the formation of a fused pyrazol-3-one.

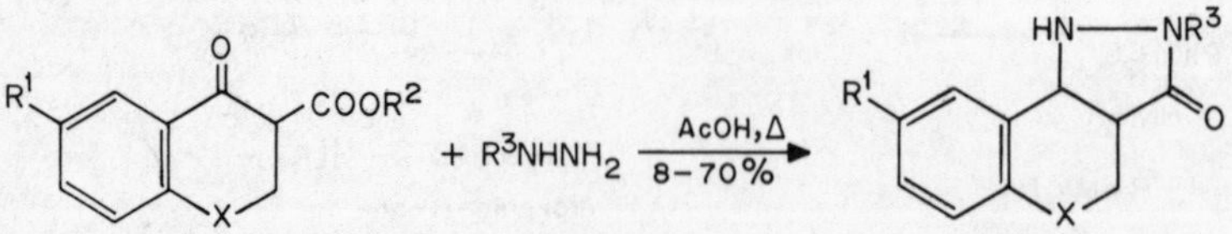

[341, 448]

R^1=H,Me,Br,Cl; R^2=H,Me;

R^3=H,Ph,Cl-,NO_2-,Me-,$MeSO_2$-C_6H_4

X=O,S

[1]Benzopyrano[4,3-*c*]-pyrazol-3-one

[1]Benzothiopyrano[4,3-*c*] pyrazol-3-one

Another ring system synthesized similarly:

[417]

Pyrazolo[3',4': 4,5]thiopyrano-
[3,2-c][1,2]-benzothiazin-1-one
5,5-dioxide

II. FORMATION OF A SIX-MEMBERED RING

1. Pyridin-2-one

Reaction of a keto-ester with ammonia or methylamine results in the formation of a new pyridin-2-one ring.

[1027]

R = H,Me

Furo[3,4-c]pyridin-4-one

Other ring systems synthesized similarly:

[185]

Pyrido[3,4-b]quinoxalin-1-one

[161]

1-Isoquinolinone

[163]

3-Isoquinolinone

[2004]

Thieno[3,4-c]pyridin-4-one

2. Pyridazin-3-one

One of the commonest reagents for ring closure of keto-acids or keto-esters is hydrazine or its salt in boiling ethanol. Monosubstituted hydrazines have also been used. When thiosemicarbazide ($R^3 = H_2NCS$) is used, the thiocarboxamide group is lost and the tautomeric phthalazine (**26.1**) is obtained.

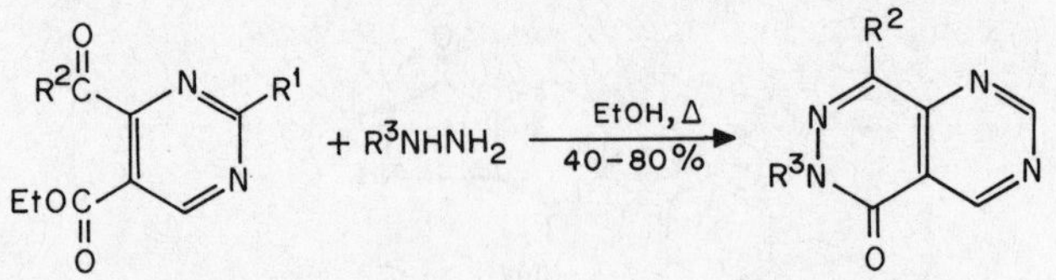

R¹=H,Me,HO,SH,NH₂,Ph;
R²=H,Me; R³=H,Ph.

Pyrimido[4,5-*d*]-
pyridazin-5-one

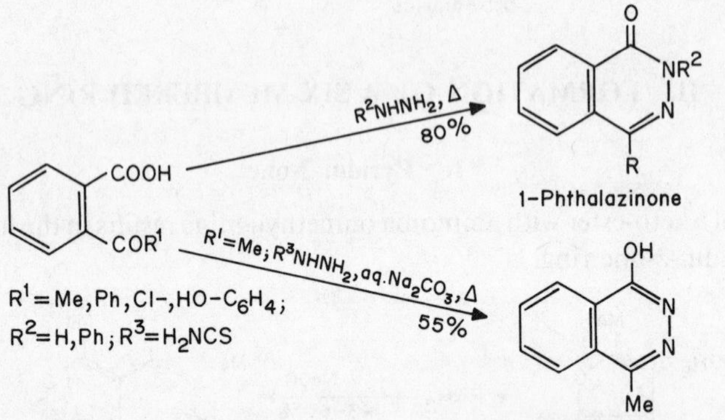

[749]

R¹=Me,Ph,Cl-,HO-C₆H₄;
R²=H,Ph; R³=H₂NCS

1-Phthalazinone   [160]

(**26.1**) Phthalazine   [1685]

Other ring systems synthesized from a keto-acid [1110] or a keto-ester [419, 771, 996, 1020, 1095, 1247] or a formyl-ester [2006]:

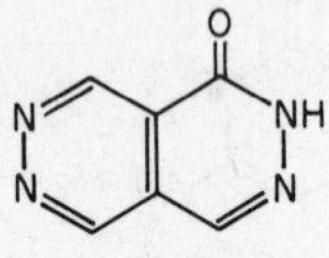

Pyridazino[4,5-*d*]-
pyridazin-1-one   [1095]

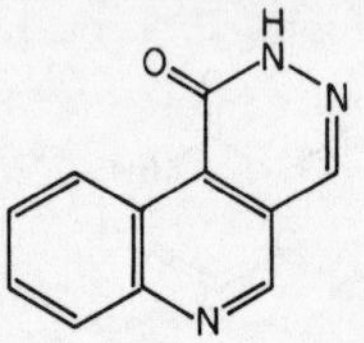

Pyridazino[4,5-*c*]quinolin-1-one   [996]

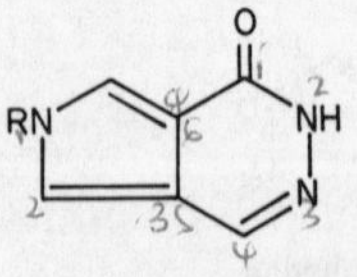

Pyrrolo[3,4-*d*]pyridazin-1-one   [2006]

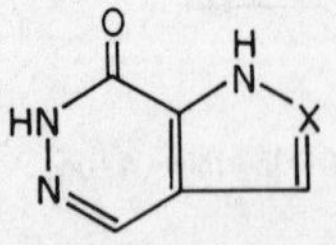

X=CH,N
Pyrrolo[2,3-*d*]-
pyridazin-7-one
Pyrazolo[3,4-*d*]-
pyridazin-7-one   [419, 771]

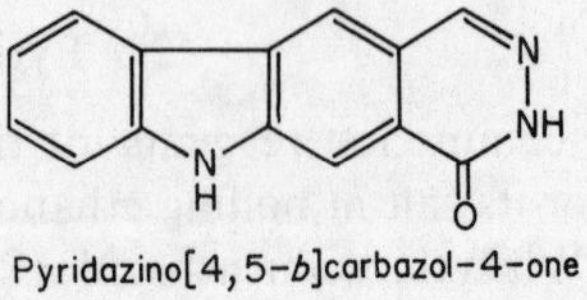

Pyridazino[4,5-*b*]carbazol-4-one   [1110]

3. Pyrimidin-4-one

A 2-aminopyridine condenses with a keto-ester to form a doubly fused pyrimidinone in good yield.

[135]

Pyrido[1,2-*a*]thieno[3,4-*d*]
pyrimidin-10-one

4. Pyran-2- or -4-one

Reaction of one of two ester groups of (26.2) with an enolized carbonyl group leads to the formation of a lactone but a more complex mechanism may account for the cyclization of a 3-azido-5-oxopyridazine carboxylic ester which gives a good yield of the pyran-2-one on treatment with sulphuric acid.

[185]

(26.2)

Pyrano[3,4-*b*]quinoxalin-1-one

[382]

Pyrano[2,3-*d*]pyridazine-2,5-dione

A Baker–Venkataraman rearrangement [B-12] in a base followed by cyclization of the diketone is a commonly used synthesis of flavones.

[981]

R = H, 3,5-$(NO_2)_2$;
Ar = 2-AcC_6H_4, 2-Ac-3,5-$(NO_2)_2C_6H_2$

1-Benzopyran-4-one, 2-aryl

(review [B-32])

With the carbonyl and carboxy groups appropriately separated, reaction between them can be effected by heating with thionyl chloride or propanoyl chloride-aluminium chloride.

2-Benzopyran-1-one   [1113]

Other ring systems synthesized similarly:

Pyrano[3,4-c]-
pyrazol-7-one   [162]

Pyrano[3,2-b]pyridin-2-one   [342]

III. FORMATION OF A SEVEN-MEMBERED RING

1. 1,2-Diazepin-3-one

When the carbonyl and ester groups have three carbon atoms separating them, a 1,2-diazepinone ring is formed in high yield. Spectral evidence later showed that the corresponding [5,4-b]isomer existed in the enol form.

Note added in proof: A recent paper [1679] shows that the product of the latter reaction is a pyrido[4,3-b]indol-3-one (26.1).

[1,2]Diazepino[4,5-b]indol-1-one   [23]

[1,2]Diazepino[5,4-b]indole   [1081]

(26.1) Pyrido[4,3-b]indol-3-one   [1679]

CHAPTER 27

Aldehyde or Ketone and Ether or Thioether

I. FORMATION OF A FIVE-MEMBERED RING

1. Pyrazole

Many thioether groups may be displaced by a nucleophile such as hydrazine hydrate and the presence of a second function which reacts with the other nitrogen of the hydrazine results in cyclization. Some pyran-4-one rings are cleaved by nucleophiles but the high reactivity of the methylthio group in some compounds allows it to be displaced without appreciable opening of the ring.

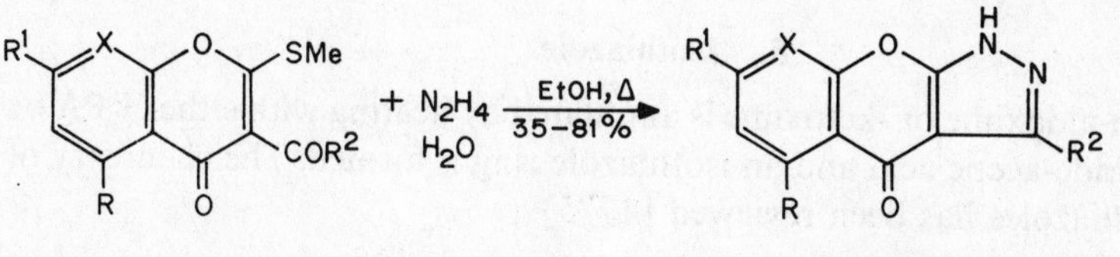

$R^1 = H, Me; R^2 = Me, Ph$

$+ N_2H_4 \cdot H_2O$ $\xrightarrow[\text{35-81\%}]{\text{EtOH, }\Delta}$ [245, 1822]

X = CH
[1]Benzopyrano[2,3-c]pyrazol-4-one

X = N
Pyrazolo[4′,3′:5,6]pyrano[2,3-b]-pyridin-4-one

2. Furan

Demethylation of an oxo-ether and heating the mixture gives a doubly fused furan in favourable substrates.

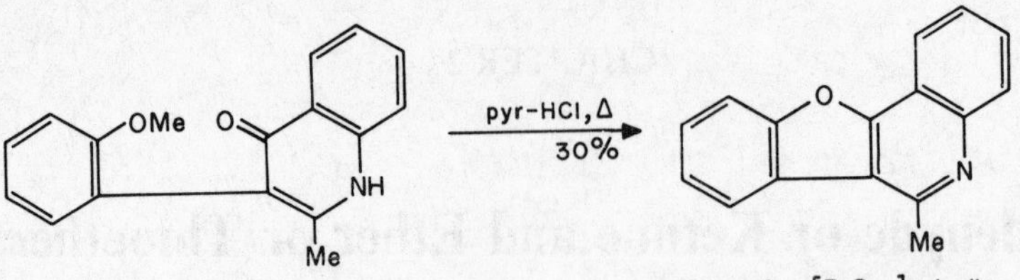

[431]

Benzofuro[3,2-c]quinoline

3. Thiophene

A methylthio group may be demethylated and may then react with chloroacetic acid to produce a fused thiophene ring.

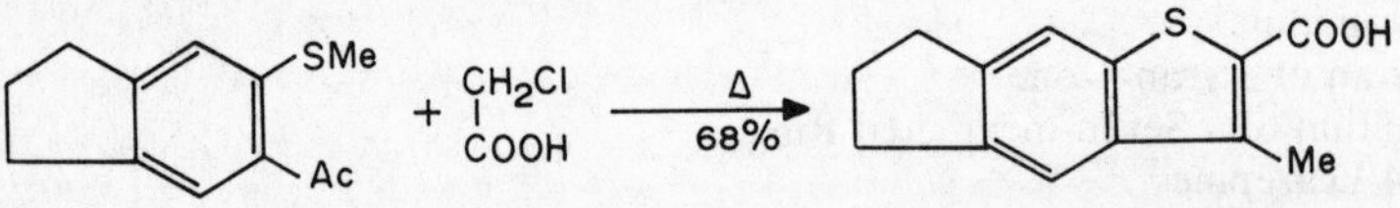

[1272]

Indeno[5,6-b]thiophene

4. Isoxazole

Similarly, hydroxylamine forms a fused isoxazole ring.

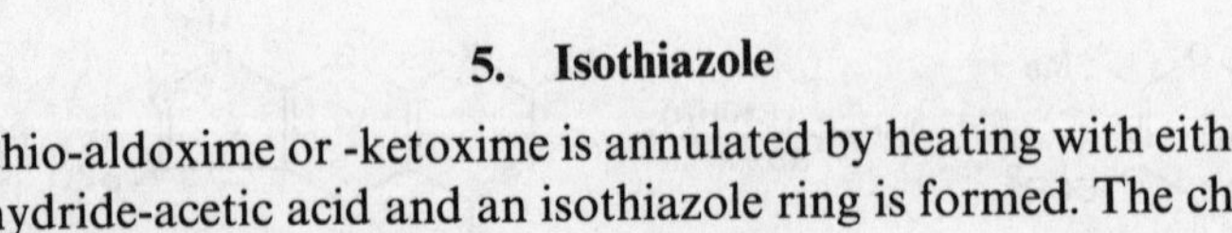

[245]

[1]Benzopyrano[2,3-c]-
isoxazol-4-one

5. Isothiazole

A 2-alkylthio-aldoxime or -ketoxime is annulated by heating with either PPA or acetic anhydride-acetic acid and an isothiazole ring is formed. The chemistry of 1,2-benzisothiazoles has been reviewed [1775].

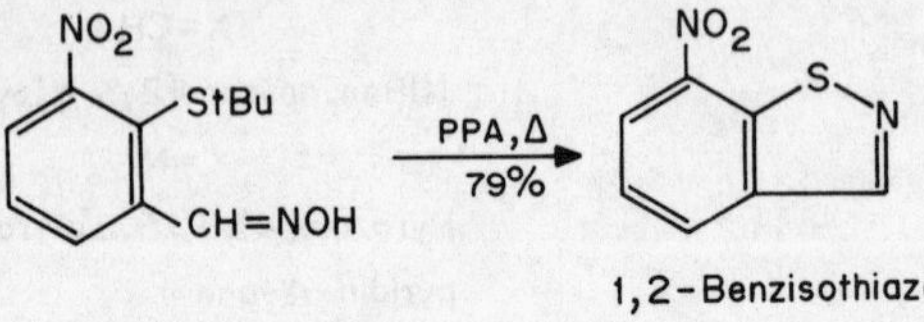

[896]

1,2-Benzisothiazole

[737]

Thieno[2,3-d]isothiazole

Another reaction in which the sulphur atom is retained is the cyclization of a 2-acetyl-thiol by chloramine.

[737]

Thieno[3,2-d]isothiazole

II. FORMATION OF A SIX-MEMBERED RING

1. Pyridine

An alkoxy group attached to an ethylideneamino function is displaced by an aminoacetaldehyde acetal under relatively mild conditions and an adjacent carbonyl group enables cyclization to occur in boiling ethanol to give a pyrimidin-4-ol but in boiling xylene, the same reactants yield the 2-aminoquinoline.

R^1 = alkyl; R^2 = H, Cl;
R^3 = (EtO)$_2$CHCH$_2$

EtOH, Δ
51–70%

Quinazoline

[539]

xylene, Δ
56–76%

Quinoline

[539]

2. Pyrimidine

An amidine reacts (in the presence of a base) with an o-oxoketene dithioketal to form a new pyrimidine ring.

R = 4-pyribyl

DMF, NaH,
PhH, Δ
57%

Benzo[h]quinazoline

[745]

An amidine also cyclizes a 2-(methylthio)-aldehyde or -ketone in good yield.

$R^1 R^2 = H, Me; R^3 = Me, Ph, NH_2$

[245, 1822]

$X = CH, N$

[1]Benzopyrano[2,3-*d*]pyrimidin-5-one
Pyrido[3′,2′:5,6]pyrano[2,3-*d*]-
pyrimidin-5-one

The reaction of an aminoacetaldehyde acetal with an alkoxyethylidene-amino ketone [539] to form a quinazolinol is described in Section II.1

3. Pyran or Pyran-4-one

1,3-Diketones cyclize on to a phenolic function, generated by dealkylation of an ether, to give a fused pyran-4-one in good yield.

Ar = 3- or 4-pyridyl

[750]

$X = CH, Y = N$
Pyrano[3,2-*c*]pyridin-4-one
$X = N, Y = CH$
Pyrano[2,3-*b*]pyridin-4-one

Attempts to generate the hydroxy group of the cinnamic aldehyde (**27.1**) were unsuccessful but a high yield of the pyran was obtained directly. It appears that the mercury (II) ion is essential for this cyclization.

(**27.1**)

MeCN, HgCl$_2$, Δ
97%

[1087]

2*H*-1-Benzopyran

III. FORMATION OF A SEVEN-MEMBERED RING

1. 1,4-Diazepine

o-Phenylenediamine cyclizes a 2-mercapto-aldehyde to form a doubly fused diazepine.

[1]Benzopyrano[2,3-*b*][1,5]-
benzodiazepin-13-one

CHAPTER 28

Aldehyde or Ketone and Halogen

I. FORMATION OF A FIVE-MEMBERED RING

1. Pyrrole

A bifunctional reagent such as hydrazine may sometimes behave as a monofunctional compound, for example, when reacting with a halogen and adjacent carbonyl group to form a 1-aminopyrrole ring.

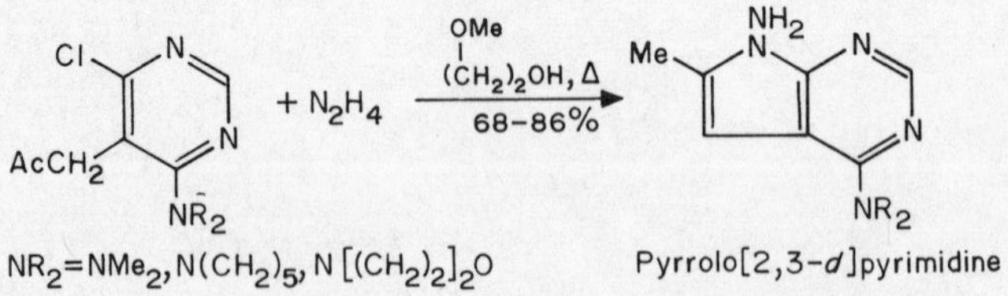

$NR_2 = NMe_2, N(CH_2)_5, N[(CH_2)_2]_2O$ Pyrrolo[2,3-d]pyrimidine

[1414]

2. Pyrazole

Hydrazine reacts in its usual difunctional role (cf. Section I.1, above) with reactive o-chloroaldehydes to produce a fused pyrazole ring, or a preformed hydrazone may be cyclized thermally, photochemically, or with DBU [1968].

130

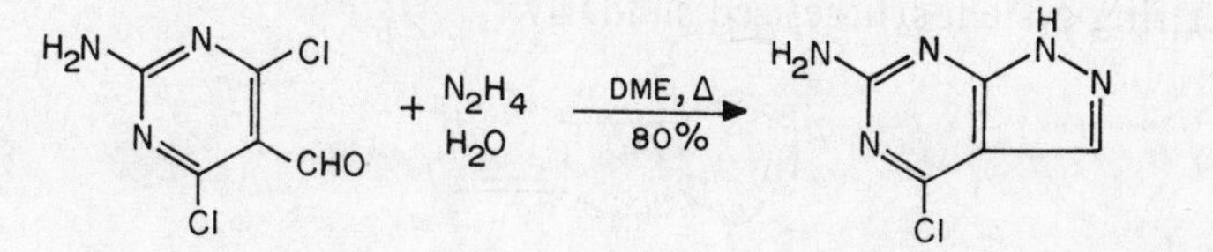

[1832]

Pyrazolo[3,4-d]pyrimidine

Other ring systems synthesized similarly:

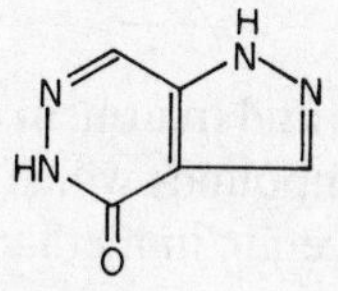

[1450, 1577]

[1578]

Pyrazolo[3,4-d]
pyridazin-4-one

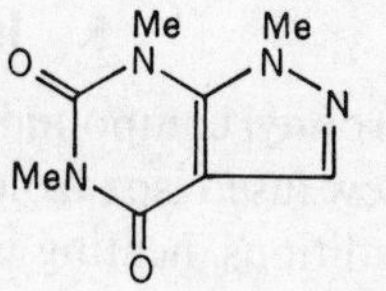

Pyrazolo[3,4-d]pyrimidine-
4,6-dione

3. Imidazole

An *N*-2-oxopropyl group adjacent to a chlorine atom reacts with ammonia to give a fused imidazole ring.

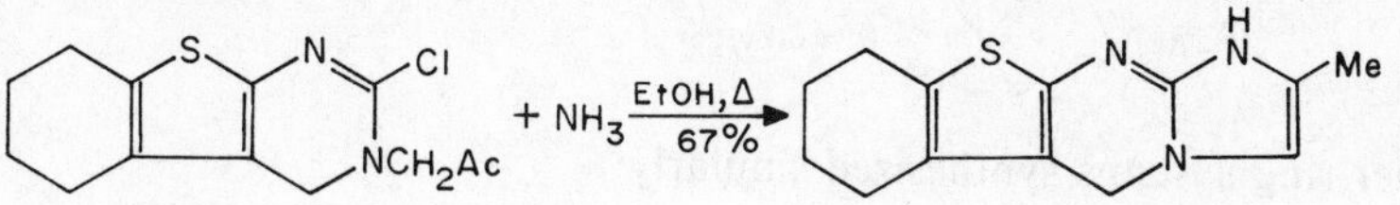

[232]

[1]Benzothieno[2,3-d]imidazo
[1,2-a]pyrimidine

4. Thiophene

Thioacetamide serves as a source of sulphur in the cyclization of some 2-bromomethyl ketones. Mercaptoacetic ester efficiently converts a 2-chloroaldehyde into a thiophene ring.

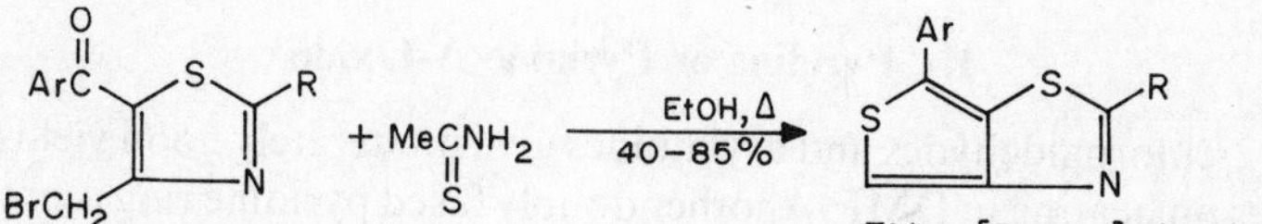

[1038]

R=H,Ph; Ar=Ph,4-MeO-,
4-MeS-C₆H₄

Thieno[3,4-d]thiazole

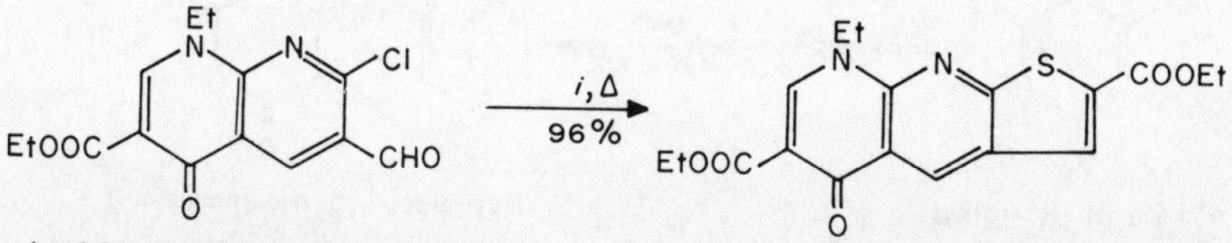

[1989]

i,HSCH₂COOEt,Na₂CO₃,EtOH

Thieno[2,3-b][1,8]naphthyridin-5-one

Another ring system synthesized similarly:

[235]

Thieno[3,4-*b*]indole

5. Isoxazole

Conversion of the carbonyl compound into its oxime and treatment of this with a strong base gives a new fused isoxazole ring. For compounds which are unstable to strongly basic conditions, heating with sodium acetate in methanol is a more promising method [1470].

$$\xrightarrow[78-99\%]{\text{1.NH}_2\text{OH 2.EtOH,KOH,}\Delta}$$

[614, 873]

$R^1 = F, Cl$

1,2-Benzisoxazole

$R^2 = Ar,$ $R^3 = $ alkyl, aryl

Other ring systems synthesized similarly:

[1168]

Furo[3,2-*f*]-1,2-benzisoxazole

[1470]

Isoxazolo[5,4-*b*]quinoline

II. FORMATION OF A SIX-MEMBERED RING

1. Pyridine or Pyridine *N*-Oxide

Reactive 2-chloroaldehydes and arylamines give moderately good yields of fused quinolines on heating in DMF. Another doubly fused pyridine ring is obtained by treating a benzyl chloride containing an aldehyde group in another ring with triethyl phosphite and base.

$+ \text{PhNHR}^2 \xrightarrow[42-62\%]{\text{DMF,}\Delta}$

[314]

$R^1 = Me, Ph; R^2 = $ alkyl

Pyrimido[4,5-*b*]quinoline-
2,4-dione

$$\text{P(OEt)}_3, \text{EtONa}, \Delta \quad 75\%$$

[1788]

OP(OEt)$_2$

Pyrrolo[1,2-a]quinoline

2-Bromo-oximes are converted under mild conditions into pyridine N-oxides by a copper-assisted cyclization using an active methylene reagent such as a malonic acid derivative. Strong alkali converts a fluoroaldoxime into a pyridine 1-oxide ring (review [B-34]).

$$\xrightarrow{\text{PhH, NaH, CuBr}} 50–79\%$$

[1418]

R^1=H, 4,5-OCH$_2$O; R^2=H, Me;
R^3=Ac, PhCO, COOMe, CN

Isoquinoline N-oxide

$$\xrightarrow{\text{aq. NaOH}} \sim 75\%$$

[1356]

R^1=H, Cl; R^2=H, Me

Dibenzo[cf][2,7]naphthyridine 8-oxide

2. Pyridazine

Hydrazine can behave as either a mono- or di-functional reagent (see Sections I.1 and I.2) and the distinction in its reaction with halogeno-ketones sometimes depends on the 6-substituent; with the 6-chloropyrimidines in ethanol at ambient temperature, it behaves as a difunctional compound to give a fused pyridazine ring.

$$+ \text{N}_2\text{H}_4 \quad \xrightarrow[\text{H}_2\text{O}]{\text{EtOH}} \quad 87\%$$

[1414]

Pyrimido[4,5-c]pyridazine

3. Pyrimidine

Guanidine annulates peri-positioned halogen and carbonyl groups in the presence of copper to form a pyrimidine ring.

Benzo[6,7]cyclohepta-
[1,2,3-*de*]quinazoline

4. 1,2-Thiazine 1,1-Dioxide

A base-generated *N*-anion from methanesulphonamide displaces a halogen and the *C*-anion (from methyl group) attacks the aldehyde to form a thiazine ring.

Benzofuro[2,3-*c*][1,2]-
thiazine 2,2-dioxide

III. FORMATION OF A SEVEN-MEMBERED RING

1. Azepine

A primary aliphatic amine (obtained *in situ* from a bromide and ammonia) condenses with a carbonyl group to give an azepine when the groups are suitably separated.

Ar = 2-ClC$_6$H$_4$

Pyrimido[5,4-*d*][2]-
benzazepine

2. 1,4-Diazepine

When the halogen and carbonyl groups are directly attached to a ring, heating the compound with a 1,2-diamine gives the diazepine ring. Ammonia supplies the second nitrogen in compounds where the halogen and ketone functions are already joined through four carbon and one nitrogen atoms.

R^1 = alkyl; R^2 = H, alkyl, halogen, CF_3;
R^3 = H, alkyl; Ar = Ph, Me-, Cl-,
MeO-, CF_3-C_6H_4

Pyrazolo[3,4-e][1,4]-
diazepine

[445]

[813]

[1,2,4]Triazolo[4,3-a]-
[1,4]benzodiazepine

Other ring systems synthesized similarly:

[1251]

Quinazolino[3,2-a][1,4]-
benzodiazepin-13-one

[983]

Thieno[3,2-e]-1,4-
diazepine

3. 1,4-Diazepin-2-one

A haloacylamine group adjacent to a ketone function is a commonly used precursor of this ring; ammonia, ammonium carbonate or hexamethylenetetramine provides the second nitrogen atom.

R^1 = Cl, NO_2; R^2 = H, Me
X = Br, Cl
i, MeCN, NaI, $(NH_4)_2CO_3$
or EtOH, hexamethylenetetramine

1,4-Benzodiazepin-2-one

[233, 976]

Other ring systems synthesized similarly:

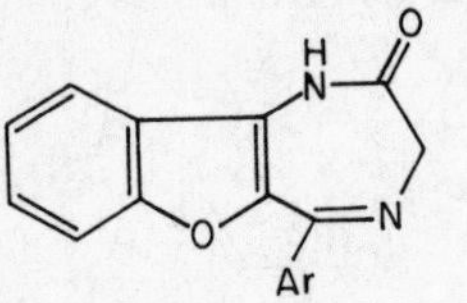

[234]

Benzofuro[3,2-e]-1,4-
diazepin-2-one

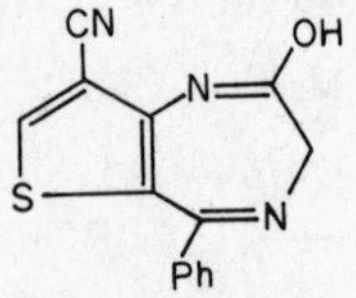

[983]

Thieno[3,2-e]-1,4-
diazepine

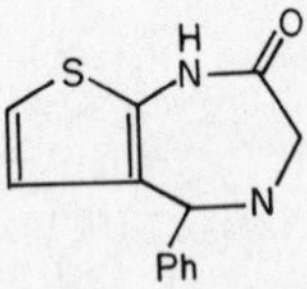

[1613]

Thieno[2,3-e]-1,4-
diazepin-2-one

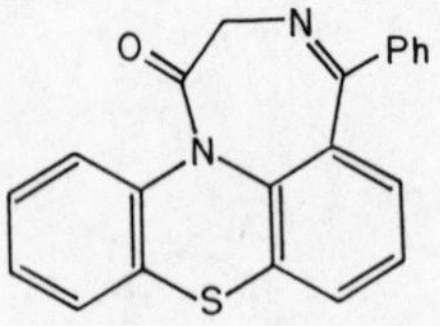

[101]

[1,4]Diazepino[6,7,1-kl]-
phenothiazin-1-one

4. 1,2-Oxazepine

Anionic displacement of a fluorine by an oxime in hot DMF gives this ring.

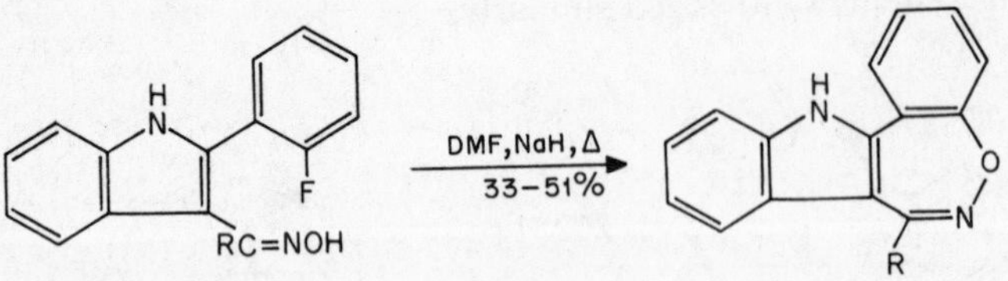

DMF,NaH,Δ
33–51%

[1459]

R=Me,Ph

Indolo[3,2-d][1,2]-
benzoxazepine

CHAPTER 29

Aldehyde and Hydroxy, Thiol or Thiocyanate

I. FORMATION OF A FIVE-MEMBERED RING

1. Isoxazole

Thermal cyclization of a 2-hydroxyaldoxime acetate in acetic anhydride gives a high yield of the isoxazole but this ring is formed under mild conditions by stirring the hydroxy-alkoxime with trichloroacetyl isocyanate.

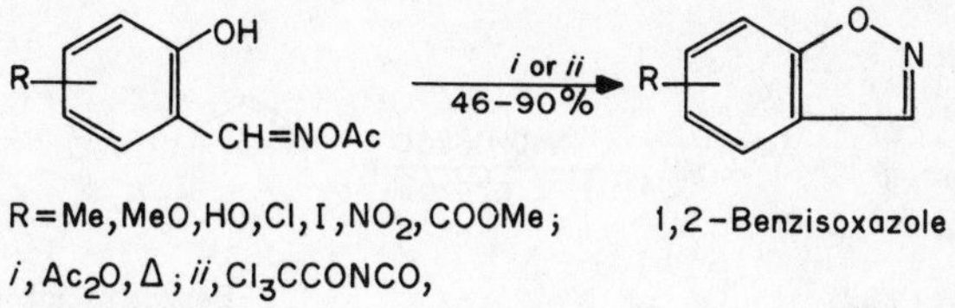

[41, 699]

R = Me, MeO, HO, Cl, I, NO_2, COOMe; *i*, Ac_2O, Δ; *ii*, $Cl_3CCONCO$, THF, K_2CO_3

1,2-Benzisoxazole

2. Furan

Salicylaldehydes react with reactive methylene compounds such as bromonitromethane or α-halogenoketones to form good yields of benzofurans. Sometimes the 3-hydroxy-2-nitrofuran derivative is first isolated and is dehydrated with hot acetic anhydride. Hydroxy, nitro or carboxylic ester-substituted salicylaldehydes are cyclized with bromonitromethane only when followed by heating with acetic anhydride.

R¹=H,Me,MeO,Cl;
R²=Me,CH₂COOEt

Benzofuran

[1738, 1881]

R=H,MeO,Br,CN,NO₂,COOEt
i,Me₂CO,K₂CO₃,Δ; ii,Ac₂O

[740, 1758]

3. Thiophene

A 2-mercaptoaldehyde on reaction with a 2-chloro-alkanoic acid or α-chloroketone in a basic medium is cyclized to the thiophene ring which will carry a 2-carboxyl or 2-acyl group.

Benzo[b]thiophene

[726]

4. Isothiazole

The nitrogen atom necessary to form this ring is usually provided by ammonia.

R=Cl,MeO

1,2-Benzoisothiazole

[726]

Another ring system synthesized similarly:

Isothiazolo[5,4-b]quinoline

[1035]

II.   FORMATION OF A SIX-MEMBERED RING

1.   2H-Pyran

2-Hydroxyaldehydes are a fruitful source of 2H-benzopyrans and many reagents have been used, depending on the nature and position of the substituents required in the product. An alkene activated by a nitro, cyano, aryl or carboxylic ester group may supply C–2 and C–3 of the pyran ring and when the alkene also has a $Me_2C=$ group, this leads to a 2, 2-dimethyl-2H-benzopyran but the reaction is inhibited when R is an OH or an electron-withdrawing group. The carboxylic ester group, however, is lost during the reaction [1893]. Allenes, and epoxides, have also been used; ultrasonic agitation shortened reaction time in reactions with β-nitrostyrene [1525].

$R^1=H,MeO,Br,Ph; R^2=H,Me;$
$R^3=H,Me,Ph,N[(CH_2)_2]_2O;$
$R^4=NO_2,CN,COOEt.$

i, Al_2O_3; *ii*, aq.NaOH, Δ;
iii, PhH, Δ.

[634, 1525, 1546, 1893, 2009]

2H-1-Benzopyran

[1425]

$R^1=H,MeO,MeOCH_2O;$
$R^2=Et,Na; Ar=4-MeO-,$
$4-MeOCH_2O-,4-PhCH_2O-C_6H_4$

[1757]

A 2-hydroxycinnamaldehyde may be cyclized in high yield by reaction with mercury(II) chloride. When saturated aliphatic nitro compounds are employed, phase transfer catalysis gives good results.

[1087]

R^1= H,MeO,Cl,NO$_2$; R^2=H,EtO

2H-1-Benzopyran

R^3= EtO,HO;

i, Bu$_2$$\overset{+}{N}H_2$$\bar{C}$l, AcOC$_5H_{11}$

ii, Et$_3$$\overset{+}{N}HTs\bar{O}$,PhMe

[839, 1480]

Acetaldehyde acetals in perchloric and acetic acids give good yields of the benzopyrylium perchlorates.

R = H,alkyl,Ph

1-Benzopyrylium

[635]

Cyanoacetamide in a basic medium gives a 2-iminobenzopyran.

R=H,MeO,Br,Me$_2$N

[26]

2. Pyran-2-one

This ring is formed from 2-hydroxyaldehydes by reaction with a carboxylic acid or one of its derivatives, for example, acid chloride, ester, or carboxamide, but the α-methylene group is further activated by a cyano, carboxylic ester or acid, alkenyl, phenyl, nitro or triphenylphosphine group. The use of a solid complex, [AlPO$_4$.Al$_2$O$_3$], enables this type of cyclization to be accomplished under mild conditions [1655]; such reactions have been reviewed [1913, 1947, B-40].

R^1 = H,Cl,MeO,NO$_2$;

1-Benzopyran-2-one

R^2= Me,Ph,NO$_2$; R^3=MeO,Et$_2$N;

i, Et$_2$$\overset{+}{N}H_2$$\bar{C}$l, PhMe; ii, POCl$_3$

[1504, 1541]

[157]

Other ring systems synthesized similarly.

[395, 473]

Furo[3,2-*g*][1]benzopyran-7-one

[881, 895, 1092, 1655]

1-Benzopyran-2-one

(review [B-40])

[502]

Pyrano[3',2':5,6]benzofuro-
[3,2-*c*]-pyridin-2-one

The imine of salicylaldehyde (readily prepared from the aldehyde and a primary amine [B-25]) reacts with carbon suboxide (review of reactions [100]) to give 1-benzopyran-2-ones.

R=alkyl,aryl

[34]

3. 4*H*-Pyran and Pyran-4-one

Salicylaldehyde and dimethyl acetylenedicarboxylate in a basic environment give a low yield of a 4*H*-benzopyran. Other derivatives of this ring system are prepared in good yield by an alumina-induced reaction of the hydroxyaldehyde with an excess of a malononitrile derivative.

[30]

4*H*-1-Benzopyran

R^1=H,MeO; R^2=CN,COOMe

[763]

Reaction with an enamine followed by an oxidant gives a rather low yield of a pyran-4-one.

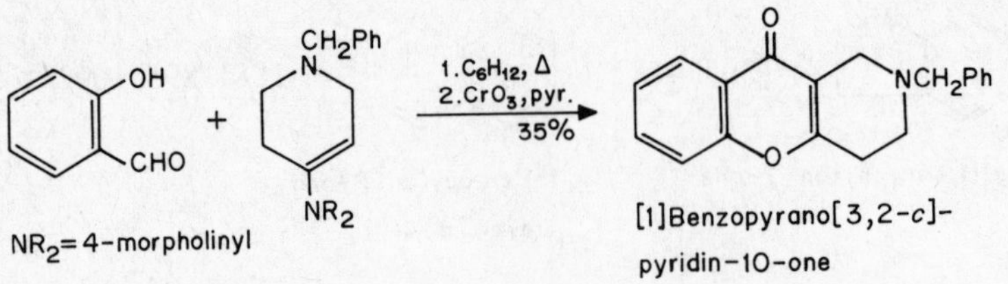

[1219]

NR$_2$= 4-morpholinyl

[1]Benzopyrano[3,2-c]-
pyridin-10-one

4. 1, 2, 3-Oxathiazine 2, 2-Dioxide

The difunctional reagent, chloromethyl isocyanate, gives high yields of benz-oxathiazine derivatives on heating with salicylaldehyde.

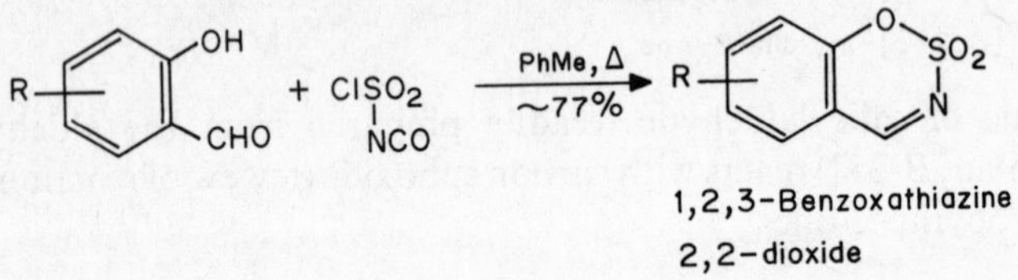

[875]

1,2,3-Benzoxathiazine
2,2-dioxide

III. FORMATION OF A SEVEN-MEMBERED RING

1. 1, 6, 2-Oxathiazepine

The oxime of a 2-thiocyanato-aldehyde is converted by a weak base into an oxathiazepine ring in good yield.

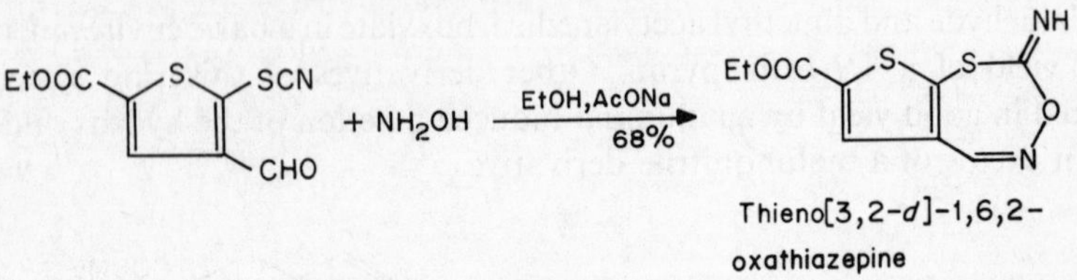

[737]

Thieno[3,2-d]-1,6,2-
oxathiazepine

CHAPTER 30

Aldehyde and Ketone; Dialdehyde or Diketone

This chapter contains cyclizations of the following combinations of carbonyl groups: aldehyde and ketone, 1,2- or 1,4-dialdehyde, 1,2-, 1,3-, 1,4- or 1,5-diketone; in addition, cyclizations of a few monohydrazones and mono-thiosemicarbazones of diketones are described.

I.   FORMATION OF A FIVE-MEMBERED RING

1.   Pyrrole

Treatment of a 1,2-dialdehyde with a primary amine (with or without bisulphite) leads to the formation of a fused pyrrole ring.

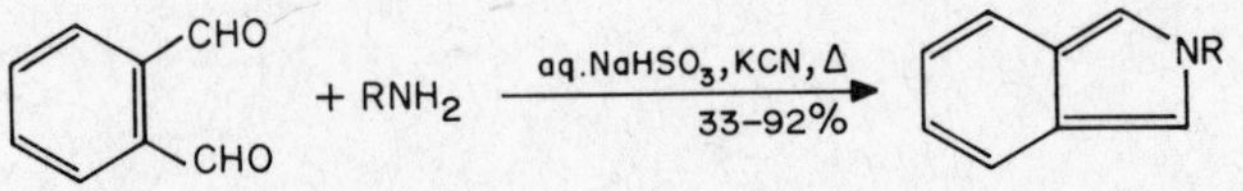

[867, 1825]

R = alkyl, aryl   Isoindole

2. Pyrazole

Hydrazines cyclize 1,3-dicarbonyl compounds and form a fused pyrazole ring.

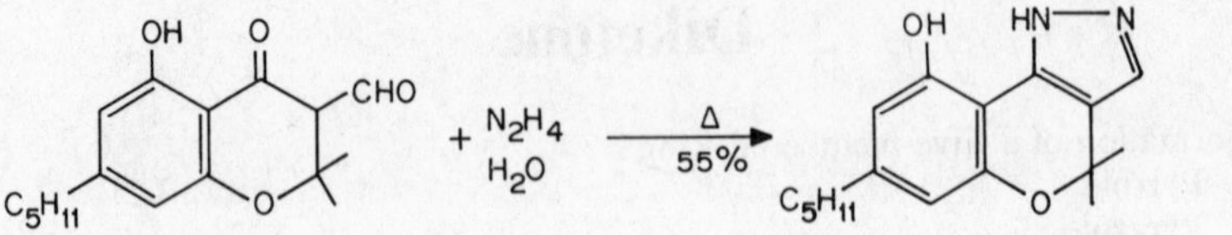

[554]

[1]Benzopyrano[4,3-c]pyrazole

[225]

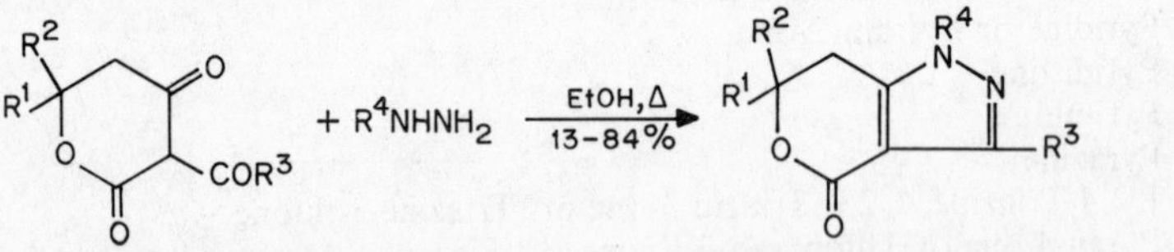

$R^1, R^2 = H, Me; R^3 = Me, Et;$
$R^4 = H, Me, Ph$

Pyrano[4,3-c]pyrazol-4-one

Other ring systems synthesized similarly:

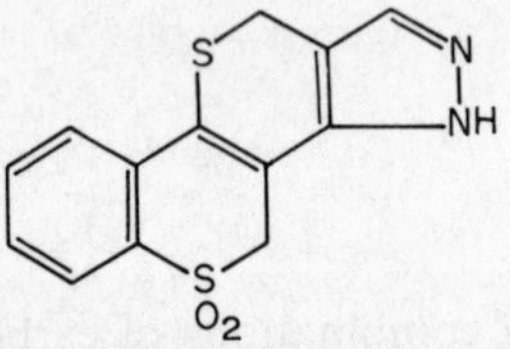

[501]

[1]Benzothiopyrano[3',4':5,6]-
thiopyrano[4,3-c]pyrazole 5,5-
dioxide

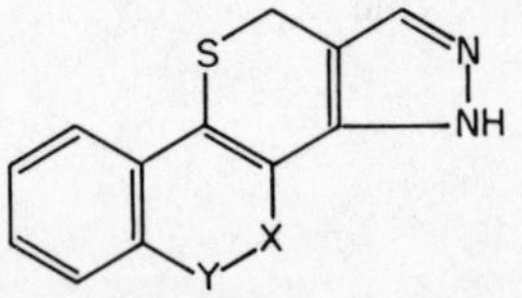

[229, 417]

$X = SO_2, Y = NH$
Pyrazolo[3',4':4,5]thiopyrano-
[3,2-c][2,1]benzothiazine 4,4-
dioxide

$X = NH, Y = SO_2$
Pyrazolo[3',4':4,5]thiopyrano-
[3,2-c][1,2]benzothiazine 5,5-
dioxide

Monohydrazones of diketones are readily cyclized to a fused pyrazole ring by heating in an acidic medium.

R = Me, Ph   Indeno[1,2-c]pyrazol-4-one   [966]

Other ring systems synthesized similarly:

Pyrazolo[4′,3′:3,4]pyrrolo[2,1-b]   [239]
benzothiazol-10-one

Pyrazolo[4′,3′:3,4]pyrrolo[1,2-a]-   [239]
quinazoline-5,10-dione

Cyclohepta[c]pyrazole   [1234]

3. Imidazole

One of the classical syntheses of an imidazole ring is the reaction of a 1,2-dione with an aldehyde and ammonia (or a source of ammonia); this gives good yields when applied to phenanthrenequinone.

R = 2- or 3-thienyl, 2-naphthyl   Phenanthro[9,10-d]imidazole   [3]

Another ring system synthesized similarly:

[1]Benzopyrano[3,4-d]imidazole   [554]

4. Furan

A 1,4-diketone in the presence of a strong base reacts with ethanolamine to give isobenzofuran as an unexpected product in low yield.

[1121]

Isobenzofuran
(review [1691])

An aldehyde group (in this example, generated by acid hydrolysis of its acetal) separated from an endocyclic ketone by two carbon atoms, may be converted in the same medium into a fused furan ring. Simultaneously, phenylsulphinic acid is eliminated.

[488]

Furo[2,3-g]quinoline

5. Thiophene

1,4-Diketones react with phosphorus pentasulphide-pyridine with the formation of a fused thiophene ring in good yield.

[1632]

Thieno[3,4-c]thiophene

Another ring system synthesized similarly:

[1633]

Thieno[3,4-c]pyrazole

6. Isoxazole

Treatment of 1,3-diketones (or 3-oxoaldehydes [554]) with hydroxylamine leads to the formation of a fused isoxazole ring. The mono-oxime may sometimes be isolated and then heated with an acid.

Indeno[1,2−d]isoxazol−4−one [309]

Other ring systems synthesized similarly:

[229, 417, 501]

X=CH₂,Y=SO₂
[1]Benzothiopyrano[3′,4′:5,6]thiopyrano−
[3,4−d]isoxazole 5,5−dioxide
X=SO₂,Y=NH
Isoxazolo[5′,4′:4,5]thiopyrano[3,2−c]−
[2,1]benzothiazine 4,4−dioxide
X=NH,Y=SO₂
Isoxazolo[5′,4′:4,5]thiopyrano[3,2−c]−
[1,2]benzothiazine 5,5−dioxide

[554]

[1]Benzopyrano[3,4−d]isoxazole

II. FORMATION OF A SIX-MEMBERED RING

1. Pyridine or Pyridin-2-one

A 1,2-dialdehyde may be converted into a pyridine-2-one ring by reaction with ethyl azidoacetate or to a pyridine with ethyl glycinate.

Pyrido[4,3−b]indol-1-one [717]

Pyrido[4,3−b]indole [717]

When a primary methylamine is heated with a 1,4-diketone and DBU, a fused pyridine ring is obtained but diketones which may be labile in basic media are cyclized by heating in butanol.

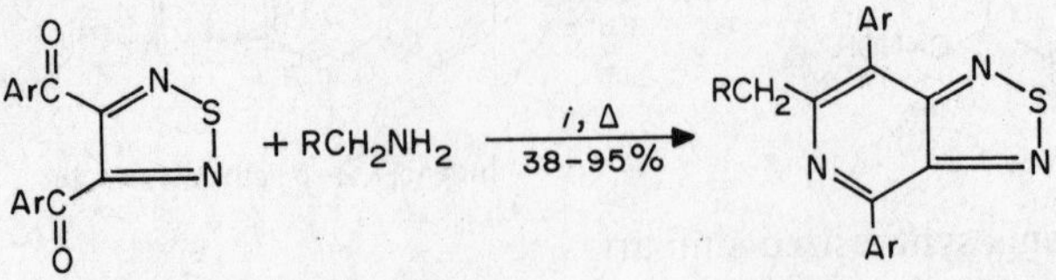

R=PhCH$_2$,NCCH$_2$,CN,COOMe;
Ar=Ph,4-Me-,4-Cl-C$_6$H$_4$
i, DBu or BuOH

[1,2,5]Thiadiazolo[3,4-*c*]-
pyridine

[1058, 1128]

Other ring systems synthesized similarly:

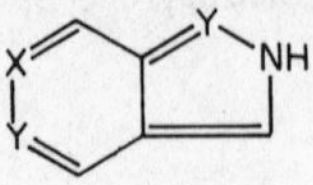

X=N,Y=CH
Pyrrolo[3,4-*c*]pyridine
X=CH,Y=N
Pyrazolo[4,3-*c*]pyridine

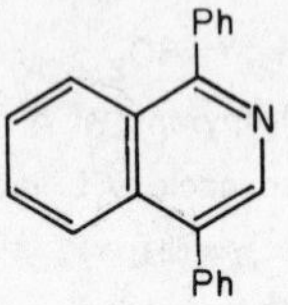

[603]

Isoquinoline

[1121]

2. Pyridazine

Hydrazine annulates 1,4-diketones (on heating in ethanol or propanol) with the formation of a fused pyridazine ring which may also be obtained in good yield by a double Wittig reaction on phthalaldehyde. A 2-acyl-aldehyde behaves similarly [1036, 1196].

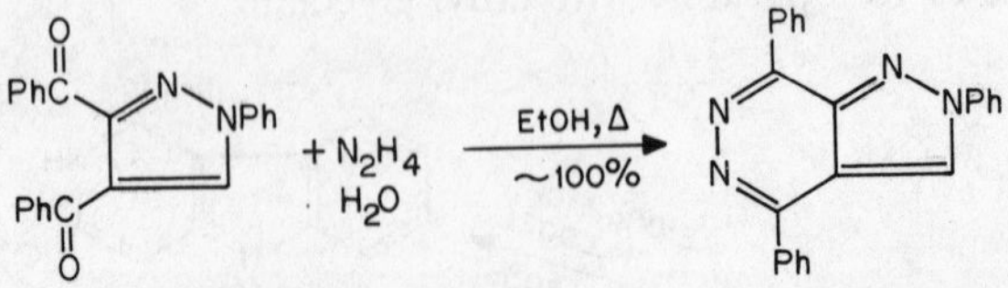

Pyrazolo[3,4-*d*]pyridazine

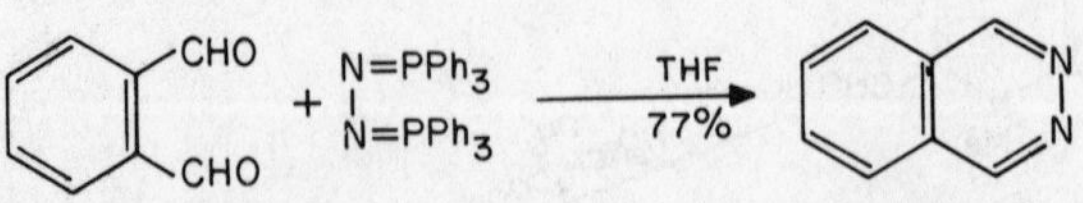

[610]

Phthalazine

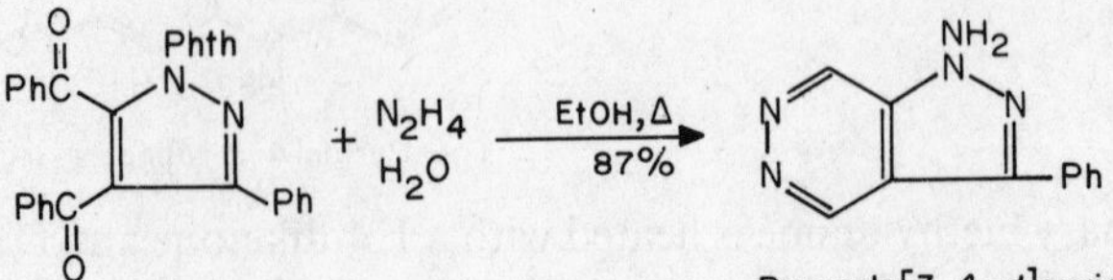

[727]

Pyrazolo[3,4-*d*]pyridazine

[2005]

Other ring systems synthesized similarly from diketones or oxo-aldehydes:

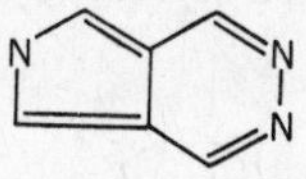

[1143, 1966]

Pyrrolo[3,4-d]pyridazine

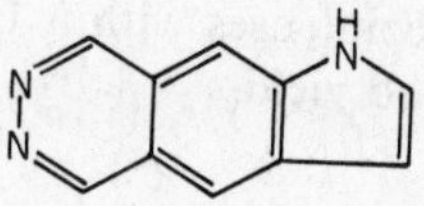

[877]

Pyrrolo[2,3-g]phthalazine

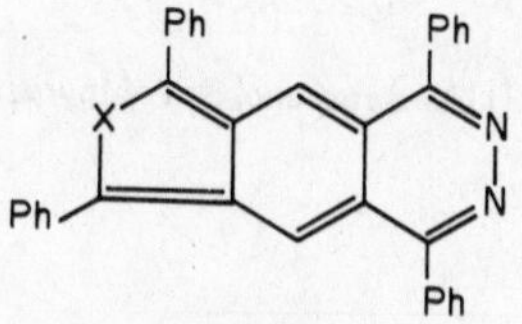

[877, 1019]

X=MeN,O,S

Pyrrolo[3,4-g]phthalazine
Furo[3,4-g]phthalazine
Thieno[3,4-g]phthalazine

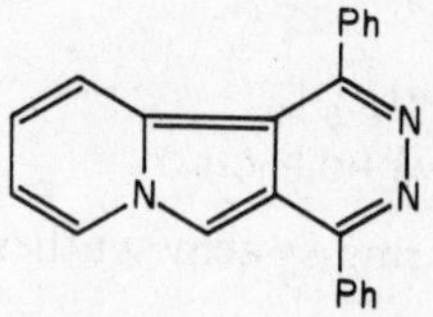

[1063]

Pyridazino[4,5-a]indolizine

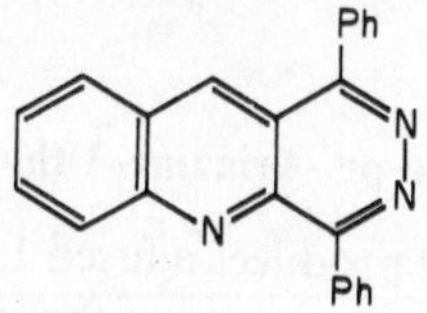

[610]

Pyridazino[4,5-b]quinoline

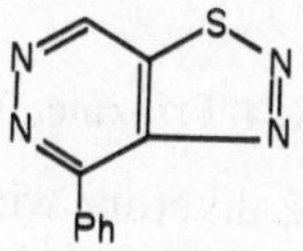

[1196]

1,2,3-Thiadiazolo-
[4,5-d]pyridazine

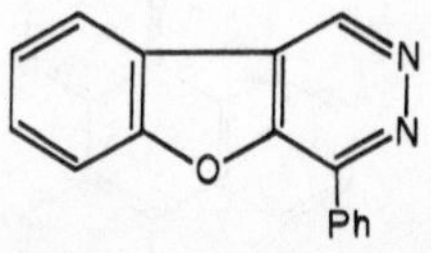

[1036]

Benzofuro[2,3-d]pyridazine

3. Pyrimidine

Guanidine or an amidine cyclizes a 1,3-dicarbonyl compound to form a new pyrimidine ring; the reaction is base-catalysed.

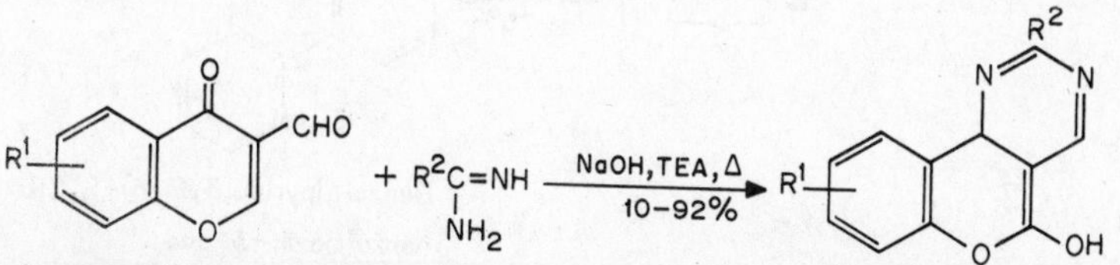

R^1=H,Me; R^2=Me,CCl$_3$,Ph
PhCH$_2$,NH$_2$

[796, 1407]

[1]Benzopyrano[4,3-d]-
pyrimidine

4. Pyrazine

A 1,2-diamine condenses with a 1,2-diketone to give a fused pyrazine ring in moderate to good yield.

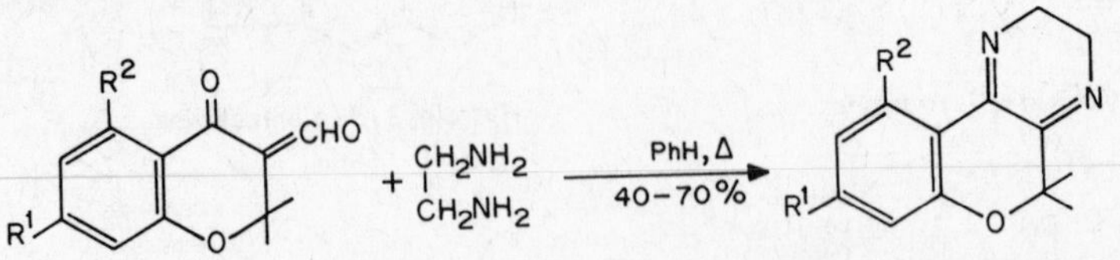

$R^1 = HO, C_5H_{11};$
$R^2 = Me, HO, PhCH_2O$

[554]

[1]Benzopyrano[3,4-b]pyrazine

Another ring system synthesized similarly:

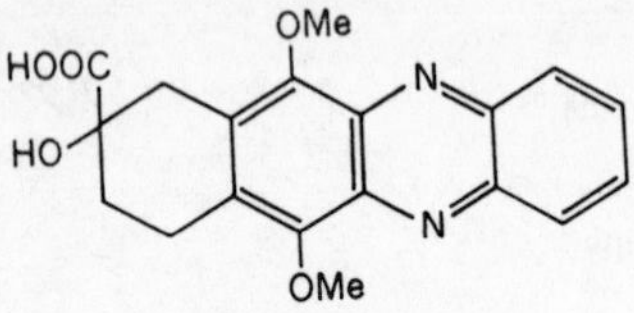

[604]

Benzo[b]phenazine

5. 1,2,4-Triazine, 1,2,4-Triazin-3-one or -Triazine-3-thione

Reaction of a 1,2-diketone with aminoguanidine produces a fused 1,2,4-triazine in high yield. It is sometimes advantageous to prepare the intermediate semicarbazone (or thiosemicarbazone).

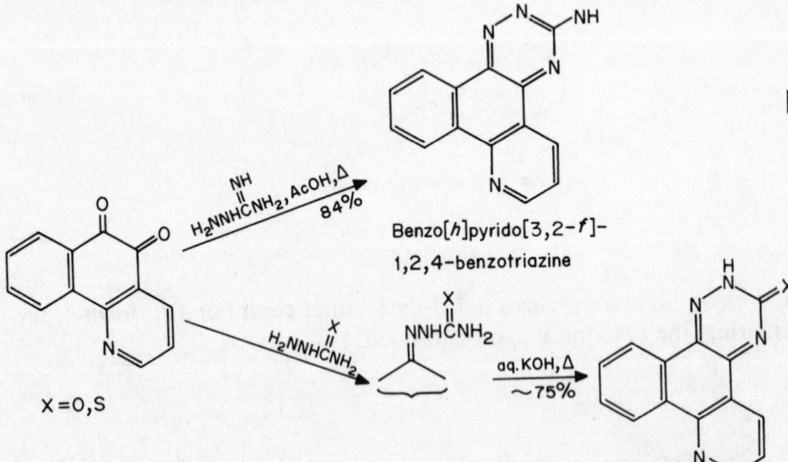

[868]

Benzo[h]pyrido[3,2-f]-1,2,4-benzotriazine

[868]

$X = O, S$

Benzo[h]pyrido[3,2-f]-1,2,4-benzotriazin-3-one
Benzo[h]pyrido[3,2-f]-1,2,4-benzotriazine-3-thione

Another ring system synthesized similarly:

[120]

X = O, S

[1]Benzopyrano[4,3-*e*][1,2,4]triazin-2-one

[1]Benzopyrano[4,3-*e*][1,2,4]triazine-2-thione

6. Pyran-4-one or Thiopyran-2-thione

A well-known synthesis of chromones depends on intramolecular reaction of phenolic and carbonyl groups [B-12]; this is illustrated in the cyclization of a (tautomeric) 1,5-diketone by hydrogen chloride-methanol.

$$\xrightarrow[\text{45-93\%}]{\text{HCl—MeOH}}$$

[299]

X = O, S

Pyrano[3,2-*c*]-1-benzopyran-4-one

[1]Benzothiopyrano[4,3-*b*]pyran-4-one

Treatment of some 1,3-diketones with phosphorus pentasulphide and acetonitrile gives simultaneous thiation and cyclization.

$$\xrightarrow[\text{45\%}]{\text{NaHCO}_3, \Delta}$$

+ P$_2$S$_5$

+ MeCN

[756]

Thiopyrano[3,4-*b*][1]benzothiophene-3-thione 9,9-dioxide

III. FORMATION OF A SEVEN-MEMBERED RING

1. Azepine

When two aldehyde groups separated by four carbon atoms are treated with a primary amine, an azepine ring may be formed. In this example, the imine is reduced with dithionite.

R=Ph,PhCH$_2$ Dithieno[2,3-c:3',2'-e]azepine   [666]

2. 1,2- or 1,4-Diazepine

A 1,3-diketone reacts with o-phenylenediamine to form a fused 1,4-diazepine
ring and a 1,5-diketone is cyclized with hydrazine to a 1,2-diazepine.

                                        $\xrightarrow[47\%]{PrOH, \Delta}$ [988]

Pyrido[2',3':4,5]cyclopenta[1,2-b]
[1,5]benzodiazepin-12-one

$\xrightarrow[29-73\%]{PrOH, \Delta}$ [974]

R^1=Ph,4-ClC$_6$H$_4$;   Indeno[1,2-c]-1,2-diazepin-6-one
R^2=Me,Ph,4-NO$_2$-,4-AcNH-C$_6$H$_4$

3. Thiepine or its 1,1-Dioxide

A double Wittig reagent containing CH$_2$SCH$_2$ group cyclizes a 1,3-dialdehyde
to give a low yield of a fused thiepine.

                + S(CH$_2$PPh$_3$)$_2$ $\xrightarrow[9-13\%]{BuLi}$ [194]
                    2Cl$^-$

X=O,S   Furo[3,4-d]thiepine
                               Thieno[3,4-d]thiepine

Condensation of an aldehyde with an ester is the basis of another synthesis of
this ring.

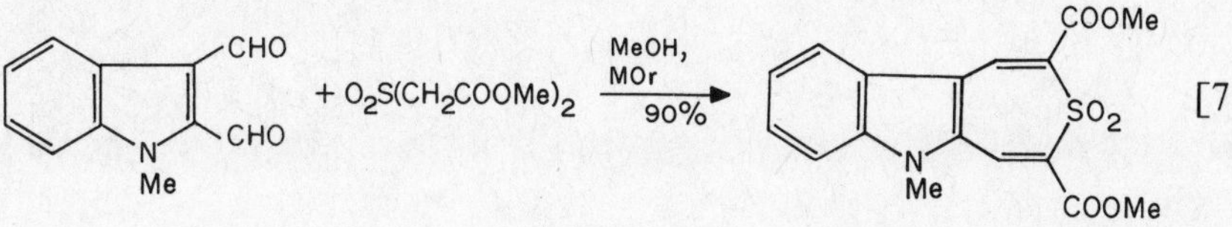

[717]

Thiepino[4,5-*b*]indole 3,3-dioxide

Other ring systems synthesized similarly:

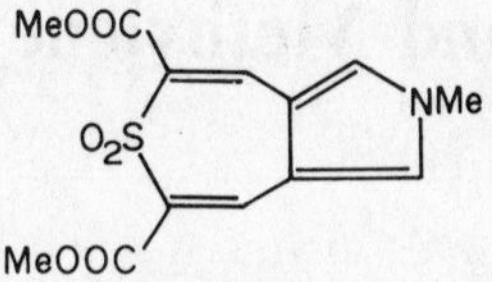

[866]

Thiepino[4,5-*c*]pyrrole
6,6-dioxide

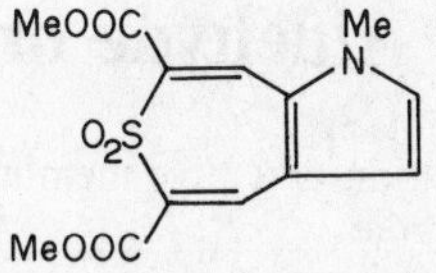

[866]

Thiepino[4,5-*b*]pyrrole
6,6-dioxide

IV. FORMATION OF AN EIGHT-MEMBERED RING

1. 1,4-Diazocine

This ring is formed by heating an *o*-diketone with *o*-phenylenediamine in the presence of an acid.

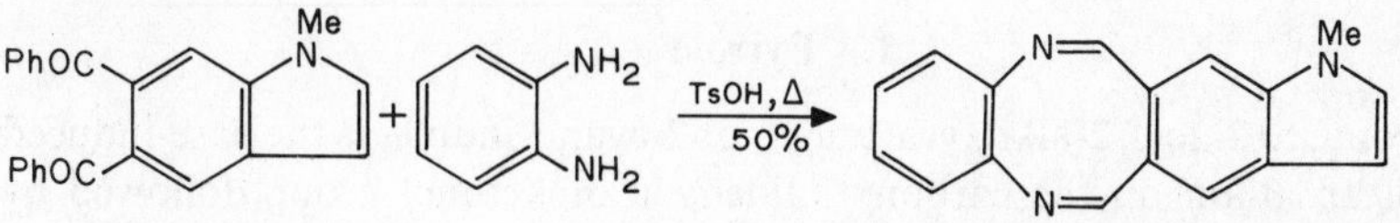

[877]

Benzo[*b*][1,4]diazocino[7,6-*f*]indole

CHAPTER 31

Aldehyde or Ketone and Methylene

'Methylene' covers reactive methyl, methylene and methine groups.

I.   FORMATION OF A FIVE-MEMBERED RING

1.   Pyrrole

A good route to 2-acyl, 2-alkoxycarbonyl- or 2-cyano-indoles is the base-induced attack of an anion on the carbonyl (aldehyde or ketone) group, followed by dehydration of the hydroxy pyrrole. Activation of a methylene group (by addition of a base or by the presence nearby of an electron-deficient centre, such as a positively charged nitrogen) and its attack on a carbonyl forms the basis of several other syntheses. The reactivity of N-substituted pyridinium salts has been reviewed [1870].

$R^1 = Ac, CN, PhCO, COOMe;$
$R^2 = H, Me, Ph; R^3 = Me, 4-MeC_6H_4$

Indole

[989]

$Ar^1, Ar^2 = Ph, 4-BrC_6H_4$

Indolizine

[1686]

154

[1686]

Ar^1=Ph,4-BrC$_6$H$_4$; Ar^2=Ph,
4-Me-,4-Cl-C$_6$H$_4$

Other ring systems synthesized similarly:

[1323, 1635]

X=NMe

Pyrrolo[1,2-*a*]imidazole

Imidazo[1,2-*a*]benzimidazole

X=S

Pyrrolo[2,1-*b*]thiazole

2. Thiophene

Anion formation from a 2-SCH$_2$COOR side-chain of a benzaldehyde gives high yields of a benzothiophene-2-carboxylic acid or ester.

[555, 726]

R^1=MeO,HO; R^2=H,Me,

i, aq.NaOH or AcONa-Ac$_2$O

Benzo[*b*]thiophene

[795]

Thieno[2,3-*b*]pyrazine

II. FORMATION OF A SIX-MEMBERED RING

1. Pyridine

3-Nitroquinolines are readily synthesized by base-catalysed ring closure of a 2-(β-nitroethylideneamino)acetophenone.

[31, 1054, 1271]

R^1=H,Me; R^2=Me, C$_5$H$_{11}$

i, aq.NaOH or Al$_2$O$_3$,Me$_2$CO

Quinoline

2. Pyridin-2-one

A cyanoacetamido group contains a highly activated methylene which reacts at ambient temperatures with a ketone to form a fused pyridin-2-one ring.

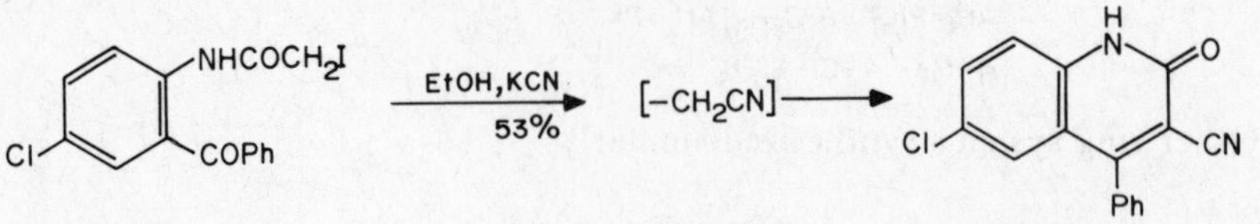

2-Quinolinone

[2]

III.   FORMATION OF A SEVEN-MEMBERED RING

1.   Azepine

Aldehyde and methylene groups on different rings can give a doubly-fused azepine ring.

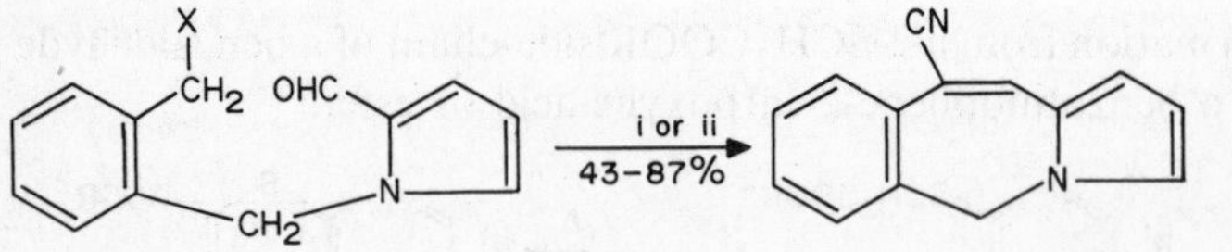

i, X=Cl; KCN, DMSO, Δ

ii, X=CN; aq. KOH

Pyrrolo[1,2-b][2]benzazepine

[932]

CHAPTER 32

Aldehyde or Ketone and Nitrile

I. FORMATION OF A FIVE-MEMBERED RING

1. Pyrrole-2-one

Ammonia or a primary or secondary non-benzenoid amine on stirring with 2-cyanobenzaldehyde at ambient temperature gives an isoindol-1-one in very high yield; arylamines give a low yield or no cyclic product.

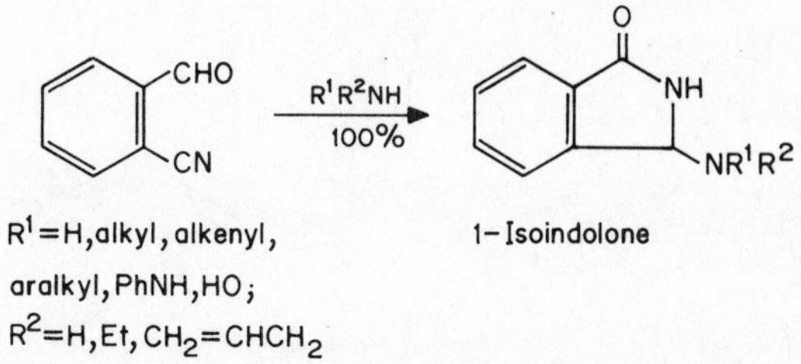

CHO
CN
R¹R²NH
100%
O
NH
NR¹R²
R¹=H,alkyl,alkenyl,
aralkyl,PhNH,HO;
R²=H,Et,CH₂=CHCH₂
1-Isoindolone

[1106]

2. Pyrazole

Under mildly basic conditions, hydrazine converts an oxonitrile into a fused pyrazole in good yield even when the ketone is a vinylogous ester.

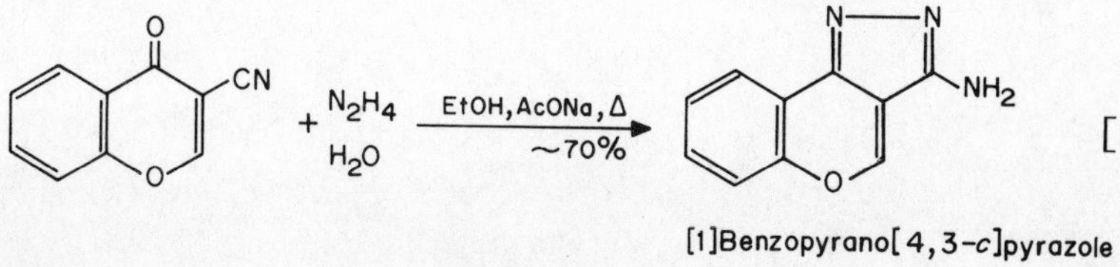

O
CN
O
+
N₂H₄
H₂O
EtOH,AcONa,Δ
~70%
N
N
NH₂
O
[1]Benzopyrano[4,3-c]pyrazole

[1689]

II. FORMATION OF A SIX-MEMBERED RING

1. Pyridin-2-one

When the carbonyl and cyano groups are separated by three carbon atoms, heating with sulphuric acid results in cyclization and the nitrile becomes a lactam.

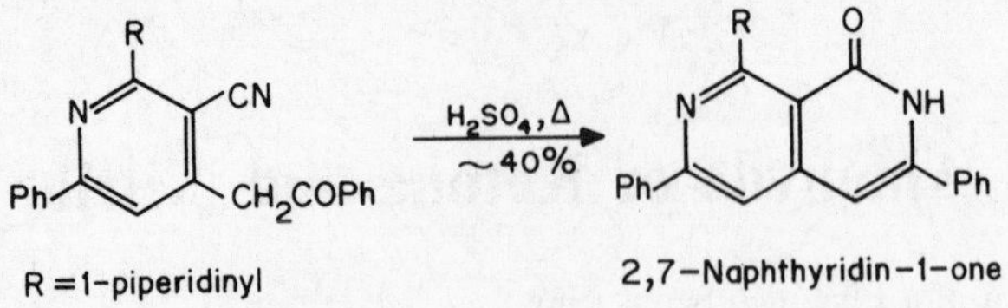

R = 1-piperidinyl   2,7-Naphthyridin-1-one

[960]

2. Pyran-2-one

In contrast to the formation of a new pyridin-2-one just mentioned, the reaction takes a different course at ambient temperature and in the presence of hydrogen bromide gas or solution; a 2-aminopyrylium salt is first formed and this, on hydrolysis, gives an isocoumarin.

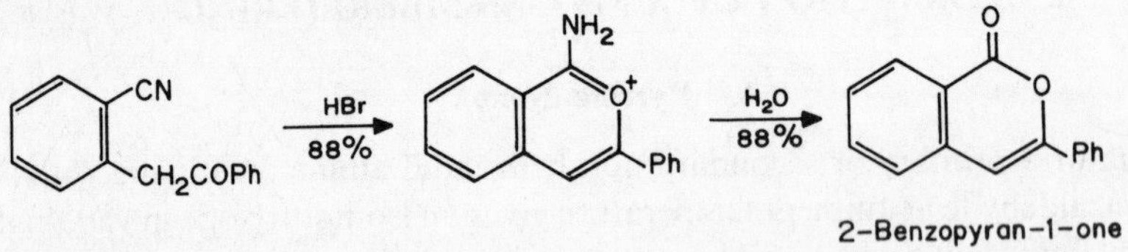

2-Benzopyran-1-one

[142]

CHAPTER 33

Aldehyde or Ketone and Nitro, Nitroso or N-Oxide

I. FORMATION OF A FIVE-MEMBERED RING

1. Isoxazole

Deoxygenation of a nitro group of a nitro-ketone with triethyl phosphite at a fairly high temperature gives moderate yields of a benzisoxazole. Alternatively, reduction of a nitroso (by phenylhydrazine) leads to high yields of the benzisoxazole at ambient temperature but the ratio of reactants and the order in which they are added have a considerable effect on the yield.

$R^1 = Me, Ph; R^2 = H, Cl$

2,1-Benzisoxazole

[1451]

[870]

II. FORMATION OF A SIX-MEMBERED RING

1. Pyridine 1-Oxide

Partial reduction of a nitro group and subsequent reaction with a neighbouring side-chain carbonyl or nuclear group yields a fused pyridine oxide ring (review

[B-34]).

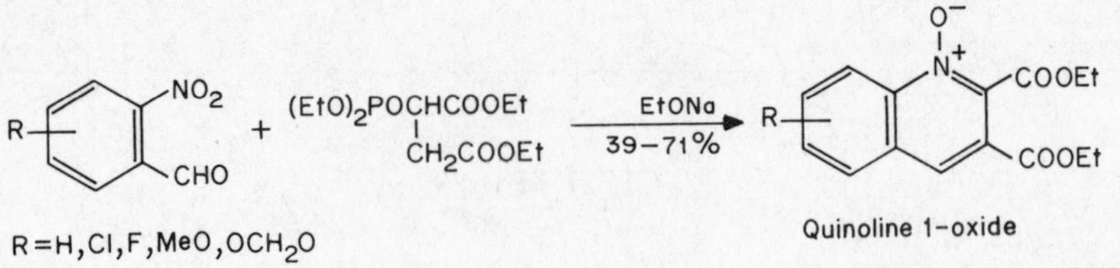

R = H, Cl, F, MeO, OCH₂O

Quinoline 1-oxide

[1404]

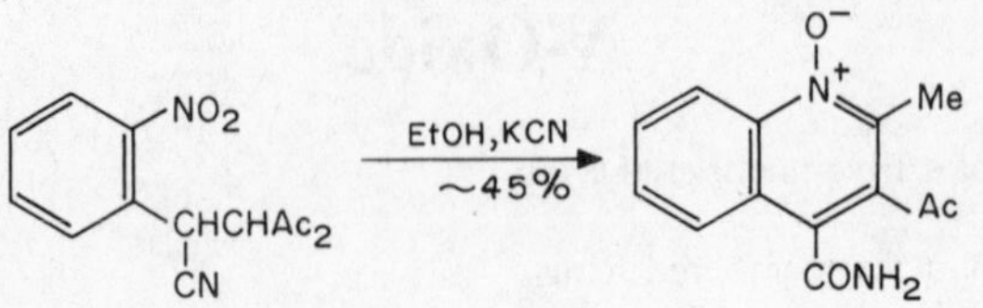

[1588]

2.   Pyridazine 1-Oxide

Treatment of a 2-methylazoxybenzophenone with hot ethanolic alkali leads to a good yield of a cinnoline 2-oxide (review [B-34]).

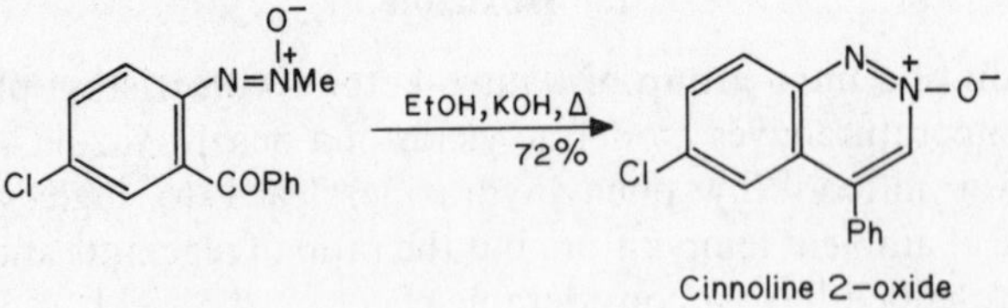

Cinnoline 2-oxide

[1273]

CHAPTER 34

Aldehyde or Other Carbonyl and Phosphorane

Reactions of the carbonyl group of aldehydes, esters, ketones, imides and lactams with phosphorane and phosphonate are covered in this chapter. The synthesis of heterocycles using phosphorus-containing groups has been reviewed [1724].

I. FORMATION OF A FIVE-MEMBERED RING

1. Pyrrole

An intramolecular Wittig–Horner–Emmons reaction is a useful way of forming a carbon–carbon bond and thus a new heterocyclic ring where a hetero atom is present between the reacting groups. For example, one carbonyl group of an N-substituted imide may react with a phosphorane or phosphonate to give a fused pyrrole ring.

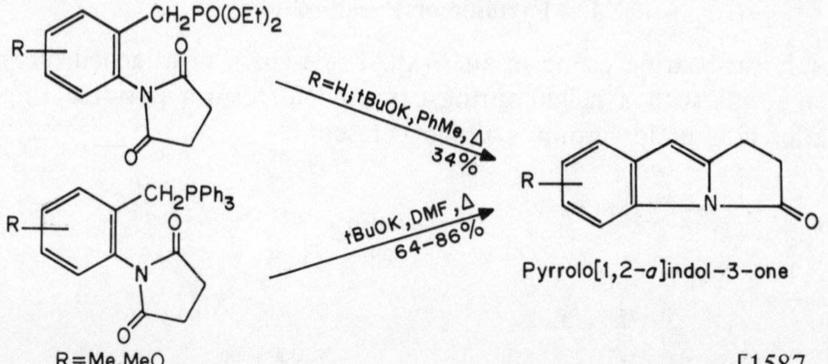

Pyrrolo[1,2-a]indol-3-one

[1587, 1944]

161

Other ring systems synthesized similarly:

Pyrrolizine   [1770]

[1943]

Isoindolo[2,1-a]indol-6-one

[1572]

(34.1) X = CH$_2$
1-Azabicyclo[3.2.0]heptan-7-one

2. Thiazole

Compound (34.1) where X = S is synthesized by the Wittig–Horner–Emmons reaction on the azetidinone with triphenyl phosphonate.

$R^1 = CCl_3, COOCH_2CCl_3;$
$R^2 = OCNH_2; R^3 = AcOCH_2$
         ‖
         O

4-Thia-1-azabicyclo[3.2.0]-
hept-2-en-7-one

[1572]

II. FORMATION OF A SIX-MEMBERED RING

1. Pyridine or Pyridin-2-one

A phosphorus-bearing group in an N-alkyl side-chain or attached directly to nitrogen reacts with a neighbouring carbonyl to form a pyridine ring or a pyridinone if an amide group is already present.

R = MeO, Ph

Quinoline

[1382]

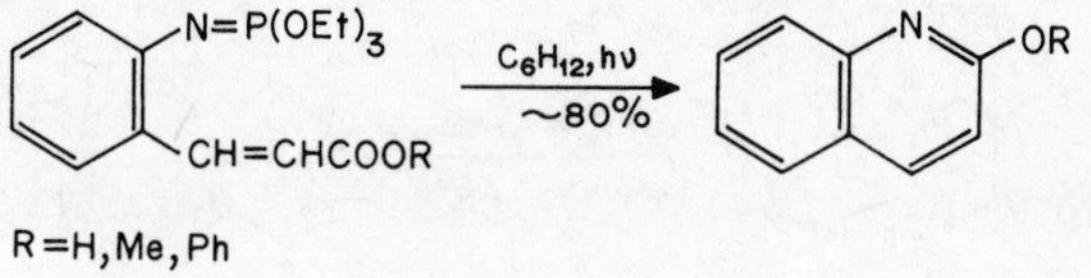

[112]

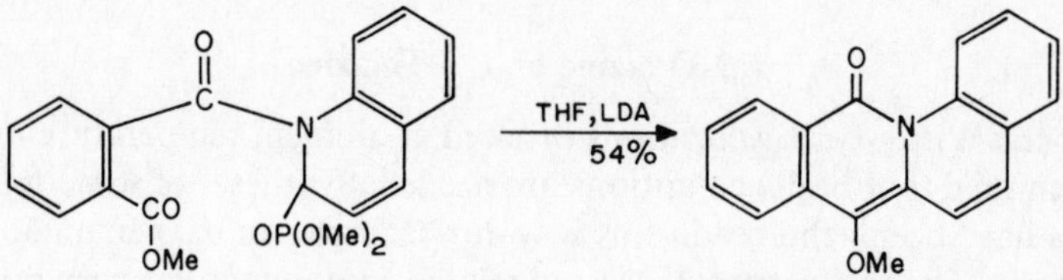

R=H,Me,Ph

[1945]

Dibenzo[bf]quinolizin–12–one

2. Pyran or Pyran-2- or -4-one

2-Formylphenoxyalkylphosphoranes cyclize on warming with a strong base to give an oxygen-containing ring.

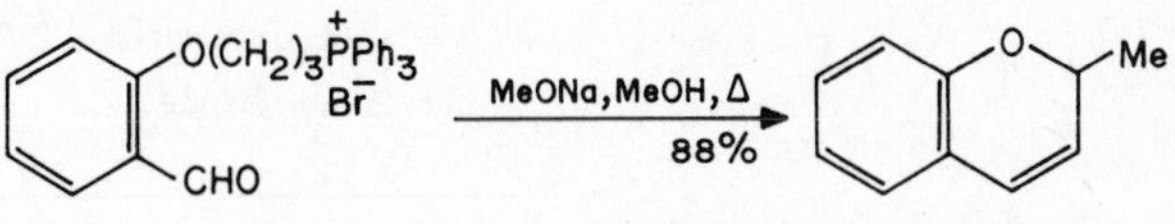

[1364]

2H–1–Benzopyran

Both variants (phosphorane and phosphonate) of the Wittig reaction have been used to synthesize benzopyran-2-ones.

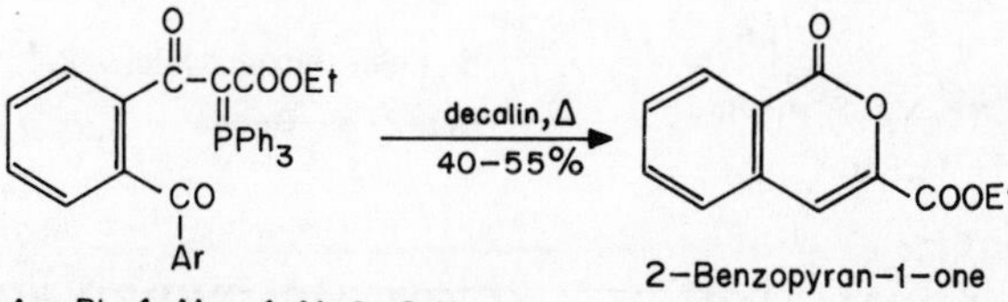

[1520]

Ar=Ph,4-Me-,4-MeO-C₆H₄

2–Benzopyran–1–one

R=H,Me,MeO

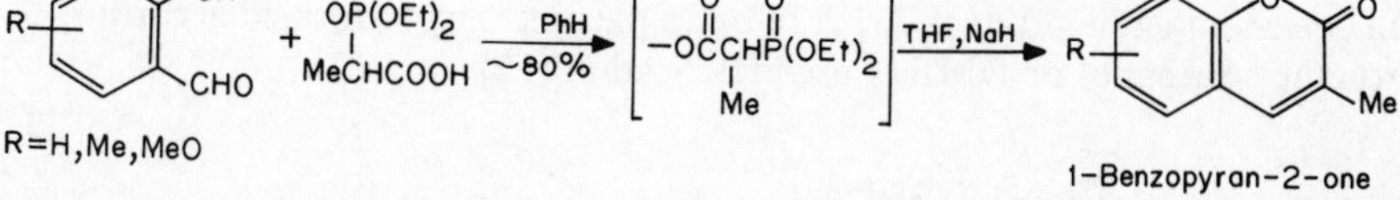

1–Benzopyran–2–one

[876]

A carbonyl group in the phosphorane-carrying side-chain can lead to the formation of a chromone.

[1705]

Ar = 4-MeOC$_6$H$_4$

1-Benzopyran-4-one

3. 1,3-Oxazine or 1,3-Thiazine

Several of the Wittig-type cyclizations proceed at ambient temperature and are useful when mild (but basic) conditions are needed. Syntheses of some β-lactam antibiotics have been achieved in this way, for example, an oxygen analogue of cephalosporin. The chemistry of these and related compounds has been reviewed [B-26, B-27].

[1924]

5-Oxa-1-azabicyclo[4.2.0]-
oct-2-en-8-one

[1427]

R^1=H,Me; R^2=H,Ph;
R^3=Me, PhCH$_2$

5-Thia-1-azabicyclo[4.2.0]-
oct-2-en-8-one

III. FORMATION OF A SEVEN-MEMBERED RING

1. Oxepine

The solvent used in the cyclization of a phosphorane can play an important role in determining the product. A pyran or an oxepine may be formed according to whether methanol or DMF is used (cf. Section II.2).

[1364]

1-Benzoxepine

IV. FORMATION OF AN EIGHT-MEMBERED RING

1. Oxocin

When the reaction described in the previous paragraph is extended to an ω-phosphoranylbutoxybenzaldehyde an oxocin ring is obtained in moderate yield.

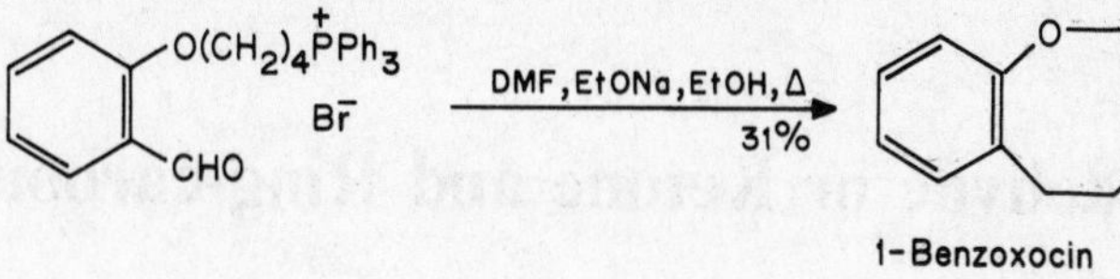

[1365]

1-Benzoxocin

CHAPTER 35

Aldehyde or Ketone and Ring-carbon

I. FORMATION OF A FIVE-MEMBERED RING

1. Pyrrole

The elements of water may be removed from a side-chain aldehyde by molecular sieve in boiling toluene or by use of PPA (review [B-21]).

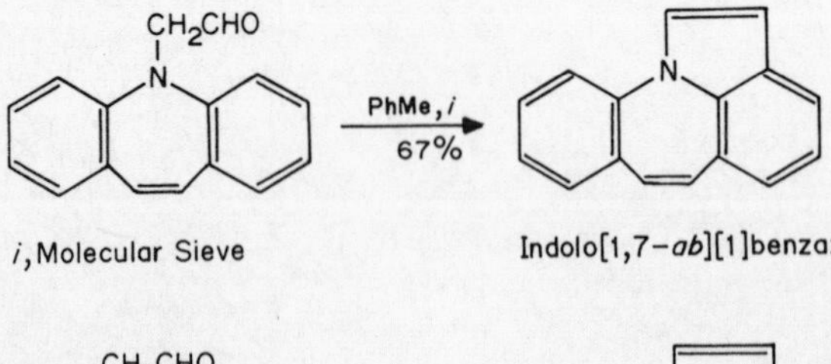

CH$_2$CHO

$\xrightarrow[\text{67\%}]{\text{PhMe, }i}$

[736]

i, Molecular Sieve

Indolo[1,7−*ab*][1]benzazepine

CH$_2$CHO

$\xrightarrow[\text{87\%}]{\text{CHCl}_3\text{, PPA}}$

[1015]

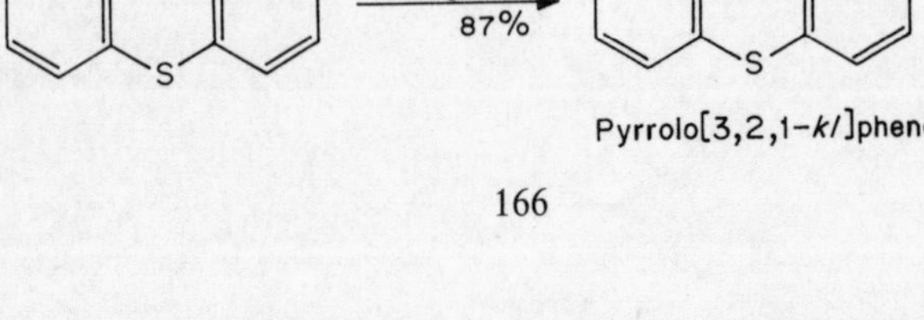

Pyrrolo[3,2,1−*kl*]phenothiazine

Quinones are able to attain the aromatic state by donating electrons to form a bond with a carbon or nitrogen. Heating in a suitable neutral solvent (nitromethane, diphenyl ether or DMF) is sometimes effective but acetic acid may also be useful. In the Nenitzescu reaction, a quinone is heated with a 3-amino-2-propenoate ester to give a 5-hydroxyindole; several variations of this reaction are known [1756].

Pyrrolo[1,2-a]indole-5,8-dione
(review [921])

[520]

Indole

[1761]

i,6-NH$_2$-1,3-Me$_2$-uracil,Δ

Pyrimido[4,5-b]indole-2,4-dione

[1771]

Pyrrolo[1,2-a]indole
(review [921])

[648]

In the Bischler indole synthesis, an oxoalkylamino chain attached to a reactive benzene ring can be cyclized in high yields to 3- or 2,3-substituted indoles; a photochemical variant of this reaction avoids the use of aluminium chloride.

R^1=H,Me; R^2=alkyl,PhCH$_2$;
R^3=H,Me,MeO

[552, 1274]

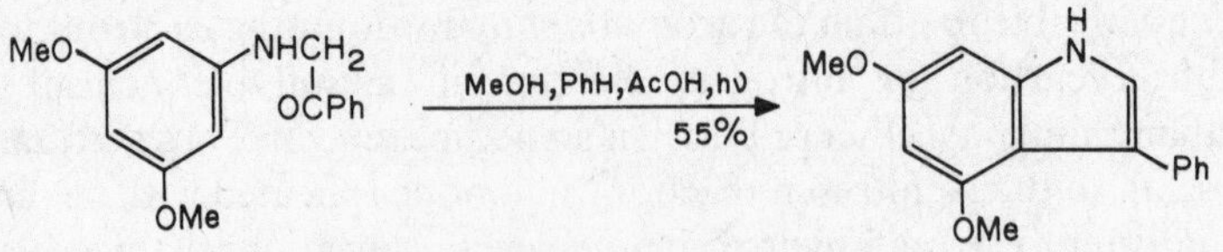

[1916]

A non-enolizable ketone in alkali can form a doubly fused pyrrole ring in good yield. An N-acetoacetylaridine cyclizes with acid to a 2-indolone [724].

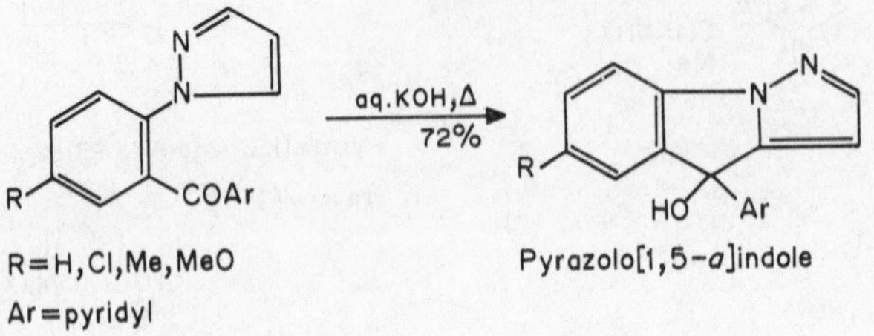

R=H,Cl,Me,MeO
Ar=pyridyl

Pyrazolo[1,5-a]indole

[370]

2. Furan

As in the last example, a ketone under hot alkaline conditions can form a C–C bond with an aromatic ring; a 2-oxoalkoxy chain can thus give a fused furan ring in the synthesis of a cannabinoid compound.

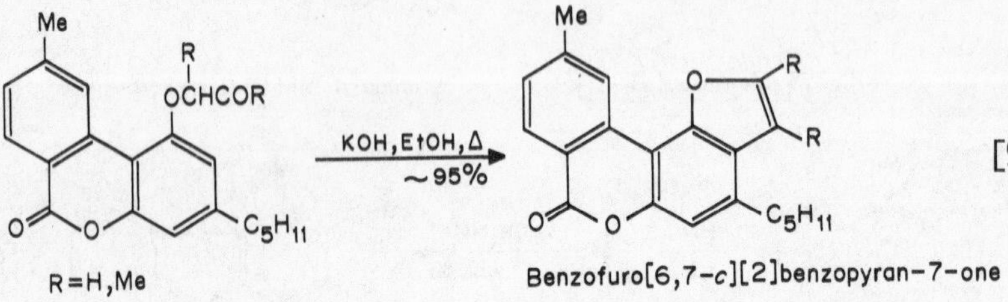

R=H,Me

Benzofuro[6,7-c][2]benzopyran-7-one

[922]

3. Thiophene

A thioether ketone similar to that used in the previous section is cyclized by heating with PPA. Benzo[b]thiophene may also be synthesized [833].

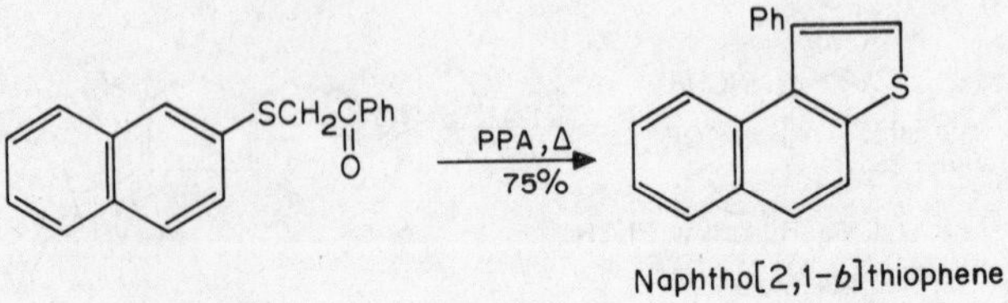

Naphtho[2,1-b]thiophene

[300]

Another ring system synthesized similarly:

Phenanthro[9,10-*b*]thiophene

[300]

4. Oxazole

In the Nenitzescu indole synthesis (Section I.1), the oxygen of a quinone is lost but at a lower temperature, the reaction takes a different course with the formation of an oxazole ring.

Pyrrolo[2,1-*b*]benzoxazole

[520]

5. Thiazole

Cyclohexane-1,3-diones on monobromination and reaction with thiourea yield a fused 2-aminothiazole in good yield.

5-Benzothiazolone

[301]

II. FORMATION OF A SIX-MEMBERED RING

1. Pyridine, Pyridin-2-one or Pyridine *N*-oxide

Acid-catalysed cyclocondensation of a cyclic ketone (containing an α-CH_2) with anthranilic acid produces a new fused pyridine ring. An ω-formylalkylamino chain or one which contains a suitably placed carbonyl group forms a pyridine ring with mineral acid or thermally. 3-Alkyl (or phenyl) quinolines are formed when 3-arylaminoacroleins are heated with an excess of aluminium bromide.

$R^1 = Me, Br, Cl, HO, EtO;$
$R^2 = Me, Ph$

EtOH
13–88%

[519]

Quino[3,2-c][1,8]naphthyridine

PhMe, TFA, Δ
80%

[556]

Indolizino[1,2-b]quinolin-9-one

H_2SO_4
~100%

[151]

Pyrido[1,2,3-de]-1,4-
benzothiazin-5-one

$Ph_2O, Δ$
~75%

[516]

[1]Benzopyrano[3,4-b]-
cyclopenta[d]pyridine-
6,8-dione

R^1—NHCH=CR^2
               |
              CHO

$AlBr_3, Δ$
51–89%

[1932]

$R^1 = H, Me, Cl; R^2 = Me, Et, Ph$ Quinoline

ω-Acetals or aldoximes cyclize under acidic conditions or sometimes thermally when the ring is reactive. Beckmann rearrangement before ring closure enables an isoquinoline to be synthesized from the ketoxime (35.1) but Lewis acids such as zinc chloride or tin(IV) chloride convert (35.1) into a quinoline.

R—⟨ ⟩—CHO   + ClCOOEt

+ $H_2NCH_2CH(OMe)_2$

1.P(OMe)$_3$ 2.TiCl$_4$
25–75%

[1287]

Isoquinoline

$R = Me, Br, HO, MeO$

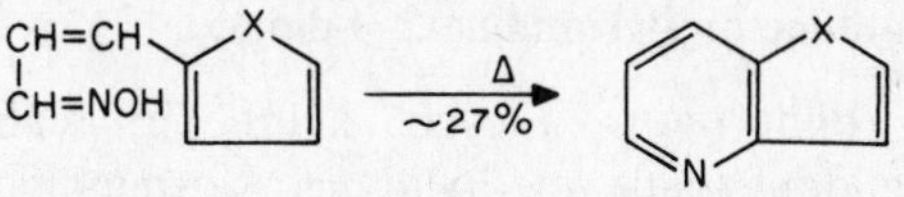

[949]

X = O, S
Furo[2,3-b]pyridine
Thieno[2,3-b]pyridine

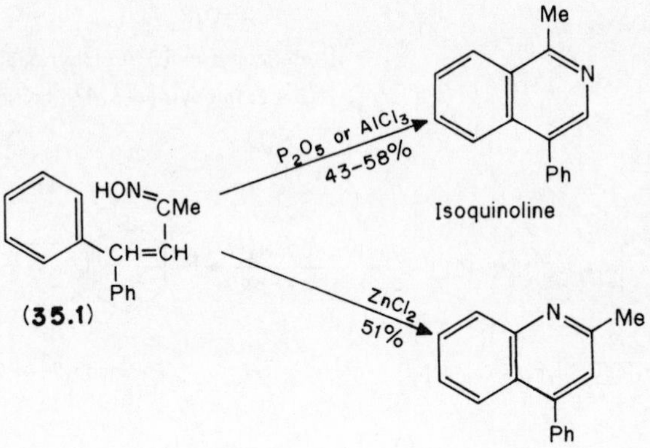

[1640]

Isoquinoline

[1640]

Quinoline

Under alkaline conditions, an oxime displaces a nuclear fluorine to produce a pyridine N-oxide ring (review [B-34]).

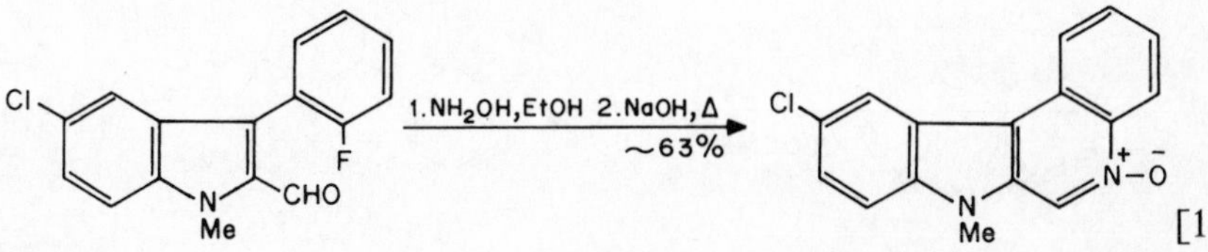

[1330]

Indolo[2,3-c]quinoline 5-oxide

2. Pyridazine

αβ-Unsaturated ketones cyclize on heating in acetic acid and can form a pyridazine ring when two nitrogen atoms are present in the ketone.

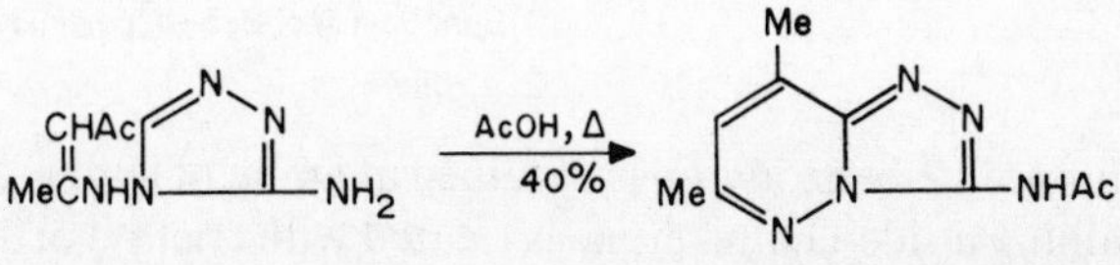

[567]

1,2,4-Triazolo[4,3-b]pyridazine

3. Pyrimidine or Pyrimidine-2, 4-dione

A pyrimidine ring may be built on to a cyclic ketone by reaction with cyanoguanidine or triformamidomethane (methylidynetrisformamide).

R≡H,Me

X=O,S

[1Benzopyrano[3,4-d]pyrimidine]
[1Benzothiopyrano[3,4-d]pyrimidine]

[1175]

R^1=H,Me,Et ; R^2=H,Me,MeO,Cl,F

Pyrimido[5,4-c]carbazole

[773]

4. Pyran or Thiopyran

Acetylenedicarboxylic ester annulates a thioxo group to a ring-carbon with the formation of a thiopyran ring in high yield.

2-Benzothiopyran

[1642]

An inter-ring bridge is formed by heating a 2-phenoxyaldehyde with PPA; a pyran fused at two of its sides is thus formed.

R=H,Me,Cl

[1Benzopyrano[2,3-d]pyrimidine - 2,4-dione]

[316]

An unusual quinonoid 2-benzopyran is obtained at ambient temperature when a resorcinol containing a side-chain ketone is treated with triethyl orthoformate.

[42]

2–Benzopyran–6–one

5. Pyran-4-one or Thiopyran-4-one

In the 1,2-addition of a phenylpropynoic ester across a cyclic carbonyl in the presence of a base and at a low temperature, both alkyne and ester groups react (cf. addition of acetylenedicarboxylate ester in preceding section) and a pyran-4-one ring is fused to the substrate.

R=Me,MeO

[517]

Naphtho[1,2–b]pyran–4–one

Fries rearrangement of the *S*-aryl ester (**35.2**) gives a 2-mercaptoacetophenone (which is not isolated); ring closure to the thiochromone follows.

(**35.2**)

R=H,Me,MeO,Br,Cl

[1894]

1–Benzothiopyran–4–one

III. FORMATION OF A SEVEN-MEMBERED RING

1. 1,4-Diazepine

On heating chromone-3-aldehyde ethyleneacetal with *o*-phenylenediamine, a benzodiazepine is obtained. Although the chromone ring tends to be opened by bases, good yields of this diazepine are reported.

R=H,Me,Cl

[1]Benzopyrano[2,3–b][1,5]–

benzodiazepin–13–one

[439]

2. 1, 3, 4-Thiadiazepine

This ring is formed by a phosphorus oxychloride-induced intramolecular acylation of a pyrrole ring by an acyl group attached to a triazole.

[911]

Pyrrolo[1,2-*d*][1,2,4]triazolo-
[3,4-*b*][1,3,4]thiadiazepine

CHAPTER 36

Aldehyde or Ketone and Ring-nitrogen

I. FORMATION OF A FIVE-MEMBERED RING

1. Pyrrole

The reaction of a Wittig reagent with an *N*-heterocyclic α-aldehyde provides a valuable method of annulating a pyrrole ring at the 1,2-bond.

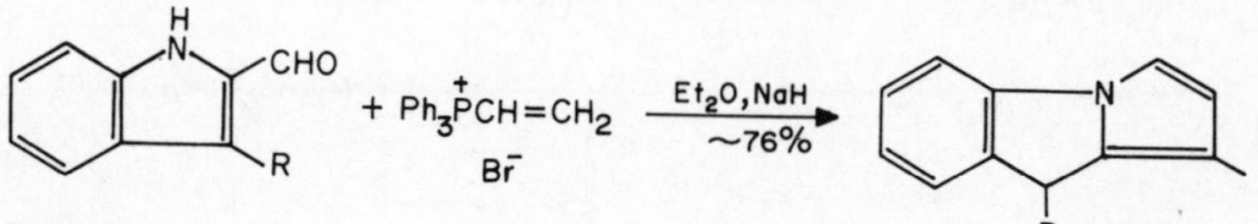

R=H,Me

Pyrrolo[1,2-*a*]indole

Other ring systems synthesized similarly:

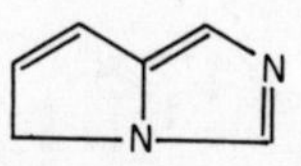

Pyrrolo[1,2-*c*]imidazole   [1190]

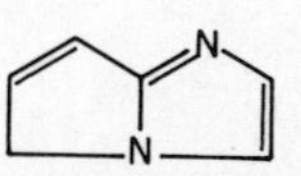

Pyrrolo[1,2-*a*]imidazole
(review[1663])   [1190]

Intramolecular *N*-acylation by the aldehyde group in the presence of a Lewis acid sometimes gives high yields of an *N*-bridgehead system.

[261]

Indolizine

$R^1, R^2 = H, Me, Ph; NR_2^3 = NEt_2,$
$N[(CH_2)_2]_2O, N(CH_2)_5$

An alkenylphosphonate reacts with an appropriately placed aldehyde to form a fused pyrrole ring.

[1557]

$R^1 = H, Ph, COOMe, PhSO_2;$
$R^2 = H, Ph; R^3 = H, PhSO_2$

Pyrrolizine
(review[1397])

2. Pyrazole

N-Amination of a 2-acylmethylpyridine followed by condensation gives a new fused pyrazole ring. The additional amino group may be provided by mesitylhydroxylamine.

[1329, 1484]

$R = Me, Ph, 2,4,6-Me_3C_6H_2$

Pyrazolo[1,5-*a*]pyridine

3. Imidazole

Under acidic conditions, a side-chain keto group reacts with a ring-nitrogen to form an imidazole, usually in good yield.

[89]

Imidazo[1,2-*a*]pyrazine

Other ring systems synthesized similarly:

[136]

Imidazo[1,2-*a*]pyridine

[649]

Imidazo[1,2-*a*][1,4]-
benzodiazepine

4. 1,2,3-Triazole or 1,2,4-Triazole

Oxidative cyclization of the pyrimidine-2-aldehyde hydrazone yields a fused
1,2,3-triazole ring but when the preformed phenylhydrazone of pyridine-2-
aldehyde was treated with lead tetra-acetate in acetic acid the reaction took a
different course to give the 1,2,4-triazole (cf. the reaction of ketone phenylhydrazones, see Chapter 95).

[1830]

[1,2,3]Triazolo[1,5-*a*]-
pyridine

[1518]

1,2,4-Triazolo[4,3-*a*]-
pyridine

5. Thiazole

An oxomethylthio side-chain is cyclized to a thiazole ring by heating with
phosphorus oxychloride.

[662]

Thiazolo[3,2-*b*][1,2,4]triazole

II. FORMATION OF A SIX-MEMBERED RING

1. Pyrimidine or Pyrimidin-4-one

Reaction between an alkylamino side-chain carbonyl and an NH in the ring
occurs usually on refluxing the compound in ethanol or toluene but a pyridine-

type (i.e. double bonded) nitrogen requires heating with a strong acid. An unexpected product (**36.1**) obtained under these conditions presumably resulted from an isomerization of the substrate [1275].

R^1 = Ac, COOEt; R^2 = CN, CONH$_2$,
CONHMe

Pyrazolo[1,5-*a*]pyrimidine   [371]

R^1 = H, HO; R^2, R^3 = H, Me, Cl   [1275]

(**36.1**)
Pyrido[1,2-*a*]pyrimidin-4-one

Other ring systems synthesized similarly:

Imidazo[1,2-*a*]pyrimidine   [372]

Pyrrolo[1,2-*a*]pyrimidine
(review [1659])   [723]

Pyrimido[1,2-*a*]indole   [779]

1,3,4-Thiadiazolo[3,2-*a*]-
pyrimidin-7-one   [1178]

2. 1, 2, 4-Triazin-3-one or 1, 2, 4-Triazine-3-thione

Thermal cyclization of the ethoxycarbonylhydrazone usually gives good yields of the fused triazines.

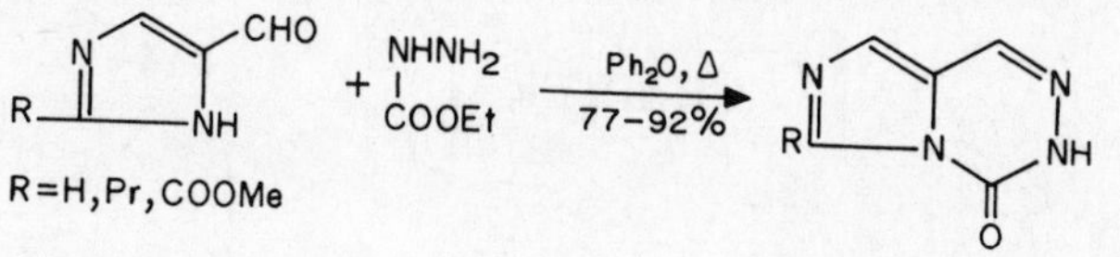

R=H,Pr,COOMe

[380]

Imidazo[1,5-d][1,2,4]–
triazin–4–one

Another ring system synthesized similarly:

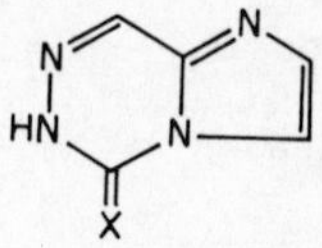

[380]

X=O,S

Imidazo[1,2-d][1,2,4]–
triazin–5–one

Imidazo[1,2-d][1,2,4]–
triazine–5–thione

III. FORMATION OF A SEVEN-MEMBERED RING

1. 1, 3-Diazepine

A carbonyl may be joined to an NH in a ring by reaction with formaldehyde.

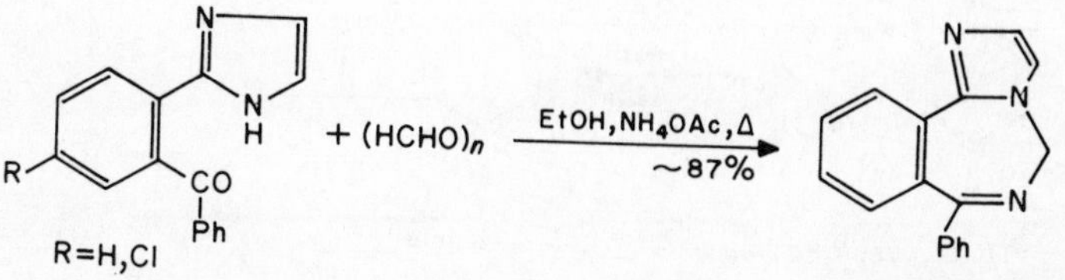

R=H,Cl

[1243]

Imidazo[2,1-a][2,4]–
benzodiazepine

CHAPTER 37

Alkene or Alkyne and Amine or Nitro

I. FORMATION OF A FIVE-MEMBERED RING

1. Pyrrole or Pyrrol-3-one *N*-oxide

A key intermediate (4-bromoindole) in the synthesis of ergot alkaloids may be obtained from palladium-mediated cyclization of a 2-vinylaniline. A similar synthesis has been used to prepare other indoles.

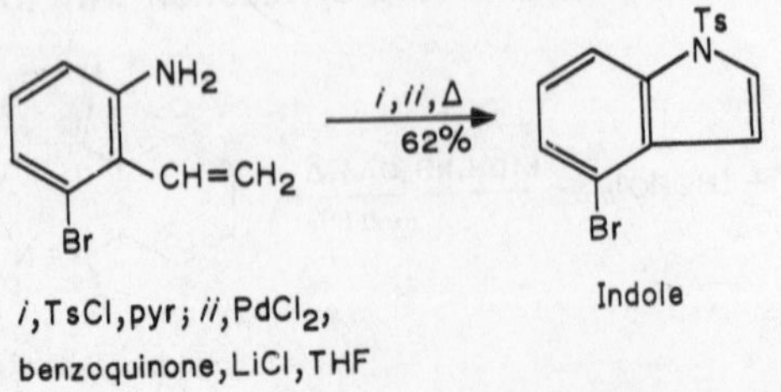

NH2
CH=CH2
Br
i,ii,Δ
62%
Ts
N
Br
Indole
i,TsCl,pyr; ii,PdCl2,
benzoquinone,LiCl,THF

[863]

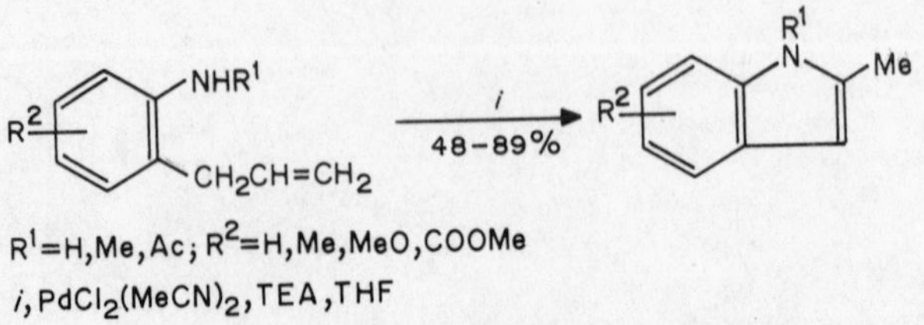

NHR1
R2
CH2CH=CH2
i
48–89%
R1
N
Me
R2
R1=H,Me,Ac; R2=H,Me,MeO,COOMe
i,PdCl2(MeCN)2,TEA,THF

[1108]

Deoxygenation of a nitro group placed ortho to a vinyl group resulted in cyclization and the formation of a pyrrole ring.

R=H,Me ; Ar=Ph,Me-,MeO-,

Cl-,Me$_2$N-C$_6$H$_4$, 2-furyl

[1]Benzopyrano[3,2-*b*]-

pyrrol-9-one

[90]

Compounds containing adjacent nitro and alkynyl groups are cyclized to 3-indolones (review [1662]) simply by heating in pyridine.

3-Indolone 1-oxide

[397]

Amino and alkyne groups in the presence of a mercury(II) salt cyclize in high yield on heating in acetic acid while carbamate and alkynyl groups cyclize efficiently in the presence of a strong base.

Pyrrolo[2,3-*b*]quinoxaline

[744]

R=H,Bu,Ph

Indole

[1555]

2. Pyrazole

An *N*-amino-2-alkynylpyridinium mesitylenesulphonate is converted into a pyrazolopyridine on treatment with a mild base; an *N*-ylide may be used [837].

R=H,alkyl,CH$_2$OH,Ph

Pyrazolo[1,5-*a*]pyridine

[296]

Other ring systems synthesized similarly:

[296, 837]   [296, 837]

Pyrazolo[1,5-a]quinoline   Pyrazolo[5,1-a]isoquinoline

[296]

Pyrazolo[1,5-b]isoquinoline

II. FORMATION OF A SIX-MEMBERED RING

1. Pyridine

Moderate to good yields of a fused pyridine ring are obtained by heating a 2-vinyl-imine in diphenyl ether.

$Ar^1CH=CH$
$Ar^2CH=N$ — Me
$\xrightarrow{Ph_2O, \Delta \atop 25-60\%}$
Ar^1
Ar^2 — Me   [847]

Ar^1=Ph,4-MeC$_6$H$_4$;
Ar^2=Ph,4-Me-,4-Cl-C$_6$H$_4$

Isoxazolo[4,5-b]pyridine

When the palladium-assisted cyclization mentioned in Section I.1 is applied to the 3,3-dimethylallyl homologue, the 2,2-dimethylquinoline derivative is obtained.

NH$_2$
CH$_2$CH=CMe$_2$
$\xrightarrow{PdCl_2(MeCN)_2, THF, TEA \atop \sim54\%}$
[1108]

Quinoline

III. FORMATION OF A SEVEN-MEMBERED RING

1. 1,4-Thiazepin-3-one

Thioglycolic acid adds across the C–C double bond and on heating, the intermediate cyclizes in good yields.

Ar^1 = Cl-, F-, NO_2-C_6H_4;
Ar^2 = Ph, 4-FC_6H_4;
R = Ar^1, Ar^1SO_2

Pyrazolo[3,4-*e*][1,4]-
thiazepin-7-one

[381]

CHAPTER 38

Alkene or Alkyne and Carboxylic Acid or its Derivative

The following derivatives of carboxylic acids are included in this chapter: carbamate, carboxamide, carboxylic ester and nitrile.

I. FORMATION OF A SIX-MEMBERED RING

1. Pyridine or Pyridin-2-one

An N-alkenyl group adjacent to a nitrile on a ring allows a base-induced reaction to occur with the result that an aminopyridine ring is produced in high yields. A 2-pyridinone ring is obtained when an alkyne and a carboxamide group react in a basic medium. Several naphthyridines may be synthesized in this way (review [1648]).

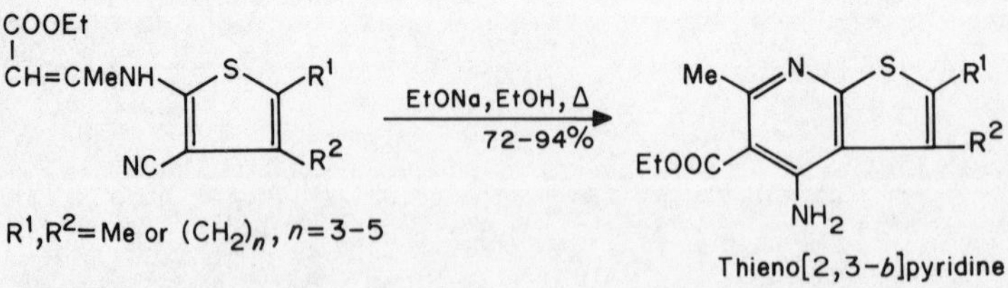

R^1,R^2=Me or $(CH_2)_n$, n=3-5

Thieno[2,3-b]pyridine

[396]

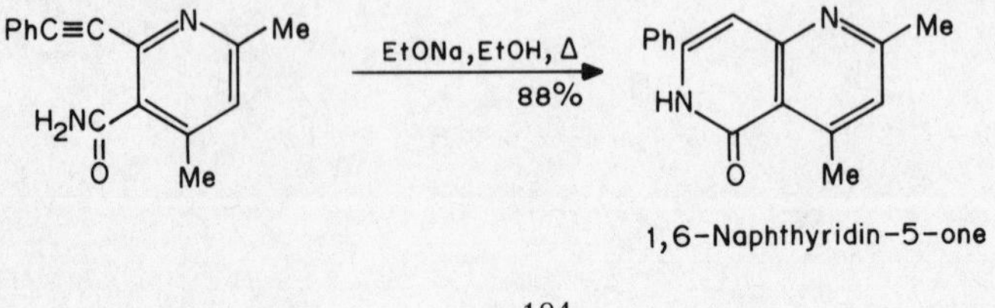

1,6-Naphthyridin-5-one

[1708]

184

Other ring systems synthesized similarly:

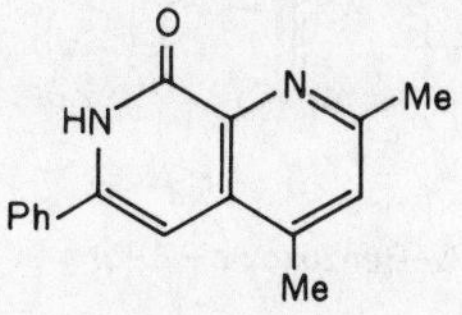

1,7-Naphthyridin-8-one   2,6-Naphthyridin-1-one

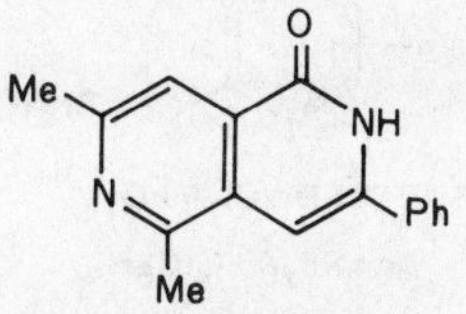

[1708]   [1708]

A 2-*N*-allylaminobenzonitrile derivative may be converted into a 4-amino-quinoline by this method or by Friedel–Crafts reaction conditions.

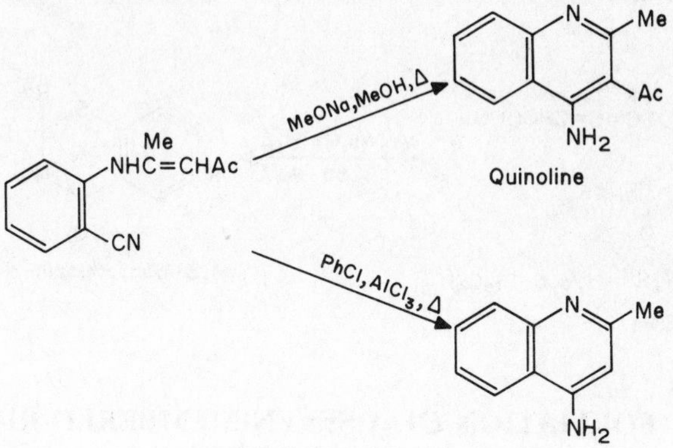

Quinoline

[1129]

[1129]

2. Pyran-2-one

Under the influence of PPA or mercuric acetate, an *O*-alkenyl or -alkynyl carboxylic ester undergoes cyclocondensation to form a pyran-2-one ring.

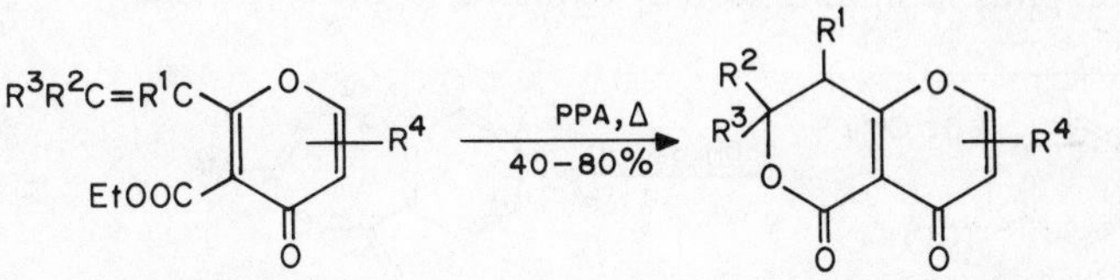

R^1–R^4=H,Me   Pyrano[4,3-*b*]pyran-4,5-dione

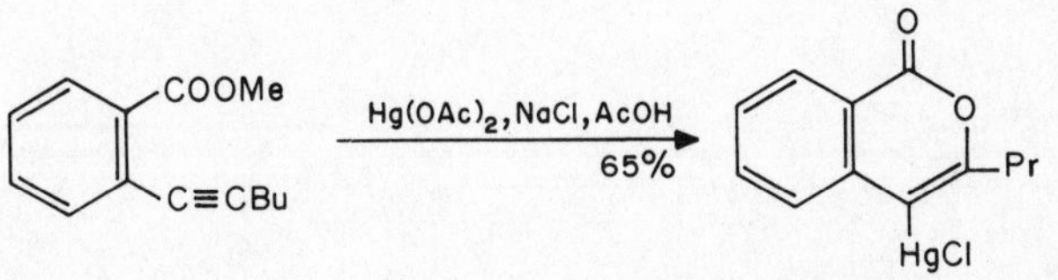

[995]

2-Benzopyran-1-one

[1500]

3. 1,3-Oxazin-2-one

A nitrovinyl-carbamate in a basic medium cyclizes to a fused oxazinone in good yield.

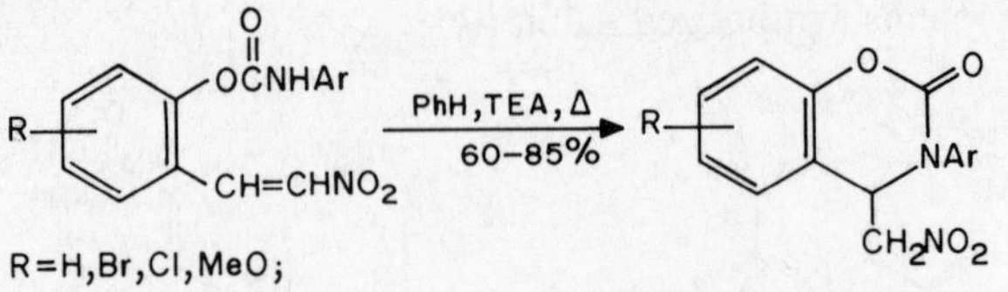

R=H,Br,Cl,MeO;
Ar=Ph,4-ClC$_6$H$_4$

1,3-Benzoxazin-2-one

[1116]

4.   1, 3-Thiazin-4-one

Thioethers which also contain a carboxamide and an olefinic double bond cyclize in a basic medium to the thiazinone ring. In this example, the double bond is part of an allene.

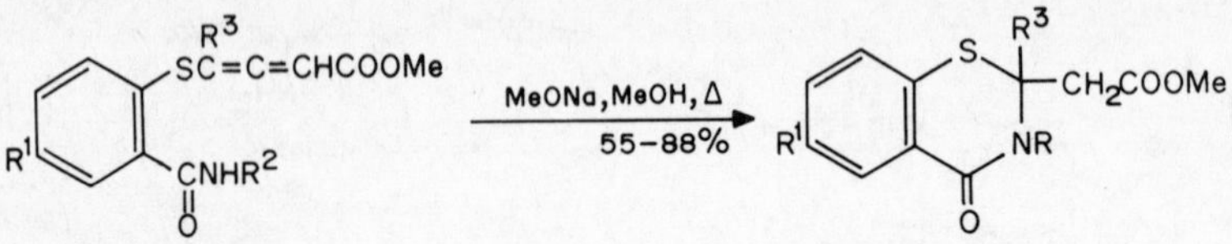

R^1=H,Br,Cl; R^2=H,3,5-Cl$_2$C$_6$H$_5$;
R^3=H,COOMe

1,3-Benzothiazin-4-one

[973]

II.   FORMATION OF A SEVEN-MEMBERED RING

1.   1, 4-Oxathiepin-7-one

Oxidative cyclization of a 2-alkynylthiobenzoic acid gives this seven-membered ring and the sulphur is simultaneously oxidized to the sulphone.

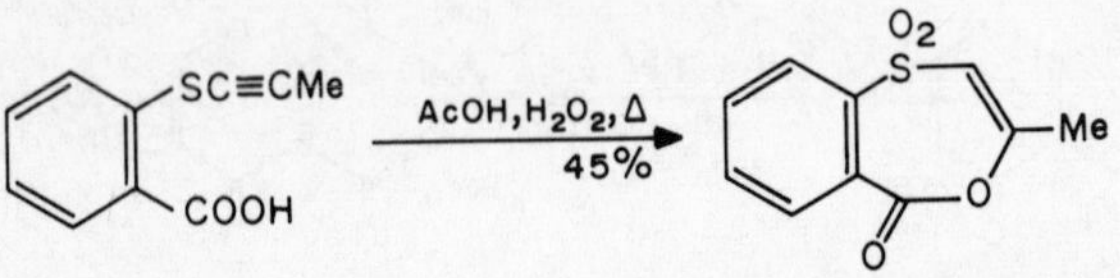

4,1-Benzoxathiepin-5-one

[957]

CHAPTER 39

Alkene or Alkyne and Halogen

I. FORMATION OF A FIVE-MEMBERED RING

1. Pyrrole or Pyrrol-2-one

Primary amines effect ring closure of a reactive halogeno-alkyne.

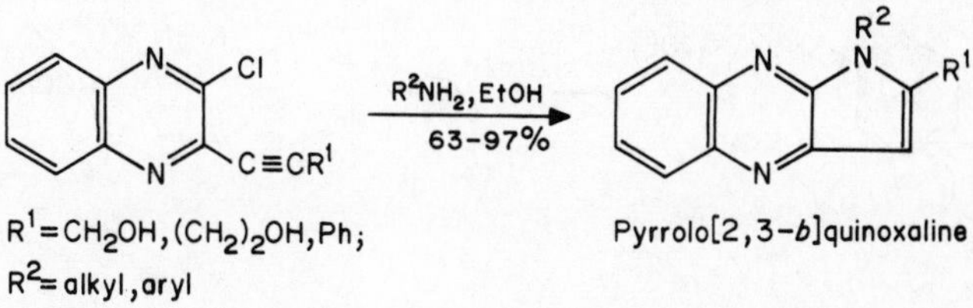

Cl
C≡CR¹
R²NH₂, EtOH
63–97%
R²
R¹
R¹=CH₂OH, (CH₂)₂OH, Ph;
R²= alkyl, aryl
Pyrrolo[2,3-b]quinoxaline

[744]

A nickel complex prepared from bis(acetylacetonate)Ni(II), aluminium triethyl and triphenylphosphine converts an *o*-chloro-allylamine into an indole but small amounts of several other products are produced simultaneously.

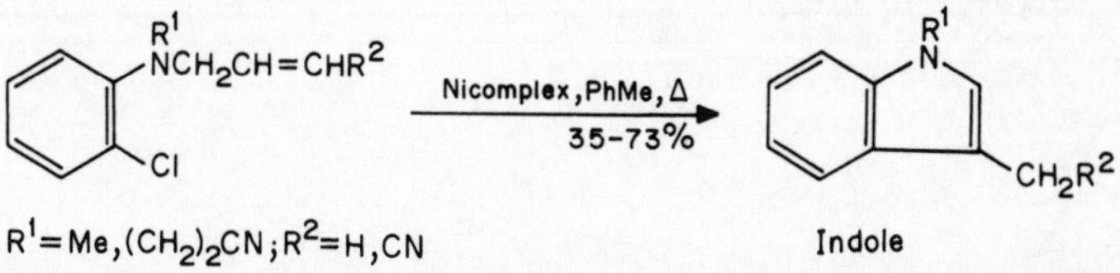

R¹
NCH₂CH=CHR²
Cl
Nicomplex, PhMe, Δ
35–73%
R¹
CH₂R²
R¹=Me, (CH₂)₂CN; R²=H, CN
Indole

[1806]

Nickel-complex-assisted cyclization of a 2-chloroalkene can give high yields of 2-indolones.

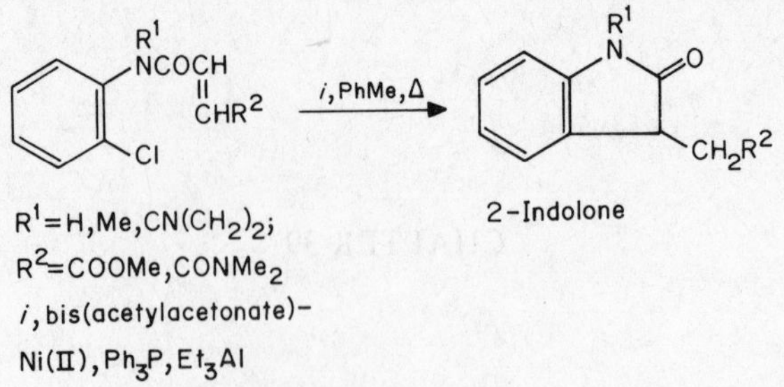

i, PhMe, Δ
2-Indolone
R1=H,Me,CN(CH2)2;
R2=COOMe,CONMe2
i, bis(acetylacetonate)-
Ni(II), Ph3P, Et3Al

[1926]

II. FORMATION OF A SEVEN-MEMBERED RING

1. 1, 2-Diazepine

The vinyl and halogenohydrazone groups react in a boiling tertiary amine, with phosphorus pentoxide-benzene or by reaction with sodium azide under phase transfer catalysis.

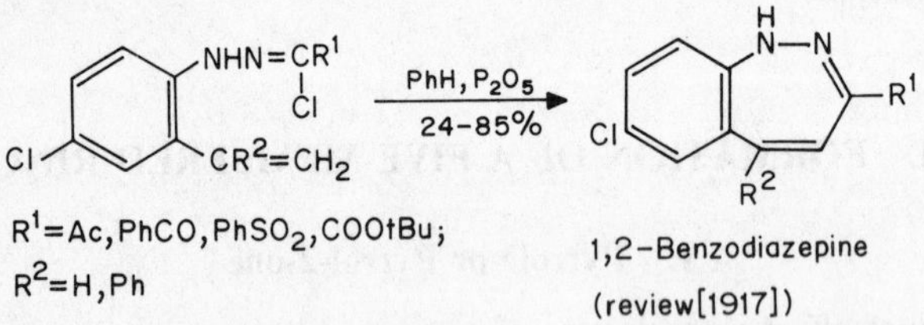

PhH, P2O5
24–85%
R1=Ac,PhCO,PhSO2,COOtBu;
R2=H,Ph
1,2-Benzodiazepine
(review[1917])

[980]

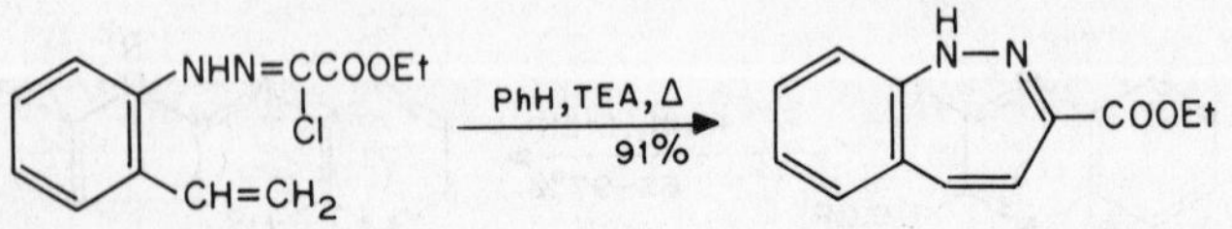

PhH, TEA, Δ
91%

[533]

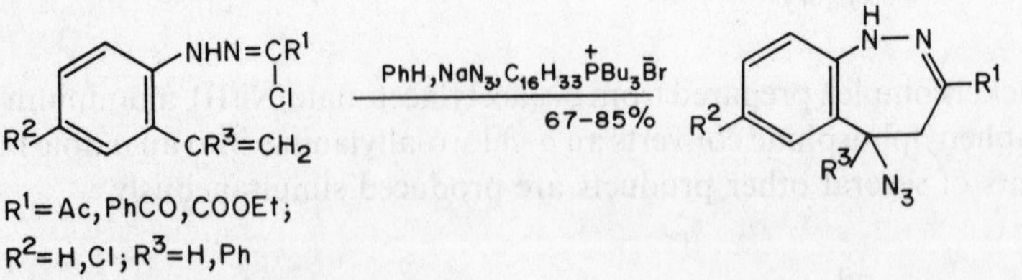

PhH, NaN3, C16H33PBu3Br
67–85%
R1=Ac,PhCO,COOEt;
R2=H,Cl; R3=H,Ph

[731]

CHAPTER 40

Alkene or Alkyne and Hydroxy, Thiol or Ether

I. FORMATION OF A FIVE-MEMBERED RING

1. Furan

Many methods have been described of converting an alkenyl or alkynyl phenol into a benzofuran; whereas the alkyne gives a furan directly, the alkene usually needs either an additional dehydrogenation (or oxidation) step or a reagent which combines both cyclizing and oxidative steps. Osmium tetroxideperiodate and 3-chloroperbenzoic acid are examples of reagents which generate 2,3-dihydrofuran-2-methanol while N-iodosuccinimide yields the furan ring in one step and usually in high yield. With the former type of reagent, the dihydrofuran may be oxidized *in situ* and without purification. Benzofurans have been reviewed [1622].

Furo[2,3-*g*]quinoline   [488]

R = Me,CHO   Benzofuran   [531]

189

Furo[3,2-*g*][1]benzopyran-7-one

[644]

Ar = 3,4-OCH$_2$OC$_6$H$_3$

[764]

o-Methoxyphenylalkynes may be dealkylated by pyridine hydrochloride, or a lithium or mercury(II) salt to give the benzofuran or the mercury chloride derivative; the mercury can be replaced by iodine or hydrogen. All methoxy groups in such compounds are demethylated.

R = H,Me; Ar = Me-, MeO-C$_6$H$_4$

[1428]

R^1 = alkyl, Ph; R^2 = H,Me,
MeO,HO,NO$_2$
i, Li, 2,4,6-Me$_3$pyridine,
R^3 = H or Hg^{2+}, AcOH, NaCl, Δ,
R^3 = HgCl

Benzofuran

[213, 1500]

2. Thiazole

1-Allyl-2-mercaptopyrimidin-4-one is cyclized by silver acetate-iodine but in the absence of silver acetate, 2-iodomethylthiazolo[3,2-*a*]pyrimidin-7-one hydro-iodide salt was isolated.

Thiazolo[3,2-*a*]pyrimidin-7-one

II. FORMATION OF A SIX-MEMBERED RING

1. Pyran

A 3-substituted allyl side-chain reacts with a hydroxy group in the presence of DDQ (review [B-39]), *N*-iodosuccinimide or sodium acetate-acetic anhydride to form a fused 2*H*-pyran ring. The allyl group is reported to resist cyclization [626]. 3-Chloroperbenzoic acid as a cyclizing agent produces a 3,4-dihydro-3-hydroxypyran ring.

R^1=Me,Ph,Me$_2$C=CH(CH$_2$)$_2$;
R^2=Me,MeO,HO,CHO
i, DDQ,Et$_2$O or NIS,CH$_2$Cl$_2$

2*H*-1-Benzopyran

Other ring systems synthesized similarly:

2*H*-Naphtho[2,3-*b*]pyran

[630]

2*H*-Naphtho[1,2-*b*]pyran

[630]

R=MeO,HO,OCH$_2$O

1-Benzopyran
(review [B-16])

[1408]

2. Pyran-4-one

To form this ring, a 2-alkenylcarbonyl hydroxy or ether moiety is needed and may be dehydrogenatively cyclized with one of several reagents. Orthophosphoric acid [1568] or a Ni–Zn–KI mixture [1284] yields a 2,3-dihydropyran-4-one while alkaline hydrogen peroxide leads to the formation of a 3-hydroxypyran-4-one. DMSO-iodine gives a good yield of 3-iodopyran-4-one derivative [1547]. Mercury(II) acetate (review [B-38]) cyclizes a 2-methylalk-ynylcarbonylbenzene to a 3-mercury substituted chromone. The side-chain undergoes a rearrangement when treated with thallium(III) nitrate; this is a convenient synthesis of an isoflavone.

R^1=Ac,MeOCH$_2$; R^2=MeO,PhCH$_2$O   1-Benzopyran-4-one,2-aryl

Ar=alkyl-Br-,Cl-,MeO-C$_6$H$_4$ 1-Benzopyran-4-one,3-aryl

R=COOH,MeO; Ar=alkyl-,
Br-,Cl-,MeO-,PhCH$_2$O-C$_6$H$_4$

R=Pr,Ph

3.   1,4-Dioxin or 1,4-Oxazine

A phenol containing a 2-alkynyloxy or 2-alkynylamino substituent cyclizes under the influence of mercury(II) oxide (review [B-37]).

R=H,Ph;   X=NAlk,NAc,O
                 1,4-Benzoxazine
                 1,4-Benzodioxin

CHAPTER 41

Alkene, Methylene, Ring-carbon, or Ring-nitrogen and Lactam Carbonyl

The term lactam carbonyl includes the thione analogue; both of these functions are able to tautomerize in many compounds and exist as either —CO—NH— or —C(OH)=N— or the corresponding sulphur equivalent. No distinction is made between these two forms in this book.

I. FORMATION OF A FIVE-MEMBERED RING

1. Pyrrole

When a lactam containing an ω-carboxyalkyl side-chain is heated with soda lime, decarboxylation and spontaneous cyclization occur. This reaction should be applicable to side-chains of various lengths but this example leads to the formation of a pyrrole ring.

Pyrrolizine
(review [1397])

[432]

2. Imidazole

The carbonyl and nitrogen of a lactam group may be linked through an imidazole ring by reaction with ethyl isocyanoacetate and a base.

193

$R = H, Me$

Imidazo[5,1-c][1,4]-
benzothiazine

[830]

3. Furan or Thiophene

Lactam carbonyl and vinyl groups combine to form a thiophene ring when such a compound is heated with phosphorus pentasulphide in pyridine; it is likely that the carbonyl is sulphurized first. Actam carbonyl and methylene react with DMFDMA to give a fused furan [1066].

$Ar = Ph, 4-Me-, 4-MeO-, 4-Cl-C_6H_4$

Thieno[2,3-e]-1,2,4-
triazine-3-thione

[320]

II. FORMATION OF A SIX-MEMBERED RING

1. Pyran or Thiopyran

Formation of a pyran or thiopyran ring between a lactam carbonyl (or its hydroxy tautomer) and a neighbouring ring-carbon atom is achieved by reaction with a malonic acid or malononitrile derivative, usually by heating in dioxan or ethanol.

$Ar = 4-Br-, 4-F-C_6H_4$

Pyrano[2,3-d]pyrimidine-
2,4-dione

[754]

Other ring systems synthesized similarly:

Pyrano[2,3-c]pyrazole

[528, 1737]

$X = O, S$

Thiopyrano[2,3-d]thiazol-2-one
Thiopyrano[2,3-d]thiazole-2-thione

[1098]

Malononitrile reacts with an alkene-containing lactam under basic conditions to form a 4*H*-pyran. Acrylonitrile in acetic acid, on the other hand, annulates the corresponding thiolactam.

$$\text{+ CH}_2\text{(CN)}_2 \xrightarrow[\sim 57\%]{\text{EtOH,TEA,}\Delta}$$

[910]

Pyrano[2,3-*d*]imidazole-2-thione

$$\text{+ CH}_2\text{=CHR} \xrightarrow[\sim 70\%]{\text{AcOH}}$$

[1858]

R=CN,COOEt;
Ar=Ph,4-Cl-,4-MeO-C$_6$H$_4$

Thiopyrano[2,3-*d*]imidazol-2-one

Another ring system synthesized similarly:

[1569]

Pyrano[2,3-*c*]pyrazole

1,3-Diketones react with lactams or their tautomers in refluxing acetic acid with the formation of a fused pyran ring.

$$\text{+ }\begin{array}{l}\text{CH}_2\text{COR}^1\\\text{COR}^2\end{array} \xrightarrow[\sim 50\%]{\text{AcOH,}\Delta}$$

R^1,R^2=Me,Ph

[880]

Pyrano[2,3-*c*]pyrazol-3-one

2. Pyran-2- or -4-one

3-Oxocarboxylic esters on heating with a 3-pyrazolone give good yields of pyranopyrazol-6-ones. To prepare the isomeric pyranopyrazol-4-ones, the keto ester is replaced by an acrylic acid chloride.

$$\text{+ }\begin{array}{l}\text{R}^3\text{CHCOR}^2\\\text{COOEt}\end{array} \xrightarrow[31-99\%]{\Delta}$$

R^1=H,Me,Ph;R^2=Me,Ph;
R^3=H,Me

[298, 632]

Pyrano[2,3-*c*]pyrazol-6-one

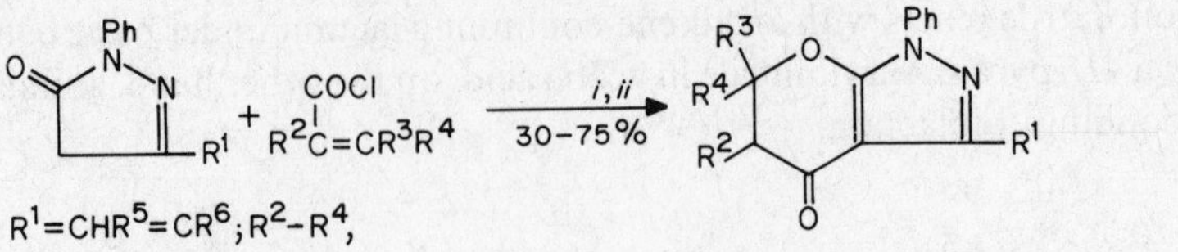

$R^1 = CHR^5 = CR^6; R^2 - R^4,$
$R^6 = H, Me; R^5 = H, Me, Ph$
$i, PhMe, Mg(OEt)_2, \Delta;$
$ii, HCl-EtOH$

[21]

Pyrano[2,3-c]pyrazol-4-one

CHAPTER 42

Alkene or Alkyne and Methylene, Ring-carbon or Ring-nitrogen

I. FORMATION OF A FIVE-MEMBERED RING

1. Pyrrole

Under acidic conditions, an alkynylamino group forms a pyrrole ring in high yield.

R^1 = Me, Et; R^2 = H, Me   Pyrrolo[2,3-d]pyrimidine-2,4-dione   [1289, 1890]

Another ring system synthesized similarly:

Pyrrolo[3,2-d]pyrimidine-2,4-dione   [1890]

197

Ring formation from an alkene and a ring-nitrogen means a loss of hydrogen atoms and may be effected under oxidizing conditions but oxalyl chloride or malononitrile is also effective.

Pyrrolo[1,2-a]quinoxaline   [443]

R^1=Me,Et; R^2=MeO,OCH$_2$O

Pyrrolo[2,1-a]isoquinoline-
2,3-dione   [1455]

Pyrrolo[1,2-c]imidazol-
1-one,3-thioxo-   [910]

The increased reactivity of quaternized pyridines is well-known (review [1870]) and in the presence of DBU, cyclization occurs between the anion and the alkyne to form a fused pyrrole (cf. [1585]).

R=Me,Bu,Ph

Indolizine
(review [1670])   [1654]

2. Pyrazole

Pyridinium aminides containing a terminal ester group when heated give a rather low yield of pyrazolopyridines.

R^1=H,Me; R^2=H,Me,Ac,COOEt

Pyrazolo[1,5-a]pyridine   [1582]

3. Furan

Alkynyloxy side-chains can be cyclized to either a fused methylfuran or pyran ring depending partly on the conditions and partly on the substituents on the reactant ring. Cyclization in hot NN-diethylaniline depends on the presence of a nitro group and the absence of a 2-methoxy group in the benzene ring (see also Section II.2) [1170], but other workers subjected variously substituted 3-phenoxy-3-methylbut-1-yne to a similar reaction and obtained the pyran [631, 822]. Cyclization in a polyethyleneglycol (PEG) gave furan or pyran depending on substituents on the benzene ring [1010]. Under acidic conditions and ambient temperature, a furan ring may be obtained [1289, 1890].

R=NO$_2$,Ac

Benzofuran

[1010]

R$^1 \neq$ MeO; R^2=2-,4-NO$_2$

[1170]

R=Me,Et

Furo[2,3-d]pyrimidine-
2,4-dione

[1289, 1890]

4. Thiophene

Alkynes react with sulphur dioxide and can incorporate the sulphur atom in the thiophene ring at temperatures below 0 °C.

R^1=COOH,SO$_2$NH$_2$;
R^2=H,MeO

Benzo[b]thiophene

[970, 1817]

5. Thiazole

An S-allene cyclizes onto a ring-nitrogen anion to form a fused thiazole ring.

$Ar = Ph, Cl-, Me-, MeO-C_6H_4$

Thiazolo[3,2-*b*]-
[1,2,4]triazole

[948]

II. FORMATION OF A SIX-MEMBERED RING

1. Pyridine

N-Alkynylanilines (which must not be tertiary amines) cyclize in a copper-mediated reaction (preferably in dioxan) to a pyridine ring. This reaction is applicable to ethers and thioethers. A similar palladium-mediated cyclization occurs at ambient temperature; a combined copper-palladium assisted reaction has also been described.

$R = H, Me, MeO, Cl$

$X = NH, O, S$
Quinoline
2*H*-1-Benzopyran
2*H*-1-Benzothiopyran

[1044]

Pyrido[3,2-*d*]pyrimidine-
2,4-dione

[1890]

$R^1 = Me, Et; R^2 = H, Me;$
$R^3 = H, Me, Et$

Pyrido[2,3-*d*]pyrimidine-
2,4-dione

[1055]

Friedel–Crafts alkylation of a nearby ring by an alkenyl group attached to another ring results in the formation of a new pyridine ring. In a basic medium, the reactive cyanoacethydrazide adds onto a ring-conjugated alkene to give a highly substituted fused pyridine.

[1537]

Pyrazolo[1,5-*a*]quinoline

[1569]

Thiazolo[3,2-*a*]pyridin-3-one

Another ring system synthesized similarly:

[1863]

Pyrido[1,2-*a*]benzimidazole

2. Pyridazine

A hetero-Diels–Alder reaction in which azodiformate ester is the dienophile proceeds at ambient temperature to give a high yield of a fused pyridazine.

[1302]

R^1=H,Me; R^2=Me,tBu

[1]Benzopyrano[3,4-*c*]pyridazine

3. Pyran

The method described in Section II.1 is supplemented by several others, especially the thermal cyclization of alkynyl ethers in *NN*-diethylaniline, polyethylene glycol, xylene or (at lower temperatures) with mercury(II) trifluoroacetate-magnesium oxide-THF [1500] or A_gBF_4 [624].

[629, 631, 633, 822, 1010]

R^1=H,Me; R^2=H,Me,Cl

2*H*-1-Benzopyran

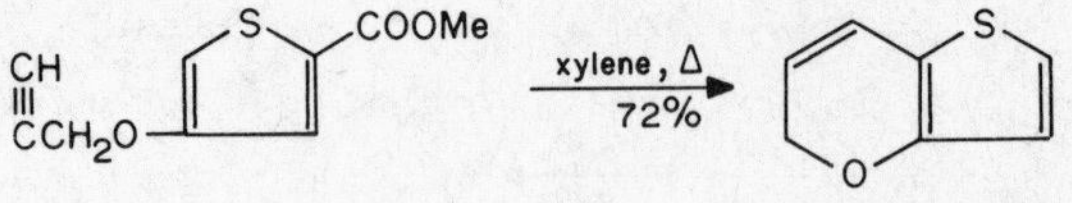

Thieno[3,2-b]pyran

[1283]

4. Thiopyran

One method of building a thiopyran ring fused to benzene has been mentioned in Section II.1 and a similar substrate cyclizes on heating in pyridine and toluene.

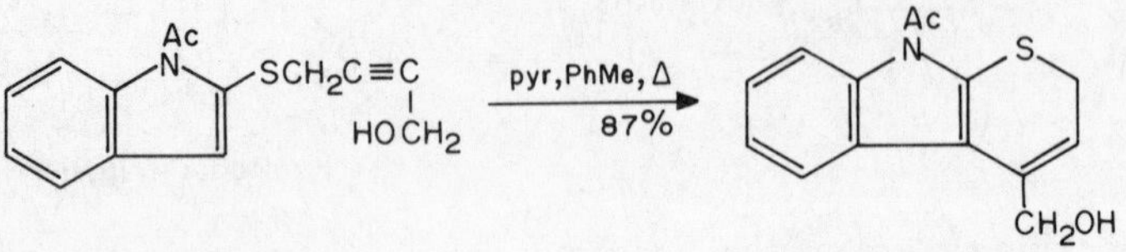

Thiopyrano[2,3-b]indole

[1100]

III.   FORMATION OF A SEVEN-MEMBERED RING

1.   1,4-Oxazepine

A doubly fused oxazepine ring is formed by a base-promoted reaction between a terminal alkyne and the NH of a triazole, the terminal methine becoming an exocyclic methylidene group.

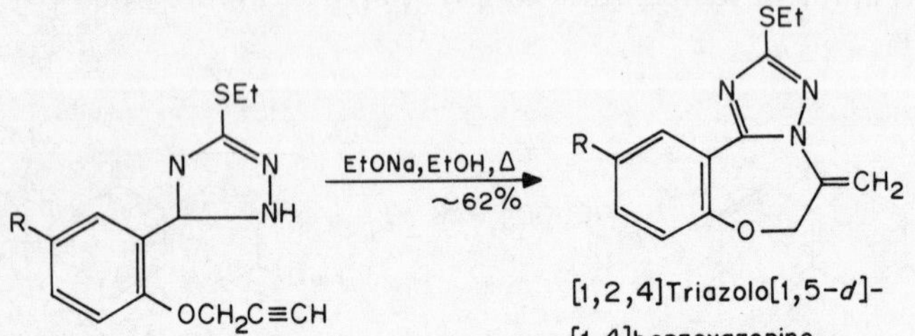

[1,2,4]Triazolo[1,5-d]-
[1,4]benzoxazepine

[933]

CHAPTER 43

Amidine and Amine, Carboxylic Acid, Ester, Hydroxy, Methylene or Nitro

I. FORMATION OF A FIVE-MEMBERED RING

1. Imidazole

N-Phenacyl quaternary heterocycles which have a neighbouring amidine group cyclize to form a fused imidazole.

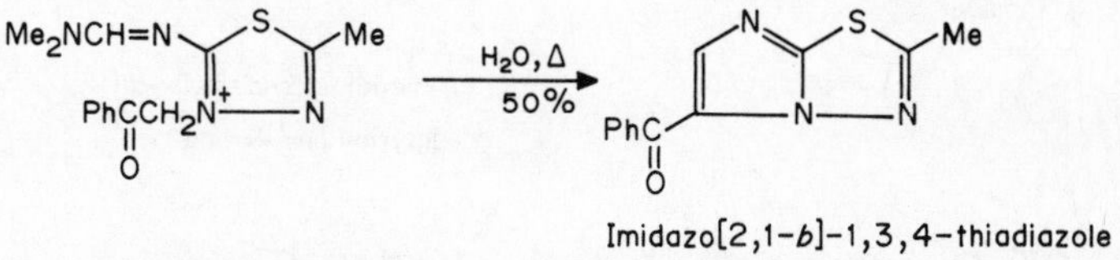

[347]

Imidazo[2,1-b]-1,3,4-thiadiazole

2. Oxazole

An o-amidinophenol undergoes ring closure thermally or in acid to form a benzoxazole in good yield.

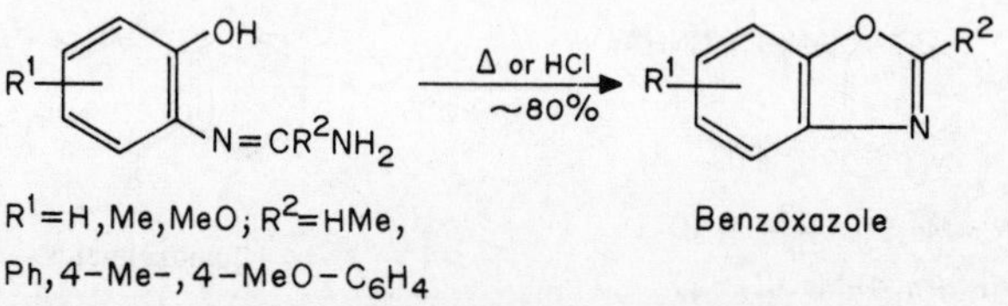

[1713]

R^1=H,Me,MeO; R^2=H,Me,
Ph,4-Me-,4-MeO-C_6H_4

Benzoxazole

203

II. FORMATION OF A SIX-MEMBERED RING

1. Pyrimidine or Pyrimidin-4-one

Thermal dehydration of an *o*-acylamidino-amine gives high yields of a fused pyrimidine.

xylene, Δ
~99%

[1735]

Quinazoline

Other ring systems synthesized similarly:

[1735]

1,2,3-Triazolo[4,5-*d*]-
pyrimidine

[1735]

Pyrimido[4,5-*d*]pyrimidine

A fused pyrimidin-4-one ring may be constructed from either an *o*-amidinocarbonyl-amine or (under different conditions) an *o*-amidino-acid or -ester or nitrile [1981].

AcOH, Δ
64%

[1153]

Thieno[2,3-*d*:4,5-*d'*]-
dipyrimidin-4-one

EtOH, KOH
79-96%

[846]

R = H, Me, Ph, Me$_2$N, MeO,
N(CH$_2$)$_5$

Oxazolo[5,4-*d*]pyrimidin-4-one

Ac$_2$O, Δ
40-80%

[1248]

R^1 = Me, EtO; R^2 = H, NO$_2$;
Ar = Ph, 4-MeOC$_6$H$_4$

4-Quinazolinone

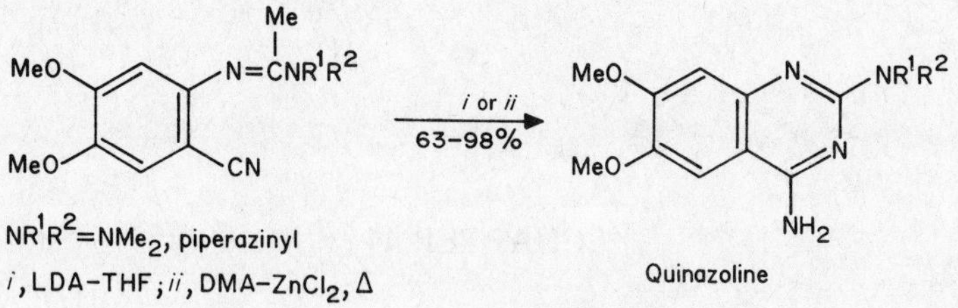
Me
MeO
N=CNR¹R²
MeO
CN
i or ii
63–98%
MeO
N
NR¹R²
MeO
N
NH2
NR¹R²=NMe2, piperazinyl
i, LDA–THF; ii, DMA–ZnCl2, Δ
Quinazoline

[1981]

2. Pyrazine *N*-oxide

Under basic conditions, an *o*-nitro-amidine cyclizes to form a fused pyrazine *N*-oxide in good yield (review of heterocyclic *N*-oxides [B-34]).

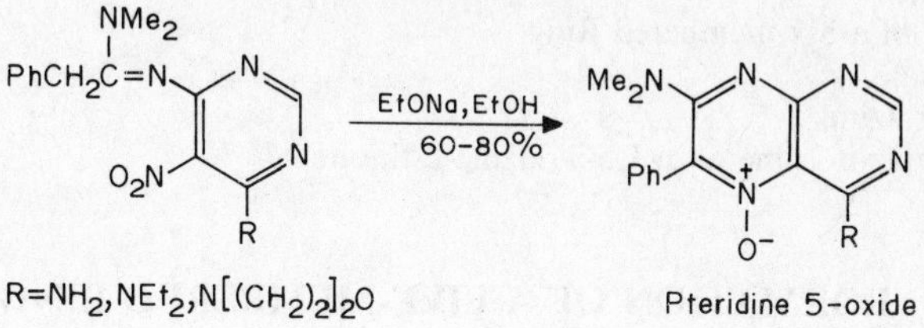
NMe2
PhCH2C=N
N
N
O2N
R
EtONa, EtOH
60–80%
Me2N
N
N
N
Ph
N
O⁻
R
R=NH2, NEt2, N[(CH2)2]2O
Pteridine 5-oxide

[913]

CHAPTER 44

Amidine and Ring-carbon or Ring-nitrogen

I. FORMATION OF A FIVE-MEMBERED RING

1. Imidazole

N-Phenylamidines contain the atoms necessary to form benzimidazoles and this conversion may be effected by sodium hypochlorite. 3-R^2-Benzamidines (**44.1**) give a mixture of 4- and 6-substituted benzimidazoles which may be separated on an alumina column. LTA is also effective [658, 1957].

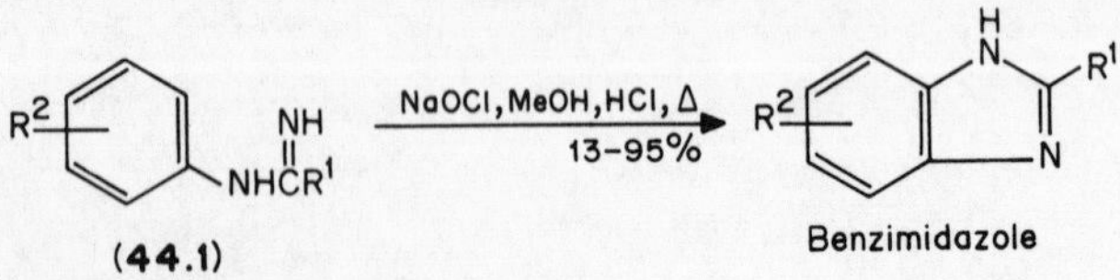

(**44.1**)

R^1=Ph,pyridinyl;
R^2=H,Me,Cl,MeO,NO_2,Ph

[334, 924]

Benzimidazole

II. FORMATION OF A SIX-MEMBERED RING

1. Pyrimidine

DMFDEA supplies one carbon atom to enable cyclization of an amidine to a ring-carbon atom to proceed on heating the reactants.

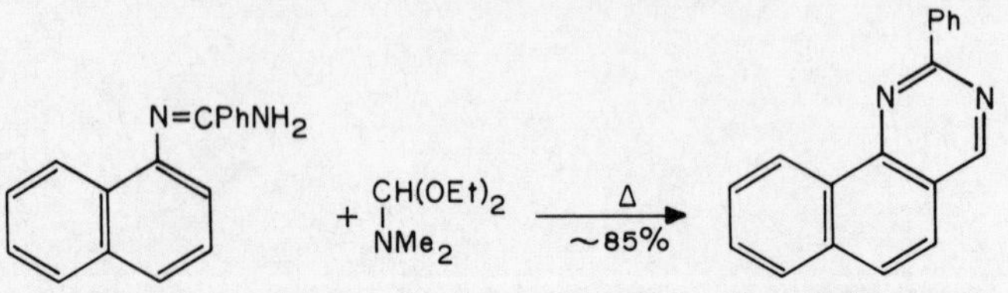

[692]

Benzo[*h*]quinazoline

2.   1, 3, 5-Triazine

When the amidine group is adjacent to the ring nitrogen atom of a heterocycle, annulation is brought about by means of a carbonyl compound, cyanogen bromide or an orthoester, depending on the kind of substituent desired in the product.

[1104]

1,3,5-Triazino[1,2-a]-
benzimidazole

[1104]

R^1=H,Pr,Ph,MeO-, Me₂N-,
HO-,NO₂-C₆H₄; R²=H,Me,or
R¹R²=(CH₂)ₙ,n=4-6

[147]

Another ring system synthesized similarly:

[1448]

Pyrazolo[1,5-a]-1,3,5-triazine

3.   1, 3, 5-Triazin-2-one or 1, 3, 5-Triazine-2-thione

Cyclization of an amidine to the triazinone is conveniently effected by heating with an azodicarboxylate ester.

[694]

i,(=NCOOEt)₂,EtOH

1,3,5-Triazino[1,2-a]-
benzimidazol-4-one (X=O)

The thioxo analogue (X = S) of this product is conveniently synthesized by reaction of the amidine with carbon disulphide in pyridine (48 per cent yield) [1104].

CHAPTER 45

Amine and Azo or Diazo

I. FORMATION OF A FIVE-MEMBERED RING

1. Imidazole

Reductive cyclization of a 2-azo-amine in the presence of formic acid eliminates one half of the azo group and a fused imidazole ring is formed.

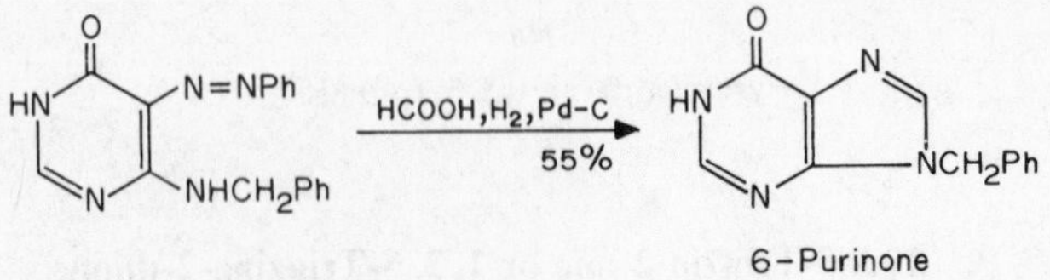

[1161]

6-Purinone

2. 1, 2, 3-Triazole

Oxidation of a similar substrate with lead tetra-acetate usually gives high yields of this fused ring. The use of this reagent in organic chemistry has been reviewed [B-36].

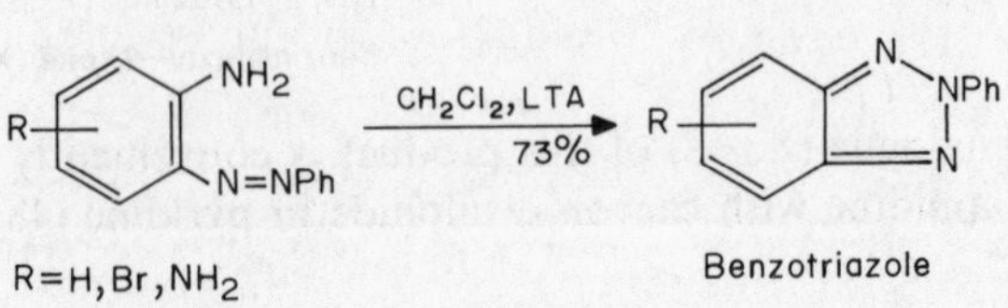

[1595]

R=H, Br, NH₂

Benzotriazole

208

Another ring system synthesized similarly:

1,2,3-Triazolo[4,5-c]-
[1,2,6]thiadiazine 2,2-dioxide

[1220]

A 1, 2, 3-triazole ring is also formed spontaneously, if rather slowly, by allowing an amino-diazo compound to stand for several days.

1,2,3-Triazolo[4,5-c]-
[1,2,6]thiadiazine 5,5-
dioxide

[1220]

II. FORMATION OF A SIX-MEMBERED RING

1. Pyrazine

The pyrazine ring of a pteridinedione may be obtained by treating an amino-azopyrimidine with dimethyl acetylenedicarboxylate.

2,4-Pteridinedione

[323]

2. 1, 2, 4-Triazin-3-one

In some of the reactions of azo groups, the PhN= part is lost but when an amino-azopyrimidine is treated with a source of one carbon atom (as a carbonyl), both nitrogen atoms are retained in the product. Carbonyldi-imidazole (CDI) gives high yields in this reaction but urea is also effective [475].

Ar = Ph,Me-,Cl-,F-C$_6$H$_4$

Pyrimido[5,4-e]-1,2,4-
triazine-3,6,8-trione

[475]

CHAPTER 46

Amine and Carboxamide or Thiocarboxamide

I. FORMATION OF A FIVE-MEMBERED RING

1. Pyrazol-3-one

Reduction of a nitro group under alkaline conditions induces cyclization with an adjacent carboxamide.

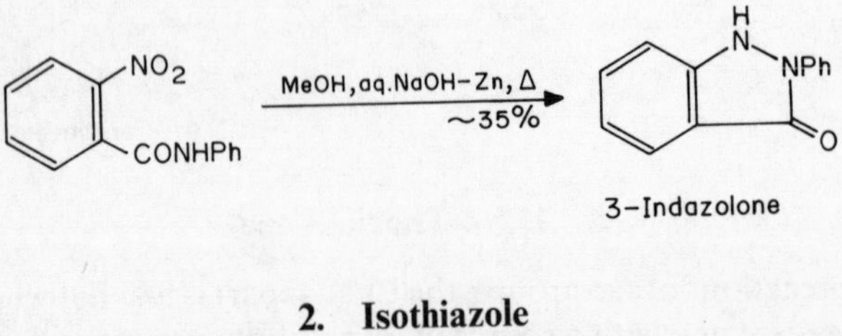

NO2
CONHPh
MeOH, aq.NaOH–Zn, Δ
~35%
H
N
NPh
O
3–Indazolone

[1449]

2. Isothiazole

Oxidative cyclization of an *o*-amino-thiocarboxamide gives good yields of a fused isothiazole.

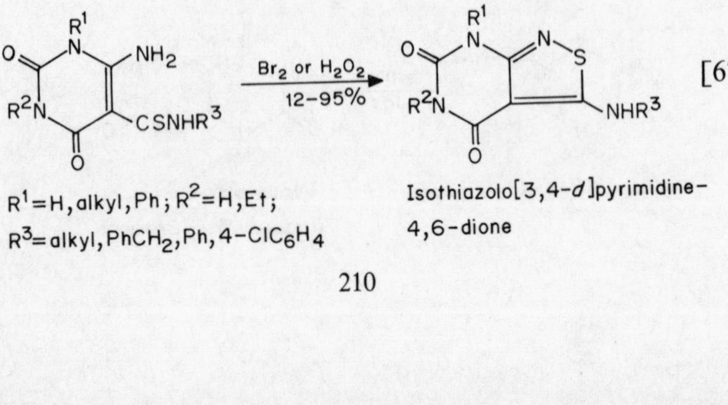

R1
O
N
NH2
R2N
CSNHR3
O
Br2 or H2O2
12–95%
N
S
NHR3
R1=H, alkyl, Ph; R2=H, Et;
R3=alkyl, PhCH2, Ph, 4–ClC6H4
Isothiazolo[3,4-d]pyrimidine–
4,6–dione

[672, 1701]

$R = MeS, Ph, N(CH_2)_4$

Thiazolo[4,5-c]isothiazole

[1641]

II. FORMATION OF A SIX-MEMBERED RING

1. Pyrimidine or Pyrimidine 1-Oxide

Activation of the carboxamide group by conversion into its chloromethine salt facilitates its reaction with pyridines. The oxime gives rise to a fused pyrimidine 1-oxide ring when it is heated with an orthoformate.

$R = H, Me, CHO, COOMe, CN$

Pyrido[1',2':1,2]pyrimido-[4,5-d]pyrimidine-2,4-dione

[510]

Pteridine 3-oxide

[244]

2. Pyrimidin-4-one or Pyrimidine-4-thione

Many examples of this cyclization exist (review [874]) and the ring-closing reagent is usually a carboxylic acid, its acyl chloride, anhydride, ester, orthoester, amide or amidine. One of the most reactive is the (ethoxycarbonyl)chloromethyleneiminium salt [451].

Pyrazolo[4,3-g]quinazolin-5-one

[454]

4-Quinazolinone   [59]

[1312]

$R^1 = Me, Ph, (CH_2)_2COOEt;$
$R^2 = 2-Me-, 2-Cl-C_6H_4;$
$R^3 = H, Cl, AcNH, NO_2;$
$X = bond, CH_2, MeCH$

$R = Ph, 4-ClC_6H_4, pyridyl$ Pyrido[2,3-d]pyrimidin-4-one   [44]

[451]

$Ar = 2-CONH_2C_6H_4$ [399]

4-Quinazolinone

$R^1 = H, MeO, MeS, Br, Cl, Ph;$ Pyrimido[4,5-b]quinolin-4-one   [38]
$R^2 = Me, Et, CF_3$

Carboxylic esters may be condensed with an amino-carboxamide under Dean and Stark conditions [38] or more efficiently in the presence of sodium ethoxide [182] while the more reactive orthoesters react on heating with or without acetic anhydride. A seleno-ester, RCSeOEt, has been used to convert anthranilamides or their thioamides into the 4-quinazolinones or 4-quinazolinethiones [1039].

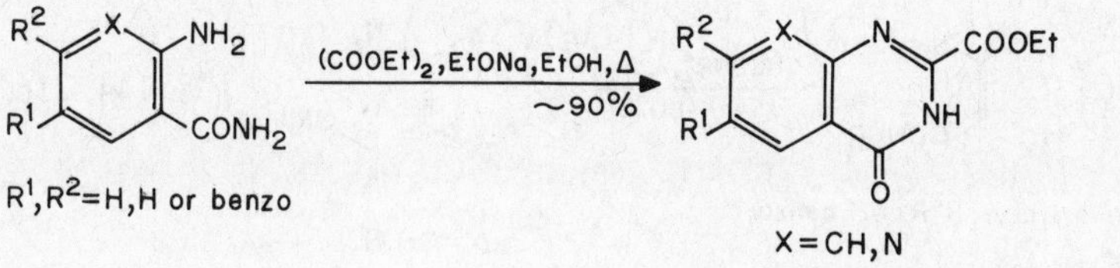

$R^1, R^2 = H, H$ or benzo

(COOEt)$_2$,EtONa,EtOH,Δ

~90%

[182]

X = CH, N
4-Quinazolinone
Pyrido[2,3-d]pyrimidin-4-one
Pyrimido[4,5-b]quinolin-4-one

Other ring systems synthesized similarly:

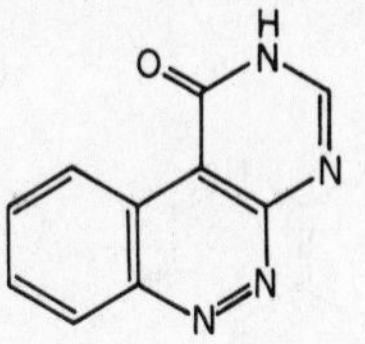

[893]

Pyrimido[4,5-c]cinnolin-1-one

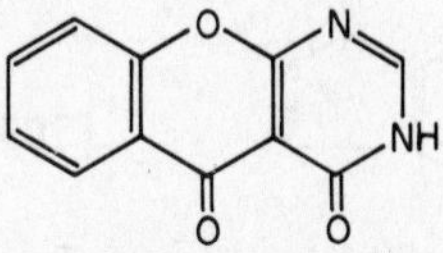

[1779]

[1]Benzopyrano[2,3-d]-
pyrimidine-4,5-dione

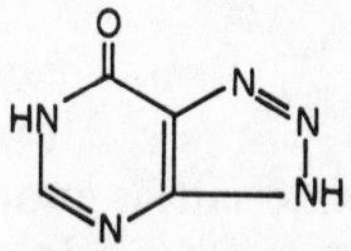

[1940]

1,2,3-Triazolo[4,5-d]-
pyrimidin-7-one
(review[2023])

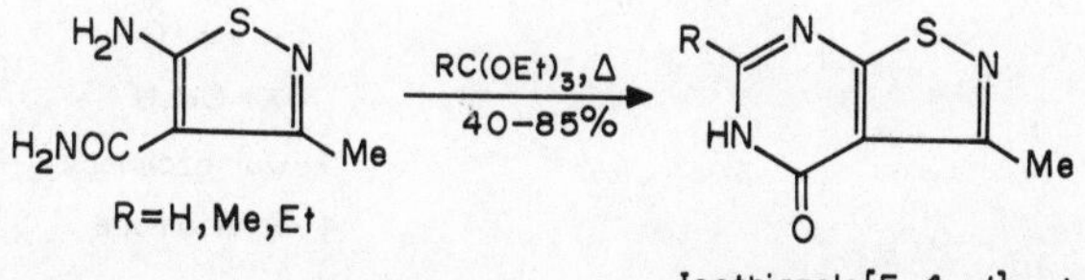

RC(OEt)$_3$,Δ

40-85%

R = H,Me,Et

[1355]

Isothiazolo[5,4-d]pyrimidin-4-one

Another ring system synthesized similarly:

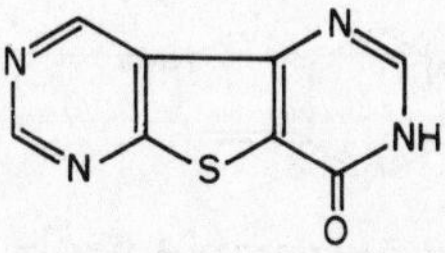

[1153]

Thieno[2,3-d:4,5-d']dipyrimidin-4-one

In several syntheses, formamide or one of its derivatives provides C-2 of the new pyrimidin-4-one ring. Diethoxymethyl acetate [1763] and potassium ethyl xanthate [1764] are also effective.

R^1=H,aryl; R^2R^3=H,benzo;
R^4=H,Me
i,EtONa,EtOH or HCl-EtOH

X=CH,N
4-Quinazolinone
4-Pteridinone
Pyrimido[4,5-b]quinolin-4-one

[38, 144, 216, 1301]

Other ring systems synthesized similarly using formamide, DMF or DMFDMA:

[709, 1091]

X=CH,N
Pyrazolo[4,3-d]pyrimidin-7-one
1,2,3-Triazolo[4,5-d]pyrimidin-7-one

[1683]

X=N,Y=CH
Pyrido[2,3-d]pyrimidin-4-one
X=CH,Y=N
Pyrido[3,2-d]pyrimidin-4-one

An amidine (preferably as its acetate salt) is another source of the C-2 carbon atom; yields vary greatly as do reaction times (4–48 h).

[216, 691]

X=CH,N
4-Quinazolinone
4-Pteridinone

Other ring systems synthesized similarly:

[215]

1,2,3-Triazolo[4,5-d]-
pyrimidin-7-one
(review [2023])

Alkyl or aryl nitriles cyclize amino-carboxamides in acid solution in very good yield.

R^1=H,Me; R^2=Me,PhCH$_2$,aryl

[63]

[1]Benzothieno[2,3-d]-
pyrimidin-4-one

Aryl aldehydes (sometimes as their bisulphite adducts) in this cyclization give 2-arylquinazolin-4-ones; 2-(3-pyridyl)purin-6-one has been prepared in this way [45]. A weak Lewis acid is sometimes helpful but the 1, 2-dihydropyrimidin-4-one ring is then formed. α-Oxo- or α-alkynoic acid esters on prolonged heating yield 2, 3-dihydroquinazolin-4-ones in good yields.

R=Ph, 4-ClC$_6$H$_4$

[93]

4-Quinazolinone

R=H,Me

[45]

Pyrido[2,3-d]pyrimidin-4-one

[62]

[62]

4-Quinazolinone

2-Arylsulphonylaminoquinazolin-4-ones are obtained by heating the amino-carboxamide with chloro(methylthio)methylenesulphonamides while quinazolin-4-ones may be prepared using ethoxymethylenemalonitrile or ethyl 2, 4-dioxo-3-ethoxymethylenepentanoate.

[305]

R^1=Ph,4-Me, 4-MeO-C$_6$H$_4$,
R^2=R^3=CN or R^1=H,R^2=Ac,
R^3=COCOOEt

[52, 398]

4-Quinazolinone

3. Pyrimidine-2, 4-dione or 2-Thioxopyrimidin-4-one

The additional carbonyl group (at C-2) may be derived from one of several different reagents [874], for example, urea, oxalyl chloride or phenyl isocyanate. In the last-named, apparently elimination of aniline (from the 5-NHCONHPh derivative) takes precedence over loss of ammonia whereas isothiocyanates behave normally to give the 3-substituted-2-thione. Diethyl carbonate gives purine-2, 6-dione [1386].

[46]

2, 4-Quinazolinedione

R^1=MeO,PhCH$_2$O;
R^2=H,Ph; R^3=H,Me

[13]

[1153]

Thieno[2,3-d:4,5-d']-
dipyrimidine-2,4-dione

[1153]

Thieno[2,3-d:4,5-d']
dipyrimidin-4-one,2-thioxo-

Other ring systems synthesized similarly:

[883]

[292]

Thieno[2,3-*d*]pyrimidin-4-one,
2-thioxo-

4-Quinazolinone, 2-thioxo

4. 1, 2, 3-Triazin-4-one

Diazotization of 2-amino-carboxamides is the most commonly used method of converting these compounds into 1, 2, 3-triazin-4-ones; 3-substituted products are obtained when the diazotized substrate is treated with a primary amine. When a conversion under weakly acidic conditions is desirable, *N*-nitrosodiphenylamine should be considered [51]. The chemistry of 1, 2, 3-benzotriazines has been reviewed [1623].

[1153]

Pyrimido[5',4':4,5]thieno-
[3,2-*d*]-1,2,3-triazin-4-one

R^1=H,Cl; R^2=alkyl,allyl

[1502]

1,2,3-Benzotriazin-4-one

Other ring systems synthesized by diazotization:

[882]

Thieno[2,3-*d*]-1,2,3-
triazin-4-one

[1958]

Pyrazolo[3,4-*d*]-
1,2,3-triazin-4-one

5. 1, 3, 5-Triazin-2-one

A 1, 3, 5-triazin-2-one ring is formed from compounds in which the carboxamide is attached to an endocyclic nitrogen—formally, a urea function but is conveniently considered in this section.

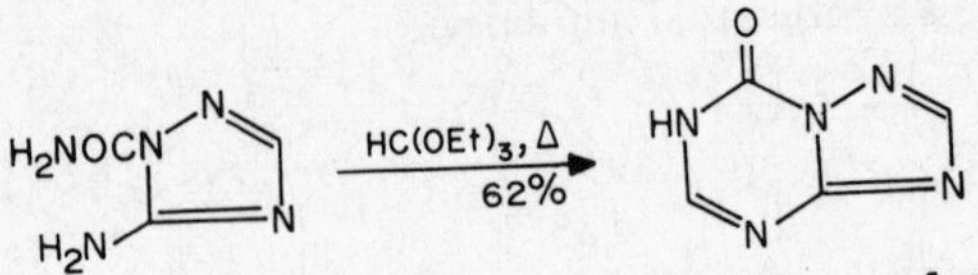

[58]

[1,2,4]Triazolo[1,5-a]
[1,3,5]triazin-7-one

6. 1, 2, 6-Thiadiazin-3-one or its 1-Oxide

The reaction of a 2-amino-carboxamide with thionyl chloride gives this thiadiazinone in high yield but when the 'amino' group is part of a cyclic imino-ether, a sulphoxide is formed.

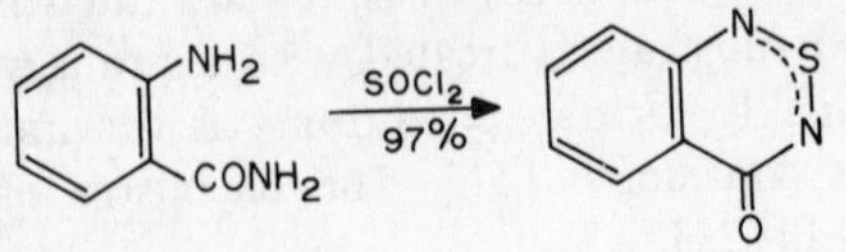

[1503]

2,1,3-Benzothiadiazin-4-one

[1099]

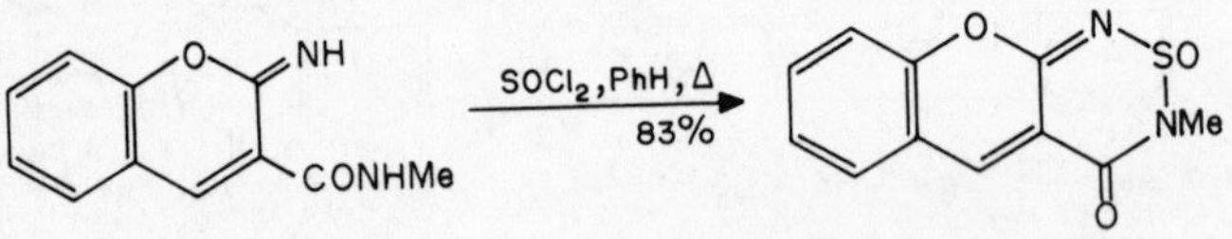

[1]Benzopyrano[2,3-c][1,2,6]-
thiadiazin-4-one 2-oxide

CHAPTER 47

Amine and Carboxylic Acid

I.   FORMATION OF A FIVE-MEMBERED RING

1.   Pyrrole

Reaction of an α-halogenoketone with a 2-amino-carboxylic acid yields an indole.

$R^1 = H, Me, Cl$;
$R^2 = Me, Et, PhCH_2$

2.   1, 2, 3-Dithiazole

Sulphur chloride and a 2-amino-carboxylic acid react on heating in benzene to give a fused dithiazolium ring.

[1658]

Thieno[2,3-*d*]-1,2,3-
dithiazol-2-ium

II. FORMATION OF A SIX-MEMBERED RING

1. Pyridine

1-Nitro-2,2-bis(methylthio)ethene is a useful synthon in heterocyclic chemistry (review [1801]). One of the methylthio groups may be replaced by a phenylamino function and the resulting compound reacts with an amino-acid to form a new pyridine ring.

[174]

Quinoline

2. Pyrimidin-4-one

Conversion of an *o*-amino-carboxylic acid into a fused pyrimidin-4-one ring is a very commonly met reaction (review [874]) and many reagents have been used. Amongst these are formamides, α-bromohydrazones, α-chloro- or -methoxy-*N*-heterocycles, formamidine acetate or a mixture of an amine and carboxylic acid (or anhydride).

[74, 84]

$R^1 = H, Me, Cl, HO, MeO, MeS;$
$R^2 = H, Me$

X = CH,N
4-Quinazolinone
Pyrido[2,3-*d*]pyrimidin-4-one

[64]

Formation of a Six-membered Ring

R$\overset{\cdot\cdot\cdot}{\bigcirc}$ (with NH$_2$ and COOH) + Cl$-$thiazole$-$COOMe $\xrightarrow[\text{31–74\%}]{i,\Delta}$ R$-$ Thiazolo product COOH [83]

R = H, alkyl, MeO;

i, KI or HCOOH, MeO(CH$_2$)$_2$OH

Thiazolo[2,3-*b*]quinazolin-5-one

Other ring systems synthesized similarly:

[1073]

Pyrido[2,1-*b*]quinazolin-6-one

[1461]

Pyrrolo[2,1-*b*]quinazolin-9-one

R$\overset{\cdot\cdot\cdot}{\bigcirc}$ (with NH$_2$ and COOH) + HC=NH$_2$ (with NH$_2$, AcO$^-$) $\xrightarrow[\text{77–89\%}]{\overset{OMe}{(CH_2)_2OH,\,\Delta}}$ R$-$ product [691]

R = H, Me, Br, Cl, I, NO$_2$

4-Quinazolinone

Another ring system synthesized similarly:

[691]

Pyrazolo[3,4-*f*]quinazolin-9-one

anthranilic acid (NH$_2$, COOH) $\xrightarrow[\text{70–90\%}]{i,RNH_2,\Delta}$ quinazolinone (R, NPh) [60, 476]

R = Me, Ph

i, RCOOH, P(OPh)$_3$–pyr or (RCO)$_2$O

Ketenimines react with anthranilic acids under neutral conditions to give 2,3-disubstituted quinazolin-4-ones.

anthranilic acid (NH$_2$, COOH) + Ph$_2$C=C=NAr $\xrightarrow[\text{32–75\%}]{\text{xylene},\Delta}$ product (CHPh$_2$, NAr) [1]

Ar = Ph, 4-BrC$_6$H$_4$

3. Pyrimidine-2, 4-dione

Heating anthranilic acids with *N*-aryldithiocarbamate or urea gives a pyrimidinedione; the former leads to a 3-aryl-dione. t-Butyl carbazate (methyl or
ethyl ester gives lower yield) forms a 3-amino-dione on heating in quinoline but
alkyl isothiocyanates produce a 2-thioxopyrimidin-4-one.

$X=N, R^2=H;$
$CO(NH_2)_2, \Delta$
44–60%

[802]

Pyrido[2,3-*d*]pyrimidine-
2,4-dione

$X=CH, R^1=H;$
$ArNHCSSMe, i$
60–96%

[684]

$R^1=$alkyl, MeCH=CHCH$_2$,–2,4–
Me$_2$C$_6$H$_3$; $R^2=$H, Me, Cl;
Ar = Ph, Me-, MeO-C$_6$H$_4$;
i, HgO, DMF, Δ

Quinazoline-2,4-dione

$+ \begin{array}{c} NHNH_2 \\ | \\ COOtBu \end{array}$

$\xrightarrow[\sim 60\%]{\text{quinoline}, \Delta}$

[994]

R=H, Me, Cl

+ MeCNS

$\xrightarrow[69–95\%]{\text{TEA}, \Delta}$

[394]

R=Me, Pr, Ph

Pyrimido[4,5-*c*]quinolin-1-one,
3-thioxo-

4. Pyrazin-2-one

2-Aminophenylglycines cyclize spontaneously on warming to give quinoxalin-2-
ones.

$\xrightarrow[57\%]{\text{EtOH, H}_2, \text{Pd-C}}$

[1853]

2-Quinoxalinone

5. Pyran-2-one

Benzyne, formed by reaction of anthranilic acid with pentyl nitrite, reacts with a 3-pyridinol to form a fused pyran-2-one ring.

R=H,Me

C₅H₁₁ONO,Δ ~25%

[2]Benzopyrano[4,3-b]-
pyridin-6-one

[1280]

6. 1,3-Oxazin-6-one or 1,3-Oxazine-2,4-dione

2-Amino-carboxylic acids react with dithiocarbamate esters, acid chlorides (or anhydrides) or N-dichloromethylenesulphamic esters to give a fused 1,3-oxazin-6-one ring.

DMF,HgO 52–90%

+ArNHCSSMe
R=H,Me,Cl; Ar =Ph,
Me-, MeO-C₆H₄

3,1-Benzoxazin-4-one

[684]

R=Ph,Me

PhCOCl or Ac₂O   pyr,Δ 46–90%

[1]Benzothieno[3,2-d]-
[1,3]oxazin-4-one

[558]

+ Cl₂C=NSO₂R   PhMe,Δ 79–88%

R=Ph,4-MeC₆H₄

[305]

Other ring systems synthesized similarly:

Pyrazolo[3,4-d][1,3]-oxazin-4-one

[304]

3,1-Benzoxazin-4-one

[1462]

Phosgene converts an *o*-amino-carboxylic acid into a fused 1,3-oxazine-2,4-dione.

[1331]

i, H₂, Pd-C
ii, COCl₂-PhH

3,1-Benzoxazine-2,4-dione
(review [1673])

Another ring system synthesized similarly:

[1951]

Thieno[2,3-*d*][1,3]-
oxazine-2,4-dione

III. FORMATION OF A SEVEN-MEMBERED RING

1. Azepin-2-one

When the amino and carboxyl groups are appropriately separated by carbon atoms, heating the compound in DMSO-xylene produces an azepinone ring. The chemistry of azepines has been reviewed [1620].

[807]

1,3-Dioxolo[4,5-*h*][3]benzazepin-6-one

2. 1,4-Diazepin-2- or -5-one

Reaction between an aliphatic carboxyl and an aromatic amino group may be brought about either by hot PPA or spontaneously during hydrogenation of a nitro group.

[770]

R¹=H,Cl,MeO; R²=H,Me

1,5-Benzodiazepin-2-one
(review [1619])

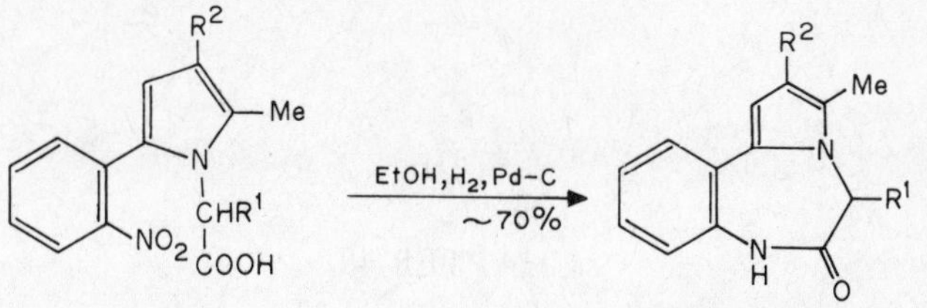

Pyrrolo[1,2-*d*][1,4]-
benzodiazepin-6-one

[203]

3. 1, 4-Oxazepin-5-one

Both erythro and threo stereoisomers of the amino-ester (**47.1**) were hydrolysed
and cyclized separately by stirring with alkali.

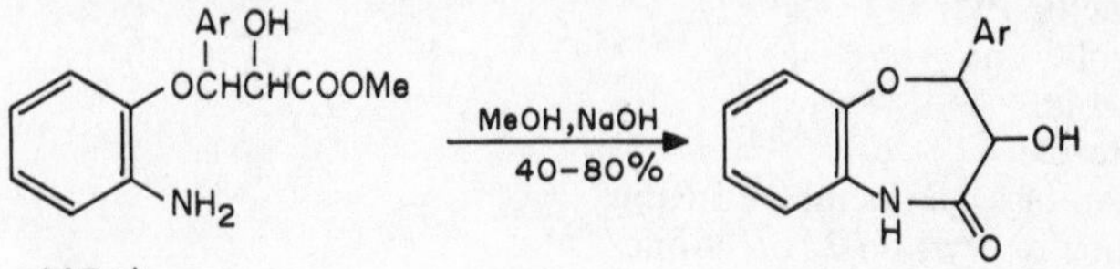

(**47.1**)
Ar = 4-MeOC₆H₄

1,5-Benzoxazepin-4-one

[1709]

IV. FORMATION OF AN EIGHT-MEMBERED RING

1. 1, 4-Thiazocin-5-one

Amino and carboxyl groups separated by five carbons and one sulphur react
under the influence of DCC to form a fused thiazocinone ring.

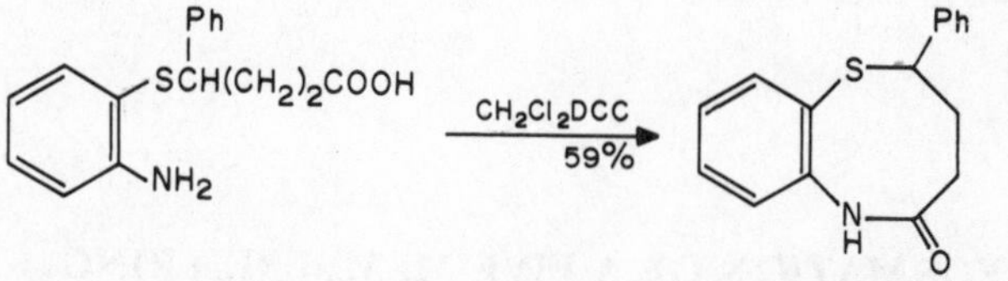

1,6-Benzothiazocin-5-one

[829]

CHAPTER 48

Amine and Carboxylic Ester

I. FORMATION OF A FIVE-MEMBERED RING

1. Pyrrole or Pyrrol-2-one

The generation of an amino group (by reduction of a nitro function) spontaneously causes interaction between it and a neighbouring carboxylic ester to form a pyrrol-2-one ring in good yield, but the nature of the ring in the substrate may account for the different products obtained under similar conditions. When an α-cyano-ester group replaces the malonate ester in this cyclization, the cyano group reacts and a pyrrole ring is formed. Heating preformed amino-ester gives a pyrrolone ring.

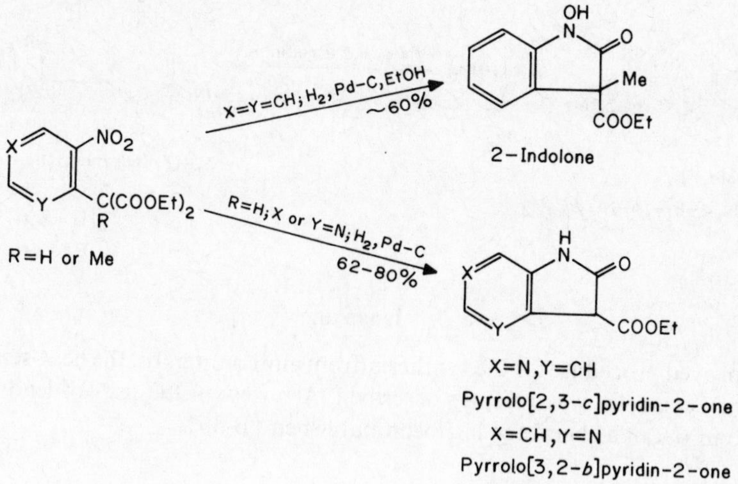
X=Y=CH; H₂, Pd–C, EtOH
~60%
2–Indolone
R=H or Me
R=H; X or Y=N; H₂, Pd–C
62–80%
X=N, Y=CH
Pyrrolo[2,3-c]pyridin-2-one
X=CH, Y=N
Pyrrolo[3,2-b]pyridin-2-one

[348]

[562]

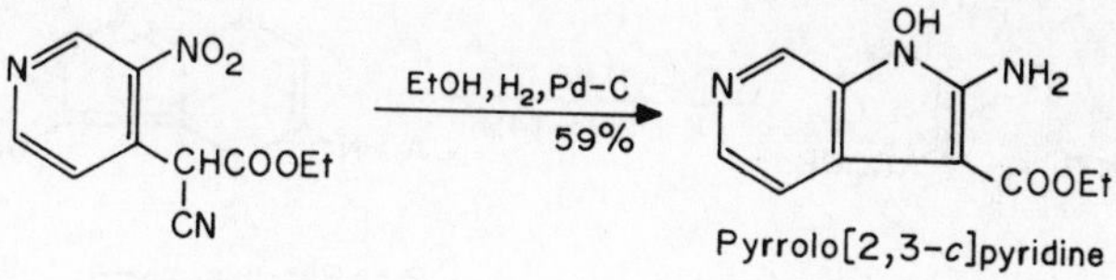
EtOH, H₂, Pd–C
59%
Pyrrolo[2,3-c]pyridine

[562]

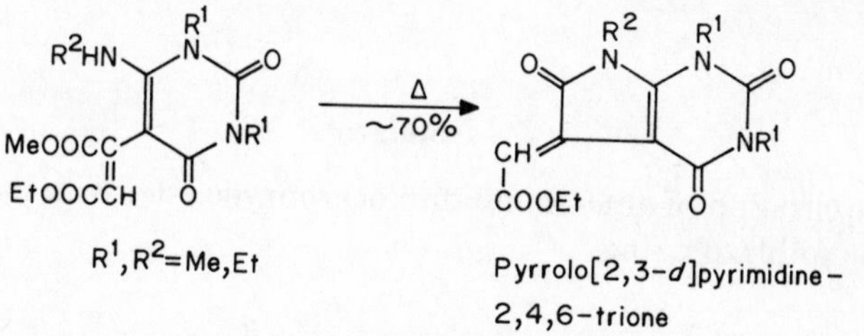
Δ
~70%
R¹, R²=Me, Et
Pyrrolo[2,3-d]pyrimidine–2,4,6–trione

[1949]

2. Pyrazole-3-one

An *N*-amino function can be produced *in situ* by hydrolysis of a hydrazone or a transhydrazonation as in the following example where the newly-generated NH_2 reacts with the ester group to form the pyrazolone.

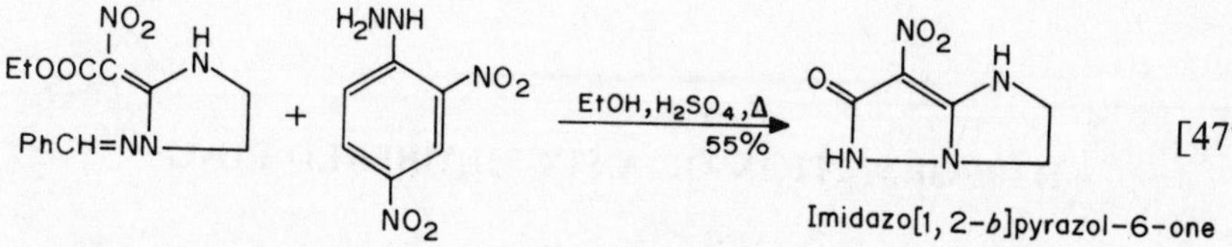
EtOH, H₂SO₄, Δ
55%
Imidazo[1, 2-b]pyrazol-6-one

[477]

A hydrazide formed *in situ* may then, in a strongly basic medium, displace a neighbouring amino group and cause cyclization.

R=H,Me,Ph;
Ar=Ph,4-Br-,4-NO$_2$-C$_6$H$_4$

3,4,7-Indazoletrione

[1226]

3. Isoxazole

An improved procedure for the synthesis (from amino-esters) of the heat-sensitive quinone isoxazoles was recently described. A review of the uses of lead tetra-acetate in organic chemistry has been published [B-36].

R=Me,Et,CD$_3$; Ar=Ph, 4-MeC$_6$H$_4$

2,1-Benzoxazole-4,7-dione

[1786]

4. Isothiazole

Oxidative cyclization of an *o*-amino-dithiocarboxylic ester with iodine-DMSO gives a new isothiazole ring.

R=Me,Ph

Isothiazolo[3,4-*d*]-
pyrimidin-4-one

[654]

II. FORMATION OF A SIX-MEMBERED RING

1. Pyridine or Pyridin-2- or -4-one

A fused 2,4-dihydroxypyridine-3-carboxylate ester is formed when an amino-ester is cyclized by reaction with diethyl malonate and a base.

R=H,Me

+ CH₂(COOEt)₂ → EtONa, EtOH, Δ / 21–95%

$X = NMe, O$

[1138, 1222]

Pyrazolo[3,4-*b*]pyridine
Isoxazolo[5,4-*b*]pyridine

Suitably positioned and separated amino and ester groups react to form a pyridin-2-one, usually induced by heat or a base.

RCH₂NH, Me, NMe, CH=CH, COOEt → DBN, TEA, Δ / 71–90%

R = Ph, 4-MeO-, 4-Cl-C₆H₄

[1862]

Pyrido[2,3-*d*]pyrimidine– 2,4,7-trione

Me, O, Me, COOMe, NO₂ → MeOH, H₂, Pd–C / ~92%

R = H, MeO, OCH₂O

[821]

Furo[3,4-*c*]quinolin-4-one

H₂N, R, CH=CH, COOEt, Me → EtONa, EtOH, Δ / ~82%

R = H, Me, iPr, Ph

[306]

Pyrido[2,3-*d*]pyrimidin-7-one

Other ring systems synthesized similarly:

Me, MeO

[212]

2-Quinolinone

[1889]

W, X, Y or Z = N, others = CH
Z = N: 1,5-Naphthyridin-2-one
Y = N: 1,6-Naphthyridin-2-one
X = N: 1,7-Naphthyridin-2-one
W = N: 1,8-Naphthyridin-2-one
(review [1648], 2009)

1,1-Bis(methylthio)-2-nitroethene is a versatile synthon (review [1801]) which reacts with amino esters to give a pyridin-4-one ring which is also formed in an acid-catalysed reaction between anthranilate ester and a resorcinol.

4-Quinolinone

[174]

9-Acridinone

[1567]

2. Pyrimidin-4-one

The combination of amine and ester groups lends itself to cyclization to pyrimidin-4-one by several reagents [874]. Perhaps the most frequently used is formamide. This has the added convenience of acting as solvent as well as reactant but occasionally either formic acid or sodium ethoxide-DMSO is added.

[69]

[1]Benzothieno[3,2-d]pyrimidin-4-one

Other ring systems synthesized similarly:

[665]

Pyrazolo[4,3-d]-
pyrimidin-7-one

[1360]

Pyrazino[2',3':4,5]thieno-
[3,2-d]pyrimidin-4-one

[1153]

Thieno[2,3-d:4,5-d']-
dipyrimidin-4-one

[1528]

Pyrimido[4',5':4,5]pyrrolo-
[2,3-c]azepine-4,6-dione

[1]Benzothieno[2,3-*d*]-
pyrimidin-4-one

4-Quinazolinone

[1]Benzothieno[3,2-*d*]-
pyrimidin-4-one

In the presence of phosphorus oxychloride, cyclic amides (lactams) also react.

Indolo[2′,3′:3,4]pyrido[2,1-*b*]-
quinazolin-5-one

Nitriles (including cyanamides), amidines or isocyanates cause ring closure under relatively mild conditions but only the lower alkyl amidines are effective [1224]. Chloroacetonitrile reacts with two moles of aminoester.

R^1, R^2 = alkyl, aryl, $(CH_2)_4$
R^3 = Cl, aryloxy, arylthio

Thieno[2,3-*d*]pyrimidin-4-one
[1]Benzothieno[2,3-*d*]pyrimidin-4-one

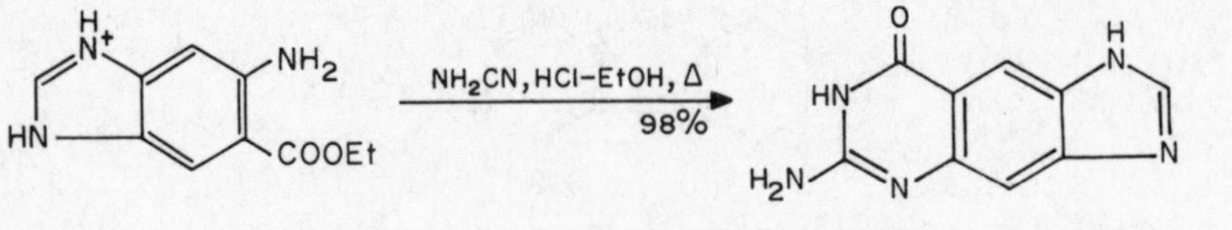

[75]

Imidazo[4,5-g]quinazolin-8-one

Other ring systems synthesized similarly:

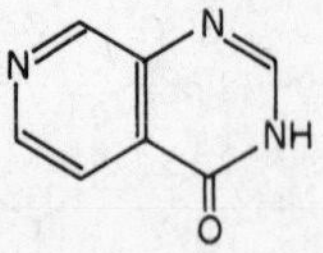

[401]

Pyrido[3,4-d]pyrimidin-4-one

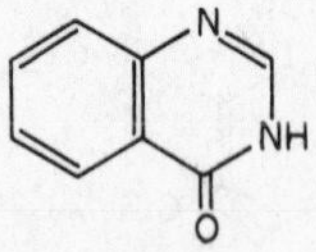

[75, 504, 691]

4-Quinazolinone

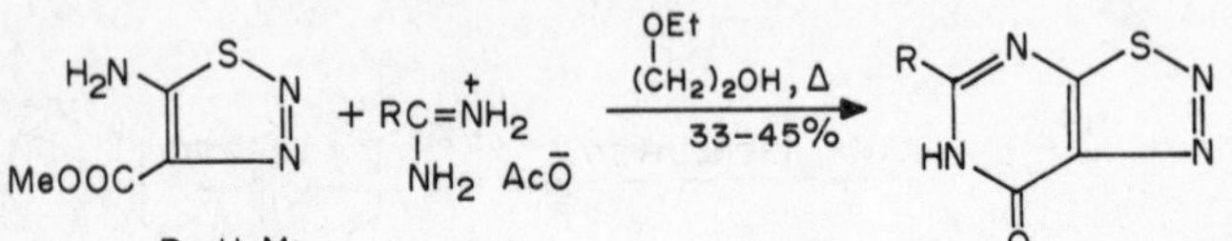

[55, 1224]

R = H, Me

1,2,3-Thiadiazolo[5,4-d]-
pyrimidin-4-one

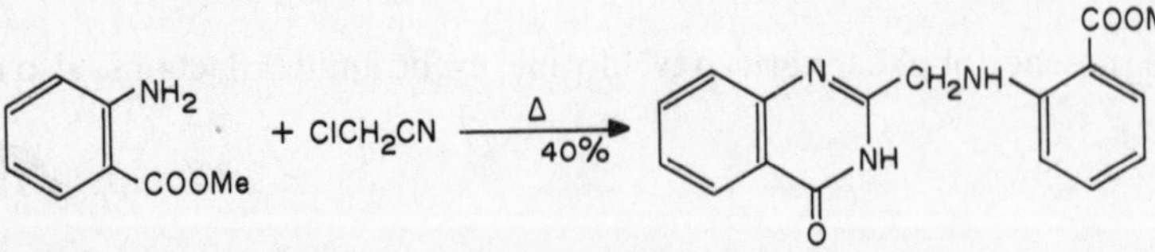

[1873]

4-Quinazolinone

A mixture of a primary arylamine, an orthoester and an amino-ester reacts on heating in decalin to form a 3-N-arylpyrimidinone ring as also does an amino-ester and a (cyclic or acyclic) chloroimine.

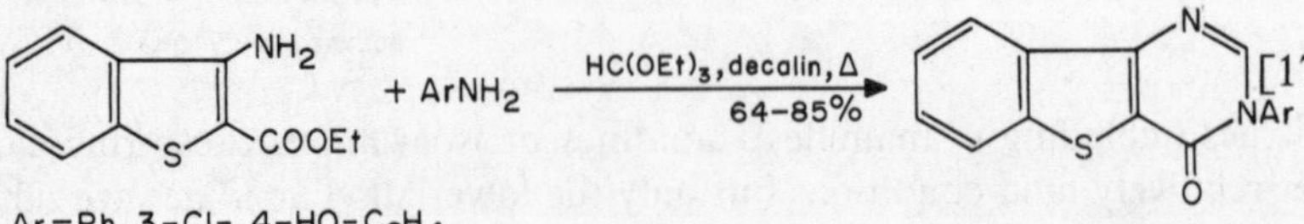

[1714]

Ar = Ph, 3-Cl-, 4-HO-C_6H_4

[1]Benzothieno[3,2-d]-
pyrimidin-4-one

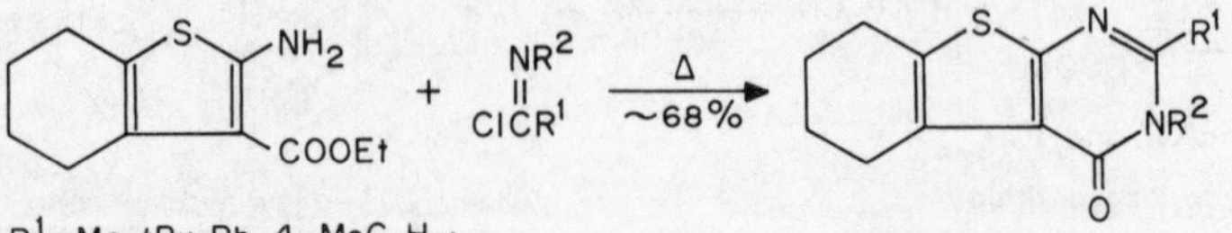

[1223]

R^1 = Me, tBu, Ph, 4-MeC_6H_4;
R^2 = PhCH$_2$CH$_2$, Ph, 3-MeC_6H_4

[1]Benzothieno[2,3-d]-
pyrimidin-4-one

Heterocycles containing a reactive halogen atom or a lactim ether group cyclize an aminoester to give a doubly fused pyrimidinone.

Pyrido[1,2-a]thieno[2,3-d]-

pyrimidin-4-one

Other ring systems synthesized similarly:

[72]

Benzothiazolo[2,3-b]quinazolin-12-one

[1950]

n=1-3

Pyrido[2,1-b]pteridin-11-one

Azepino[2,1-b]pteridin-12-one

Azocino[2,1-b]pteridin-13-one

3.   Pyrimidine-2,4-dione or 2-Thioxopyrimidin-4-one

Aminoesters react with isocyanates (or isothiocyanates) to give first the ureides which, on treatment with a base, cyclize to the pyrimidinedione (or thioxo-one).

[13, 74]

i, X=CH; NaH-DMF

X=N; pyr., Δ

X=CH

2,4-Quinazolinedione

X=N

Pyrido[2,3-d]pyrimidine-2,4-dione

Other ring systems synthesized similarly:

[66]

Furo[2,3-d]pyrimidine-2,4-dione

[114]

4-Quinazolinone, 2-thioxo

[560]

[1]Benzothieno[2,3-d]pyrimidin-4-one

The highly reactive chlorosulphonyl isocyanate (for a review of its chemistry, see [324]) behaves differently from other isocyanates in this reaction in that the nitrogen is unsubstituted.

Pyrazolo[3,4-d]pyrimidine-
4,6-dione

[1751]

A ureido (or thioureido) group, whether formed *in situ* from an amine or present in the substrate, reacts with an ester group to give a pyrimidinedione (or thioxo-one).

R=H,Me

Pyrazino[2,3-g]quinazoline-2,4-dione

[559]

Other ring systems synthesized similarly:

Pyrimido[5,4-b]indol-
4-one,2-thioxo-

[511]

Thieno[2,3-d]pyrimidin-
4-one,2-thioxo-

[883]

A ureide may also be prepared by reaction of an amine and an azide and in this way, a 3-substituent may be introduced.

Pyrido[2,3-d]pyrimidine-
2,4-dione

[68]

4. Pyrazin-2-one or Pyrazine-2,3-dione

A glycine ester side-chain on annulation to a neighbouring amino group yields a new pyrazin-2-one ring which can also arise when the two groups are attached to separate rings.

RCHCOOEt
|
NH

R=H,Me

electrochem. reduction
~58%

[77]

[1]Benzopyrano[3,4-b]-
pyrazine-3,5-dione

EtOH,H₂,Pd-C
51%

[78]

Pyrrolo[1,2-a]quinoxalin-4-one

The amino group may be converted into its ethoxalyl derivative and this cyclizes readily on to an amino group to form a pyrazinedione ring. Catalytic N-debenzylation (in one reaction) enables ring closure to proceed.

DMF,H₂,Pd-C
84%

[1584]

2,3-Quinoxalinedione

H₂,Pd-C,EtOH,AcOH
96%

[828]

Pyrazino[1,2-a]pyrrolo[2,1-c]-
[1,4]benzodiazepine-3,4-dione

5.   1,3-Oxazin-2- or -6-one or 1,3-Oxazine-2,4-dione

When a 2-imino carboxylate ester is heated in pyridine, an oxazin-2-one is obtained and also when an aminoester is treated at ambient temperature with an isocyanate in an acid medium but phosgene in alkaline solution yields a 2,4-dione.

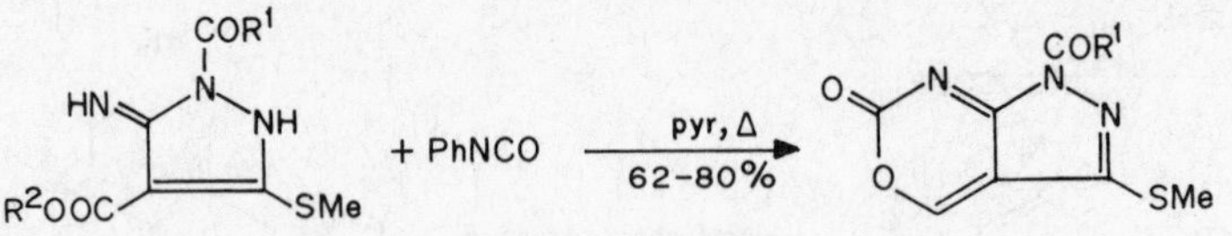

$R^1 = Ph, 4-MeO-, 4-Cl-C_6H_4;$
$R^2 = Me, Et$

[1860]

Pyrazolo[3,4-d][1,3]-
oxazin-6-one

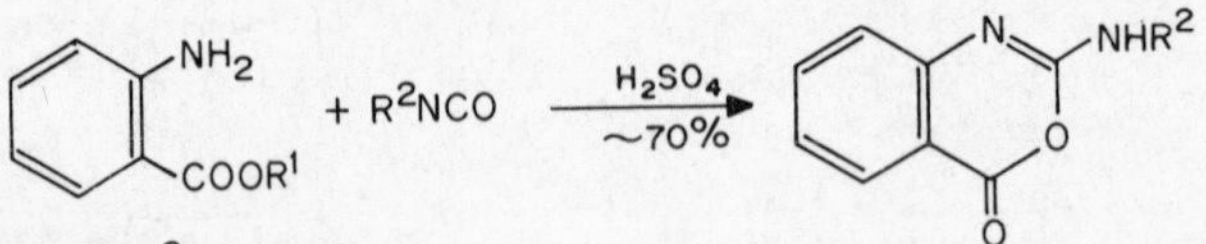

$R^1 = Me, Et; R^2 = (CH_2)_2Cl,$
CH_2COOEt, Ph

[73]

3,1-Benzoxazin-4-one

[214]

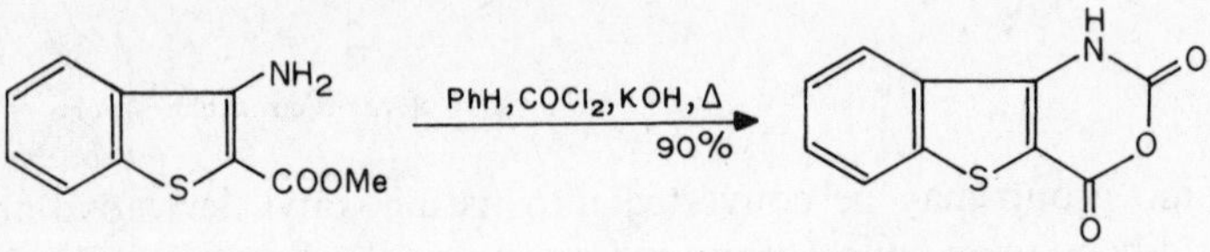

[1]Benzothieno[3,2-d][1,3]-
oxazine-2,4-dione

6. 1,4-Thiazin-3-one

With a sulphur atom in the side-chain, a thiazinone may be obtained by heating the substrate under acidic conditions. 1,4-Benzothiazines have been reviewed [942].

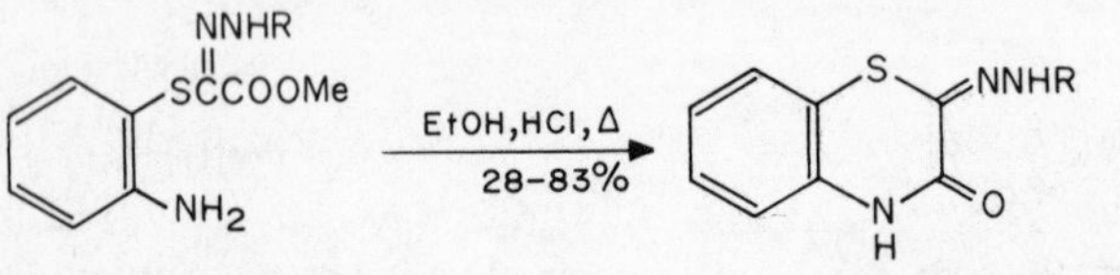

$R = Bu, Ph, Cl-, CF_3-,$
$NO_2-C_6H_4$

[825]

1,4-Benzothiazin-3-one

7. 1,2,6-Thiadiazin-3-one 1,1-Dioxide

Sulphamoyl chloride in alkali cyclizes an aminoester to a thiadiazinone dioxide.

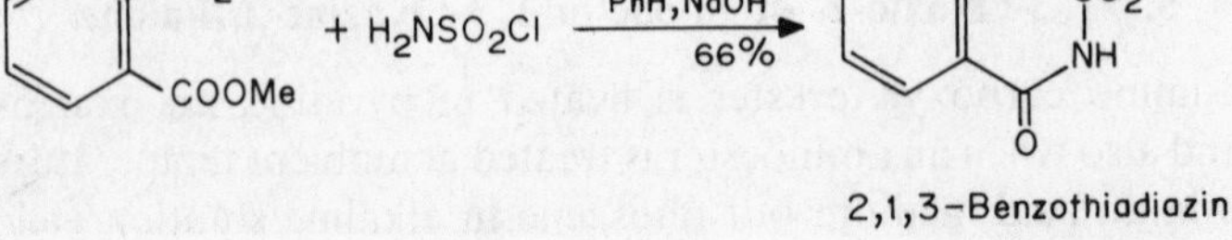

[67]

2,1,3-Benzothiadiazin-
4-one 2,2-dioxide

III. FORMATION OF A SEVEN-MEMBERED RING

1. Azepin-2-one or Azepine-2,5-dione

Amino (or potential amino) and ester groups separated by five carbon atoms (one of which may be a carbonyl group) cyclize to benzazepines thermally or under the influence of a base. Their chemistry has been reviewed [1620].

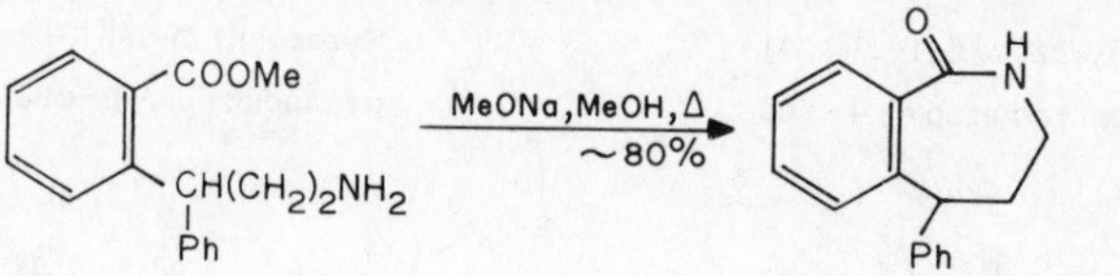

[1225]

2-Benzazepin-1-one

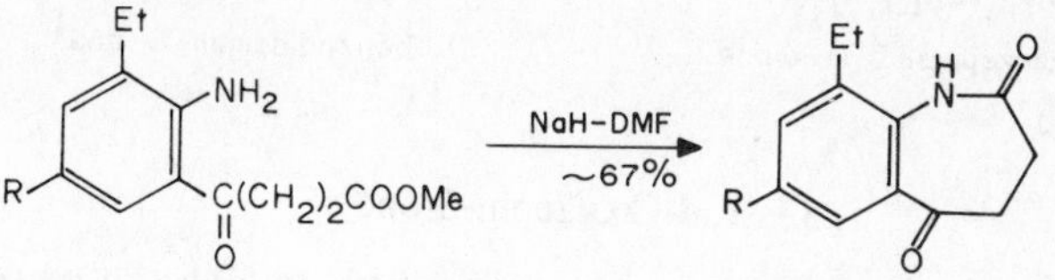

R=Br,COOMe

[914]

1-Benzazepine-2,5-dione

Other ring systems synthesized similarly:

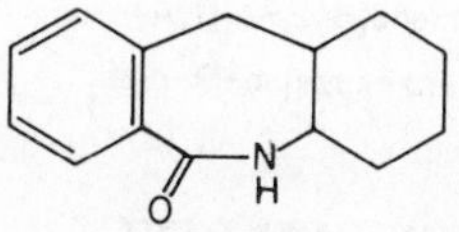

[169]

Dibenz[be]azepin-6-one

[990]

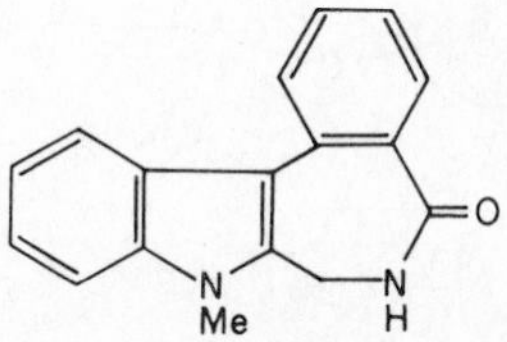

Indolo[2,3-d][2]benzazepin-5-one

2. 1,4-Diazepin-5-one or 1,4-Diazepine-2,5-dione

These rings are formed by reaction between an amino and an ester group, sometimes aided by a basic medium (see [1895] for a review).

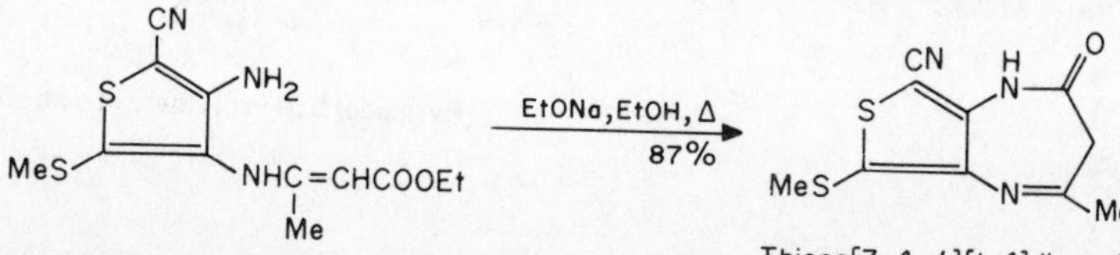

[191]

Thieno[3,4-b][1,4]diazepin-2-one

Other ring systems synthesized similarly:

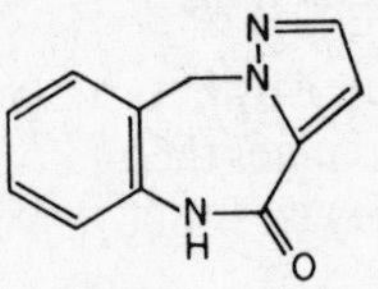

[561]

Pyrazolo[5,1-c][1,4]-
benzodiazepin-4-one

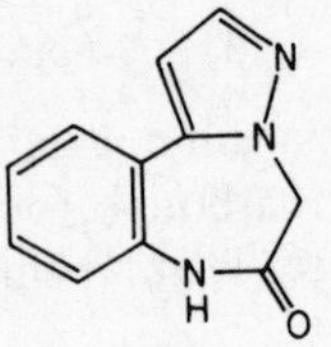

[79]

Pyrazolo[1,5-d][1,4]-
benzodiazepin-6-one

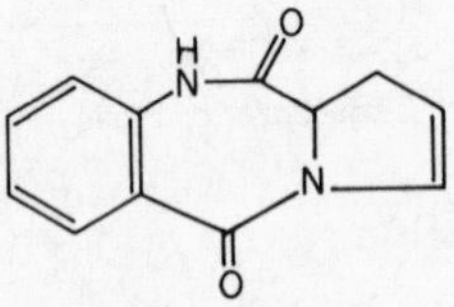

[85]

Pyrrolo[2,1-c][1,4]-
benzodiazepine-5,11-dione

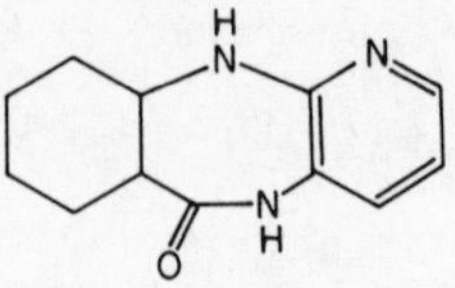

[1990]

Pyrido[2,3-b][1,4]-
benzodiazepin-6-one

3. 1,4-Oxazepin-5-one

Thermal cyclization of a suitable aminoester substrate gives an oxazepinone.

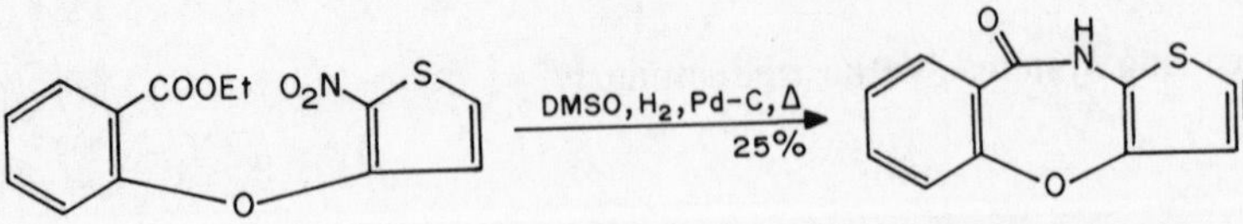

Thieno[3,2-b][1,4]-
benzoxazepin-9-one

IV. FORMATION OF AN EIGHT-MEMBERED RING

1. Azocin-2-one

Base-induced reaction of suitably placed groups gives a moderate yield of this eight-membered ring.

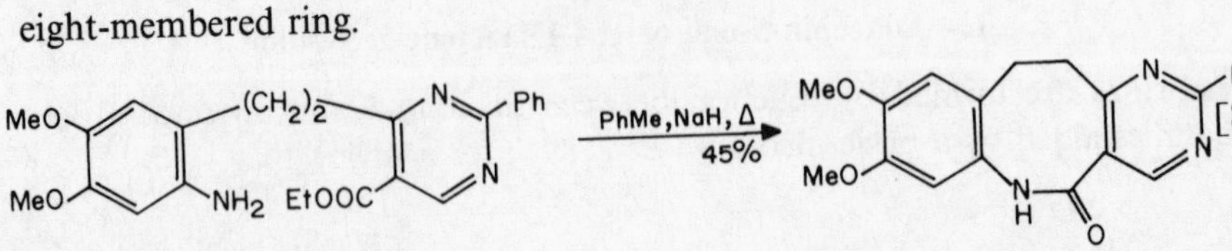

Pyrimido[5,4-c][1]benzazocin-5-one

CHAPTER 49

Amine and Enamine

I.   FORMATION OF A FIVE-MEMBERED RING

1.   Pyrrole

o-Amino-enamines (prepared *in situ* from the corresponding nitro compounds) readily cyclize to a fused pyrrole ring. This synthesis is often called the Leimgruber–Batcho cyclization (review [1439]) and has been frequently used to synthesize indoles and azaindoles. Originally, the nitro compounds were reduced chemically using iron-acetic acid [441] or titanium(III) chloride [456] but catalytic reduction has been found to be more efficient [857]; lithium aluminium hydride may be used but may reduce other groups [1446]. Dinitroenamines to be converted into a nitroindole may be more conveniently cyclized after partial reduction with titanium chloride [907, 1441] while catalytic reduction is quite satisfactory to prepare aminoindoles [1443]. The conditions of hydrogenation are important if the amount of 1-hydroxyindoles is to be kept low [39]. It is not necessary to isolate the enamine and good yields have been recorded in the conversion of methyl 2-methyl-3-nitrobenzoate into the indole by successive treatment with DMFDMA and a reducing agent.

$R^1, R^2 = H, MeO, Cl, CN, PhCH_2O, COOMe$;   Indole
$NR_2^3 = NMe_2, N(CH_2)_5$
i, $TiCl_3, Me_2CO, NH_4OAc$; *ii*, $PhH, H_2, Pd–C$

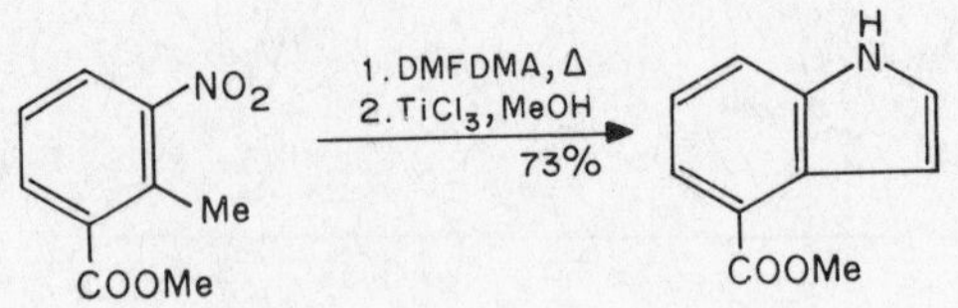

[907]

Indole

Other ring systems synthesized similarly:

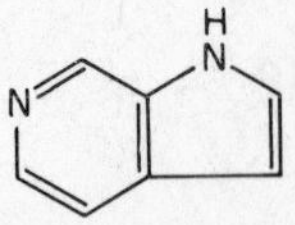

[441]

Pyrrolo[2,3-c]pyridine

[17]

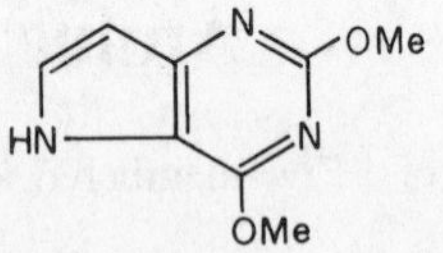

Pyrrolo[3,2-d]pyrimidine

In one variation of the Leimgruber–Batcho cyclization the nitro dimethyl-amine enamine is converted into its semicarbazone analogue; its lower solubility reduces the tendency to form dimeric products.

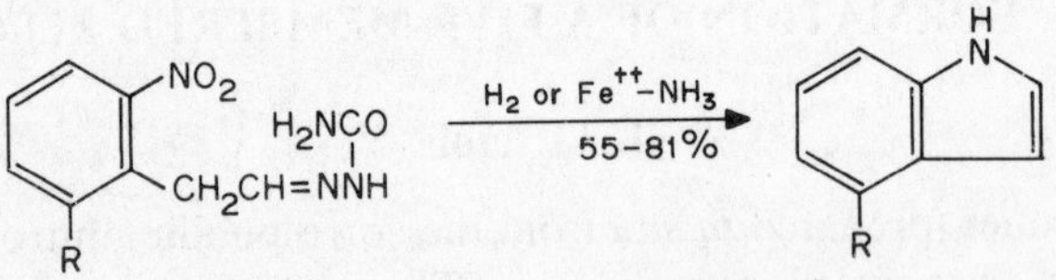

[1443]

R=Me,HO,MeO,PhCH$_2$O

Indole

II. FORMATION OF A SIX-MEMBERED RING

1. Pyridine

Cyclization of a C-enamine-N-enamine may be mentioned here. It gives a fused pyridine ring on mild treatment with acetic acid.

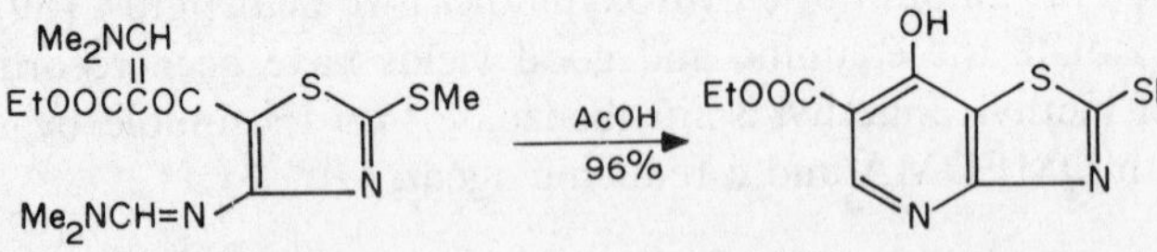

[1464]

Thiazolo[4,5-b]pyridine

(review [1865])

Amine and Hydrazide or Hydrazine

In addition to the combination of groups shown in the title of this chapter, examples of acylamine or carbamate and hydrazide and of hydrazide and hydrazine groups are included.

I. FORMATION OF A FIVE-MEMBERED RING

1. Pyrrol-2-one

Although a 2-hydrazino-hydrazide can in theory cyclize to a pyridazinone, heating in toluene gives a pyrrolone as the sole product.

[1267]

Pyrrolo[3,2-e]-1,2,4-triazolo-
[4,3-b]pyridazin-7-one

2. Pyrazole

Diazotization of a 2-amino-hydrazide results in the formation of a pyrazole ring and loss of nitrogen.

R = 4-Me-, 4-Cl-C$_6$H$_4$ Indazole

[2031]

II. FORMATION OF A SIX-MEMBERED RING

1. Pyridin-2-one

When a tosylhydrazide is heated with a base, an aldehyde group is formed (McFadyen–Stevens reaction); this reacts with the methylene of ethyl 2-pyridineacetate to give a new pyridin-2-one ring.

+ ArCH$_2$COOEt
Ar = 2-pyridyl

1,8-Naphthyridin-2-one
(review [1648])

[1187]

2. Pyridazin-3-one

An amino group reacts with an NN'-diacylhydrazine in acid to yield a pyridazinone ring.

Benzofuro[2,3-d]pyridazin-4-one

[935]

3. Pyrimidin-4-one

3-Aminoquinazolin-4-ones are prepared by heating a 2-amino-hydrazide with an orthoester but the course of the cyclization is dependent on the solvent used, reaction time, the proportion of reactants and the orthoester employed [1562–1564]. When these factors are controlled, quinazolinones rather than the isomeric triazepinones (see Section III) are obtained in good yield. 2-Acylaminobenzoic acid hydrazide reacts with hydrazine hydrate to yield a 3-aminoquinazolin-4-one.

$R^1, R^2 = H, Me, (CH_2)_5;$
$R^3 = H, Me, Cl, Br, NO_2; R^4 = H, Me$

[826, 1562, 1564]

4-Quinazolinone

Other ring systems synthesized similarly:

[1714]

[1]Benzothieno[3,2-*d*]-
pyrimidin-4-one

[271, 1802]

6-Purinone

[271]

Pyrazolo[3,4-*d*]pyrimidin-4-one

[1379]

4-Quinazolinone

4. Pyrimidine-2, 4-dione

The additional carbonyl group required to form this ring may be provided by ethyl chloroformate, preferably in benzene although pyridine has been used [1472].

$R^1 = H, alkyl, CH_2COOH;$
$R^2 = H, alkyl; R^3 = H, alkyl, Ac$

[1680]

2,4-Quinazolinedione

Phosgene reacts with N^2-alkoxycarbonylamino-hydrazides to give the dione in good yield. The free 3-amino group is obtained when the t-butyloxycarbonylhydrazide is cyclized.

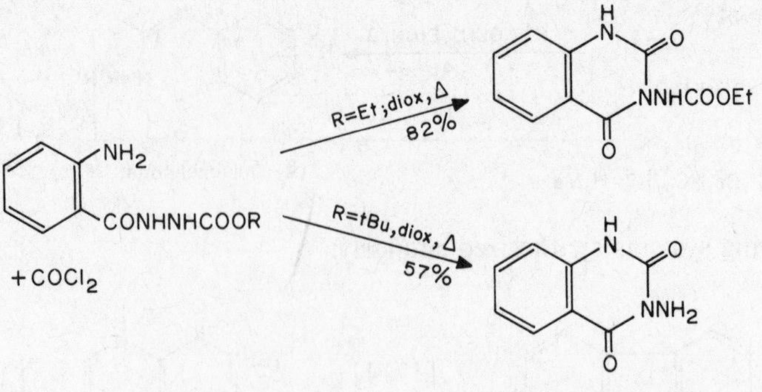

[978]

[978]

2,4-Quinazolinedione

5. 1,2,3-Triazin-4-one

Diazotization of the amino group of an unsubstituted amino-hydrazide causes cyclization to the triazinone in good yields (cf. diazotization of an amino-N^2-arylhydrazide, Section I.1).

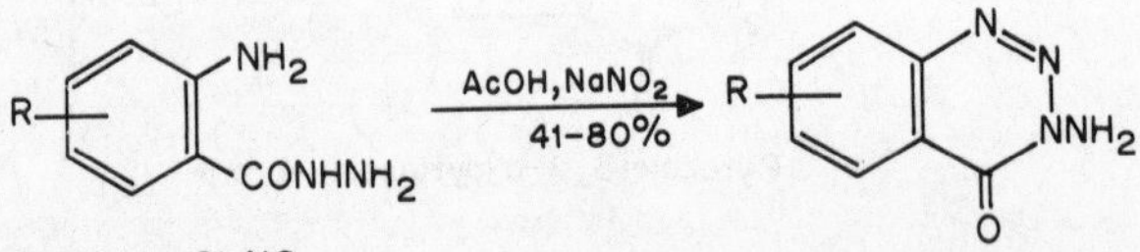

[1496]

R = H, Me, Cl, NO$_2$

1,2,3-Benzotriazin-4-one

Another ring system synthesized similarly:

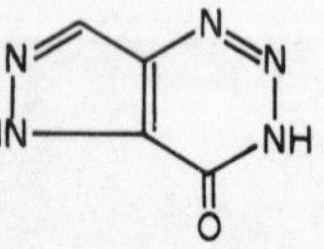

[967]

Pyrazolo[4,3-d]-1,2,3-
triazin-4-one

6. 1,2,4-Triazin-3-one

Thermal cyclization of an ethoxycarbonylaminohydrazine gives the isomeric 1,2,4-triazin-3-one ring in good yield.

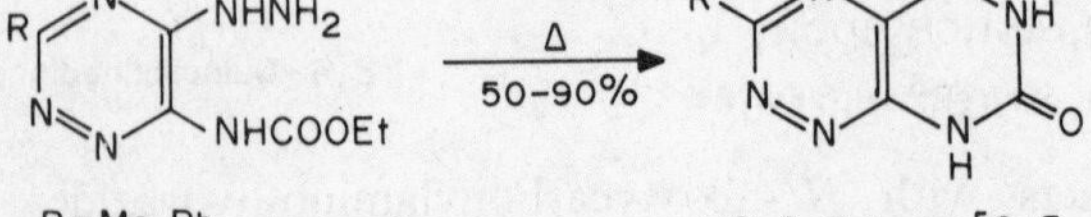

[683]

R = Me, Ph

1,2,4-Triazino[6,5-e]-
[1,2,4]-triazin-3-one

III. FORMATION OF A SEVEN-MEMBERED RING

1.   1, 2, 4-Triazepin-7-one

As mentioned in Section II.3, *o*-amino-hydrazides can react with orthoformates to give either 3-aminoquinazolin-4-ones or the isomeric 1, 3, 4-benzotriazepin-5-ones. Factors affecting the course of the cyclization are discussed [1562, 1564, 1802]. Early work in which the synthesis of the triazepinone was claimed should therefore be approached cautiously [94, 1230]. A 2-(hydrazinomethyl)aniline, however, cyclizes in acid solution to the 5-methylene analogue.

R = H, Me, Ph

1,3,4-Benzotriazepin-5-one

[1379, 1563]

R¹ = H, Me; R² = Me, Et

1,3,4-Benzotriazepine

[1797]

Cyanogen bromide or an isothiocyanate efficiently cyclizes an amino-hydrazide to a triazepinone in good yield.

R¹ = H, Br, Cl, NO₂;
R² = H, Me

1,3,4-Benzotriazepin-5-one

[1166]

R = Bu, PhCH₂, CH₂=CHCH₂,
Ph, 4-MeC₆H₄

1,3,4-Benzotriazepin-5-one

[936]

CHAPTER 51

Amine and Hydrazone or Imine

I. FORMATION OF A FIVE-MEMBERED RING

1. Pyrazole

An *o*-amino-imine (produced *in situ* in this example from a nitro-imine) may be cyclized by prolonged heating in a high boiling solvent with a deoxygenating agent.

$$\text{PhtBu, P(OEt)}_3, \Delta \quad 15-27\%$$

Ar = Ph, 4-Me$_2$NC$_6$H$_4$

Thieno[3, 2-*c*]pyrazole

[321]

2. Imidazole

In anaerobic conditions, 2--aminophenylhydrazones cyclize to benzimidazoles but in the presence of oxygen (or a mild oxidant) a mixture of the benzimidazole **(51.1)** (as major product) and 1, 2, 4-benzotriazine is obtained.

[1124]

Benzimidazole
(51.1)

R=H,Me,PhCH₂,Ph

(51.1) +

[1124]

1,2,4-Benzotriazine

Oxidative cyclization of *o*-amino-imines yields a fused imidazole in high yield. Benzimidazole may be synthesized in 89 per cent yield by the oxidation of the mono-Schiff's base of *o*-phenylenediamine with plumbophosphates [1849].

[1209]

2,6-Purinedione
(review [B-7])

3.   1, 2, 3-Triazole

Diazotization of an amino group of an amino-imine results in the formation of a triazole ring in high yield.

[1501]

Benzotriazole

II.   FORMATION OF A SIX-MEMBERED RING

1.   Pyridin-2-one

One synthetic use (review [100]) of carbon suboxide is its cyclization of 2-aminobenzaldehyde imine (or oxime or oxime *O*-methyl ether) to a 2-quinolinone.

R = HO, MeO, Ph, 4-Me-,   2-Quinolinone
4-MeO-C$_6$H$_4$

[1045]

2. Pyrazine

2-Amino-Schiff's bases may be cyclized into a pyrazine ring, the additional carbon being supplied by an orthoformate ester.

R^1, R^2 = H, Me; R^3 = aryl, heteroaryl   2,4-Pteridinedione

[902]

3. 1, 2, 3-Triazine

In a reaction similar to that described in the previous section, the same groups in the peri positions of naphthalene yield a fused triazine ring.

Naphtho[1,8-*de*]-1,2,3-triazine

[47]

Oxidative treatment of a 2-amino-hydrazone results in cyclization to the triazine at ambient temperature.

R = Me, Ph, 4-MeOC$_6$H$_4$ 1,2,3-Benzotriazine

[1697]

4. 1, 2, 4-Triazine

Amino-hydrazones or -amidrazones cyclize on warming to form a fused 1, 2, 4-triazine.

[1820]

$R^1=H,Me,Br,Cl,NO_2$;
$R^2=EtO,NH_2$

1,2,4-Benzotriazine
(review [B-14])

Dehydrogenative cyclizations by bromine or azodiformate ester also give rise to a 1, 2, 4-triazine; the example [1124] mentioned in Section I.2 is also of this type.

[766]

$R^1=Ph,4-Cl-,4-MeO-C_6H_4$;
$R^2=Me,Et$
$i,(=NCOOEt)_2$

Imidazo[4,5-e]-1,2,4-
triazin-6-one

[683]

1,2,4-Triazino[6,5-e]-
[1,2,4]triazine

5. 1, 2-Oxazine

Methylation of a tertiary amine and treatment of the adjacent oxime with butyllithium gives an oxazine ring (and expulsion of trimethylamine).

[1727]

2,3-Benzoxazine

6. 1, 3, 4-Thiadiazine 1, 1-Dioxide

When a sulphone group is present in the side-chain, ring closure of an amino-hydrazone can yield a fused thiadiazine ring.

[710]

4,1,2-Benzothiadiazine 4,4-dioxide

III. FORMATION OF A SEVEN-MEMBERED RING

1. 1, 2, 4-Triazepine or 1, 2, 4-Triazepin-3-one

An amino-hydrazone is converted into a fused triazepine or triazepinone ring in moderate yield by treatment with paraformaldehyde or phosgene.

1, 3, 4-Benzotriazepine   [457]

1, 3, 4-Benzotriazepin-2-one   [457]

CHAPTER 52

Amine and Ketone

I. FORMATION OF A FIVE-MEMBERED RING

1. Pyrrole

For pyrrole formation, the presence of a 2-(β-oxoalkyl)-aniline-type molecule is one requirement. Such compounds cyclize spontaneously or with some heating but when a γ-carboxylic ester group is also present, the expected fused pyrrole may be accompanied by the corresponding fused pyridin-2-one [565].

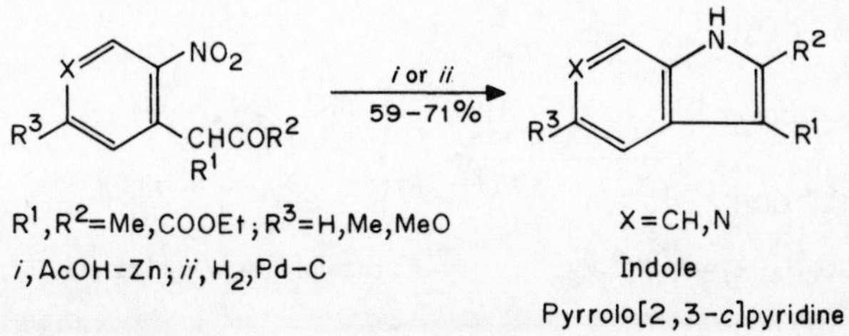
NO2
CHCOR2
R1
R3
X
i or ii
59–71%
H
N
R2
R1
R1, R2=Me, COOEt; R3=H, Me, MeO
i, AcOH–Zn; ii, H2, Pd–C
X=CH, N
Indole
Pyrrolo[2,3-c]pyridine

[550, 565, 1377]

The nitro group may be in an alkyl chain and may need more drastic conditions for its reduction. When both amine and carbonyl are attached to a ring, an additional carbon has to be supplied by another reactant, for example, an α-bromoketone.

[495]

Pyrrolo[2,3-*f*]quinoline

[20]

R^1, R^4 = Me, Ph; R^2 = H, Me;
R^3 = H, Cl, MeO

Indole

2. Pyrazole

Displacement of an amino (preferably primary or tertiary) group placed between two carbonyl groups is effected by heating with hydrazine hydrate in methanol. The hydrazine then reacts with one of the carbonyl groups to give a fused pyrazole ring. A similar reaction is possible when an amino and one carbonyl group are attached to a π-deficient ring. The iminium salt is more reactive than a carbonyl group and so the amino-ketone needs a higher reaction temperature.

[1234]

Cyclohepta[*c*]pyrazol-8-one

[1234]

R^1 = H, Me; R^2 = H or
$R^1 R^2$ = $(CH_2)_4$

Cyclohepta[*c*]pyrazole

[1887]

R^1 = 4-ClC$_6$H$_4$; R^2 = Me, PhCH$_2$;
X = Br, I

Pyrazolo[3,4-*d*]pyridazine

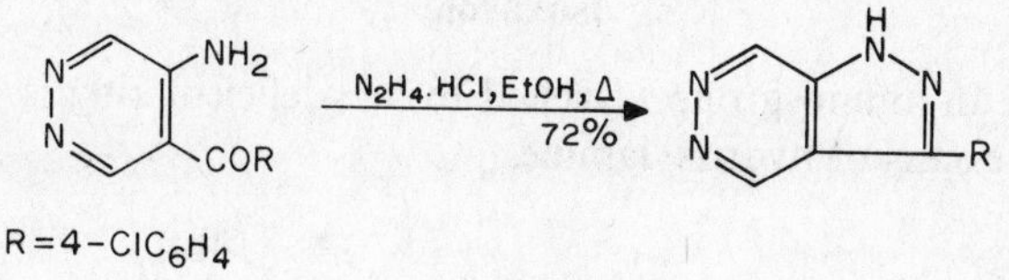

[1887]

R = 4-ClC₆H₄

3. Imidazole

Potassium thiocyanate converts an amino-ketone into an imidazole-2-thiol.

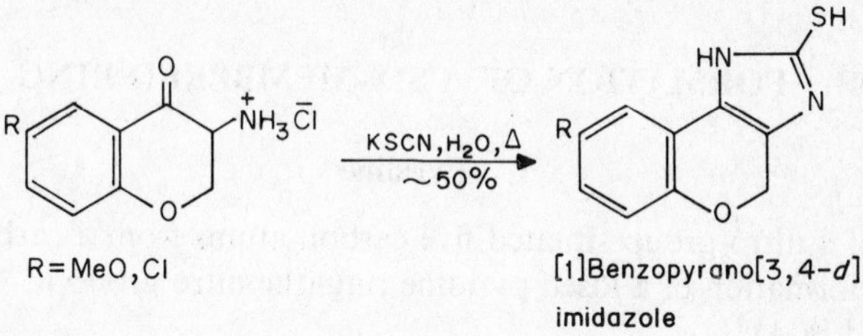

[102]

R = MeO, Cl

[1]Benzopyrano[3,4-*d*]-
imidazole

A quaternized 2-aminoheterocycle containing a *β*-carbonyl *N*-substituent cyclizes on heating in ethanol or with PPA (review [B-21]) to a fused imidazole ring.

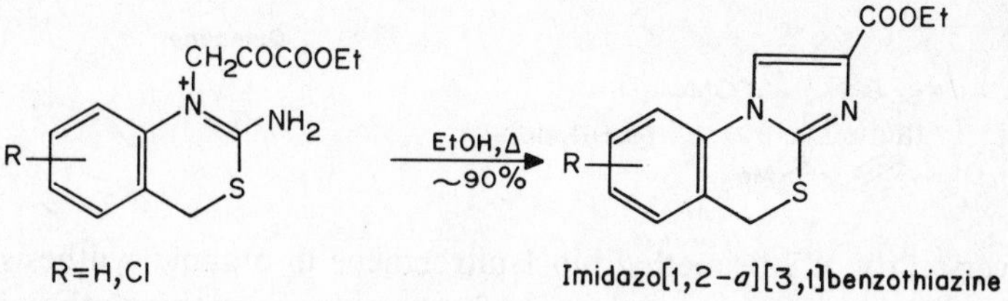

[1374]

R = H, Cl

Imidazo[1,2-*a*][3,1]benzothiazine

Another ring system synthesized similarly:

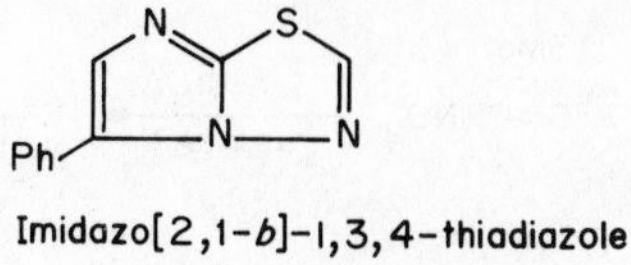

Imidazo[2,1-*b*]-1,3,4-thiadiazole

4. 1,2,3-Triazole

An *N*-aminopyridinium salt (review of reactivity [1870] containing an oxime group cyclizes to give a fused 1,2,3-triazole ring.

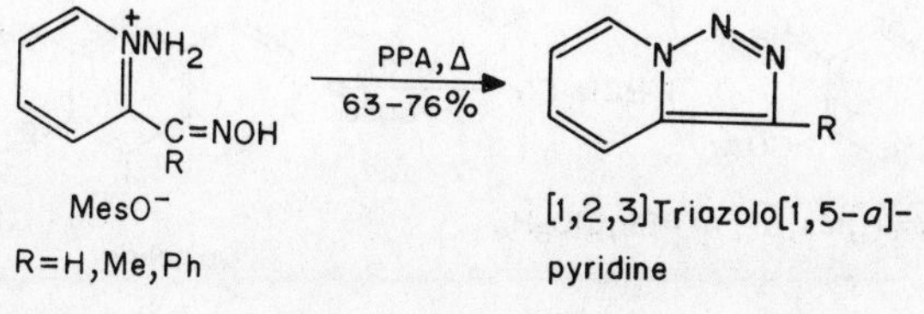

[1329]

MesO⁻
R = H, Me, Ph

[1,2,3]Triazolo[1,5-*a*]-
pyridine

5. Isoxazole

Displacement of an amino group attached to a π-deficient ring is possible by heating with an excess of hydroxylamine.

$$\text{NH}_2\text{OH, EtOH, }\Delta$$
$$36\%$$

[1887]

R = 4-ClC$_6$H$_4$

Isoxazolo[4,5-*d*]-
pyridazine

II. FORMATION OF A SIX-MEMBERED RING

1. Pyridine

Reduction of a nitro group situated five carbon atoms from a carbonyl group leads to the formation of a fused pyridine ring (the nitro group in Ar^1 has also been reduced in Ar^2).

$$\text{Na}_2\text{S}_2\text{O}_4\text{, MeOH, }\Delta$$
$$\sim 90\%$$

[1232]

Quinoline

Ar^1 =

R = HO, MeO

The versatility of 2-bismethylthio-1-nitroethene in organic synthesis (review [1801]) is illustrated in two syntheses of 3-nitroquinolines; the synthon has been modified in one synthesis by reaction with aniline.

$$\text{AcOH, }\Delta$$
$$42\text{--}76\%$$

R^1 = Me, Ph, 4-MeC$_6$H$_4$;
R^2 = H, Br, Cl

[174, 1799]

Two methods which proceed at low temperatures in high yields are worth attention. An aliphatic nitrile in the presence of sodamide cyclizes an amino-ketone while the dilithio derivative of a ketoxime reacts with LDA in THF.

$$+ \text{ RCH}_2\text{CN} \xrightarrow[\sim 80\%]{\text{NaNH}_2}$$

[327]

R = H, Me, Et; Ar = Ph, ClC$_6$H$_4$

Quinoline

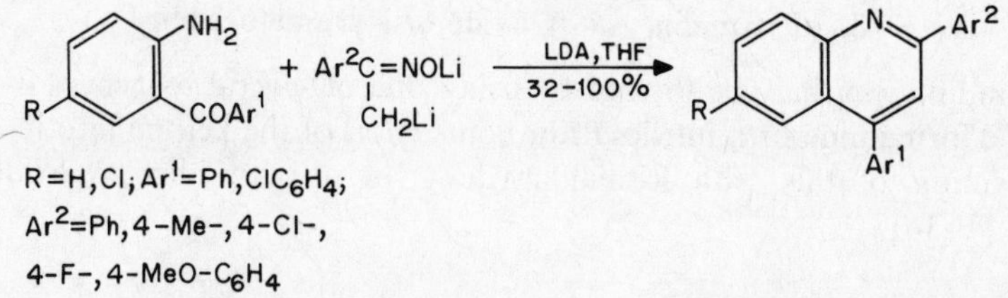

[650]

R=H,Cl; Ar¹=Ph,ClC₆H₄;
Ar²=Ph,4-Me-,4-Cl-,
4-F-,4-MeO-C₆H₄

In a Friedlander-type reaction (review [1093]), amino-ketones condense with 3-oxobutyrolactone (furan-2, 4-dione) to give a doubly fused pyridine ring. Other ketones also condense with amino-ketones in the presence of either a Lewis acid ($ZnCl_2$ or $HClO_4$) or a strong base.

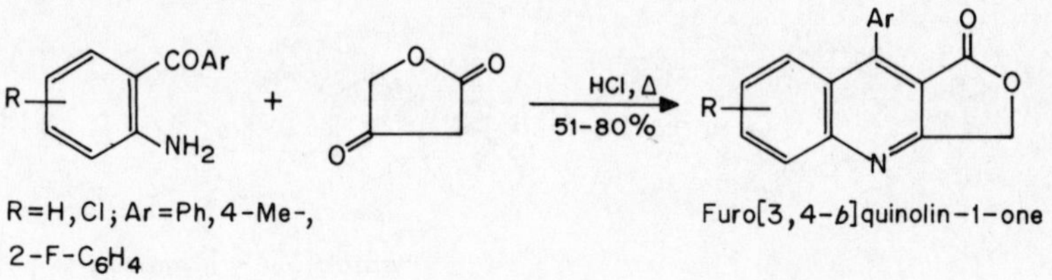

[490]

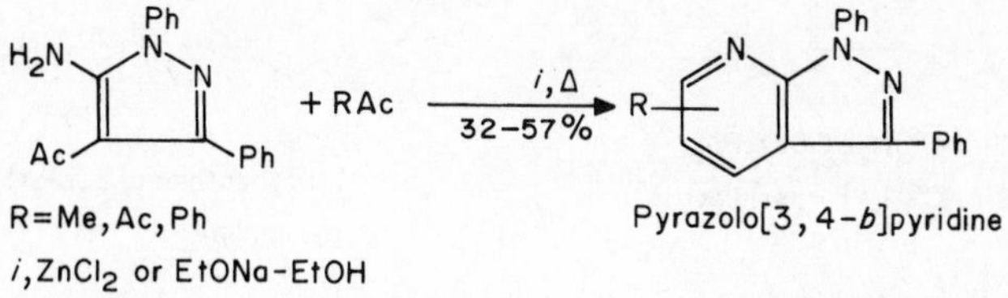

R=H,Cl; Ar =Ph, 4-Me-,
2-F-C₆H₄

Furo[3,4-b]quinolin-1-one

[1846]

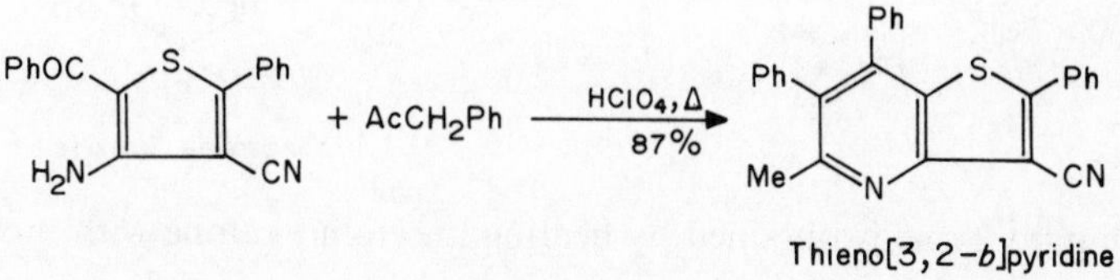

R=Me, Ac, Ph
i, ZnCl₂ or EtONa-EtOH

Pyrazolo[3, 4-b]pyridine

[392]

Thieno[3,2-b]pyridine

2. Pyridin-2-one

When an amino-ketone is heated with diketene (review [2018]), a high yield of a fused pyridinone is obtained.

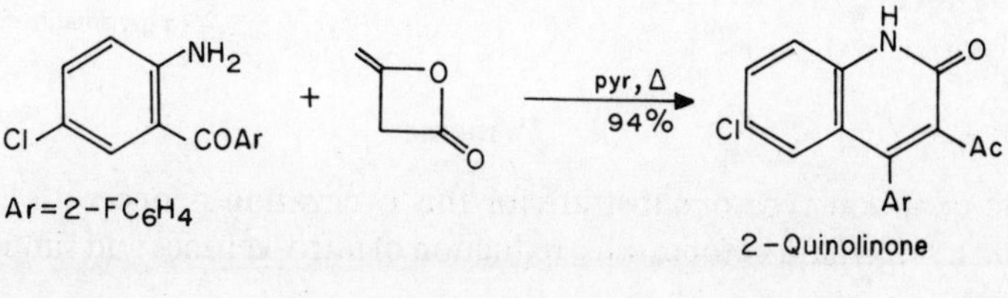

[1356]

Ar=2-FC₆H₄

2-Quinolinone

3. Pyrimidine, its *N*-oxide or Pyrimidin-2-one

A pyrimidine ring may be formed by using one of several reagents [B-5], for example, formamide or a nitrile. Prior conversion of the ketone into its oxime and reaction of this with formaldehyde gives a pyrimidine *N*-oxide ring (review [B-34]).

[211]

Quinazoline

[1846]

Pyrazolo[3,4-*d*]pyrimidine

[63]

R=Me,CH$_2$COOEt,Ph,
4-ClC$_6$H$_4$,3-pyridinyl

[1]Benzothieno[2,3-*d*]-
pyrimidine

[990]

Quinazoline 3-oxide

A pyrimidin-2-one is obtained by heating an amino-ketone with urea.

[1668]

R^1=H,Et,PhCH$_2$;
R^2=H,Me

Pyrrolo[3,4-*d*]pyrimidin-2-one

4. Pyrazine

The most common type of substrate for this cyclization is one with a nuclear amine and a side-chain ketone; thus, reduction of nitro-ketones and simultaneous cyclization is achieved readily. Sometimes, the ketone is masked or protected as part of a previous synthetic step.

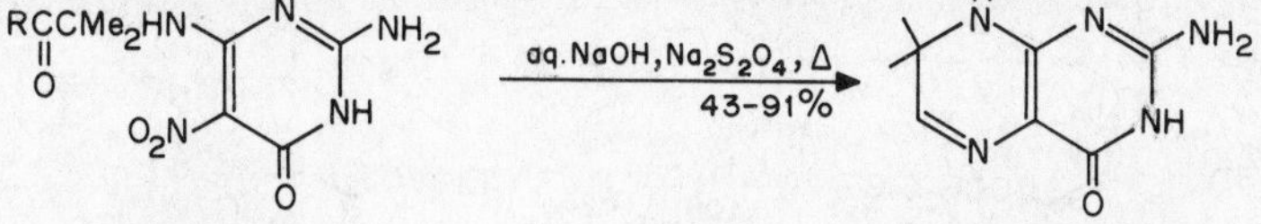

R=Me, Et

4-Pteridinone

[1829]

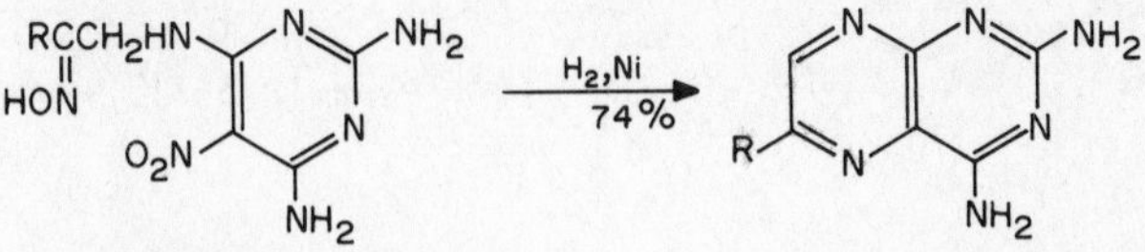

R=CH₂NMeC₆H₄-4-COOMe

Pteridine

[1375]

Other ring systems synthesized similarly:

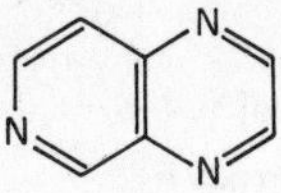

Pyrido[3,4-b]-
pyrazine

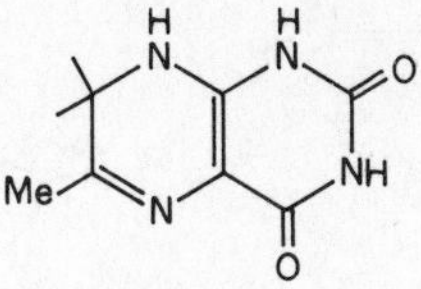

[491, 1378]

2,4-Pteridinedione

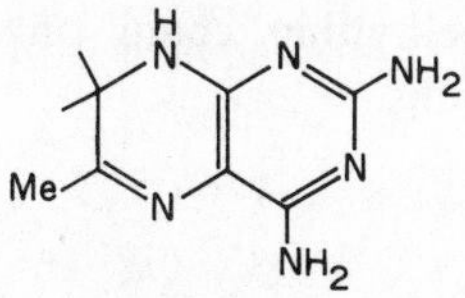

[1790]

Pteridine

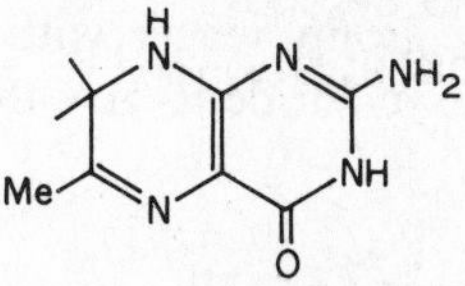

[1790]

4-Pteridinone

[1791]

5. 1,3- or 1,4-Oxazine or 1,4-Thiazine 1,1-Dioxide

Deprotection (by an acid) of an aliphatic aminoalkoxy group (as its Schiff's base) conveniently placed to react with a quinone carbonyl results in a rearrangement of the chain and the formation of a 1,4-oxazine.

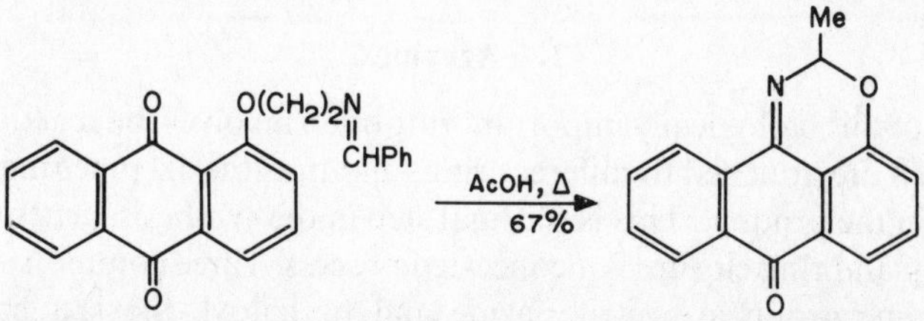

Anthra[9,1-de][1,3]-
oxazin-7-one

[566]

Suitably placed amino and ketone groups spontaneously condense to form a 1,4-oxazine and 1,4-thiazine dioxide.

[360]

Pyrano[2,3-*f*][1,4]benzoxazin-7-one

[1207]

Pyrrolo[3,4-*b*]-1,4-thiazine
1,1-dioxide

6. 1,3,4-Thiadiazine

An *N*-amino group reacts with a 2-oxoethylthio chain on heating with phosphorus oxychloride to give this fused ring.

[911]

1,2,4-Triazolo[3,4-*b*]-
[1,3,4]thiadiazine

III. FORMATION OF A SEVEN-MEMBERED RING

1. Azepine

Formation of this biologically important ring often involves the reaction between groups which are attached to different rings and thus the azepine ring is fused to two others in the product. This is the final step in the synthesis; generation of the amine group and ring closure is a concerted process. Three common sources of a primary amine group are nitrile, azide and phthaloyl, the last having been introduced earlier as a protecting group. The chemistry of the benzazepines has been reviewed [1620, B-19].

Thiazolo[5,4-d][2]benzazepine   [492]

R=phthalimido, i, MeNH$_2$EtOH;
R=N$_3$, i, EtOH, H$_2$, RaNi
Ar=Ph, 2-Cl-, 2-F-C$_6$H$_4$

1,2,3-Triazolo[4,5-d][2]-
benzazepine   [461]

Another ring system synthesized similarly:

2-Benzazepine   [2019]

Pyrimido[5,4-d]-
[2]benzazepine   [871]

2.   1,4-Diazepine or 1,4-Diazepin-2- or 5-one

Strategies for the synthesis of this medically important ring are similar to those
mentioned in the preceding section (reviews [1895, 1917]). Thermal dehydration
of an amino-ketone may also be employed.

1,4-Benzodiazepine   [990]

Pyrrolo[2,1-c][1,4]-
benzodiazepine   [828]

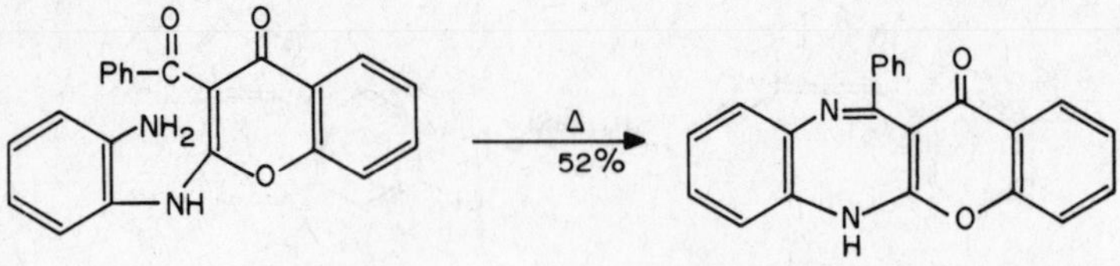

[245]

[1]Benzopyrano[2,3-b][1,5]-
benzodiazepin-13-one

Other ring systems synthesized similarly:

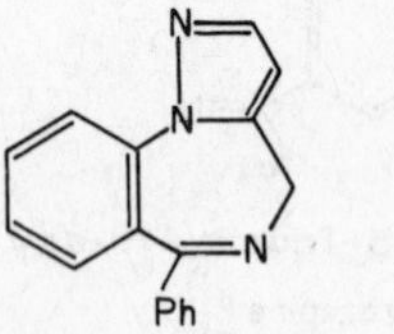

[1233]

Pyrazolo[1,5-a][1,4]-
benzodiazepine

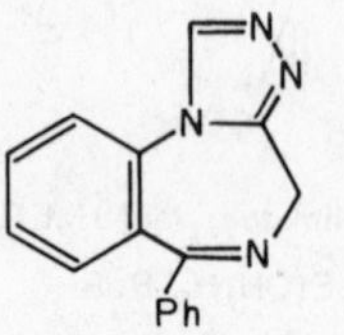

[813]

[1,2,4]Triazolo[4,3-a]-
[1,4]benzodiazepine

A 1,4-diazepin-2- or -5-one ring may be constructed from an amino-ketone which has a carbonyl in the amine-carrying side-chain.

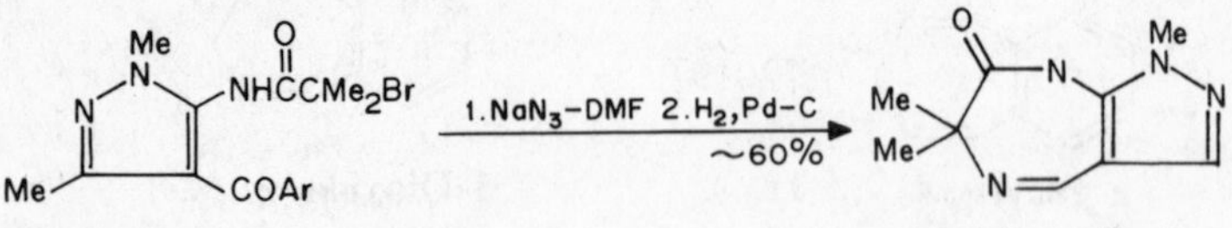

[445]

Ar = 3-ClC$_6$H$_4$

Pyrazolo[3,4-e][1,4]-
diazepin-7-one

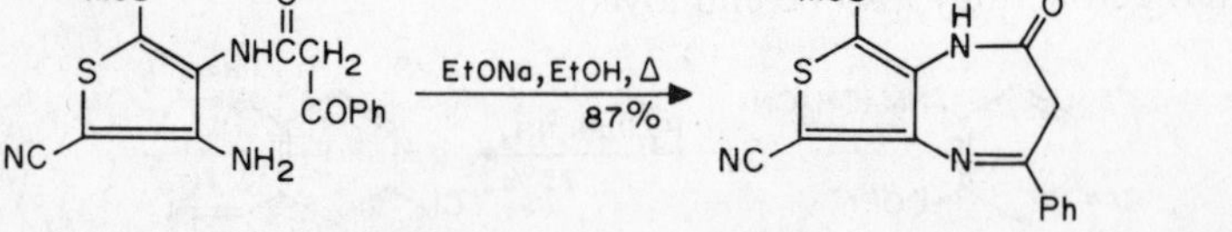

[191]

Thieno[3,4-b][1,4]diazepin-2-one

Another ring system synthesized similarly:

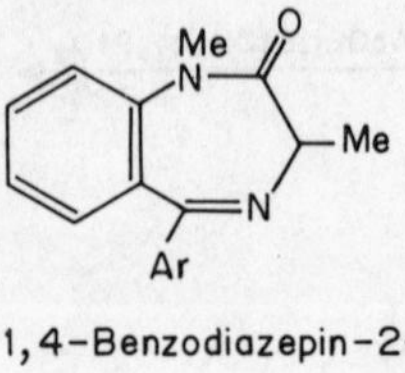

[1986]

1,4-Benzodiazepin-2-one

3. 1,4-Oxazepine

A protected 2-aminoethoxy group reacts with a quinone carbonyl under dehydrating conditions to give a fused oxazepine ring.

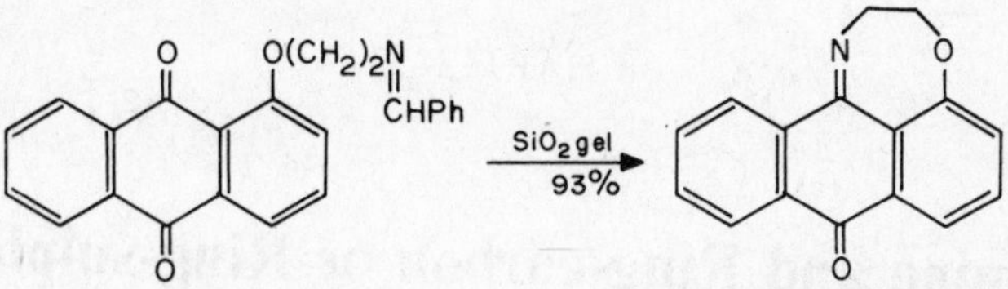

[566]

Anthra[9,1-*ef*][1,4]–
oxazepin–8–one

4. 1,2,6-Oxadiazepine

This ring may be obtained by annulating an amino and a ketoxime with formaldehyde.

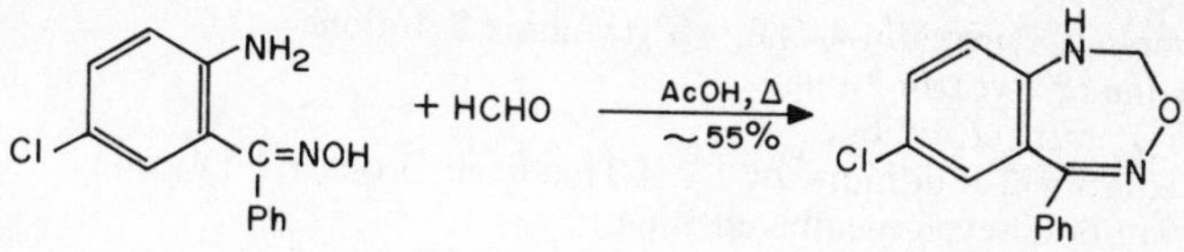

[990]

3,1,4–Benzoxadiazepine

5. 1,2,5-Thiadiazepine 1,1-Dioxide

Reaction of an amine with a ketosulphonamide in the usual way gives the benzothiadiazepine dioxide (and its tautomer) together with an equal amount of the 3,4-dihydro derivative. Reduction of the nitro group was effected by chemical means in this reaction.

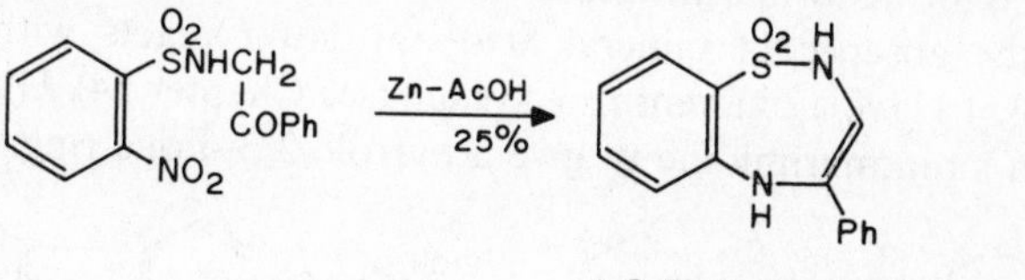

[118]

1,2,5–Benzothiadiazepine
1,1–dioxide

CHAPTER 53

Amine and Ring-carbon or Ring-sulphur

I. FORMATION OF A FIVE-MEMBERED RING

1. Pyrrole

A pyrrole ring is formed when an amine is treated with an α-halogeno-ketone or a benzoin. In the absence of mineral acid, the latter reacts with 3-amino-5-phenylpyrazole to give a different ring system (see Chapter 54). Oxalyl chloride reacts with an aminopyrimidine to give a pyrrole-2, 3-dione ring.

R = Me, 2, 3-Me₂   Carbazole   [575]

Ar = Ph, 4-MeOC₆H₄

Pyrrolo[2,3-c]pyrazole   [1305]

[408]

R = H, Me

Pyrrolo[2,3-d]pyrimidine–
2,4,5,6-tetraone

Other ring systems synthesized similarly:

[1916]

Indole
(review [2008])

[1842]

Pyrrolo[2,3-b]pyridine

2. Pyrazole or 1,2,3-Triazole

1-Aminopyridinium salts (review [1870]) react with acetylenes in a basic medium to give a pyrazolopyridine (this type of heterocycle has been reviewed [1600]).

[573]

R = Me, Ph, COOMe

Pyrazolo[1,5-a]pyridine

Transfer of a labile azide group in acid solution gives good yields of a fused 1,2,3-triazole.

1,2,3-Triazolo[4,5-b]pyridine
(review[1694])

[1120]

Other ring systems synthesized similarly:

[1120]

X = O, S, NPh
Isoxazolo[4,5-d]-1,2,3-triazole
Isothiazolo[4,5-d]-1,2,3-triazole
Pyrazolo[3,4-d]-1,2,3-triazole

[1120]

1,2,3-Triazolo[4,5-b]indole

3. Thiazole or Isothiazole

Sulphur may be supplied for these rings by an inorganic thiocyanate, elemental sulphur or (for the isothiazole) an isothiocyanate.

R^1 = Me, MeO, PhCH$_2$O, Br, Cl;
R^2 = NH$_4$, Na;
i, AcOH, Br$_2$; ii, NaBr, HCl

Benzothiazole

[88, 570]

R^1 = Cl, MeO, CF$_3$, PhO;
R^2 = 2- or 4-pyridyl

[334]

R = peracetylated sugar

Isothiazolo[3,4-d]-
pyrimidine-4,6-dione

[1111]

4. 1, 2, 4-Dithiazole

Treatment of a 3-imino-1,2,4-dithiazole with an isothiocyanate leads to the formation of a second dithiazole ring by reaction with a ring-sulphur.

R = Me, Ph

[1,2,4]Dithiazolo[1,5-b][1,2,4]-
dithiazole-4-S^{IV}

[1720]

II. FORMATION OF A SIX-MEMBERED RING

1. Pyridine

In this cyclization, the three carbon atoms required to complete the pyridine ring may be supplied by an α, β-unsaturated aldehyde or ketone, glycerol (a Skraup reaction), an enamine or triformylmethane. A modification which avoids the preparation and isolation of triformylmethane uses a mixture of phosphorus oxychloride, ethyl bromoacetate and DMF; the salt, $[CH(CH={\overset{+}{N}}Me_2)_3]_3 3Cl^-$, is believed to be formed and converted *in situ* by alkali into the trialdehyde. In the presence of a rhodium–norbornadiene complex, acetaldehyde or propionaldehyde (2 moles) provides the necessary carbon atoms while 1,3-diketones react with an arylamine under acidic conditions to form a 2,4-dialkylpyridine ring. The application of 3(5)-aminopyrazoles as intermediates has been reviewed [1123].

Pyrido[3,2-*d*]pyrimidin-4-one   [574]

R^1=alkyl, Ph; R^2=Me, COOEt;
R^3=Me, Ph

Pyrido[2,3-*d*]pyrimidine-
2,4-dione   [1530]

R=alkyl

Quinoline   [526]

Pyrazolo[3,4-*b*]pyridine
(review[1600])   [734]

A similar reaction on 5- or 6-aminoindazole gave the pyrazoloquinoline.

Pyrazolo[4,3-*f*]quinoline   [1237]

Pyrazolo[3,4-*f*]quinoline   [1237]

H_2N ... $+ \begin{array}{c} COR^1 \\ | \\ CH=CR^2NMe_2 \end{array}$ $\xrightarrow[45-100\%]{AcOH,\Delta}$ ... [884]

$R^1 = Me, Ph, 4-Cl^-, 4-Me-C_6H_4;$
$R^2 = H, Me; X = O, S$

$X = O, S$

Pyrido[2,3-*d*]pyrimidine-2,4-dione
Pyrido[2,3-*d*]pyrimidin-4-one, 2-thioxo

H_2N ... $+ HC(OEt)_3$ $\xrightarrow[77\%]{DMF, \Delta}$ ... [746]

H_2N ... NH_2 $+ [CH(CH=\overset{+}{N}Me_2)_3]$ $3Cl^-$ $\xrightarrow[82\%]{NaOH, \Delta}$ ... [1391]

Pyrido[2,3-*d*]pyrimidin-4-one

... NH_2 $+ R^2CH_2CHO$ $\xrightarrow[30-82\%]{i, \Delta}$ ... [1125]

$R^1 = H, MeO; R^2 = H, Me$
$i, [Rh(norbornadiene)Cl]_2, PhNO_2$

Quinoline

H_2N ... R^1 ... $+ Ac_2CHR^3$ $\xrightarrow[29-67\%]{H_2PO_4, \Delta}$ ... [1354]
R^2

$R^1 = iPr, PhCH_2, Ph; R^2 = H, Me, Ph;$
$R^3 = H, Cl, AcO$

Pyrazolo[3,4-*b*]pyridine
(review [1600])

In an acidic medium, 2,6-diaminopyridine is converted into a naphthyridine by 1,3-diketones [1186].

H_2N ... NH_2 $+ \begin{array}{c} PhCO \\ | \\ CH_2Ac \end{array}$ $\xrightarrow[80\%]{H_3PO_4, \Delta}$ Ph ... NH_2 [10]

1,8-Naphthyridine

In the Pictet–Spengler synthesis, a reduced pyridine ring is formed from a 2-aminoethyl group and a ketone or aldehyde (or acetal); phenylpyruvic acid also gives a good yield [297]. Although it is usually a source of isoquinolines, it has also been used to form a pyridine fused to a heteroring [1744]. When a 2-oxocarboxylic acid reacts with the ethylamine, simultaneous decarboxylation occurs.

Ar = 3,4,5-(MeO)₃C₆H₂

Isoquinoline   [263]

R¹=H,PhCH₂; R²=H,MeO;
R³=H,Me,(CH₂)₂COOH

Pyrido[3,4-b]indole   [330, 331]

R=H,Me

Pyrrolo[3,2-c]pyridine   [518]

Another ring system synthesized similarly:

Pyrido[2,3-d]pyrimidine-
2,4-dione   [1837]

Some of the atoms which become part of a newly formed pyridine ring may be supplied by another ring or by a second molecule of the substrate, the latter sometimes being cleaved in the process.

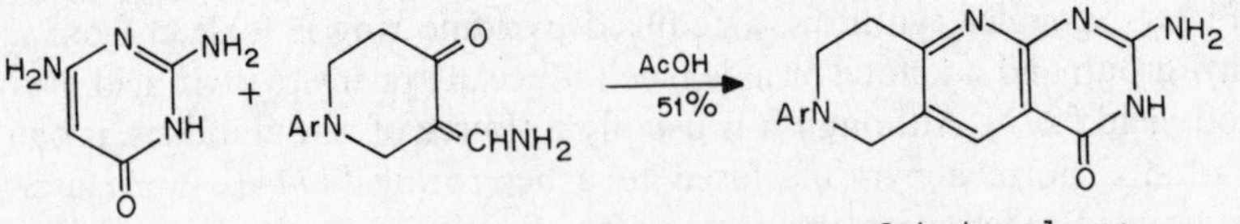

Ar = 4-tBuOOCC₆H₄

Pyrido[3′,4′:5,6]pyrido-
[2,3-d]pyrimidin-4-one

[1608]

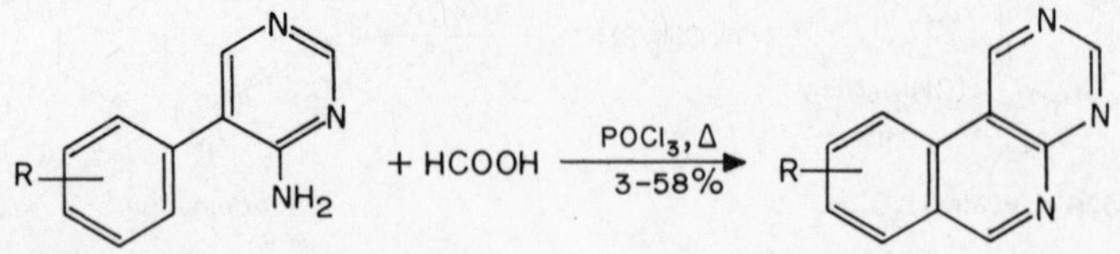

R = H, Me, MeO

Pyrimido[4,5-c]isoquinoline

[657]

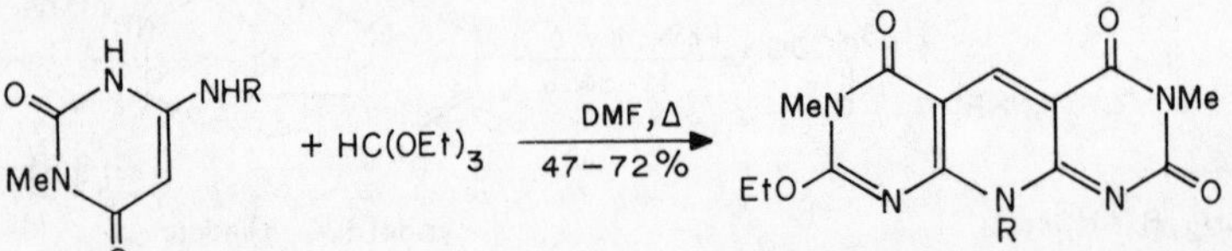

R = Me, COOMe, Ph

Pyrido[2,3-d:6,5-d′]dipyrimidine-
2,4,6-trione

[572]

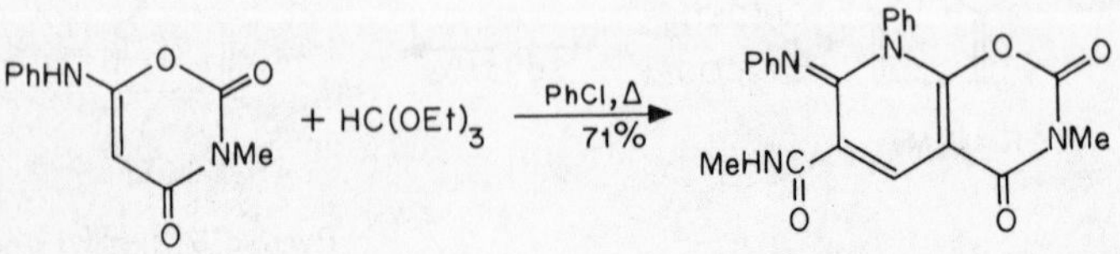

Pyrido[3,2-e]-1,3-oxazine-
2,4-dione

[1112]

Oxidative cyclization through C–N bond formation induced by diethyl azodiformate leads to high yields of a pyridine-containing ring system. Phenylacetylene was the only one of several alkynes which annulated arylamines when treated successively with butyl-lithium and tin(IV) chloride; the applicability of this method appears to be limited.

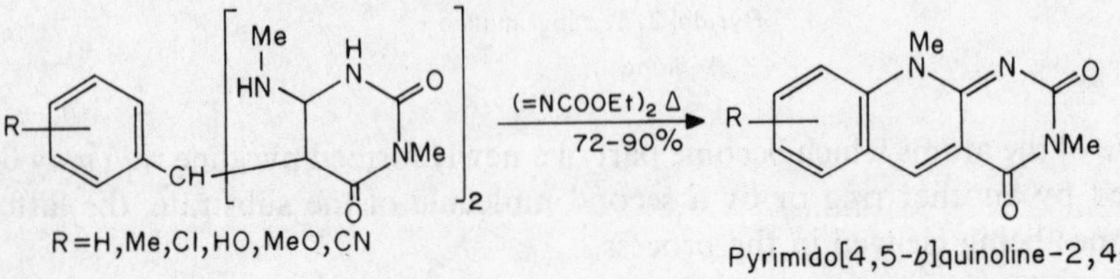

R = H, Me, Cl, HO, MeO, CN

Pyrimido[4,5-b]quinoline-2,4-dione

[738]

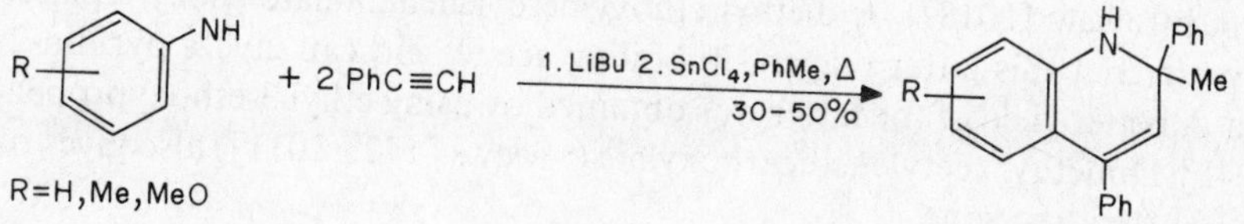

Quinoline [1107]

R=H,Me,MeO

Other ring systems synthesized similarly:

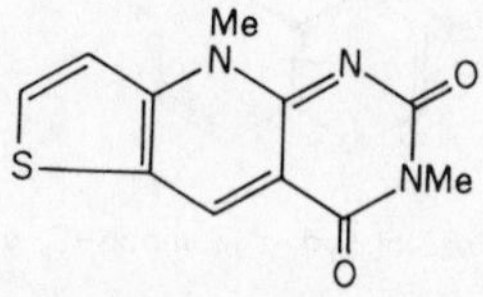

Thieno[2′,3′:5,6]pyrido[2,3-d]-
pyrimidine-2,4-dione [738]

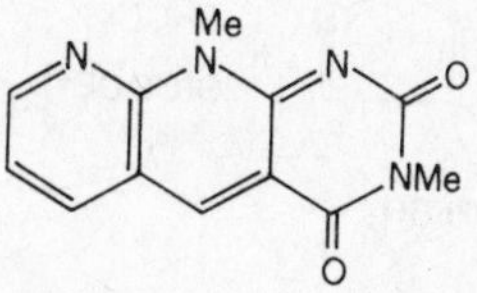

Pyrimido[4,5-b][1,8]naphthyridine-
2,4-dione [738]

2. Pyridin-2- or -4-one

Heating an aryl- or heteroaryl-amine with malonic acid and phosphorus oxychloride or with ethyl acetoacetate-acetic acid gives a tautomeric fused 4-hydroxypyridin-2-one ring in moderate yield.

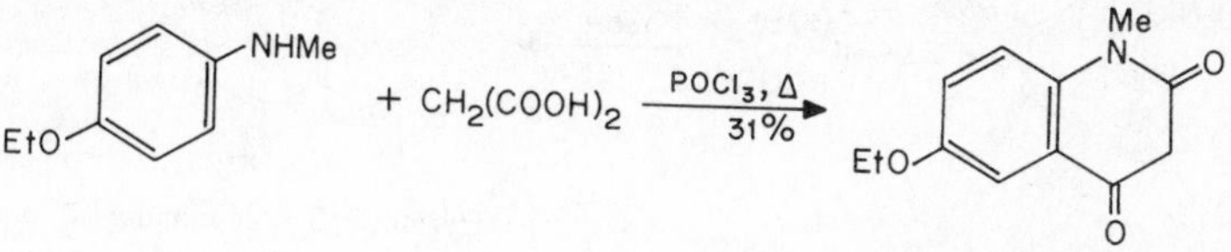

2,4-Quinolinedione [333]

Pyrazolo[3,4-b]pyridin-6-one
(review[1600]) [734, 1332]

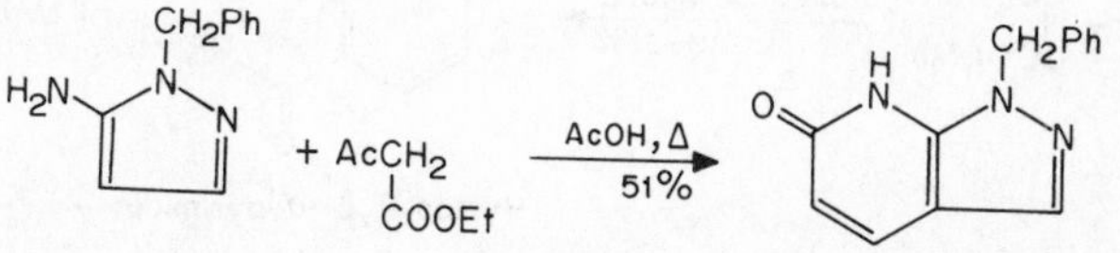

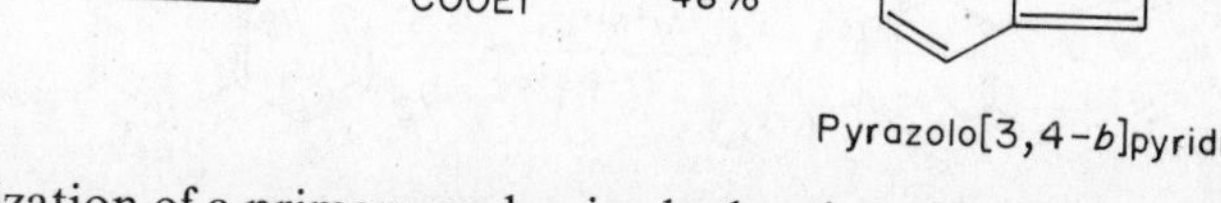

Pyrazolo[3,4-b]pyridin-6-one [1115]

Cyclization of a primary arylamine by heating with a β-keto-ester (a Conrad–Limpach reaction) yields 4-quinolinones as does a variation in which diketene

[1960] (review [2018]) or diethyl ethoxymethylenemalonate [862] replaces the ketoester but this latter reagent in boiling acetic acid can give a pyridin-2-one ring. A better yield of the latter was obtained by using ethyl 3-ethoxypropenoate [1942]. Dimethyl acetylenedicarboxylate (reviews [1725, 2011]) also gives rise to a fused pyridin-2-one.

Pyrazolo[3,4-f]quinolin-9-one   [862]

4-Quinolinone   [571]

Pyrido[2,3-d]pyrimidin-7-one   [1942]

Pyrido[2,3-d]pyrimidine-
2,4,7-trione   [1910, 1960]

2,4-Dinitrodiphenyl ketones undergo selective denitration of the 2-substituent when heated with an arylamine and when an amine group is present at the 2'-position, cyclization to an acridone occurs in moderate to good yield.

Ar^1=Ph,Me-,Cl-,MeO-,CN- C$_6$H$_4$;   9-Acridinone
Ar^2=3-MeC$_6$H$_4$ (review[B-10])   [329]

3. Pyrimidine, Pyrimidin-4-one or Pyrimidine-2, 4-dione

Hydroxylamine (or an aralkyl ether of it) reacts with an activated pyrimidine ring and formaldehyde to form another (reduced) pyrimidine ring. The trichloromethylamine derivative, $CCl_3N=CCl_2$, has been used to annulate several aminoheterocycles with the formation of a new pyrimidine ring.

R=H, PhCH$_2$, PhCH$_2$CH$_2$ Pyrimido[4,5-d]pyrimidine   [458]

R^1=Me, Ph, Ph(CH$_2$)$_2$, CH$_2$=CHCH$_2$;
R^2=H, Me, CH$_2$=CHCH$_2$, Ph(CH$_2$)$_2$

Pyrimido[4,5-d]pyrimidine–
2,4-dione   [1703]

Another ring system synthesized similarly:

[1747]

Pyrazolo[3,4-d]pyrimidine

Ethoxymethylene compounds condense on heating with an α-amino-N-heterocycle to give a pyrimidin-4-one ring.

Pyrimido[1,2-a]azepin-4-one   [1295]

Pyrazolo[3,4-d]–
pyrimidin-4-one   [1681]

Another ring system synthesized similarly.

[1681]

Pyrido[3,4-*d*]pyrimidin-4-one

A doubly fused reduced pyrimidin-4-one ring is formed by an intramolecular amination at C-2 of an indole.

$$\xrightarrow[79\%]{HCl-MeOH,\Delta}$$

[975]

Indolo[2,1-*b*]quinazolin-12-one

A 4-amino-2-methylthiopyrimidine ring is formed when dimethyl cyanocarbonimidodithionate, $(MeS)_2C{=}NCN$, is heated with an amine.

$+ (MeS)_2C{=}NCN$

$$\xrightarrow[38-70\%]{DMF,K_2CO_3,\Delta}$$

[673]

$R^1=Me,Ph; R^2=H,Me,Ph$

Pyrimido[4,5-*d*]pyrimidine-
2,4-dione

4. Pyrazine or Pyrazin-2-one

A fused pyrazine ring is formed by heating a 6-aminouracil with nitrosobenzene in acetic anhydride. Formic acid annulates an amino group to a reactive ring such as pyrrole and doubly fused pyrazine is produced.

$+ PhNO$

$$\xrightarrow[42-87\%]{Ac_2O,AcOH,\Delta}$$

[937]

$R=$alkyl,$PhCH_2$,$HO(CH_2)_2$,
Ph,4-Cl-,4-CN-C_6H_4

Benzo[*g*]pteridine-2,4-dione

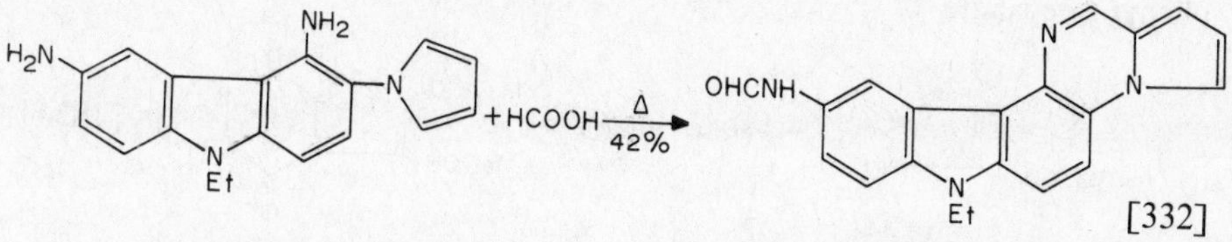

+HCOOH $\xrightarrow{\Delta}$ 42%

[332]

Pyrrolo[1′,2′:1,2]pyrazino[6,5-c]carbazole

Other ring systems synthesized similarly:

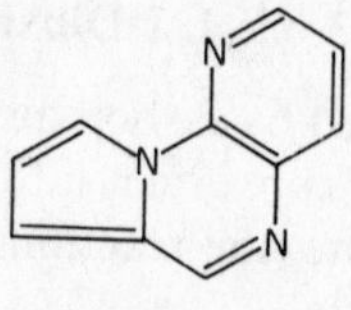

[1885]

Pyrido[3,2-e]pyrrolo-
[1,2-a]pyrazine

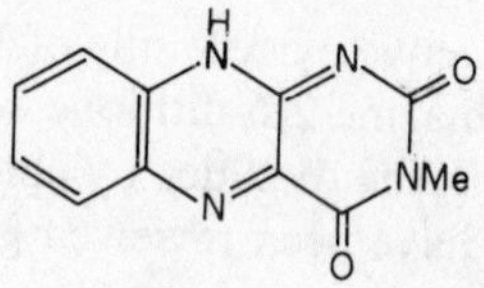

[1869]

Pyrimido[4,5-b]quinoxaline-
2,4-dione

Phosgene forms a bridge between an amine group attached to one ring and the carbon of another electron-rich ring; this can produce a doubly fused pyrazinone ring.

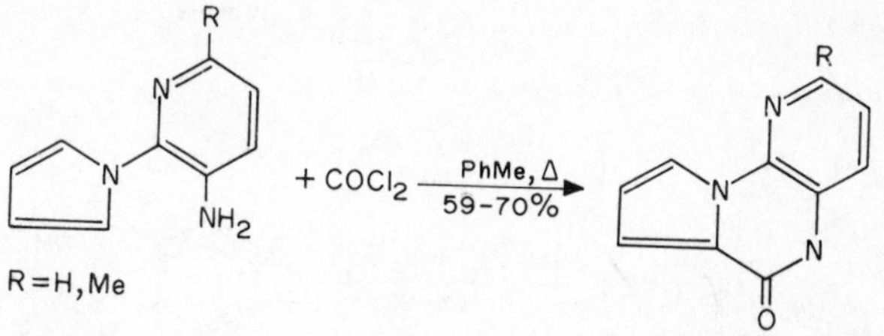

+ COCl₂ $\xrightarrow[59-70\%]{PhMe, \Delta}$

R = H, Me

[1781, 1885]

Pyrido[3,2-e]pyrrolo-
[1,2-a]pyrazin-6-one

Another ring system synthesized similarly:

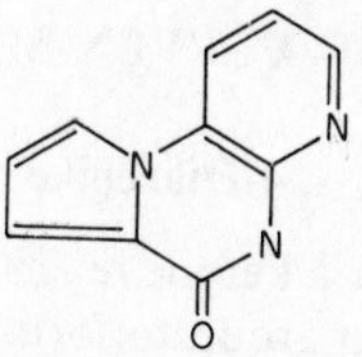

[1783]

Pyrido[2,3-e]pyrrolo-
[1,2-a]pyrazin-6-one

5. 1,3,5-Triazine-2,4-dione

A fused 1,3,5-triazinedione is obtained by reaction of a 2-aminopyrrole with phenyl isocyanate.

[1834]

Pyrrolo[1,2-a]-1,3,5-triazine-
2,4-dione

6.   1,3-Thiazine-2,6-dithione or 1,2,4-Thiadiazin-3-one 1,1-Dioxide

The very reactive pyrrole ring of 3-aminoindole is cyclized by carbon disulphide
to a 1,3-thiazine-2,6-dithione while arylamines are converted into 1,2,4-
benzothiadiazine dioxides by chlorosulphonyl isocyanate (the reactions of this
compound have been reviewed [324]).

[1639]

R=H,Br

[1,3]Thiazino[5,4-b]indole-
2,4-dithione

[119]

R^1=H,Me,Ph; R^2=H,MeO,Cl

1,2,4-Benzothiadiazin-3-one
1,1-dioxide

III.   FORMATION OF A SEVEN-MEMBERED RING

1.   1,4-Diazepine

An aldehyde (or its hemiacetal) or a ketone reacts at the 2-position of a pyrrole
ring with a suitably placed amino group to form a new 1,4-diazepine ring.

[1238]

Pyrrolo[2,1-c][1,4]benzodiazepine

$R^1 = H, Me; R^2 = alkyl, aryl, Ac$

Pyrrolo[1,2-a][1,4]-
benzodiazepine

[407]

CHAPTER 54

Amine and Ring-nitrogen

The N–C–N fragment involved in this type of cyclization may be regarded as a cyclic amidine; the reactions of such compounds have been reviewed [788].

I. FORMATION OF A FIVE-MEMBERED RING

1. Imidazole

The N–C–N linkage is well-suited to the formation of an imidazole ring and a number of reagents bring about this cyclization. α-Halogeno-carbonyl compounds are the most frequently used, especially α-halogeno-ketones (some of which are themselves cyclic), but α-halogeno-aldehydes or their acetals, α-halogeno-oximes, α-halogeno-carboxamides and α-halogeno-hydrazones derived from α-oxo-aldehydes provide further variations on this versatile cyclization. The halogen atom of α-halogeno-ketones has been replaced by a hydroxy group. α-Halogeno-hydrazones have many applications in the synthesis of fused heterocycles and these have been reviewed [1437, 1753].

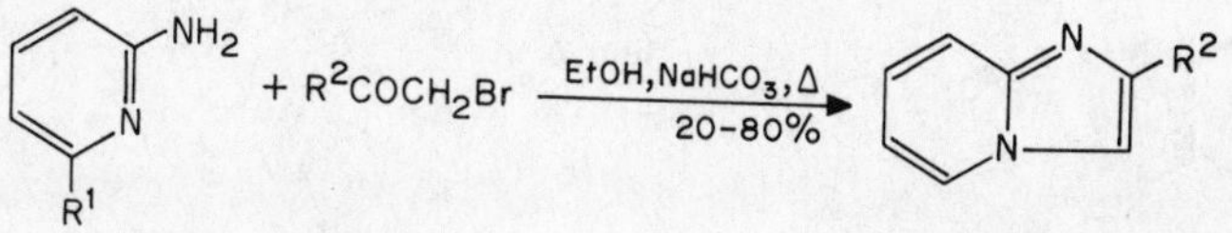

R^1=H,Me,Ph$(CH_2)_2$; R^2=Me,Ph,
3,5-tBu$_2$-4-HOC$_6$H$_2$

Imidazo[1,2-a]pyridine

[136, 1296]

Other ring systems synthesized similarly:

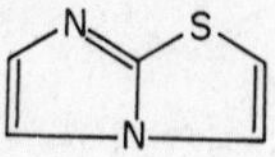

[930, 1300, 1877]

Imidazo[2,1-b]thiazole

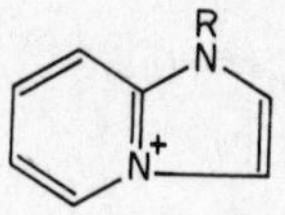

[1029]

Imidazo[1,2-a]pyridinium

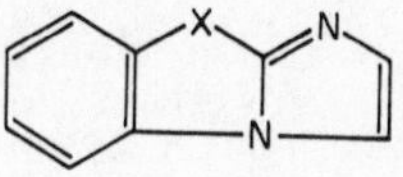

[1300, 1948]

X=O,S

Imidazo[2,1-b]benzothiazole
Imidazo[2,1-b]benzoxazole

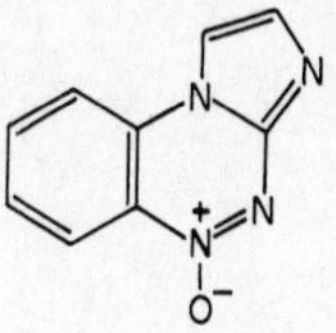

[1374]

Imidazo[1,2-c][1,2,4]-
benzotriazine 5-oxide

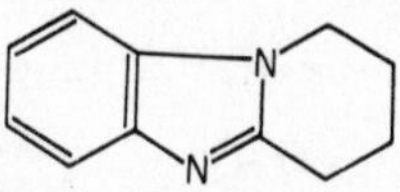

[136]

Pyrido[1,2-a]benzimidazole

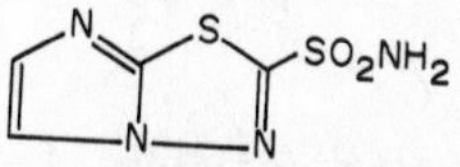

[580, 1877]

Imidazo[2,1-b]-1,3,4-
thiadiazole

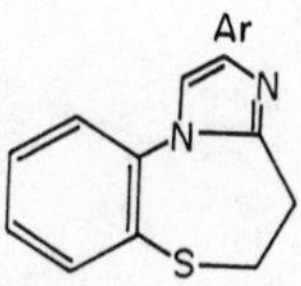

[1798]

Imidazo[2,1-d][1,5]-
benzothiazepine

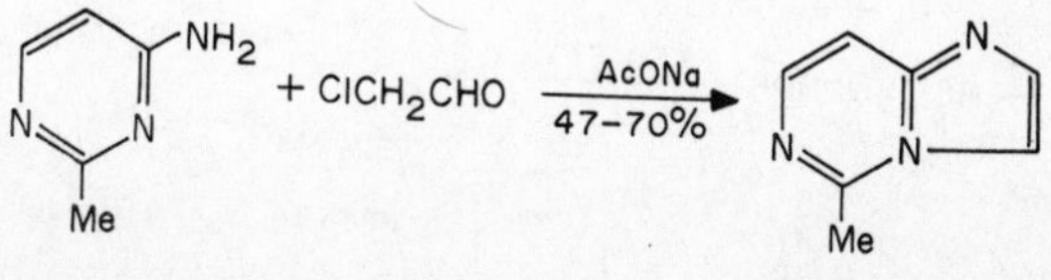

[1136]

Imidazo[1,2-c]pyrimidine

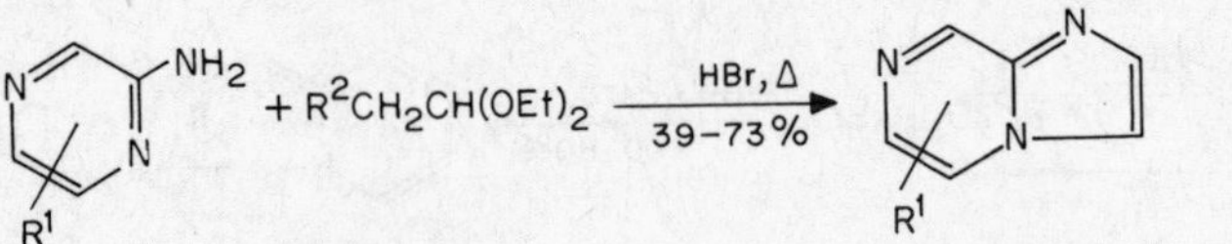

$R^1 = H, Br, Cl, Ph; R^2 = Br, Cl$

Imidazo[1,2-*a*]pyrazine

[89, 1353]

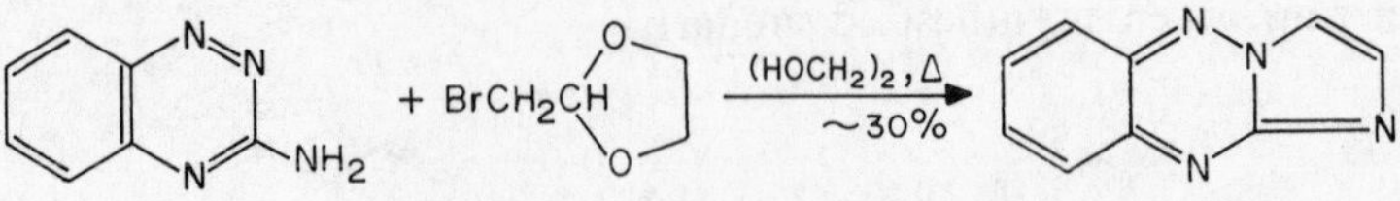

Imidazo[1,2-*b*][1,2,4]-
benzotriazine

[132]

Another ring system synthesized similarly:

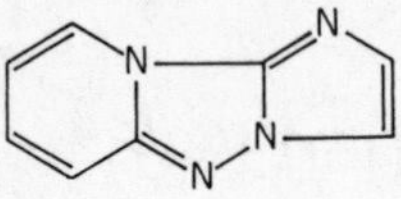

[1148]

Imidazo[1',2':1,5][1,2,4]-
triazolo[4,3-*a*]pyridine

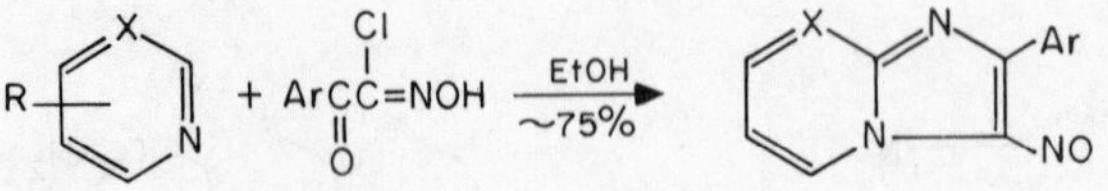

$R = H, Me; X = CH, N;$
$Ar = Ph, 4-NO_2C_6H_4$

X=CH,N
Imidazo[1,2-*a*]pyridine
Imidazo[1,2-*a*]pyrimidine

[1308]

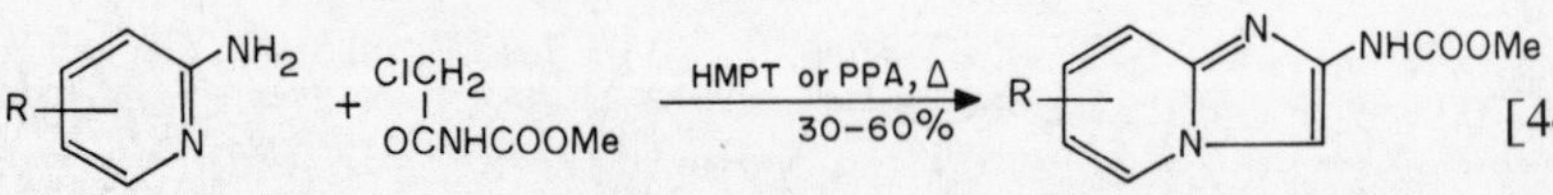

R=H,PhS,PhCONPh

Imidazo[1,2-*a*]pyridine

[447, 576]

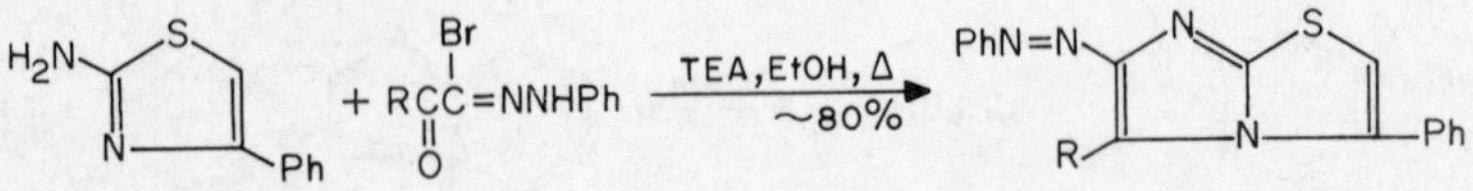

R=Me,Ph

Imidazo[2,1-b]thiazole

[527]

Other ring systems synthesized similarly:

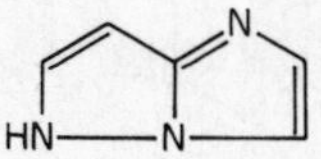

Imidazo[1,2-b]pyrazole   [64, 527]

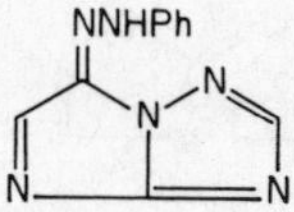

Imidazo[1,2-b][1,2,4]-   [527]
triazole

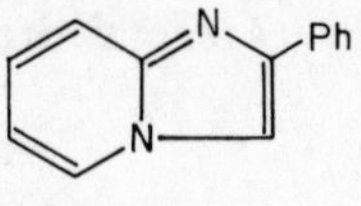

Imidazo[1,2-a]pyridine   [64]

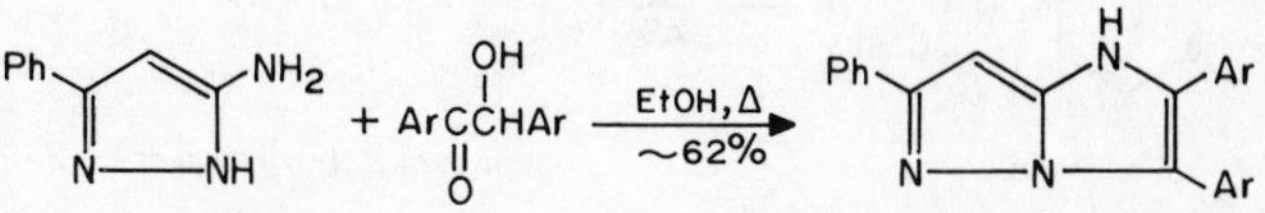

Ar = Ph, 4-MeOC$_6$H$_4$

Imidazo[1,2-b]pyrazole   [1305]
(review [1123])

Among other reagents which have been employed to annulate an amino-heterocycle to an imidazole are orthoesters, formic acid and p-quinone.

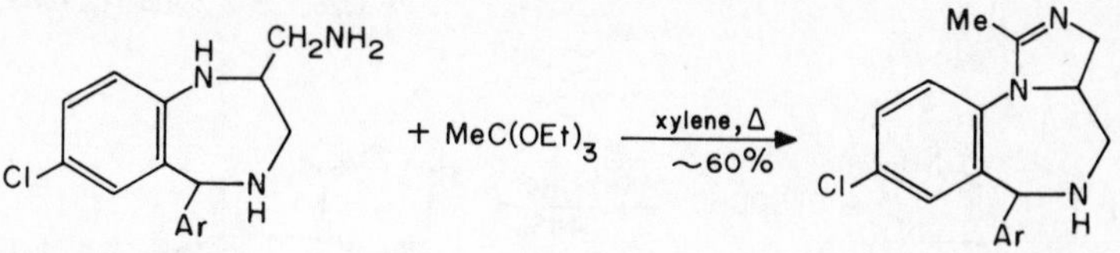

Ar = 2-FC$_6$H$_4$

Imidazo[1,5-a][1,4]-   [509]
benzodiazepine

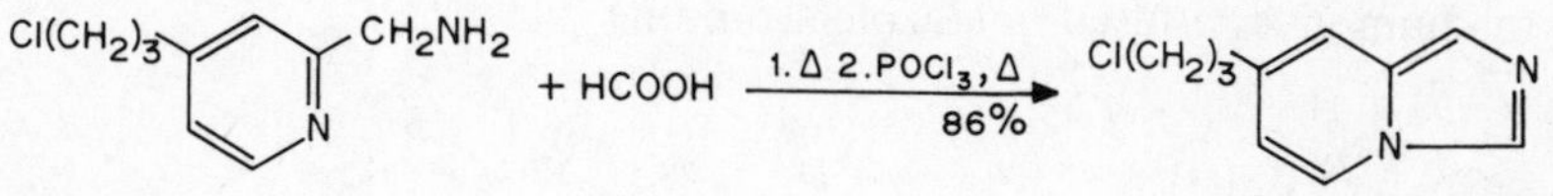

Imidazo[1,5-a]pyridine
[1575]

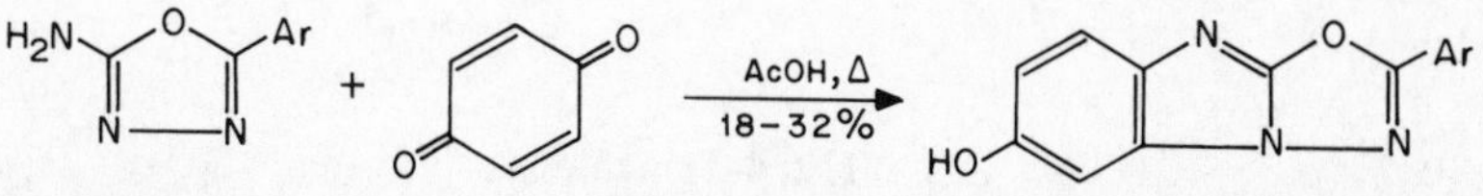

Ar = Ph, 4-Cl-, 4-Me-, 4-F C$_6$H$_4$;
3,5-Br$_2$-2-HOC$_6$H$_2$

1,3,4-Oxadiazolo[3,2-a]-
benzimidazole   [1105]

Other ring systems synthesized similarly:

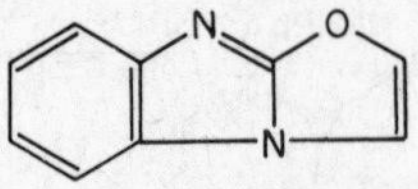

Oxazolo[3,2-α]benzimidazole [1103]

Imidazo[1,5-α]pyridine [1575]

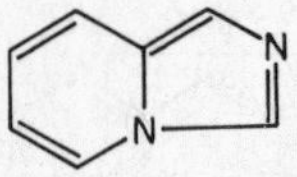

2. Imidazol-4-one or Imidazole-4, 5-dione

α-Halogeno-esters or α-oxoaldehydes react with aminoheterocycles to yield fused imidazol-4-ones.

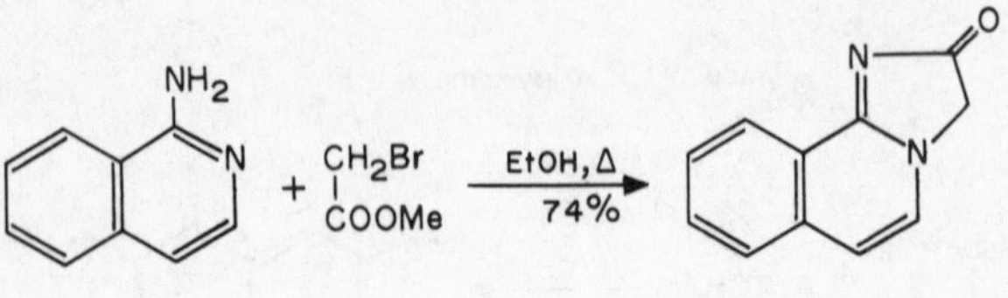

Imidazo[2,1-α]isoquinolin-2-one [1322]

Imidazo[1,2-α]pyridin-2-one [527]

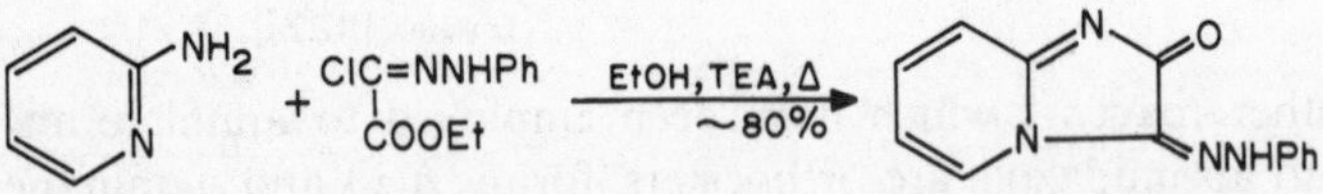

Imidazo[1,2-α]pyrazin-3-one [1071]

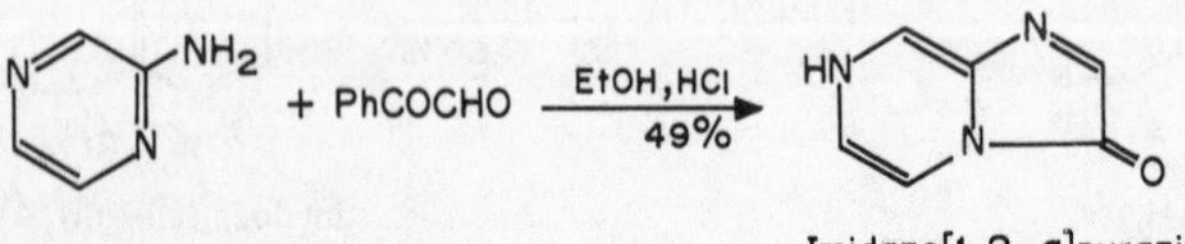

Reaction of 2-aminobenzimidazoles (review [1429]) with oxalyl chloride in a basic medium gives a fused imidazole-4, 5-dione.

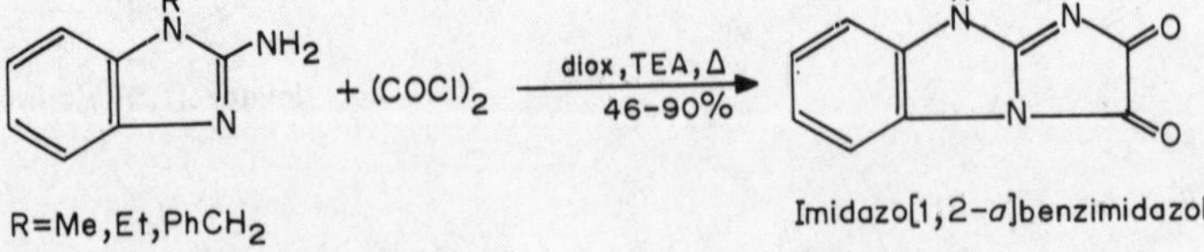

R=Me,Et,PhCH₂

Imidazo[1,2-α]benzimidazole-
2,3-dione [1321]

3. 1, 2, 4-Triazole

An amine heterocycle or its imino tautomer may be annulated by reaction with a hydrazide or an α-chlorohydrazone (reviews [1437, 1753]).

+ AcNHNH₂ →[HMPT, Δ][41%]

[652]

Thieno[3,2-f][1,2,4]triazolo-
[4,3-a][1,4]diazepin-4-one

+ PhC=NNHPh →[diox,TEA,Δ][60%]

[1819]

Pyrazolo[5,1-c]-1,2,4-triazole

Successive reaction with DMF (to form the —N=CHNMe₂), hydroxylamine (to give —NCH=NOH) and phosphorus oxychloride leads to a triazole but in moderate yields.

→[i, ii, iii][25-50%]

[1367]

R = H, Me
i, DMF; ii, NH₂OH; iii, POCl₃

1,2,4-Triazolo[1,5-b]-
[1,2,4]triazine

II. FORMATION OF A SIX-MEMBERED RING

1. Pyridazine

A 4-amino-1,2,4-triazole is cyclized by heating with a 1,3-diketone and a mild base.

+ Ac₂CH₂ →[AcONa,Δ][80%]

[567]

1,2,4-Triazolo[4,3-b]pyridazine

2. Pyrimidine

Many 1,3-diketones and 1,3-dialdehydes (sometimes protected from self-condensation as their diacetals) have been condensed azeotropically or with a base or acid in this way to form a fused pyrimidine ring. The readily prepared ethoxymethylene compounds condense in acetic acid while ketene thioacetals react on heating with a 2-aminoheterocycle such as amino-1,2,4-triazole. Although diethyl malonate yields a pyrimidin-4-one ring (see next section), the di-imidate gives a 4,6-diaminopyrimidine ring.

R=CF$_3$,alkyl
i,azeotropic distillation

Pyrimido[1,2-a]benzimidazole   [146]

R^1=H,Ph(CH$_2$)$_2$,4-COOHC$_6$H$_4$CH$_2$;
R^2=Me,Pr

[1,2,4]Triazolo[1,5-a]-
pyrimidine   [133, 134]

R=H,COOH,CN
i,EtOH,ZnCl$_2$ or HCl

Pyrazolo[1,5-a]pyrimidine
(review [1123])   [954, 1646]

R^1=H,CN,PhSO$_2$;R^2=CN,PhCO,COOMe;
R^3=NH$_2$,Ph,HO

[1,2,4]Triazolo[1,5-a]-
Pyrimidine   [1721]

R=H,Me,Ph

Pyrazolo[1,5-a]pyrimidine   [1684]

Other ring systems synthesized similarly:

X=CH,N
Pyrazolo[1,5-a]pyrimidine
[1,2,4]Triazolo[1,5-a]pyrimidine   [507, 529]

X=CH,N
[1,2,3]Triazolo[1,5-a]pyrimidine
Tetrazolo[1,5-a]pyrimidine   [529]

Imidazo[1,2-a]pyrimidine   [529]

Isoxazolo[2,3-a]pyrimidin-8-ium   [496]

Thieno[3″,2″:5′,6′]pyrido[4′,3′:3,4]
pyrazolo[1,5-a]pyrimidine

[1812]

R^1 = H, Me, NO_2; R^2 = H, alkyl,
$AcO(CH_2)_3$

Pyrazolo[1,5-a]pyrimidine

[1393]

Other ring systems synthesized similarly:

[530]

Imidazo[1,2-a]pyrimidine

[530]

Pyrimido[1,2-a]benzimidazole

In the above examples, the amino group has been attached to the *N*-containing ring but the two functions may be well-separated or be in different rings and a reagent which supplies one carbon atom, for example, an orthoester, an aldehyde [123] or cyanogen bromide [106], then forms a new pyrimidine ring.

R = H, Ph

[1351]

Imidazo[1,2-c]pyrazolo[4,3-e]-
pyrimidine (review [1542])

Other ring systems synthesized similarly:

[106, 123]

R = H_2 or NH, *n* = 2
Imidazo[1,2-c]quinazoline
R = NH, *n* = 3
Pyrimido[1,2-c]quinazoline
(review [1888])
R = NH, *n* = 4
[1,3]Diazepino[1,2-c]quinazoline

[1863]

Pyrimido[1,6-a]benzimidazole

3. Pyrimidin-4-one

The formation of this ring from a 2-aminoheterocycle has been widely applied (review [874]) and the most common type of reagent is an acyclic or cyclic 3-oxo-ester including diketene which acts as a source of acetoacetyl group. Another compound which has an active methylene group is ethyl 5-tetrazolylacetate in conjunction with triethyl orthoformate.

The chemistry of pyrido[1,2-a]pyrimidines, which are accessible by this route, has been reviewed [1693]; reactions of 3(5)-aminopyrazoles have also been surveyed [1123].

R^2—pyridine-NH_2, R^1 + OCR^3, CHR^4, $COOEt$ → $\xrightarrow[33-96\%]{i,\Delta}$ Pyrido[1,2-a]pyrimidin-4-one (R^2, R^3, R^4, O) [281, 410, 841, 1383]

$R^1 = Me, NH_2; R^2 = H, Me, Cl, HO;$
$R^3 = alkyl, aryl, MeOOCCH_2;$
$R^4 = H, Me, Cl, EtOOCCH_2$
i, PPA or $Et_2C_6H_4$

Pyrido[1,2-a]pyrimidin-4-one

pyridine-NH_2, R + $EtOOC$ cyclopentanone $\xrightarrow[46-60\%]{\Delta}$ Cyclopenta[d]pyrido[1,2-a]pyrimidin-10-one [1241]

R = Me, HO, COOH

Cyclopenta[d]pyrido[1,2-a]-pyrimidin-10-one

HN-pyrazole-R^1, H_2N, R^2 + OCR^3, CH_2COOEt $\xrightarrow[31-92\%]{i,\Delta}$ Pyrazolo[1,5-a]pyrimidin-7-one [446, 507, 845, 1305, 1785, 1819]

$R^1 = Me, NH_2; R^2 = H, Br, aryl, COOEt;$
$R^3 = H, Me, Ph$
i, EtOH or AcOH

Pyrazolo[1,5-a]pyrimidin-7-one

Other ring systems synthesized similarly:

Pyrido[1,2-b]quinazolin-6-one [409]

Pyrimido[1,2-a]benzimidazol-4-one [1395]

[1988]

[1,2,4]Triazolo[1,5-a]-
pyrimidin-7-one

[1962]

Imidazo[1,2-a]pyrimidin-5-one

[1545]

Pyrimido[1,2-b][1,2,4]-
triazin-8-one

Pyrimido[2,1-b]benzoxazol-
4-one

[579]

The use of αβ-unsaturated nitriles in the synthesis of heterocycles has been reviewed [1096] and an example is given. Ethoxymethylenemalonate condenses with cyclic amidines at low temperature to give compounds containing a pyrimidin-4-one ring. Activated alkynes have been used to cyclize several aminoheterocycles and a review [1725] of such reactions was published in 1978. Finally, a combined cyclization–chlorination using diethyl malonate, PPA and phosphorus oxychloride has given a fused chloropyrimidin-4-one ring.

[528]

Pyrazolo[1,5-a]pyrimidin-5-one

[336]

Pyrrolo[1,2-a]pyrimidin-4-one
Pyrido[1,2-a]pyrimidin-4-one
Pyrimido[1,2-a]azepin-4-one

Another ring system synthesized similarly:

[1745]

Pyrimido[2,1-*b*]benzothiazol-4-one

H_2N — (imidazole with N-Me, NO$_2$) + $\begin{array}{c} CCOOEt \\ ||| \\ CCOOEt \end{array}$ $\xrightarrow[\sim 40\%]{MeOH,\Delta}$ EtOOC / EtOOC ... NO$_2$ [337]

Imidazo[1,2-*a*]pyrimidin-5-one

H_2N ... $\begin{array}{c} X \\ Y \end{array}$ R^1 + R^{2}C≡CCOOEt $\xrightarrow[44-60\%]{\Delta \text{ or THF,18 °C}}$ ... [1682, 1961]

R^1=H,Ph,4-Cl-,4-MeO-C$_6$H$_4$
R^2=H,COOMe,Ph

X=S,Y=CH
Thiazolo[3,2-*a*]pyrimidin-7-one

X=O,Y=N
1,3,4-Oxadiazolo[3,2-*a*]pyrimidin-7-one

X=S,Y=N
1,3,4-Thiadiazolo[3,2-*a*]pyrimidin-7-one

(benzo-fused X—NH$_2$) + $\begin{array}{c} CCOOMe \\ ||| \\ CCOOMe \end{array}$ $\xrightarrow[52-70\%]{MeOH \text{ or diox},\Delta}$ ... COOMe [901, 947, 1522]

X=NMe
Pyrimido[1,2-*a*]benzimidazol-2-one

X=O
Pyrimido[2,1-*b*]benzoxazol-2-one

X=S
Pyrimido[2,1-*b*]benzothiazol-2-one

H_2N ... (thiazole with R) + CH$_2$(COOEt)$_2$ $\xrightarrow[15-60\%]{POCl_3,PPA,\Delta}$ Cl ... R [931]

R=Me,4-Me-,4-MeO-,
4-Cl-C$_6$H$_4$

Thiazolo[3,2-*a*]pyrimidin-5-one

4. Pyrimidine-4, 6-dione

When an aminoheterocycle is heated with either a malonic ester (sometimes with HMPT) or a malonyl dichloride–triethylamine mixture, a new pyrimidine-4, 6-dione ring is formed.

Imidazo[1,2-*a*]pyrimidine-
5,7-dione

[797]

R=H,alkyle

Pyrimido[1,2-*a*]benzimidazole-
2,4-dione

[685]

Other ring systems synthesized similarly:

[798]

X = NMe

Dibenzo[*bf*]pyrimido[1,2-*d*][1,4]diazepine-2,4-dione

X = S

Dibenzo[*bf*]pyrimido[1,2-*d*][1,4]thiazepine-2,4-dione

5.   1, 3, 5-Triazine

A new 1, 3, 5-triazine ring may be fused to an aminoheterocycle by reaction with ethoxymethylenecyanamide while an amino group attached to another ring may be annulated using an orthoester as a source of the necessary carbon atom. The chemistry of 4-amino-1, 2, 3-triazoles (used as substrates [1874]) has been reviewed [2027].

$$[147]$$

R = H, Me, Et, Ph

[1,2,4]Triazolo[1,5-a]-
[1,3,5]triazine

Other ring systems synthesized similarly:

[147]

1,3,5-Triazino[1,2-a]-
benzimidazole

[1784]

[1,2,3]Triazolo[1,5-a]-
[1,3,5]triazine

[1240]

Pyrazolo[1,5-a]-
1,3,5-triazine

[1729]

Pyrimido[1,2-a]-1,3,5-
triazin-6-one

[1062]

R = H, Me

Imidazo[1,2-e][1,2,4]triazolo-
[1,5-a][1,3,5]triazine

Other ring systems synthesized similarly:

[1239]

Pyrazolo[1',5':3,4][1,3,5]triazino-
[1,2-b][1,2,4]triazin-9-one

[753]

Pyrazolo[1',5':3,4][1,3,5]-
triazino[1,2-a]benzimidazole

6. 1,3,5-Triazinone or -Triazinedione

A 1,3,5-triazin-2-one ring is formed in variable yields depending on the substituents on the precursor amine by reaction with the carbamoyl chloride, PhC(Cl)=NCOCl, in boiling dry toluene.

[915]

Pyridazino[2,3-a]-1,3,5-
triazin-4-one

Alkyl, aryl, alkoxy- or phenoxy-carbonyl isocyanates or isothiocyanates are versatile cyclizing reagents and have been used on aminoheterocycles. The use of alkoxycarbonyl isothiocyanates in cyclization reactions has been reviewed [860].

[508]

Imidazo[1,2-a]-1,3,5-
triazine-2,4-dione

Other ring systems synthesized similarly:

[798]

X=S
Dibenzo[bf]-1,3,5-triazino-
[1,2-d][1,4]thiazepine-2,4-dione
X=NMe
Dibenzo[bf]-1,3,5-triazino-
[1,2-d][1,4]diazepine 2,4-dione

[1836]

Pyrrolo[1,2-a]-1,3,5-
triazine-2,4-dione

[859]

Thiazolo[3,2-a]-1,3,5-
triazin-2-one, 4-thioxo-

III. FORMATION OF A SEVEN-MEMBERED RING

1. 1,3-Diazepine or 1,3-Diazepin-2-one

Aminomethyl and ring-nitrogen functions may be linked together and form a 1,3-diazepine by either formaldehyde or phosgene.

[1243]

Imidazo[2,1-a][2,4]-
benzodiazepine

Imidazo[2,1-a][2,4]-
benzodiazepin-5-one

2. 1,4-Diazepine or 1,4-Diazepin-2-one

An amine and a ring-nitrogen in different rings react with a difunctional reagent such as bromoacetyl chloride in TEA to give a doubly fused diazepinone. A similar ring is formed in a reaction between an arylamine and an N-bromoethyl lactam carrying a displaceable ethoxy group.

R^1=H,Ph; R^2=H,MeO,CF$_3$

[1242]

Tetrazolo[1,5-d][1,4]-
benzodiazepin-6-one

R=H,MeO,Cl

[141]

Pyrrolo[1,2-d][1,4]-
benzodiazepin-3-one

3. 1,3,5-Thiadiazepine

An extension of the method of cyclizing an imidazolylbenzylamine mentioned in Section III.1 to the substrate in which the two rings are joined through a sulphur atom and the amino group is directly attached to the benzene ring gives a thiadiazepine ring.

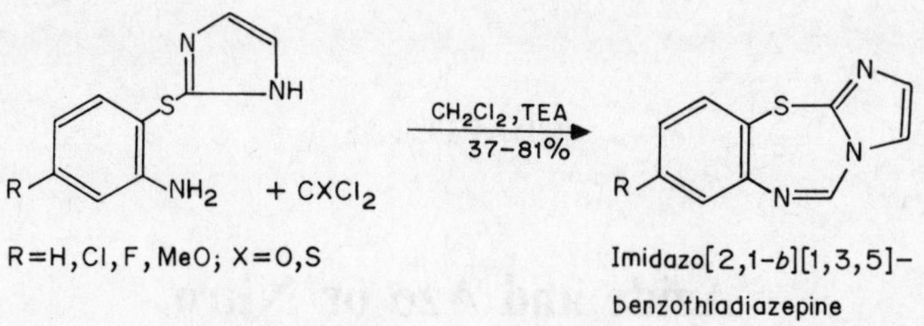
CH2Cl2,TEA
37–81%
+ CXCl2
R=H,Cl,F,MeO; X=O,S
Imidazo[2,1-b][1,3,5]-
benzothiadiazepine

[861]

IV. FORMATION OF AN EIGHT-MEMBERED RING

1. 1,3,5-Triazocin-6-one

When an amino and endocyclic nitrogen are suitably separated in different rings, a triazocine ring may be formed by reaction with an orthoester.

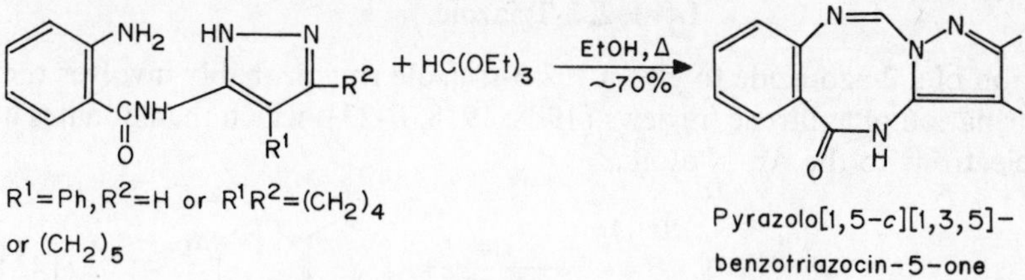
+ HC(OEt)3
EtOH, Δ
~70%
R1=Ph, R2=H or R1R2=(CH2)4
or (CH2)5
Pyrazolo[1,5-c][1,3,5]-
benzotriazocin-5-one

[1352]

CHAPTER 55

Azide and Azo or Nitro

I. FORMATION OF A FIVE-MEMBERED RING

1. 1,2,3-Triazole

Cyclization of a 2-azo-azide to give a 1,2,3-triazole ring probably involves the initial formation of a nitrene (reviews [1908, 1918, B-23]) which then donates a pair of electrons to the Ar–N atom.

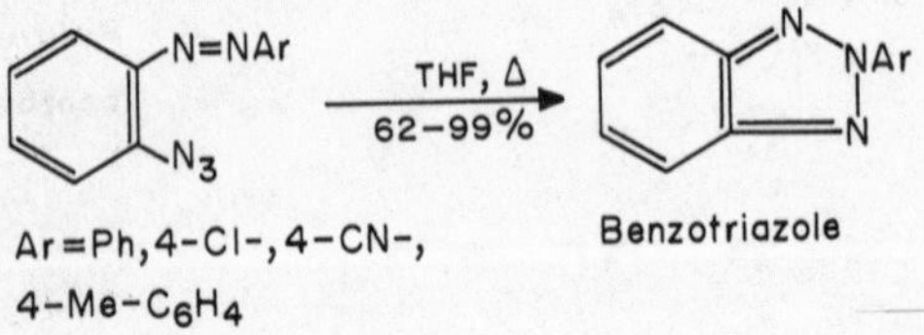

Ar = Ph, 4-Cl-, 4-CN-, 4-Me-C₆H₄

Benzotriazole [43]

2. Furazan or Furazan *N*-Oxide

Benzofurazans are conveniently prepared in high yields from 2-nitro-azides and a deoxygenating agent. Heating these substrates in benzene or DMF can also give benzofurazan 1-oxides (reviews [1690, 1726]).

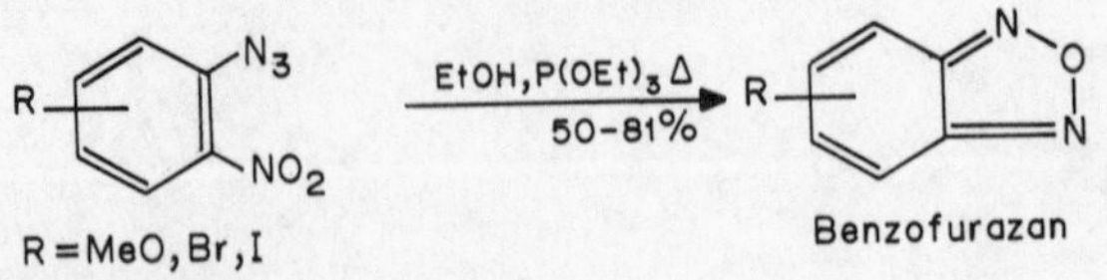

R = MeO, Br, I

Benzofurazan [1609, 1677]

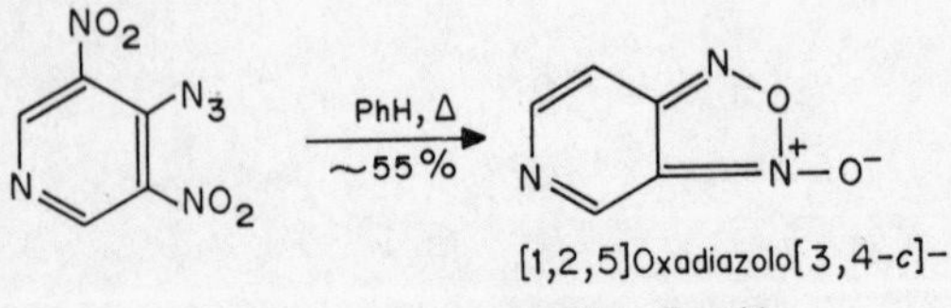

[1,2,5]Oxadiazolo[3,4-c]-pyridine 3-oxide [1495]

Another ring system synthesized similarly:

[1,2,5]Oxadiazolo[3,4-d]-
pyrimidine-5,7-dione 1-oxide

[315]

CHAPTER 56

Azide and a Carboxylic Acid or its Derivative

Carboxamide, ester and nitrile are included as derivatives of a carboxylic acid.

I. FORMATION OF A FIVE-MEMBERED RING

1. Pyrrole

Thermolysis of azides is an important reaction in the synthesis of heterocycles (reviews [1908, 1918, B-23]). When such a group is adjacent to an ester, carbazoles are obtained in low to moderate yield. A pyrrole ring is also formed from a 2-azidocinnamic acid but under reducing conditions and low temperature o-azido-acrylic esters behave differently (see Section II.1).

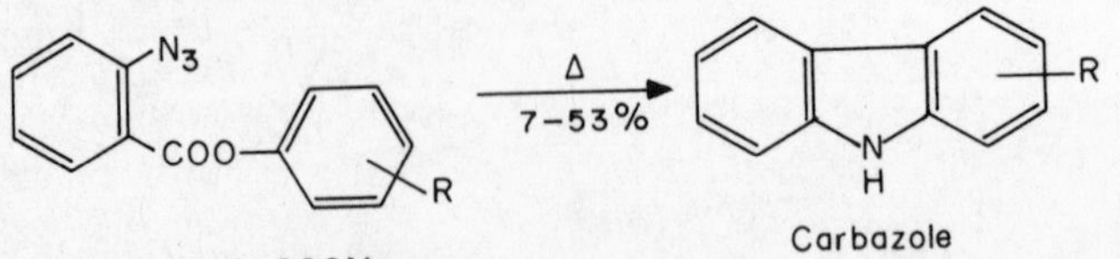

R=H,Me,Br,Cl,COOMe

Carbazole
(review [1695])

[897]

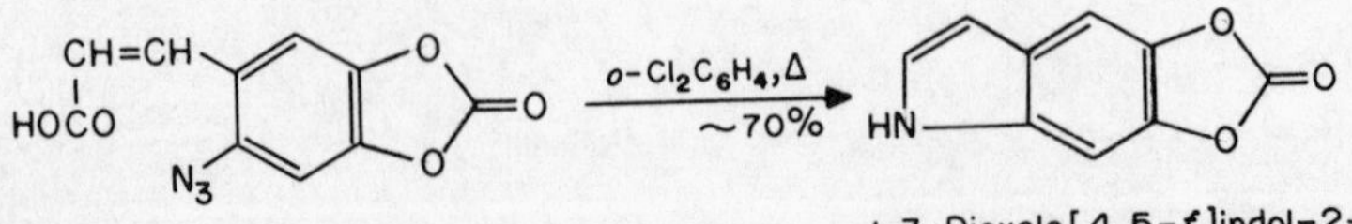

1,3-Dioxolo[4,5-f]indol-2-one

[1316]

294

2. Pyrazole

2-Azido-carboxamides are cyclized either to a fused pyrazol-2-one ring on heating with a base in DMF or to a pyrazole by hot thionyl chloride. 2-Azido-nitriles may be converted into fused pyrazoles either by conversion into the imidate followed by pyrolysis, or by heating with hydrazine in ethanol or DMF (with simultaneous elimination of RNH). Azido-esters also undergo the latter cyclization.

Indazol-3-one   [377]

$Ar = Ph, Me-, Cl-,$
$MeO-, NO_2-C_6H_4$

Indazole   [728, 729]

[728, 729]

$R = H, Me$ Indazole   [1335]

Another ring system synthesized similarly:

[217]

Imidazo[4,5-f]indazol-3-one

II. FORMATION OF A SIX-MEMBERED RING

1. Pyridin-2-one

As mentioned in Section I.1, o-azido-cinnamic esters are converted into fused pyridin-2-ones under mildly reducing conditions.

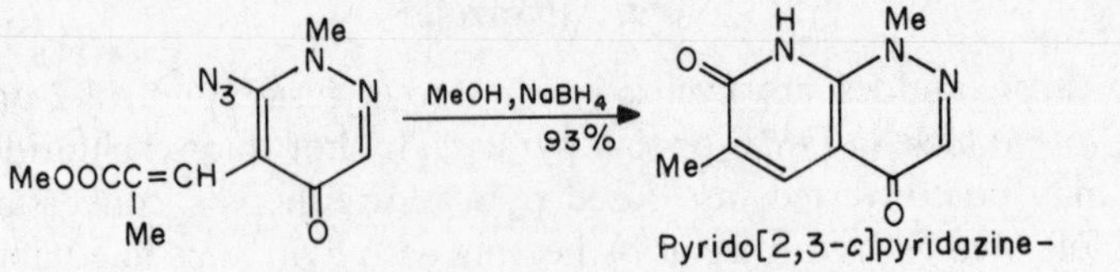

Pyrido[2,3-c]pyridazine-
4,7-dione

[382]

2. Pyrimidine or Pyrimidine-2, 4-dione

A 2-azido-nitrile reacts (in a complex fashion) with an anion (from malonic acid derivatives) to form a fused pyrimidine ring.

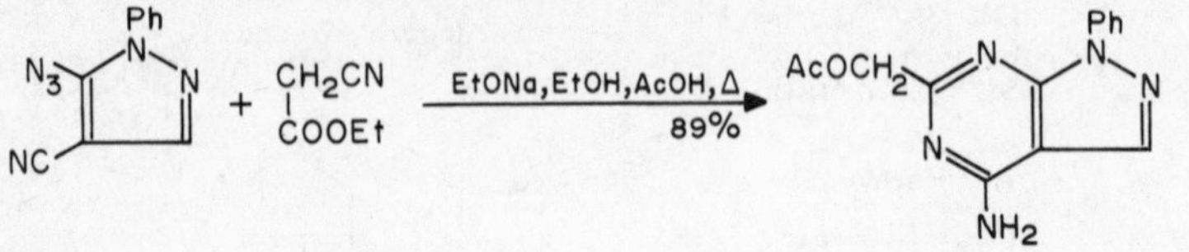

Pyrazolo[3,4-d]pyrimidine

[140]

Thermolysis of an azidocarbonyl group (through a Curtius reaction) adjacent to a carboxamide leads to the formation of a fused pyrimidinedione ring in high yield.

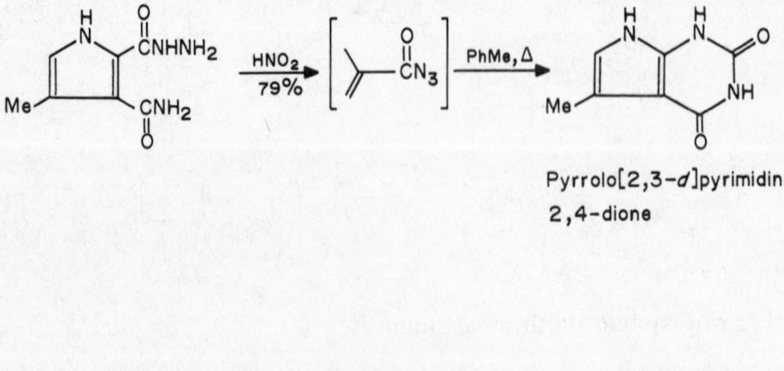

Pyrrolo[2,3-d]pyrimidine-
2,4-dione

[705]

CHAPTER 57

Azide and Methyl, Methylene or Methine

I. FORMATION OF A FIVE-MEMBERED RING

1. Pyrrole or Pyrrol-3-one

Most of the reactions described in this chapter require the azide to be heated and therefore a nitrene is a likely intermediate. This explains the retention of only one of the three nitrogen atoms of the azide group (reviews [1908, 1918, B-23]). A fused pyrrole ring may be formed from nuclear or side-chain azides and some unexpected products have been isolated.

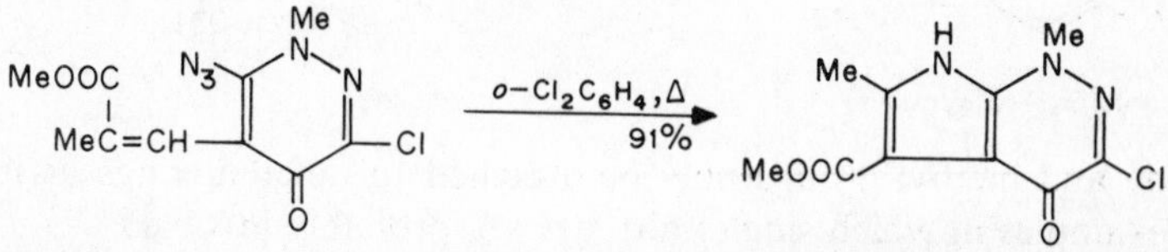

Pyrrolo[2,3-*c*]pyridazin-4-one [979]

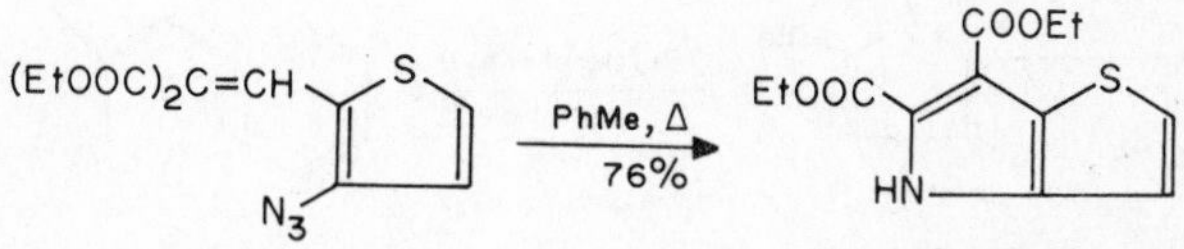

Thieno[3,2-*b*]pyrrole (review [1663]) [1008]

Another ring system synthesized similarly:

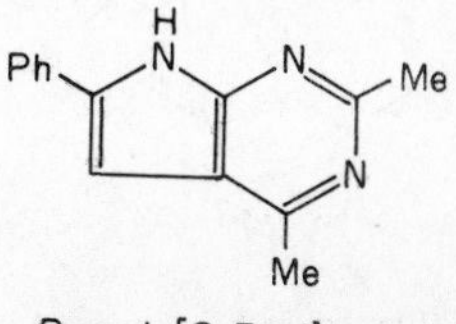

Pyrrolo[2,3-*d*]pyrimidine [317]

Loss of nitrogen does not always require thermolysis; basic conditions at ambient temperature may be sufficient to form a pyrrol-3-one ring.

R=Me,Et,(CH₂)₄   Indol-3-one   [378]

II. FORMATION OF A SIX-MEMBERED RING

1. Pyridine

A nitrene from a side-chain azide may attack an adjacent nuclear methyl to give a pyridine ring. The presence of iodine as oxidant frequently increases the yield.

Thieno[3,2-c]pyridine   [258]

Other ring systems synthesized similarly:

Pyrazolo[3,4-c]pyridine   [258]   Isoquinoline   [258, 1303]

The azide and methyl groups may be attached to different rings as in the following examples in which singlet nitrenes are probably involved.

Thieno[3,2-c]quinoline   [1937]

Another ring system synthesized similarly:

Phenathridine   [1937]

CHAPTER 58

Azide and Ring-carbon

In this chapter the reactions of both azides and acylazides with an adjacent ring-carbon (which may or may not be in the same ring) are described. Some of these proceed through the reactive nitrene and two of the three nitrogen atoms are lost. The reactions of nitrenes have been reviewed [1908, 1918, B-23]. An example of two azide groups reacting together is included.

I. FORMATION OF A FIVE-MEMBERED RING

1. Pyrrole

Pyrolysis of an azide, RN_3, frequently produces the reactive nitrene, $R\ddot{N}$: which immediately attacks a nearby position of the same or another ring. The azide is usually heated in an inert high-boiling solvent although cyclization may occur without a solvent. Photochemically induced cyclization of an azide can often give a good yield of product.

$$\text{EtOOCC=CH} \quad \xrightarrow[\sim 96\%]{\text{xylene}, \Delta} \quad \text{EtOOC} \qquad \qquad [849]$$

R = Me, Br

Furo[3,2-*b*]pyrrole

299

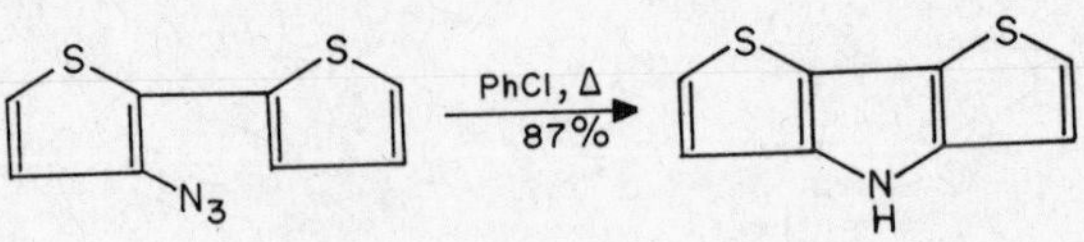

PhCl, Δ
87%

[730]

Dithieno[3,2-b:2',3'-d]pyrrole

MeOH, hν
75%

[81]

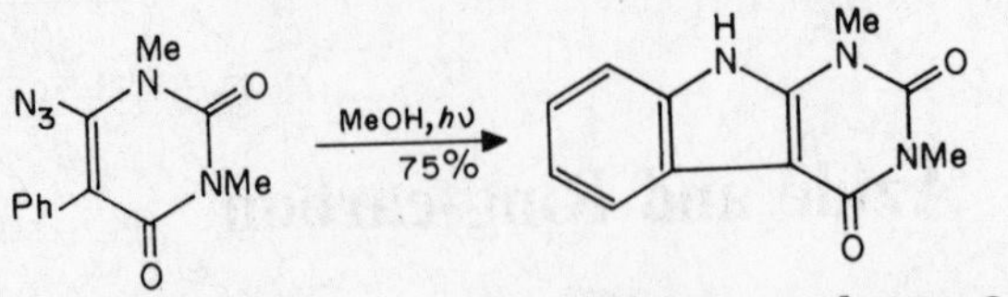

Pyrimido[4,5-b]indole-2,4-dione

Other ring systems synthesized similarly:

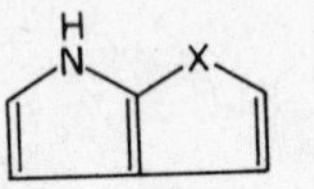

[25, 997]

X=NH,O,S
Pyrrolo[2,3-b]pyrrole
Furo[2,3-b]pyrrole
Thieno[2,3-b]pyrrole
(review [1663])

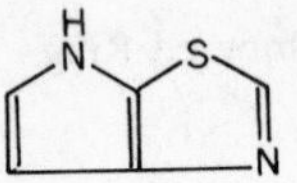

[939]

Pyrrolo[3,2-d]thiazole

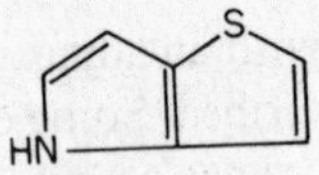

[997]

Thieno[3,2-b]pyrrole
(review [578])

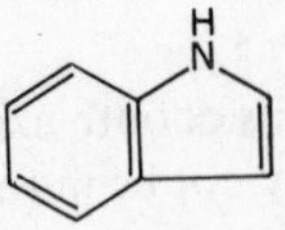

[1011]

Indole

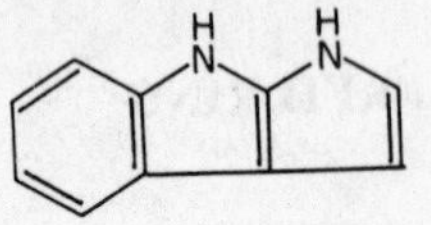

[1304]

Pyrrolo[2,3-b]indole
(review [921])

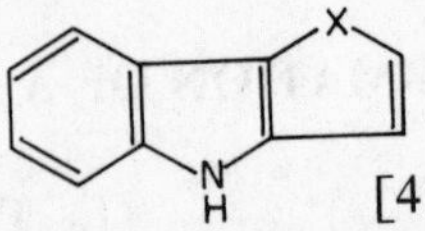

[413, 730, 1244, 1938]

X=O,S
Furo[3,2-b]indole
Thieno[3,2-b]indole

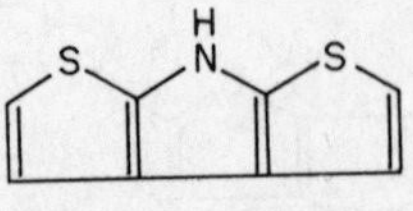

[730]

Dithieno[2,3-b:2',3'-d]-
pyrrole

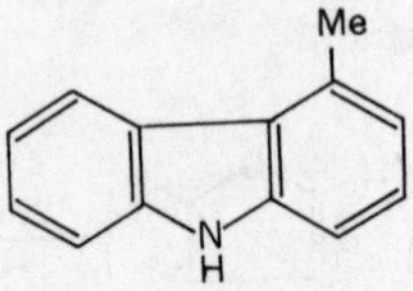

[1937]

Carbazole
(review [1695])

A detailed study of the effect of varying the substituents on the benzene ring showed that this reaction can give an indole or a mixture of indole and isoquinoline [258] but the following reaction gave indoles only:

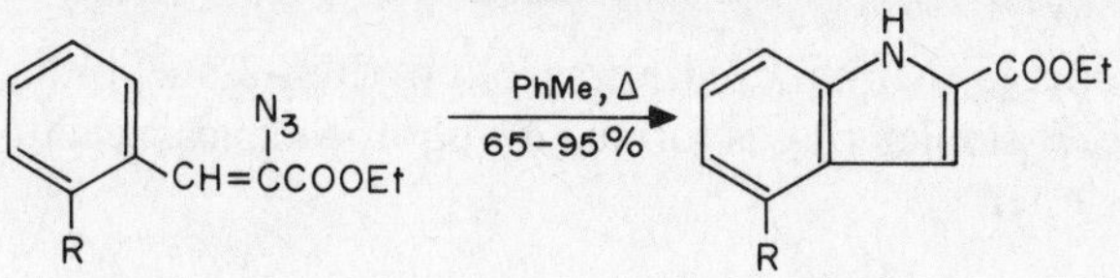

R=Me,iPr,PhCH₂O

Indole
(review [2008])

[258]

2. Isoxazole

Heating an azide with either acetic anhydride and zinc chloride or Vilsmeier reagents gives a new isoxazole ring.

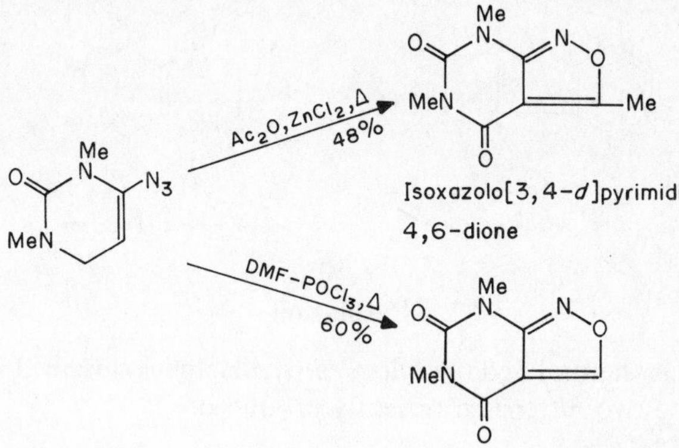

Ac₂O,ZnCl₂,Δ
48%

[Isoxazolo[3,4-*d*]pyrimidine-
4,6-dione

[81]

DMF-POCl₃,Δ
60%

[81]

3. Oxazole

Heating an azide in PPA and acetic acid gives a new oxazole ring in moderate to high yields.

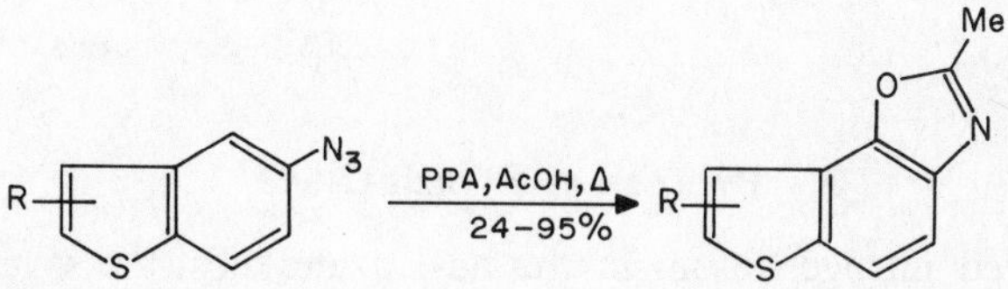

PPA,AcOH,Δ
24-95%

R=H,Me,COOMe

Thieno[2,3-*g*]benzoxazole

[1297]

Another ring system synthesized similarly:

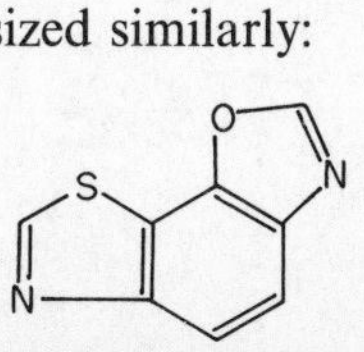

[1494]

Thiazolo[4,5-*g*]benzoxazole

II. FORMATION OF A SIX-MEMBERED RING

1. Pyridin-2-one

Acyl azides undergo thermal cyclization to a pyridin-2-one in the presence of a tertiary amine. A similar ring is formed by photochemical reaction of an azide with a glycine ester.

[5]

Furo[2,3-c]pyridin-7-one

Other ring systems synthesized similarly:

[1415]

[840]

Furo[3,2-c]pyridin-4-one

2,4,6-Pteridinetrione

2. Pyridazine

When a diazide is thermolysed (in xylene) a pyridazine is obtained, possibly by interaction of the two nitrenes transiently produced.

[1008]

X=O,S
Furo[3,2-c]pyridazine
Thieno[3,2-c]pyridazine

3. Pyrazine or Pyrazin-2-one

A suitably placed nitrogen atom in the acyl azide substrate gives rise to a pyrazinone.

[585]

Pyrrolo[1,2-a]thieno-
[3,2-e]pyrazin-5-one

Another ring system synthesized similarly:

[1783]

X = CH, N

Pyrido[2,3-e]pyrrolo[1,2-a]pyrazin-6-one
Pyrazino[2,3-e]pyrrolo[1,2-a]pyrazin-6-one

Two molecules of an azide on heating in formamide can produce a tricyclic product containing a new pyrazine ring.

[335]

Pyrimido[5,4-g]pteridine-
2,4,6,8-tetraone

4. 1, 2, 4-Triazine

Photochemical cyclization of an azide in the presence of an acid hydrazide can give good yields of a ring system containing a newly-formed 1, 2, 4-triazine.

[840]

R = H, Me, PhCH₂, Ph,
4-Me-pyridyl

Pyrimido[5,4-e]-1,2,4-
triazine-5,7-dione

5. 1, 4-Thiazine 1, 1-Dioxide

Triethyl phosphite converts a 2-azidodiphenyl sulphone into a phenothiazine dioxide.

[1298]

Phenothiazine 5,5-dioxide

CHAPTER 59

Azide and Ring-nitrogen

I. FORMATION OF A FIVE-MEMBERED RING

1. Pyrazol-3-one

Treatment of an acid hydrazide with a two-fold excess of nitrous acid gives an azide which can annulate on heating with a neighbouring ring-nitrogen to form a pyrazolone. A larger excess of nitrous acid gives a mixture of the pyrazolone and a degradation product of the hydrazide.

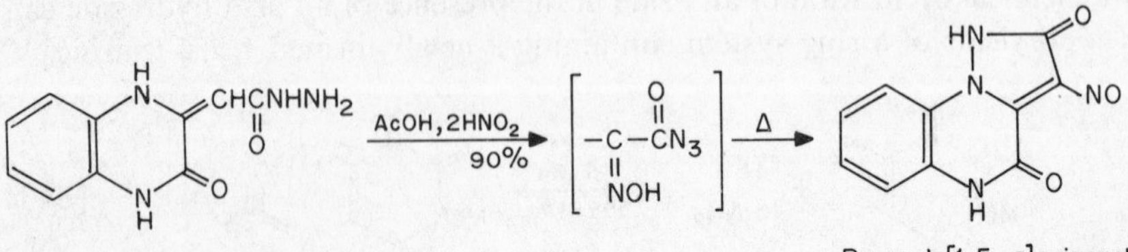

Pyrazolo[1,5-*a*]quinoxaline-2,4-dione

[1533]

2. Tetrazole

An azide group (formed by diazotization of a hydrazine group or by displacement of a reactive halogen) adjacent to an endocyclic nitrogen atom readily forms a 1,2-fused tetrazole ring. DMF or acetic acid may be added as a solvent.

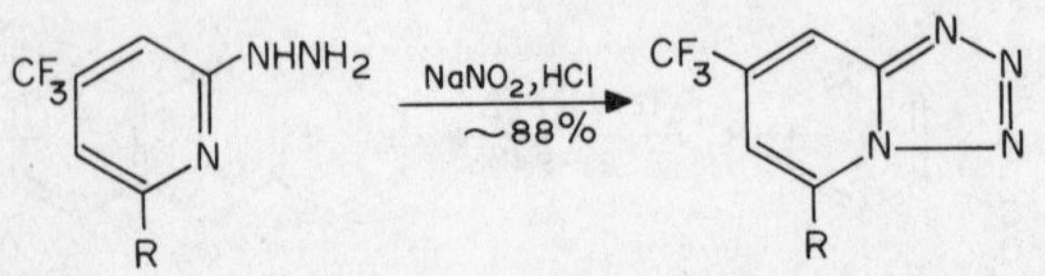

[963]

R=Me, Ph, 2-thienyl

Tetrazolo[1,5-*a*]pyridine

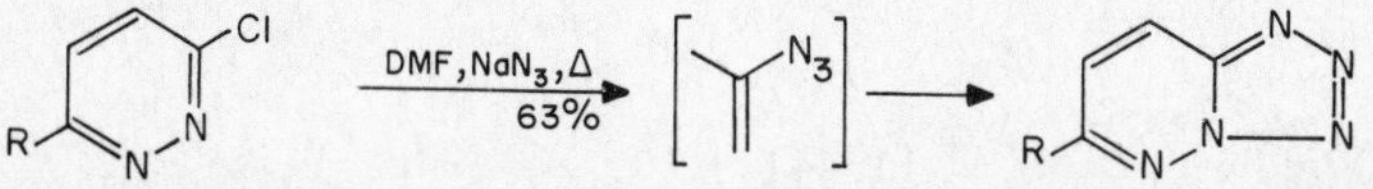

R = 5-Br-2-thienyl

Tetrazolo[1,5-b]pyridazine [618]

Other ring systems synthesized from a hydrazine:

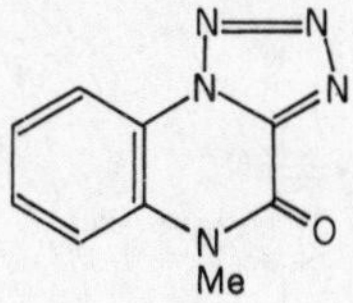

[1851]

Tetrazolo[1,5-a]quinoxalin-
4-one

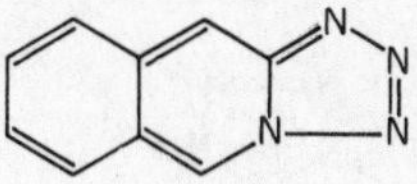

[1163]

Tetrazolo[1,5-b]isoquinoline

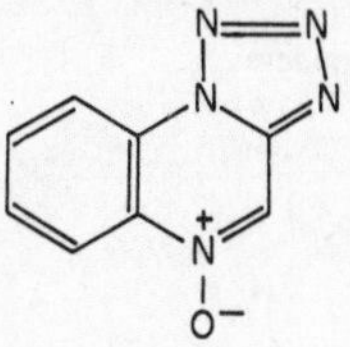

[279]

Tetrazolo[1,5-a]quinoxaline-
5-oxide

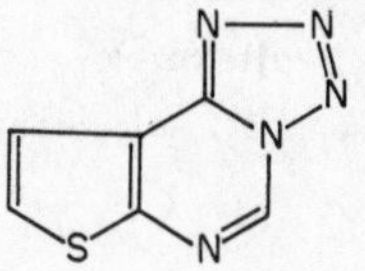

[1343]

Tetrazolo[1,5-c]thieno-
[3,2-e]pyrimidine

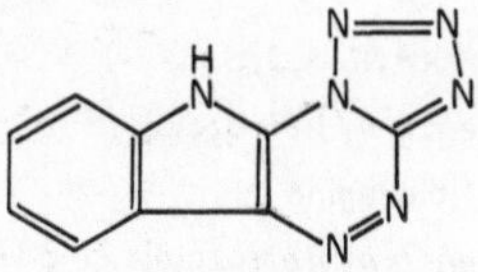

[484]

Tetrazolo[5',1':3,4][1,2,4]-
triazino[5,6-b]indole

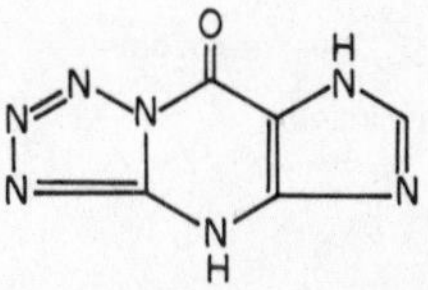

[1782]

Tetrazolo[1,5-a]purin-9-one

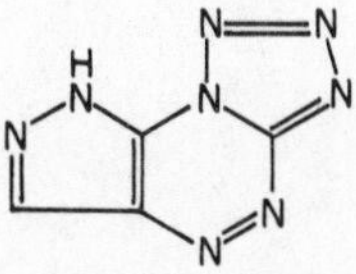

[1474]

Pyrazolo[3,4-e]tetrazolo-
[5,1-c][1,2,4]triazine

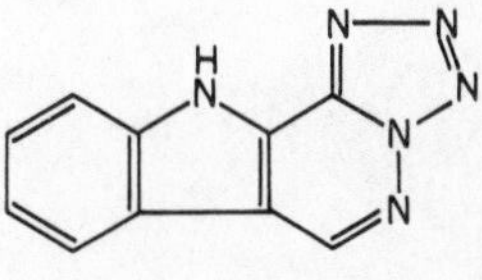

[799]

Tetrazolo[1',5':2,3]pyridazino-
[4,5-b]indole

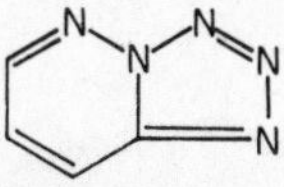

[618]

Tetrazolo[1,5-b]-
pyridazine

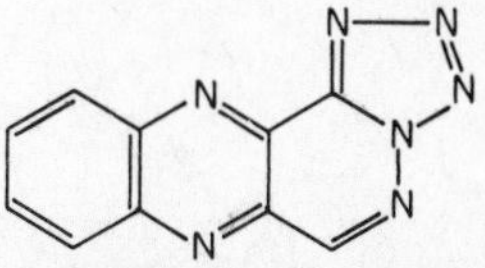

[1460]

Tetrazolo[1',5':1,6]pyridazino-
[4,5-b]quinoxaline

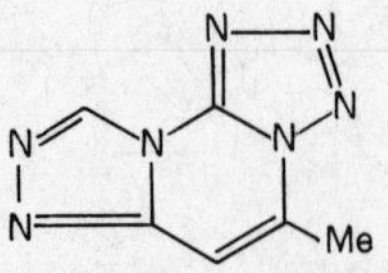

[1344]

Tetrazolo[1,5-*a*][1,2,4]-
triazolo[4,3-*c*]pyrimidine

Other ring systems synthesized from halide:

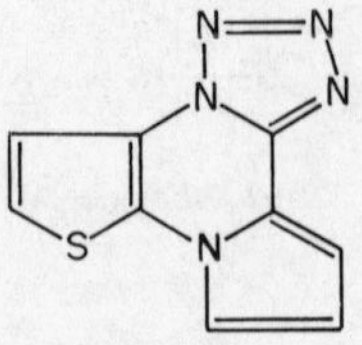

[585]

Pyrrolo[1,2-*a*]tetrazolo-
[5,1-*c*]thieno[3,2-*e*]pyrazine

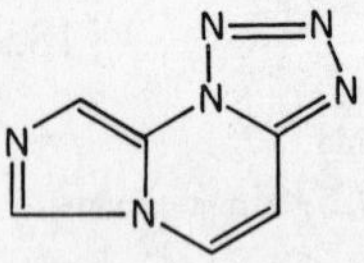

[702]

Imidazo[1,5-*a*]tetrazolo-
[1,5-*c*]pyrimidine

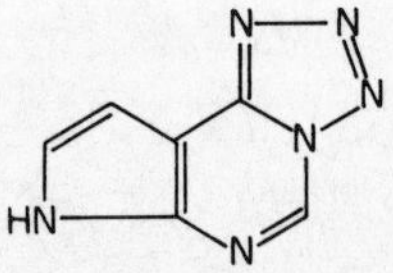

[1361]

Pyrrolo[3,2-*e*]tetrazolo-
[1,5-*c*]pyrimidine

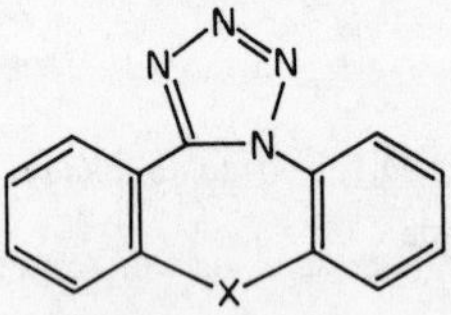

[1362]

X=NMe,O,S
Dibenzo[*bf*]tetrazolo[1,5-*d*]-
[1,4]diazepine
Dibenzo[*bf*]tetrazolo[1,5-*d*]-
[1,4]oxazepine
Dibenzo[*bf*]tetrazolo[1,5-*d*]-
[1,4]thiazepine

CHAPTER 60

Azo or Triazenyl and Carbamate, Carboxylic Acid, Ester or Nitrile

I. FORMATION OF A SIX-MEMBERED RING

1. Pyrimidin-4-one

When a 2-triazenyl-nitrile is heated in formamide, the final product is a fused pyrimidin-4-one; the mechanism proposed involves a number of intermediates.

Ar = 4-CN-, 4-NO$_2$-C$_6$H$_4$

4-Quinazolinone
(review[B-5])

[1301]

2. 1,2,3-Triazine

In contrast to the ring closure described in the preceding section, heating the same compound in ethanol gives a 4-imino-1,2,3-triazine in high yield. Replacing the nitrile by an ester group and adding piperidine results in the formation of a triazin-4-one again in high yield.

Ar = Me-, Et-, NO$_2$-C$_6$H$_4$

R=CN; EtOH, Δ ~92%

1,2,3-Benzotriazine

[704]

R=COOMe; EtOH, pip, Δ ~86%

1,2,3-Benzotriazin-4-one

[725]

3. 1, 2, 4-Triazin-3-one

Ortho-placed azo and carbamate groups react on thermolysis in decalin to produce the triazinone in high yield. Treatment successively with mineral acid and ammonia also brought about a similar cyclization.

Ar = Ph, 4-Me-, 4-MeO-,
4-Br-, 4-Cl-C_6H_4

1,2,4-Triazino[6,5-*b*]indol-3-one

[1508]

Ar = Ph, Me-, Cl-, Br-, F-,
MeO-C_6H_4

1,2,4-Benzotriazin-3-one

[790]

4. 1, 4-Oxazin-3-one

A terminal carboxylic acid adjacent to an azo group gives a high yield of a fused 1, 4-oxazin-3-one ring on heating with thionyl chloride. When the benzene ring is activated by methoxy or similar substituents, some chlorinated products are also obtained.

Ar = 2-ClC_6H_4, 2,6-Cl_2C_6H_3

1,4-Benzoxazin-3-one

[1716]

CHAPTER 61

Carbamate or Ureide and Ring-carbon or Ring-nitrogen

Compounds containing a carbamate as one of the reacting functions are usually regarded as acylamines (q.v.) but the numerous cyclizations of carbamates to neighbouring ring-carbon or -nitrogen justify a separate chapter. Ureas may be considered as the amides of carbamic acids and so cyclization of compounds which contain a ureide or thioureide adjacent to a ring carbon or nitrogen are also considered in this chapter.

I. FORMATION OF A FIVE-MEMBERED RING

1. Imidazole, Imidazol-2-one or Imidazole-2-thione

Reductive cyclization of phenyl 8-naphthylcarbamate proceeds under mild conditions to give a peri-fused imidazole ring. Alkyl carbamates fail to react in this way.

EtOH, THF, NaBH$_4$, Δ
48%

Imidazo[5,4,1-ij]-
quinolin-2-one

[338]

309

Interaction between a thioureide group and a ring-nitrogen is brought about by DCC and an imidazole ring is formed but on heating an N^3-phenylureide in toluene, the thioxo group is retained.

[811]

Imidazo[1,5-a]pyridine

R=Me,PhCH$_2$,Ph,Cl-,
MeO-C$_6$H$_4$

[664]

Imidazo[1,5-a]pyridine-
3-thione

The anion generated by the action of t-butoxide on a benzodiazepine attacks a side-chain carbamate ester to form a fused imidazol-2-one ring.

Ar = 2-ClC$_6$H$_4$

tBuOK,tBuOH,Δ
58%

[682]

Imidazo[1,5-a][1,4]-
benzodiazepin-1-one

Heating a ureide with phosphorus oxychloride causes annulation of the carbonyl to the ring-nitrogen so as to form a fused imidazole ring.

POCl$_3$,Δ
~63%

[1327]

R=Me,Ph

Imidazo[1,5-a]pyrazine

2. 1, 2, 4-Triazole

Oxidation of a thiosemicarbazide group at the 2-position of pyridine with nickel peroxide gives a high yield of a fused triazole. The [1, 5-a] isomer is obtained from the carbamate ylide [152, 613].

NiO$_2$,MeCN
89%

[757]

1,2,4-Triazolo[4,3-a]-
pyridine

3. Thiazole

Oxidative cyclization of N-arylthioureas, usually with bromine, yield fused 2-aminothiazoles [148]. Plumbophosphates have also been used as oxidants [1849].

R = Ac, PhCO, COOEt   Thieno[3,2-d]thiazole   [1142]

Another ring system synthesized similarly:

Benzothiazole   [1849]

4. 1, 2, 4-Oxadiazol-5-one, 1, 2, 4-Thiadiazole or 1, 2, 4-Thiadiazol-5-one

Carbamate and N-oxide functions interact to produce an oxadiazolone ring but the corresponding thiadiazolone is conveniently prepared from the ureide and chlorothioformyl chloride.

[1,2,4]Oxadiazolo[2,3-a]-
pyrimidin-2-one   [1162]

[1,3,4]Thiadiazolo[3,2-b]-
[1,2,4]thiadiazol-6-one   [1366]

Another ring system synthesized similarly:

[1,2,4]Thiadiazolo[2,3-b]-
pyridazin-2-one   [1366]

A fused thiadiazole ring is formed in good yield when a 2-thioureidopyrimidine is treated with bromine-acetic acid at ambient temperature.

[872]

[1,2,4]Thiadiazolo[2,3-*a*]-
pyrimidine

II. FORMATION OF A SIX-MEMBERED RING

1. Pyridin-2-one

In a variation on the Bischler–Napieralski reaction (see Chapter 17) a side-chain carbamate reacts in the presence of PPA to form an isoquinolinone.

[588, 707]

1-Isoquinolinone

2. Pyrimidin-2- or -4-one, Pyrimidine-2, 4-dione or Pyrimidine-2-thione

An *N*-ethoxycarbonylamidine is converted into a quinazolinone on heating in quinoline. Reaction of benzonitrile with the anion of an *N*-phenylcarbamate gives a high yield of a quinazolin-2-one but if the nitrile is replaced by phenyl isocyanate, the quinazolinedione is obtained. Lithiation of heterocycles has been reviewed [1549, 1550].

[865]

4-Quinazolinone

[1548]

2-Quinazolinone

[1548]

2,4-Quinazolinedione

Phenylureas or phenylthioureas are cyclized to a pyrimidinone (or thione) ring by prolonged heating with an aldehyde and methanesulphonic acid.

R^1=MeO, OCH$_2$O; R^2=alkyl;
Ar=Ph, Me-, Cl-, CF$_3$-, NO$_2^-$,
COOH-C$_6$H$_4$
X = O, S

2-Quinazolinone
2-Quinazolinethione

[288]

3. Pyrazine or Pyrazin-2-one

A doubly fused pyrazinone is obtained by annulating a carbamate group to an adjoining ring-carbon atom with phosgene while heating a ureide with phosphorus oxychloride yields a fused pyrazine ring.

Pyrido[2,3-e]pyrrolo-
[1,2-a]pyrazin-6-one

[1783]

Pyrido[2,3-h]pyrrolo[1,2-a]-
quinoxaline

[711]

4. 1,2,4-Triazin-3-one or 1,2,4-Triazine-3,6-dione

Annulation of either a C=NNHCOOEt or a CONHNHCOOEt side-chain on to a ring NH may be achieved thermally with a base or by reaction with ethyl chloroformate-pyridine.

R=H, Pr, MeOCH$_2$

Imidazo[1,5-d][1,2,4]triazin-4-one

[380]

$$EtOH, KOH, \Delta \atop 98\%$$ [589]

Pyrrolo[1,2-*d*][1,2,4]-
triazine-1,4-dione

$$pyr, ClCOOEt, \Delta \atop 35\%$$ [1171]

[1,2,4]Triazino[4,5-*a*]-
benzimidazol-1-one

Other ring systems synthesized similarly:

[678]

[1,2,4]Triazino[4,5-*b*]-
indazole-1,4-dione

[276]

Pyrrolo[1,2-*d*][1,2,4]-
triazin-4-one

[940]

[1,2,4]Triazino[4,5-*a*]-
indole-1,4-dione

[513]

Pyrazolo[1,5-*d*][1,2,4]-
triazin-7-one

5.   1, 3, 5-Triazin-2-one, 1, 3, 5-Triazine-2, 4-dione or
their Sulphur Analogues

An *N*-ethoxycarbonyl-amidine or -thiourea is a useful precursor for this ring.

$$i, \Delta \atop \sim 78\%$$ [1240]

i, Xylene or EtOH, K_2CO_3

Pyrazolo[1,5-*a*]-1,3,5-
triazin-4-one

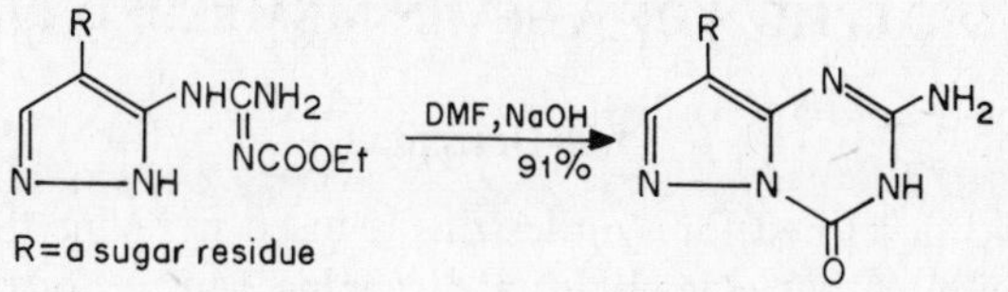

R = a sugar residue

DMF, NaOH
91%

[466]

Pyrazolo[1,5-*a*]-1,3,5-
triazin-4-one

Other ring systems synthesized similarly:

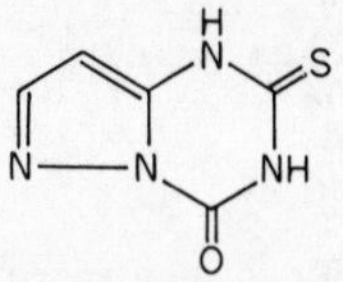

[1762, 1785]

Pyrazolo[1,5-*a*]-1,3,5-
triazin-4-one,2-thioxo

[402]

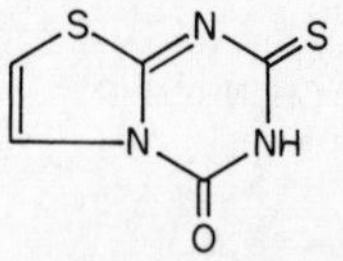

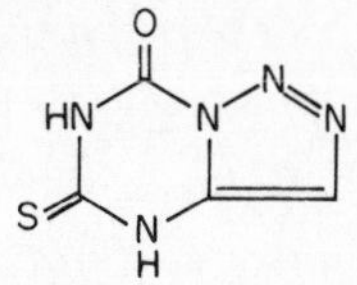

Thiazolo[3,2-*a*]-1,3,5-
triazin-4-one,2-thioxo

[1204]

[1,2,3]Triazolo[1,5-*a*][1,3,5]-
triazin-7-one,5-thioxo-

Annulation of an *N*-aminocarbonyl or thioureide onto the neighbouring ring-carbon or ring-nitrogen may be effected by treatment with an isocyanate or an isothiocyanate in a basic medium.

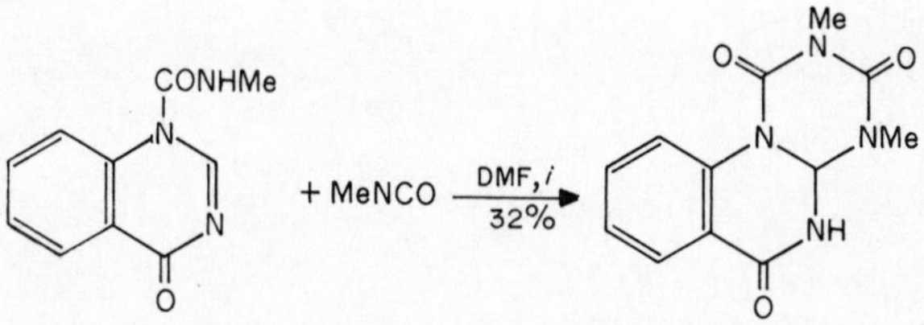

+ MeNCO

DMF, *i*
32%

[1576]

i, tetramethylguanidine

[1,3,5]Triazino[1,2-*a*]quinazoline-
1,3,6-trione

+ ArNCS

MeCN, KOH
~60%

[1325]

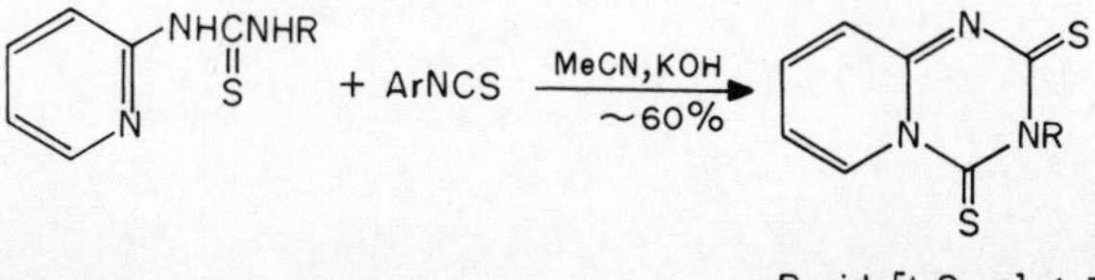

Pyrido[1,2-*a*]-1,3,5-
triazine-2,4-dithione

III.   FORMATION OF A SEVEN-MEMBERED RING

1.   1,4-Diazepine

When the method described for synthesizing a fused pyrazine (Section II.3) is applied to a ureidomethyl side-chain, a diazepine ring is formed in low to moderate yield.

Pyrrolo[1,2-*a*][1,4]-
benzodiazepine

[1571]

CHAPTER 62

Carboxamide and Another Carboxylic Acid Derivative

The second functional group in this chapter may be carboxyl, ester or nitrile.

I.   FORMATION OF A SIX-MEMBERED RING

1.   Pyridine, Pyridin-2-one or Pyridine-2,6-dione

Reaction between suitably separated nitrile and carboxamide groups in the presence of a base yields a tautomeric 2-aminopyridine ring.

Pyrrolo[3,2-c]pyridine   [1034]

Isoquinoline   [1854]

Imidazo[4,5-c]pyridin-4-one   [1131]

Strongly basic conditions promote cyclization of a carboxamide and an ester while an acidic medium facilitates reaction between a carboxamide and a nitrile.

[1246]

Pyrrolo[2,3-c]pyridine-
5,7-dione

Another ring system synthesized similarly:

[143]

Pyrrolo[3,4-c]pyridine-
1,4,6-trione

[126]

1,6—Naphthyridine-5,7-dione

2. Pyridazin-3-one

Heating a 2-cyano-carboxamide with hydrazine hydrate effects ring closure to the pyridazinone with loss of the carboxamide nitrogen.

[1314]

Pyrazolo[3,4-d]pyridazin-7-one

II. FORMATION OF A SEVEN-MEMBERED RING

1.   1,4-Diazepine-2,7-dione

Reaction between a carboxamide and a carboxylic acid may be brought about by DCC and when these groups are separated by three carbons and a nitrogen, a diazepinedione results.

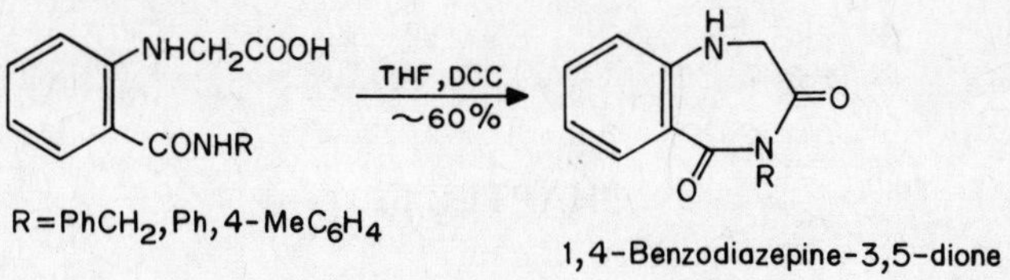

R = PhCH$_2$, Ph, 4-MeC$_6$H$_4$

1,4-Benzodiazepine-3,5-dione

[1245]

CHAPTER 63

Carboxamide or Sulphonamide and Diazonium Salt or Diazo

I. FORMATION OF A FIVE-MEMBERED RING

1. Pyrrole

When a 2-tosylamino-diazoacetophenone is subjected to acetylation conditions, the arylsulphonyl group is lost (cf. Chapter 75, Section II.1.E) and a fused pyrrole ring is formed.

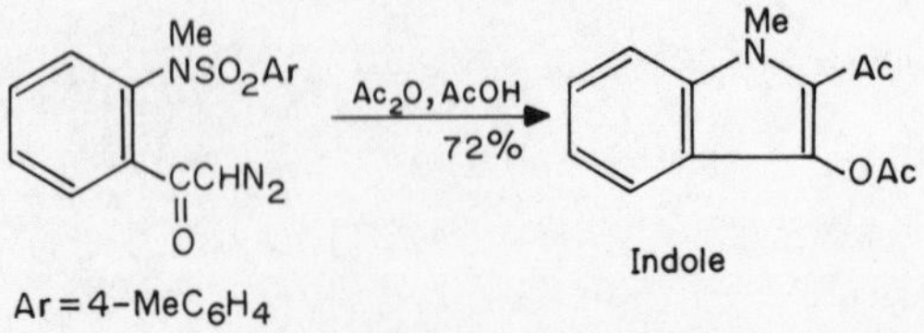

[1299]

Ar = 4-MeC$_6$H$_4$

Indole

2. Isothiazole 1,1-Dioxide

A diazo-sulphonamide gives a fused isothiazole dioxide when treated with toluenesulphonic acid.

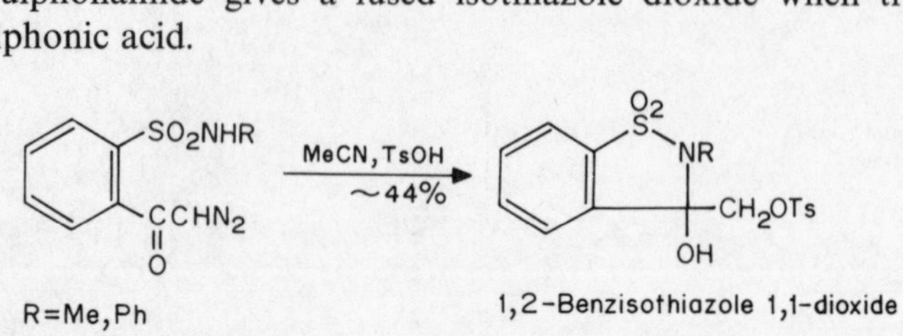

[472]

R = Me, Ph

1,2-Benzisothiazole 1,1-dioxide

II. FORMATION OF A SIX-MEMBERED RING

1. 1, 2, 3-Triazin-4-one

Diazotization of a 2-amino-carboxamide produces a fused triazinone ring in high yield.

R^1=H, PhCH$_2$; R^2=H, Me

i, AcOH or TFA, diox

[1119, 967]

Pyrazolo[4,3-d]-1,2,3-
triazin-4-one

Other ring systems synthesized similarly:

[53]

Pyrazolo[3,4-d]-1,2,3-
triazin-4-one

[49]

1,2,3-Triazolo[4,5-d]-
[1,2,3]triazin-7-one

[882]

Thieno[2,3-d]-1,2,3-
triazin-4-one

[365]

Thiazolo[4,5-d]-1,2,3-
triazin-4-one

2. 1, 2-Thiazin-4-one 1, 1-Dioxide

Formic acid promotes nitrogen elimination and induces cyclization of a 2-sulphonamido-diazoacetophenone to this ring with loss of gaseous nitrogen but heating the same substrate in chlorobenzene gives the 3-one through a Wolff rearrangement (review [1919]).

[472]

1,2-Benzothiazin-4-one
1,1-dioxide (review [1773])

[472]

1,2-Benzothiazin-3-one
1,1-dioxide (review [942])

CHAPTER 64

Carboxamide, Hydroxamic Acid, Hydrazide, Nitrile or Ureide and Hydroxy or Ether

Functions in the title in which sulphur replaces oxygen are included in this chapter. An isolated example of the cyclization of a hydroxy-isonitrile is given in Section III.1.

I. FORMATION OF A FIVE-MEMBERED RING

1. Furan or Thiophene

A 2-hydroxy-nitrile reacts with bromonitromethane in the presence of a base to form a fused 3-aminofuran ring. The corresponding thiol reacts also with an α-halogenoketone; in alkali, this cyclization proceeds at ambient temperature.

$R^1 = H, MeO; R^2 = NO_2, Ac, CN, PhCO;$

$R^3 = Br, Cl$

i, when X = O: EtOH, TEA, Δ

ii, when X = S: aq. NaOH

[366, 726, 1734]

X = O, S

Benzofuran

Benzo[*b*]thiophene

322

2. Isoxazole

Cyclodehydration of a 2-hydroxy-hydroxamic acid in the presence of carbonyldiimidazole (CDI) gives good yields of 1,2-benzisoxazoles.

[386]

R=H,Cl

1,2-Benzisoxazole
(review [1674])

3. Oxazole or Oxazol-2-one

Oxidative cyclization occurs when a 2-hydroxyphenylthiourea is treated with nickel peroxide but when the corresponding urea is heated (without solvent), the benzoxazolone is obtained.

[389]

R^1=H,Me,Ph; R^2=Me,Ph

Benzoxazole

[869]

R=H,MeO,NO_2

2-Benzoxazolone

4. Isothiazole or Isothiazol-3-one

Sodium hypochlorite converts 2-mercapto-nitriles in alkaline solution into a fused 3-aminoisothiazole in good yield. Similarly, the mercapto-carboxamide is cyclized to the 3-one by alkaline iodine at ambient temperature.

[726]

1,2-Benzisothiazole
(review [1775])

Other ring systems synthesized similarly:

[154]

W,X,Y or Z=N, others=CH
W = N : Isothiazolo[5,4-*b*]pyridin-3-one
X = N : Isothiazolo[5,4-*c*]pyridin-3-one
Y = N : Isothiazolo[4,5-*c*]pyridin-3-one
Z = N : Isothiazolo[4,5-*b*]pyridin-3-one

II. FORMATION OF A SIX-MEMBERED RING

1. Pyridin-2-one

Heating 2-ethoxyvinylbenzamide with 4-toluenesulphonic acid leads to the isoquinolinone in high yield.

[1708]

1-Isoquinolinone

2. Pyrimidine, Pyrimidine-2,4-dithione or 4-Thioxopyrimidin-2-one

A 2-methylthio-thiocarboxamide is cyclized to one or other of these rings on heating with an aryl isocyanate or isothiocyanate.

[1531]

X = O, S
Pyrrolo[2,3-*d*]pyrimidin-
2-one,4-thioxo-
Pyrrolo[2,3-*d*]pyrimidine-
2,4-dithione

Reactive methoxy-nitriles are converted into fused pyrimidines when treated with guanidine or a primary amine.

R=H,Me,Ph,2-furyl

Pyrido[2,3-*d*]pyrimidin-7-one

[1731]

1,2,3-Triazolo[4,5-*d*]pyrimidine
(review[2023])

[1591]

Another ring system synthesized similarly:

[537]

Oxazolo[5,4-*d*]pyrimidine

3. Pyran-2-one or Thiopyran-4-one

Hydrolytic cyclization of a nitrile on to a neighbouring hydroxy group leads to the formation of a pyranone ring.

[253]

Pyrano[3,2-*b*]indol-2-one

S-Debenzylation of a thioether which has a 2-cyanoacetyl group is followed by cyclization to a thiopyran-4-one in high yield at ambient temperature.

[1638]

1-Benzothiopyran-4-one

4. 1,3-Oxazin-2- or -4-one or 1,3-Oxazine-2,4-dione

Reaction of a 2-hydroxy-nitrile with reagents which provide a carbon atom at the correct oxidation level leads to an oxazinone; acetic anhydride, an isocyanate, and a phosgeniminium salt are examples.

[367]
1,3-Benzoxazinium, 4-oxo

R=H: Ac₂O, ArCHO, Δ
47–100%

R=H: PhMe, ArNCO, TEA, Δ
74–87%

[864]
1,3-Benzoxazin-2-one

Cl₂C=⁺NMe₂Cl⁻, (CH₂Cl)₂, Δ
~82%

[1041]
1,3-Benzoxazin-4-one

R=MeO, Br, Cl, NO₂

A 2-hydroxybenzamide is converted into a benzoxazinone by heating with either an aldehyde or a ketone.

[357, 1728]

R^1=H,Me; R^2=H,Me,Et,CH₂COOEt;
R^3=alkyl,aryl,CH₂COOEt,
2-thienyl or R^2R^3=(CH₂)₅
i,CHCl₃,H₂SO₄ or PhH,pyrrolidine

1,3-Benzoxazin-4-one

A source of one carbonyl group to convert the hydroxy-carboxamide (or hydroxamic acid or hydrazide) into a benzoxazinedione is provided by either ethyl chloroformate or carbonyldi-imidazole (CDI).

[1422]

i,CDI,THF; *ii*,DMF,TEA,
ClCOOEt

Pyrazolo[3,4-*e*][1,3]-
oxazine-5,7-dione

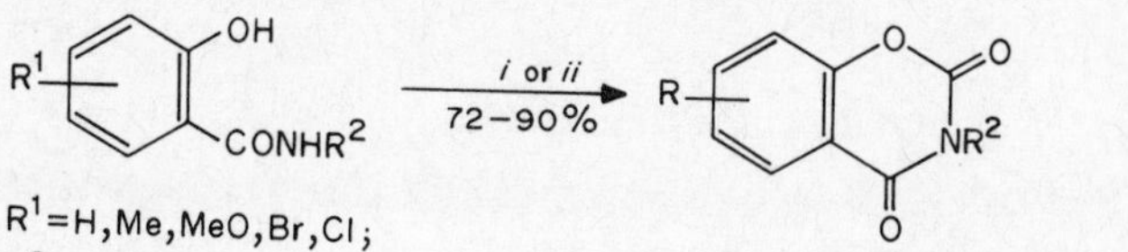

[368, 887]

R^1=H,Me,MeO,Br,Cl;
R^2=HO,Me$_2$C=N
i,pyr,ClCOOEt,Δ;
ii,CH$_2$Cl$_2$,CDI.

1,3-Benzoxazine-2,4-dione

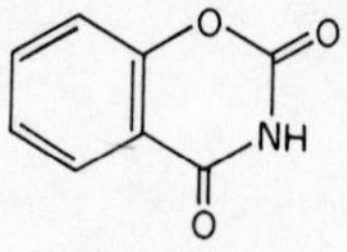

[357]

1,3-Benzoxazine-2,4-dione

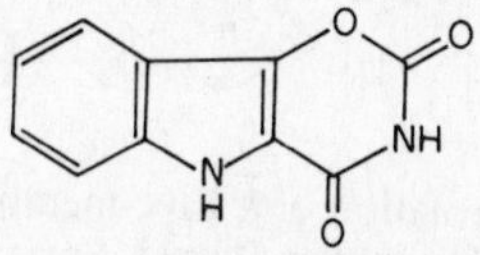

[1079]

1,3-Oxazino[5,6-*b*]-
indole-2,4-dione

5.   1,2,3-Oxathiazin-4-one 2-Oxide

Heating a salicylamide derivative with thionyl chloride gives the relatively little known benzoxathiazinone.

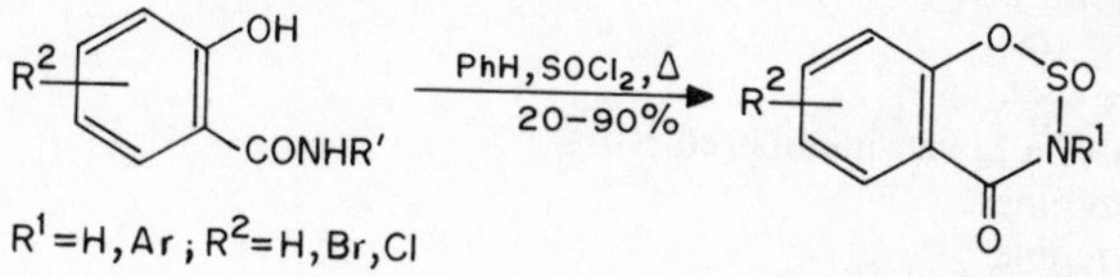

[356]

R^1=H,Ar; R^2=H,Br,Cl

1,2,3-Benzoxathiazin-
4-one 2-oxide

III.   FORMATION OF AN EIGHT-MEMBERED RING

1.   1,3-Oxazocine

A side-chain alcohol reacts with an isonitrile on heating with copper(I) oxide to give (in this example) an eight-membered ring.

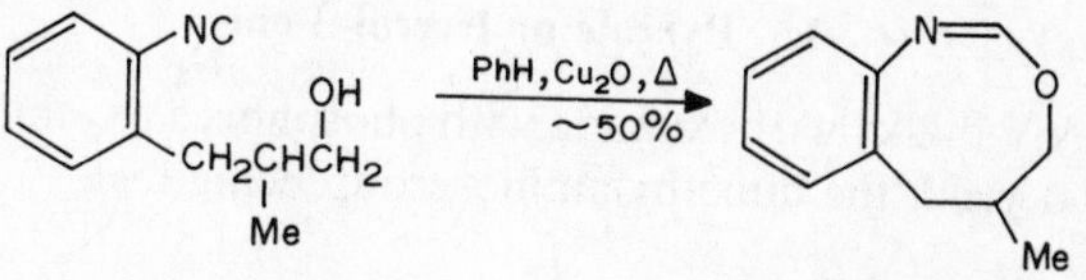

[1884]

3,1-Benzoxazocine

CHAPTER 65

Carboxamide or Nitrile and Ring-carbon or Ring-nitrogen

Cyclizations in which a hydrazide, hydroxamic acid or isonitrile reacts with a ring-carbon or ring-nitrogen are included.

I. FORMATION OF A FIVE-MEMBERED RING

1. Pyrrole or Pyrrol-3-one

Treatment of *NN*-dialkylcarboxamide with phosphorus oxychloride effects ring closure in good yield, the dimethylamino group being lost.

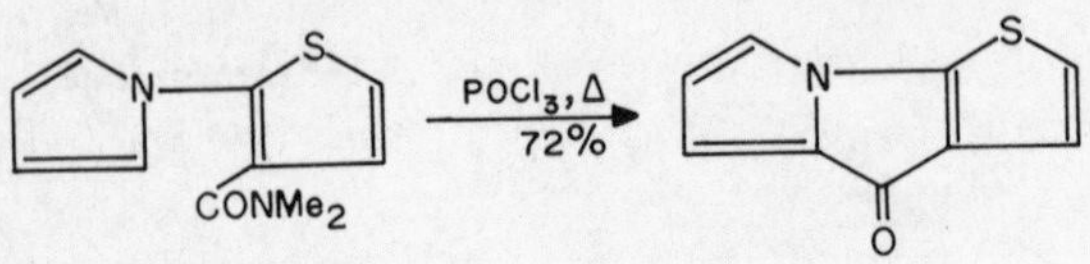

Thieno[3,2-*b*]pyrrolizin-4-one

[653]

Another ring system synthesized similarly:

Thieno[2,3-*b*]pyrrolizin-8-one   [653]

An *N*-phenylhydroxamic acid and vinyl acetate undergo palladium-mediated cyclization to produce an indole.

R^1 = alkyl, alkoxy, aryl;
R^2 = H, Me, Cl

Indole   [1515]

2. Imidazol-2-one or Imidazole-2, 4-dione

Annulation of a nitrile to an adjacent ring nitrogen is achieved by treatment with methyl isocyanate in the presence of 2-methyl-1, 4-diazabicyclo[2.2.2]octane and dibutyltin laurate.

i, 2-MeDBO,
$C_{12}H_{23}OSnBu_2$

Imidazo[1,5-*c*]imidazol-3-one   [542]

A carboxamide also reacts with an isocyanate and a base to form an imidazoledione in high yield.

R^1 = cyclohexyl, aryl; R^2 = alkyl, Ph;
Ar = Ph, 4-Me-, 4-Cl-C_6H_4

Imidazo[1,5-*a*]pyrazine-1,3-dione   [463]

3. 1, 2, 4-Triazole

A heteroarylhydrazone of a mesoxalic ester nitrile cyclizes on heating with acetic acid to a fused triazole ring.

[1293]

[1,2,4]Triazolo[5,1-c][1,2,4]-
triazine

Other ring systems synthesized similarly:

[1293]

Pyrazolo[5,1-c][1,2,4]-
triazine

[1293]

Tetrazolo[5,1-c][1,2,4]-
triazine

4. Furan

A benzofuran is obtained when a tertiary amide of a phenoxyacetic acid is heated with phosphorus oxychloride but the NN-dialkylamino group is retained in this cyclization (cf. previous section).

[590]

R = Cl, benzo   Benzofuran

5. Thiophene

On heating cyclohexylidenemalononitrile with sulphur a fused thiophene ring is formed in good yield.

[715]

Benzo[b]thiophene
(review [1649])

II. FORMATION OF A SIX-MEMBERED RING

1. Pyridine

A side-chain nitrile attacks the ring carbon under Friedel–Crafts conditions; on the other hand, a base-induced reaction with a malononitrile-type reagent can give high yields of a fused 2-aminopyridine ring. The mechanism of this may involve addition of nucleophile, ring opening and ring closing (ANRORC) [551].

R^1 = H, Me, benzo;
R^2 = H, Me; R^3 = CN, COOEt

PhCl, AlCl$_3$, Δ
34-91%

Quinoline

[1003]

Another ring system synthesized similarly:

[1094]

Thieno[2,3-*b*]pyridine

+ RCH$_2$CN

EtONa, EtOH, Δ
~80%

R = CN, COOEt, CONH$_2$

Pyrido[2,3-*d*]pyrimidine-
2,4-dione

[1473]

A side-chain isonitrile reacts under mild conditions with an acid halide to give a fused pyridine by a mechanism involving an imidoyl halide.

+ RCX

CF$_3$SO$_3^-$Ag$^+$
57-82%

R = alkyl, alkenyl, alkynyl;
X = Br, Cl, I

Isoquinoline
(review [2003])

[1824]

Cyclization of a cyanovinyl side-chain under very mild conditions is effected by stirring with a primary amine.

MeOH, RNH$_2$
40-71%

[285]

Pyrido[2,3-*b*]quinoxaline 5-oxide

2.   Pyridin-2- or -3-one or Pyridine-2,6-dione

Annulation of a carboxamide group to the ring carbon with orthoesters or DMFDEA gives a fused pyridin-2-one ring. An unexpected course (probably

through a spiro intermediate) is taken in the acid-induced cyclization of the
amino-nitrile derivative of indole which yields a pyridin-3-one.

R=H,Me,Ph

Indeno[2,1-c]pyridine-3,9-dione

[480]

Pyrido[3,4-b][1,4]-
benzothiazin-3-one

[591]

R^1, R^2=H,Me; R^3, R^4=Me,(CH$_2$)$_5$

Pyrido[3,4-b]indol-4-one

[1579]

Malonic acid derivatives react with a carbamoyl-pyridinium salt (review of
reactivity [1870]) and a base at ambient temperature to give a pyridine-2,6-dione
ring.

R=COOEt,CN; Ar=Me-,Br-,NO$_2$-C$_6$H$_4$;
X=Br,Cl

2,7-Naphthyridine-1,3-dione

[686]

3. Pyridazine

An arylhydrazone of a mesoxalonitrile derivative undergoes an intramolecular
Friedel–Crafts reaction to form a pyridazine ring.

R^1=H,Me,MeO;
R^2=CONH$_2$,CN,COOH,CSNH$_2$

Cinnoline

[1507]

Other ring systems synthesized similarly:

Benzofuro[2,3-*g*]cinnoline   Benzofuro[3,2-*g*]cinnoline

[265]   [541]

4. Pyrimidin-4-one or Pyrimidine-2, 4-dione

A carboxamide and a ring nitrogen may be joined through another carbon atom by heating the compound with triethyl orthoformate but when diethyl carbonate is used, a pyrimidine-2, 4-dione ring is formed.

BuOH, HC(OEt)$_3$, HCl, Δ
64%
Pyrimido[6,1-*a*]isoquinolin-2-one   [294]

EtONa, EtOH, Et$_2$CO$_3$
87%
Pyrimido[6,1-*a*]isoquinoline-2,4-dione   [294]

Nitrile groups react with an adjacent ring nitrogen on heating with strong acids but aqueous alkali is effective and preferable when acid-sensitive groups are present.

i or *ii*
20–77%

R = 5-tetrazolyl
i, CF$_3$COOH, H$_2$SO$_4$, HCl–EtOH
ii, aq. KOH

Pyrimido[2,1-*b*]benzothiazol-4-one   [280]

Another ring system synthesized similarly:

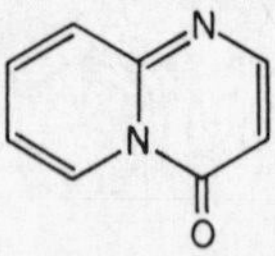

[919]

Pyrido[1,2-a]pyrimidin-4-one

5. Pyrazin-2-one

An electron-rich ring (for example, pyrrole) is attacked by an isocyanate (produced *in situ* by a Lossen rearrangement of a hydroxamic acid) and a pyrazin-2-one ring can be produced in this way.

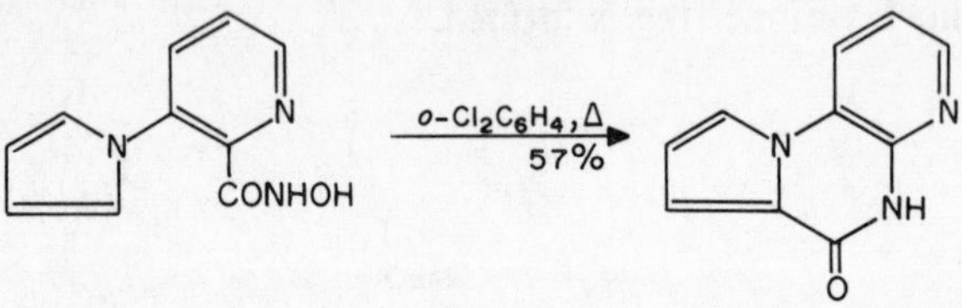

[1783]

Pyrido[2,3-e]pyrrolo[1,2-a]-
pyrazin-6-one

6. Thiopyran-4-one

Under the influence of PPA, a cyano group may acylate an adjacent ring in a Friedel–Crafts type reaction.

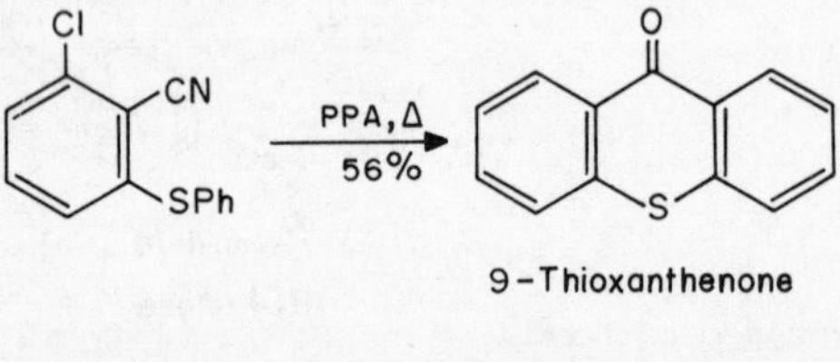

[714]

9-Thioxanthenone

III. FORMATION OF A SEVEN-MEMBERED RING

1. 1, 2-Diazepine

Treatment of a hydrazide with Vilsmeier reagents can cause ring closure on to the neighbouring ring-carbon atom.

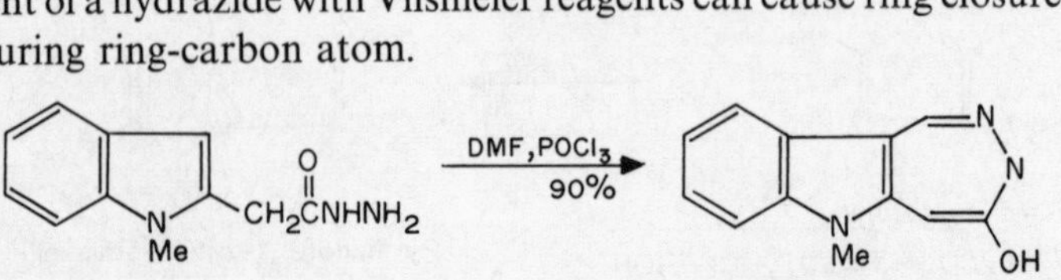

[1081]

[1,2]Diazepino[5,4-b]indole

2. 1, 4-Diazepine

Acid-catalysed reaction of *o*-phenylenediamine with a nitrile gives moderate yields of a fused benzodiazepine.

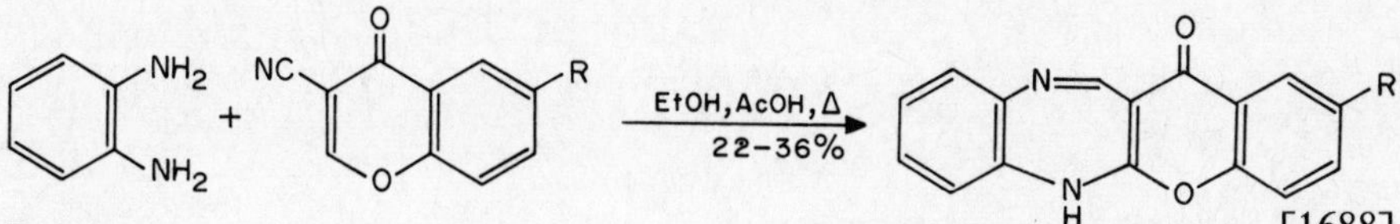

R = Me, Br

EtOH, AcOH, Δ
22–36%

[1688]

[1]Benzopyrano[2,3-*b*][1,5]–
benzodiazepin–13–one

CHAPTER 66

Carboxylic Acid or its Derivative and Halogen

In this chapter, cyclizations in which reaction occurs between a halogen and a carboxamide, a carboxylic acid, a carboxylic acid halide, carboxylic ester or nitrile, are discussed.

I. FORMATION OF A FIVE-MEMBERED RING

1. Pyrrol-2-one

Heating a 2-halogenomethyl-ester or -nitrile with a primary amine leads to the formation of a pyrrol-2-one ring. A benzenoid halogen also reacts with a carboxamide and DMF-LiF to give this ring.

336

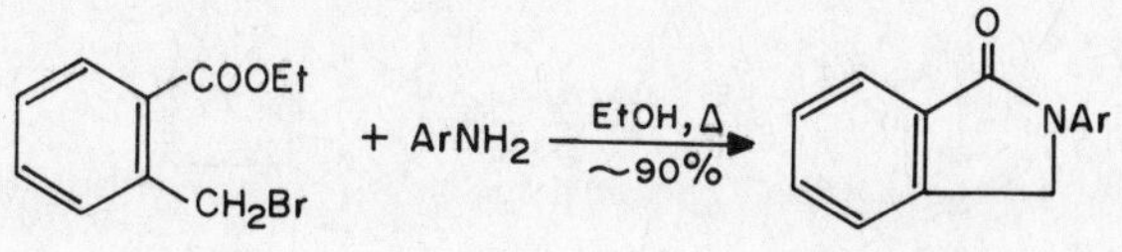

+ ArNH$_2$ $\xrightarrow[\sim 90\%]{\text{EtOH},\Delta}$ [170]

COOEt

CH$_2$Br

Ar = 4-AcC$_6$H$_4$

1-Isoindolone

(review[1615])

$\xrightarrow[64-82\%]{\text{DMF,LiF},\Delta}$ [1912]

X

CONH$_2$

2-Indolone

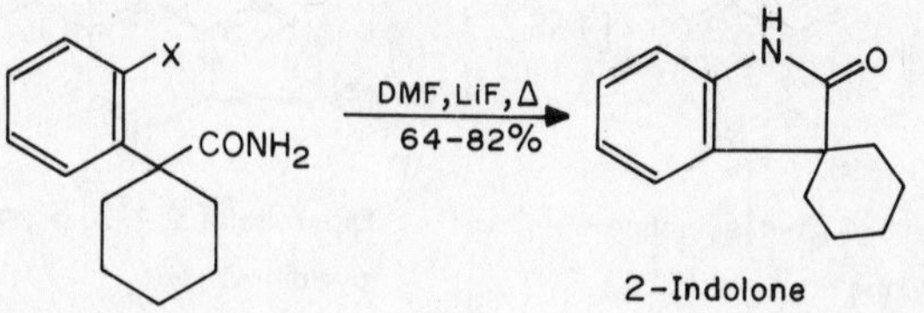

X = Br,Cl,F

Other ring systems synthesized from a halogenomethyl derivative:

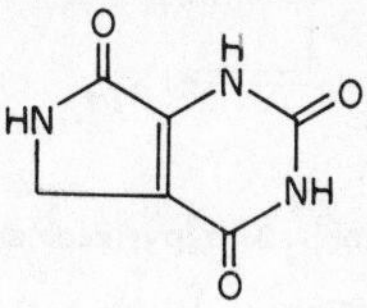

[71]

Pyrrolo[3,4-d]pyrimidine-
2,4,7-trione

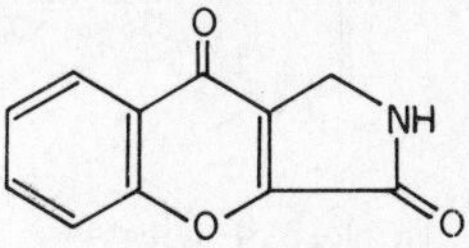

[1040]

[1]Benzopyrano[2,3-c]-
pyrrole-3,9-dione

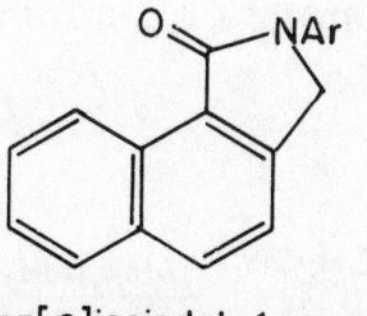

[767]

Benz[e]isoindol-1-one

2. Pyrazole or Pyrazol-3-one

An activated 2-halogeno-ester or -nitrile reacts readily with hydrazines to form a
fused pyrazolone or aminopyrazole, respectively.

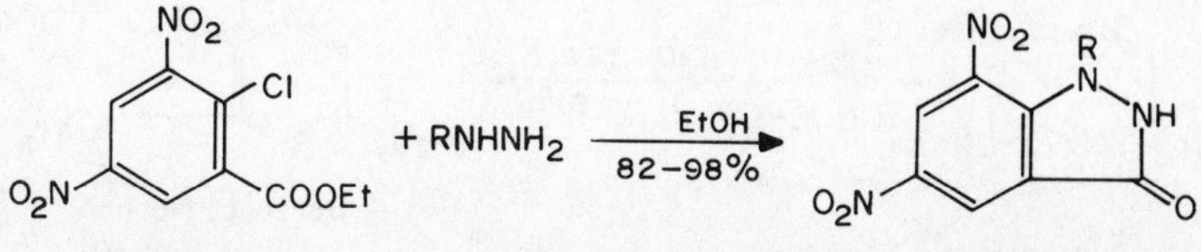

NO$_2$

Cl

O$_2$N

COOEt

R = Me,Ph

+ RNHNH$_2$ $\xrightarrow[82-98\%]{\text{EtOH}}$ [166]

NO$_2$ R

O$_2$N

3-Indazolone

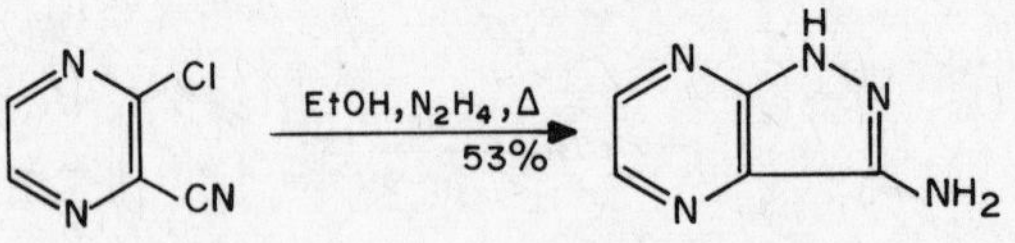

EtOH, N₂H₄, Δ
53%

$$EtOH, N_2H_4, \Delta \quad 53\%$$

[888]

Pyrazolo[3,4-b]pyrazine

Other ring systems synthesized similarly:

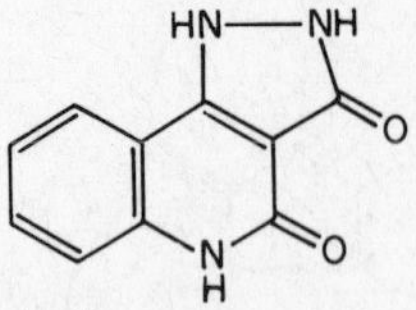

[168]

Pyrazolo[4,3-c]quinoline-
3,4-dione

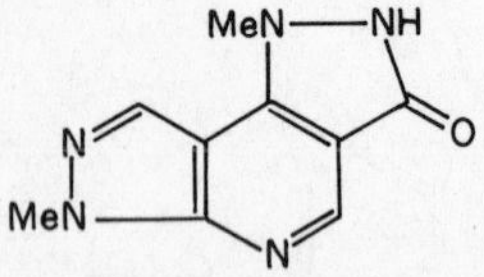

[167]

Dipyrazolo[3,4-b; 3',4'-d]-
pyridin-3-one

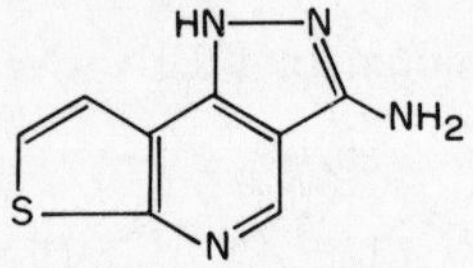

[500]

Pyrazolo[3,4-d]thieno-
[2,3-b]pyridine

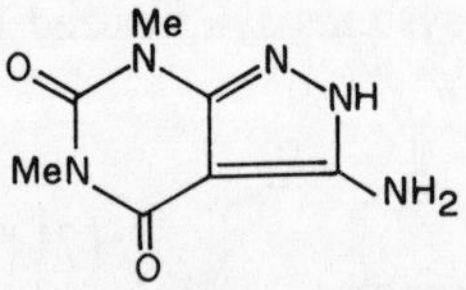

[927]

Pyrazolo[3,4-d]pyrimidine-
4,6-dione

3. Thiophene or Thiophen-3-one

In the presence of a base, mercaptoacetic ester converts a reactive halogeno-ester or -nitrile into a fused thiophenone or 3-aminothiophene ring, respectively.

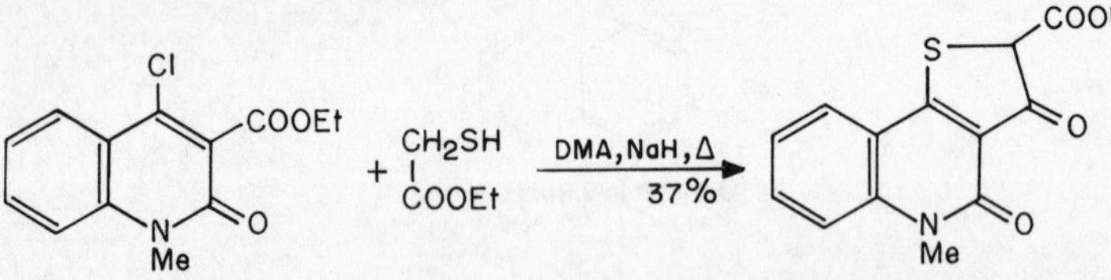

$$CH_2SH \; | \; COOEt \quad DMA, NaH, \Delta \quad 37\%$$

[168]

Thieno[3,2-c]quinoline-
3,4-dione

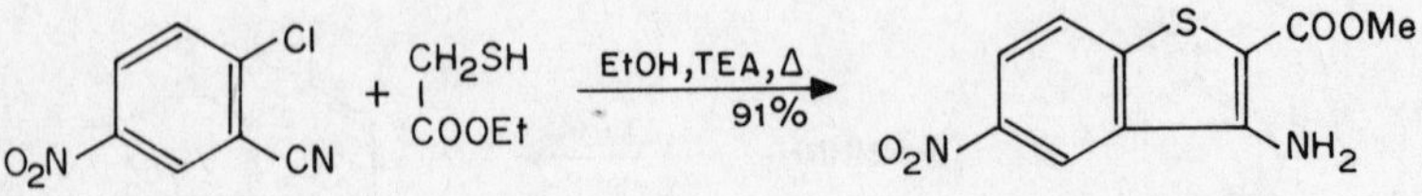

$$CH_2SH \; | \; COOEt \quad EtOH, TEA, \Delta \quad 91\%$$

[675]

Benzo[b]thiophene
(review [1649])

Other ring systems synthesized similarly:

Thieno[2,3-d]pyrimidine [1266]

Thieno[2,3-b]pyrazine [1360]

Thieno[2,3-c]pyridine [1387]

Thieno[2,3-b][1,8]naphthyridin-5-one [1989]

4. Isoxazole

A fused isoxazole ring is obtained when a 2-halogeno-nitrile is heated with hydroxylamine.

$R^1 = Cl, F; R^2 = H, Cl, CN, NO$

$$\xrightarrow[43-72\%]{THF, NaOH, \Delta}$$

2,1-Benzisoxazole
(review [1674]) [1516]

5. Isothiazole or Isothiazol-3-one

Sodium hydrogen sulphide and chloramine convert an activated 2-bromonitrile into an isothiazole ring. The 3-one derivative may be obtained from 2-chlorosulphenylbenzoyl chloride by reaction with a primary amine.

$$\xrightarrow[78\%]{Me_2CO, NaSH, ClNH_2}$$

1,2-Benzisothiazole
(review [1617]) [896]

II. FORMATION OF A SIX-MEMBERED RING

1. Pyridine or Pyridin-2-one

Heating a reactive 2-chloro-ester with a 3-aminoalkanoic ester gives good yields of a tautomeric 4-hydroxypyridine. The pyridine nitrogen may come from a primary arylamine to give a tri- or tetra-cyclic product.

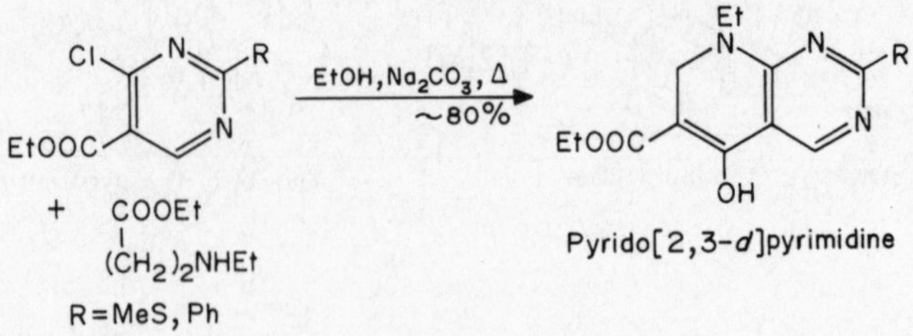

[1336]

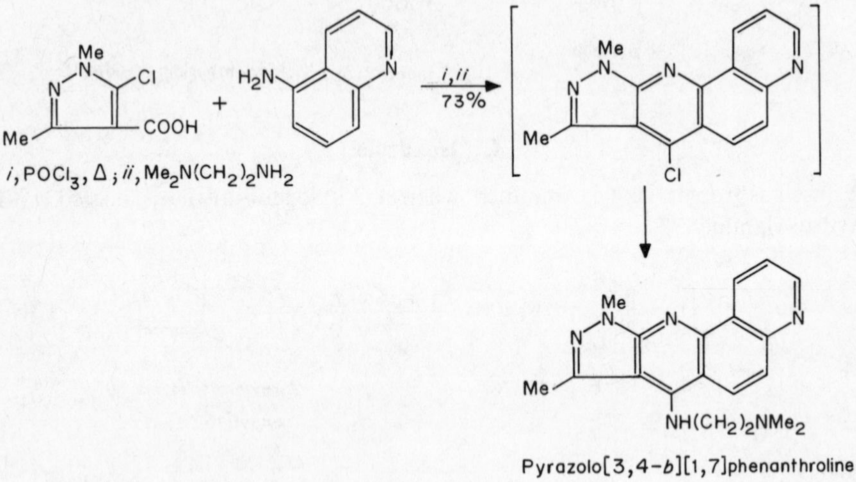

Pyrido[2,3-*d*]pyrimidine

R = MeS, Ph

i, POCl₃, Δ ; *ii*, Me₂N(CH₂)₂NH₂

Pyrazolo[3,4-*b*][1,7]phenanthroline

[1193]

A side-chain carboxylic ester on heating with ammonia or hydrazine gives a fused pyridinone in high yield.

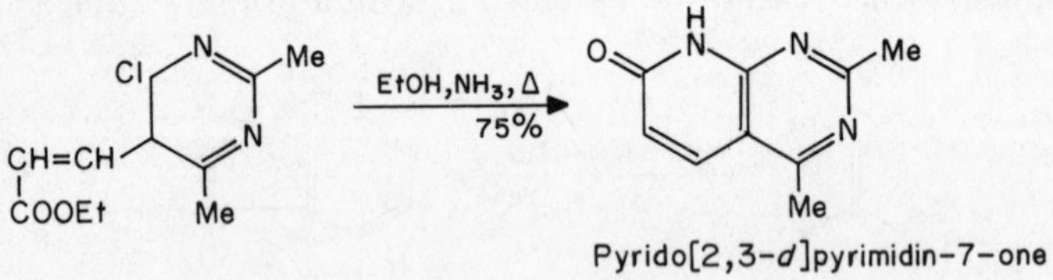

[306]

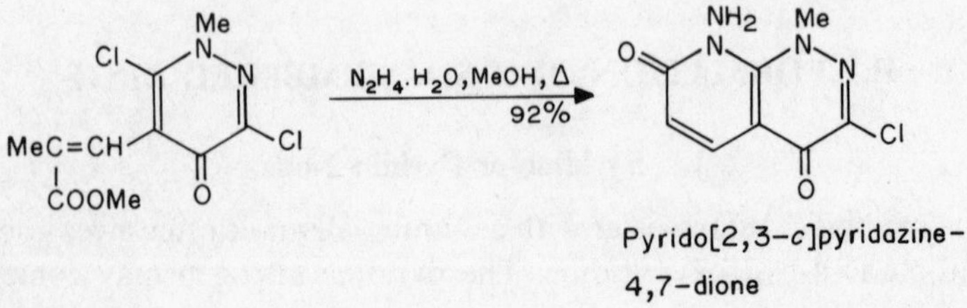

Pyrido[2,3-*d*]pyrimidin-7-one

[418]

Pyrido[2,3-*c*]pyridazine-
4,7-dione

Displacement of fluorine by an anion (from a carboxamide) can be used to form a fused pyridinone ring.

R^1 = H, Me; R^2 = H, Me, Me$_2$N,
Me$_2$N(CH$_2$)$_2$

Indolo[2,3-c]quinolin-6-one

[1330]

Reissert compounds (review [1935]) containing a suitably positioned halogen cyclize under basic conditions with elimination of hydrogen cyanide and hydrogen halide; this type of reaction is useful in the synthesis of berbines.

R = H, MeO

X = CH, N
Dibenzo[ag]quinolizin-8-one
Isoquino[3,2-a]phthalazin-8-one

[615]

2. Pyridazin-3-one

In Section I.1, the synthesis of isoindoles from halogeno-esters and amines was mentioned; when the amine is replaced by hydrazine or methylhydrazine, both nitrogen atoms become part of the new ring. The synthetic usefulness of hydrazines has been reviewed [1437].

R = H, Me

Benzofuro[2,3-d]pyridazin-
4-one

[935, 1247]

3. Pyrimidine or Pyrimidine-2-thione

Reactive 2-halogeno-nitriles and guanidine, an amidine or thiourea in the presence of a strong base, cyclize to form a pyrimidine ring.

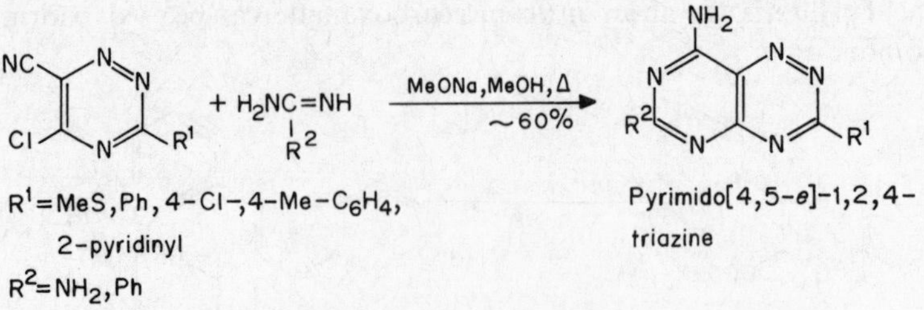
MeONa,MeOH,Δ
~60%
R¹=MeS,Ph,4-Cl-,4-Me-C6H4,
2-pyridinyl
R²=NH2,Ph
Pyrimido[4,5-e]-1,2,4-
triazine

[362, 1733]

Another ring system synthesized similarly:

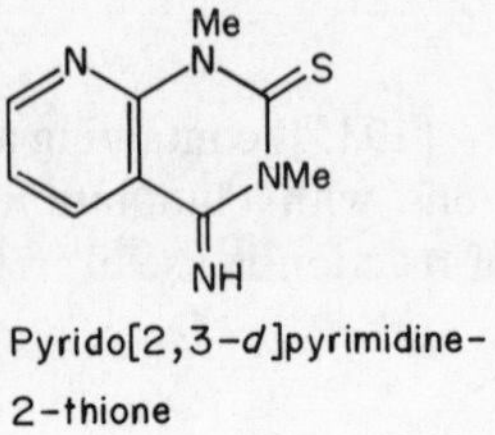
Pyrido[2,3-d]pyrimidine-
2-thione

[236]

4. Pyran-3- or -4-one

Cyclization of a reactive *o*-chloro-propenoic acid by heating with a base yields a pyran-2-one ring. A pyran-4-one ring is obtained by reaction in a basic medium of 2-fluorobenzoyl chloride with a 3-oxoester.

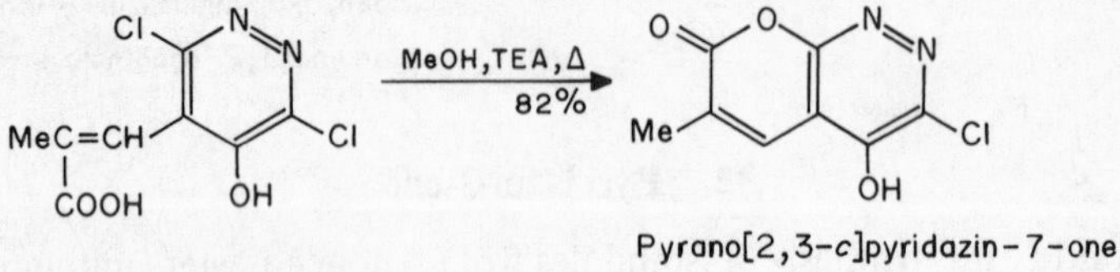
MeOH,TEA,Δ
82%
Pyrano[2,3-c]pyridazin-7-one

[416]

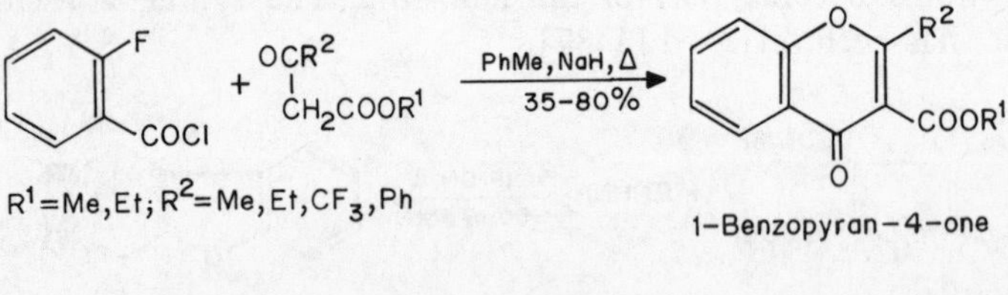
PhMe,NaH,Δ
35-80%
R¹=Me,Et; R²=Me,Et,CF3,Ph
1-Benzopyran-4-one

[982]

5. Thiopyran

A substituted ethanethiol condenses with a reactive 2-chloro-nitrile in a basic environment to form a fused thiopyran.

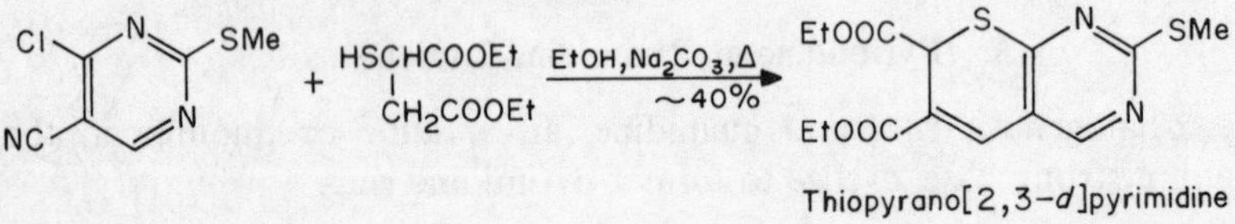
EtOH,Na2CO3,Δ
~40%
Thiopyrano[2,3-d]pyrimidine

[1266]

6. 1,3-Oxazin-4-one or 1,3-Oxazine-2,6-dione

An *NN*-disubstituted cyanamide condenses with a reactive halogeno-acid on heating to give the oxazin-4-one. The 2,6-dione ring is formed in a copper-assisted two-stage reaction of a halogeno-acid first with an amine and then with phosgene.

1,3-Benzoxazin-4-one

Another ring system synthesized similarly:

[200]

[1]Benzothieno[2,3-*e*]-1,3-oxazin-
4-one 5,5-dioxide

[1331]

R=cyclohexyl, 4-F-, 3-CF$_3$-C$_6$H$_4$
i, K$_2$CO$_3$, Cu, Δ, *ii*, COCl$_2$

3,1-Benzoxazine-2,4-dione

7. 1,3-Thiazin-4- or -6-one

The two reactive halogens in 2-chloropyridine-3-carbonyl chloride react readily with the SH and NH of mercaptobenzimidazole under mild conditions.

[1318]

Pyrido[3',2':5,6][1,3]thiazino-
[3,2-*a*]benzimidazol-5-one

A doubly fused thiazin-6-one is obtained when an *o*-chloro-nitrile reacts with a cyclic mercaptoamidine.

i,DMA,NaH; ii,HCl

Pyrido[2,3-d]pyrimido[2,1-b]-
[1,3]thiazin-5-one

[236]

III. FORMATION OF A SEVEN-MEMBERED RING

1. 1,3-Diazepin-4-one

The 2-halogenomethyl-ester is a versatile combination (see Sections I.1 and II.2) and it reacts with 2-amino-pyrimidine or -pyridine with the formation of a 1,3-diazepin-4-one.

[1]Benzopyrano[2,3-e]pyrido[1,2-a]-
[1,3]diazepine-6,14-dione

[1040]

Another ring system synthesized similarly:

[1040]

[1]Benzopyrano[2,3-e]pyrimido-
[1,2-a][1,3]diazepine-7,13-dione

2. 1,4-Diazepin-5-one or 1,4-Diazepine-2,5-dione

The combination of a bromomethyl and an ester (see preceding section) can give rise to a 1,4-diazepin-5-one ring if the groups are appropriately separated.

$R=H,Me,Et,HO(CH_2)_2,NH_2$

Quinazolino[3,2-a][1,4]-
benzodiazepine-5,13-dione

[1251]

Provided the halogen is reactive, a 2-bromo-ester reacts with *o*-phenylenediamine to form a fused 1,4-diazepine-5-one ring. α-Haloacyl-esters react with gaseous ammonia to produce a diazepinedione.

[1]Benzopyrano[3,2-*b*][1,5]-
benzodiazepine-6,13-dione

[1779]

[1242]

1,4-Benzodiazepine-
2,5-dione

3. 1,4-Oxazepin-7-one

An aminoalcohol reacts with a reactive chloro-ester to give an oxazepinone in good to high yield. Primary amines do not react in this way.

R^1=Me,MeS,Ph,piperidinyl;
R^2=Me,HO(CH$_2$)$_2$,Me$_2$N(CH$_2$)$_2$;
R^3=H,Me

Pyrimido[4,5-*e*][1,4]-
oxazepin-5-one

[1349]

Other ring systems synthesized similarly:

[1349]

Pyrido[4,3-*e*][1,4]-
oxazepin-5-one

[1349]

Pyridazino[3,4-*e*][1,4]-
oxazepin-5-one

IV. FORMATION OF AN EIGHT-MEMBERED RING

1. 1,4-Diazocin-5-one

The versatile 2-bromomethyl-ester (see Section III.2) again demonstrates its usefulness in its reaction with *o*-phenylenediamine.

[1]Benzopyrano[2,3-*c*][1,6]-
benzodiazocine-6,14-dione

[1040]

CHAPTER 67

Carboxylic Acid or Ester and Hydrazide or Hydrazine

Cyclization of an ester-hydrazone is included in this chapter.

I. FORMATION OF A FIVE-MEMBERED RING

1. Pyrazole

Heating a 2-hydrazino-carboxylic acid with mineral acid gives a fused pyrazole ring.

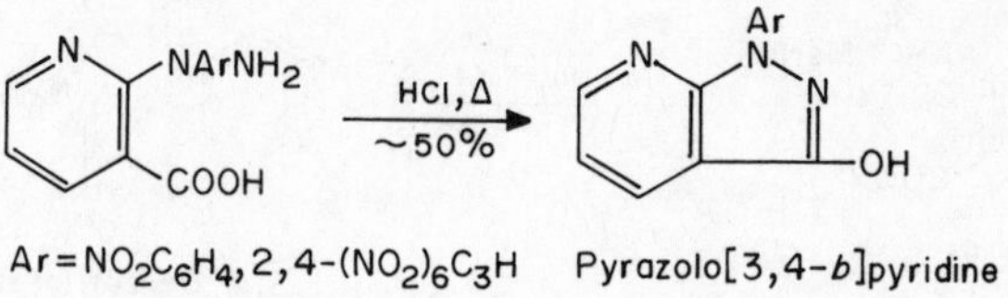

[1612]

$Ar = NO_2C_6H_4, 2,4-(NO_2)_6C_3H$ Pyrazolo[3,4-b]pyridine

II. FORMATION OF A SIX-MEMBERED RING

1. Pyridazine, Pyridazin-3-one or Pyridazine-3,6-dione

Carboxylic ester and hydrazone groups are converted into a fused pyridazine ring in a weakly basic medium.

R=Ph, 4-Me-, 4-Br-C₆H₄,
4-pyridinyl

Pyrido[1,2-b]pyridazin-9-
ium 4-hydroxy inner salt

A hydrazino group separated from a carboxyl by three carbon atoms can react with the latter to form a pyridazinone in moderate yield. A hydrazide and an ester similarly give a pyridazinedione.

Benzofuro[2,3-d]pyridazin-4-one

Pyrrolo[2,3-d]pyridazine-
4,7-dione

III. FORMATION OF A SEVEN-MEMBERED RING

1. 1,2,4-Triazepine-3,7-dione

A suitably positioned $CONMeNH_2$ group being a nucleophile, attacks an ester group to form the title ring.

R=Me, Et

1,3,4-Benzotriazepine-
2,5-dione

IV. FORMATION OF AN EIGHT-MEMBERED RING

1. 1,2,5-Triazocine-3,4,8-trione

Under basic conditions, an oxalic hydrazide side-chain behaves (as in the preceding section) as a nucleophile and reacts with the ester group.

1,4,5-Benzotriazocine–2,3,6-trione

[1031]

CHAPTER 68

Carboxylic Acid, Acyl Chloride or Ester and Hydroxy or Ether

Examples of the sulphur analogues of hydroxy or alkoxy derivatives are included.

I. FORMATION OF A FIVE-MEMBERED RING

1. Pyrazole or Pyrazol-3-one

On heating a 2-hydroxy- or 2-methylthio-ester with a hydrazine, a fused pyrazole or pyrazolone is formed depending apparently on the parent ring system.

[1158]

Pyrazolo[3′,4′:5,6]thiopyrano–
[3,4-*b*]pyridine, 3–OH inner salt

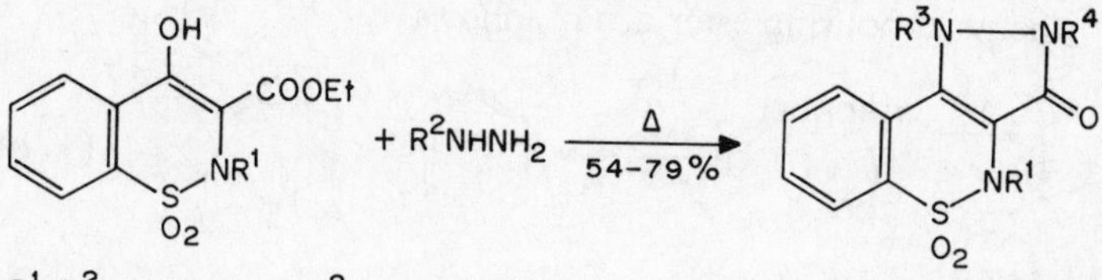

R^1,R^2=H,Me; when R^2=Me,
R^3 or R^4=Me

Pyrazolo[4,3-c][1,2]benzothiazin-
3-one 5,5-dioxide

[1346]

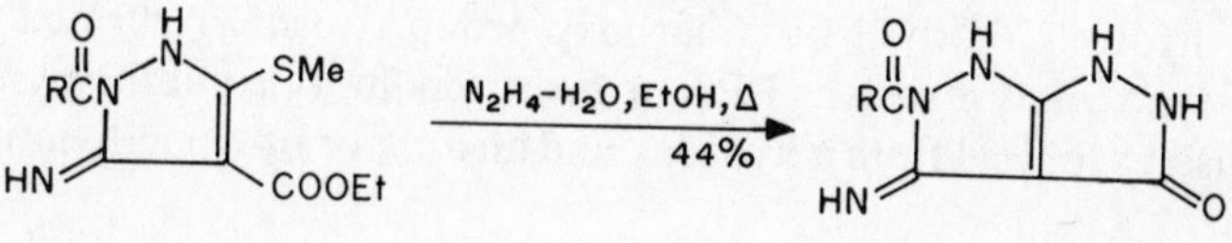

R=4-MeOC$_6$H$_4$

Pyrazolo[3,4-c]pyrazol-3-one

[1860]

2. Thiophene

When a 2-mercaptobenzoic acid is heated with chloroacetic acid and a base, a
benzo[b]thiophene is formed.

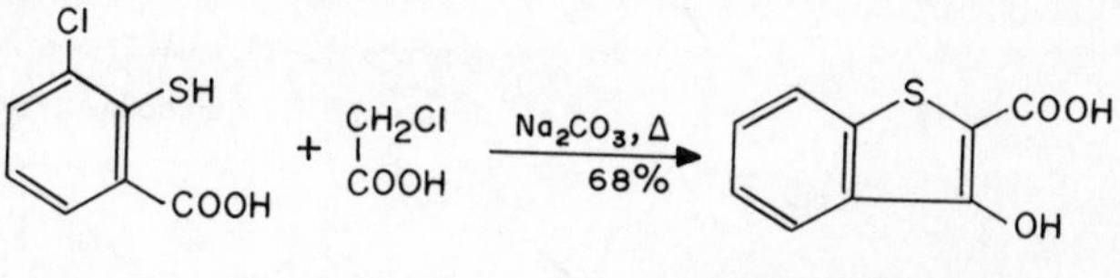

Benzo[b]thiophene

[896]

II. FORMATION OF A SIX-MEMBERED RING

1. Pyrimidine or Pyrimidin-4-one

A 2,3-diaminoquinazolinone is obtained by heating ethyl salicylate with
aminoguanidine.

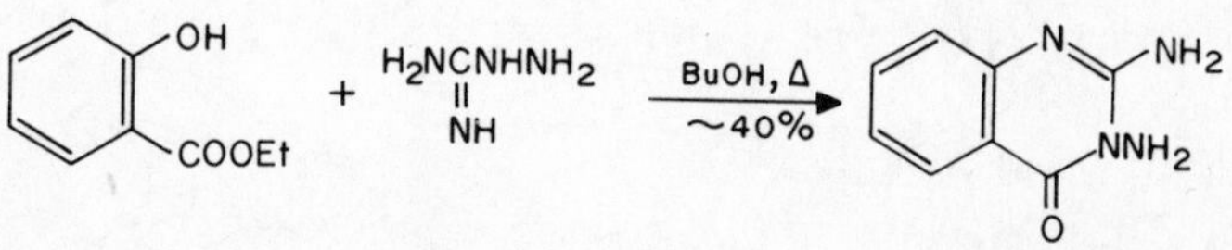

4-Quinazolinone
(review[1671])

[1457]

A reactive ethoxy group is displaced by heating with a primary amine and cyclization onto the neighbouring ester group follows.

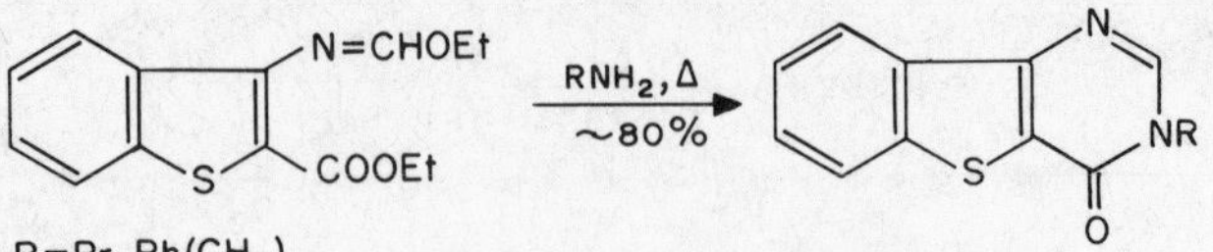

[1714]

[1]Benzothieno[3,2-d]-
pyrimidin-4-one

2. Pyran-2-one or Thiopyran-2-one

Acylation of a hydroxy or thiol by a carboxy group is usually effected by a
Friedel–Crafts-type catalyst such as PPA or hydrogen fluoride but the pyran-2-
one ring may also be formed from a hydroxy and an ester or by demethylation of
an ether-ester.

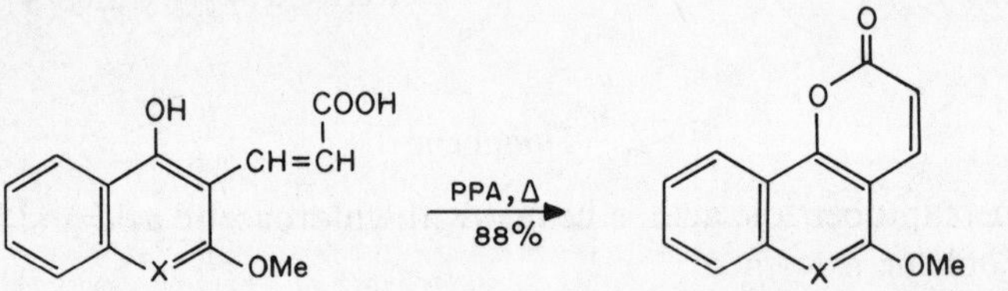

[156]

X = CH, N
Naphtho[1,2-b]pyran-2-one
Pyrano[3,2-c]quinolin-2-one

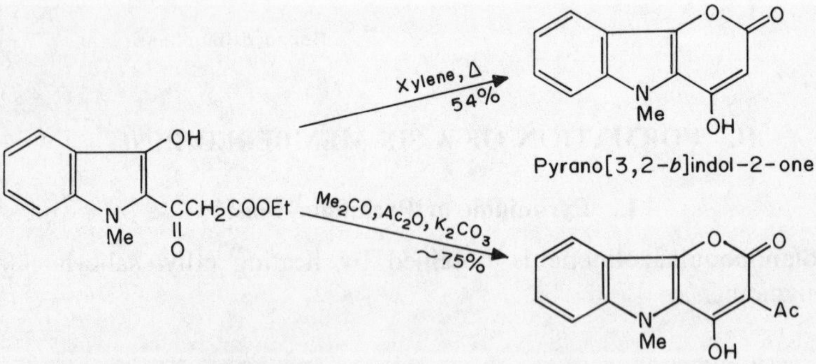

[1079]

Pyrano[3,2-b]indol-2-one

[1079]

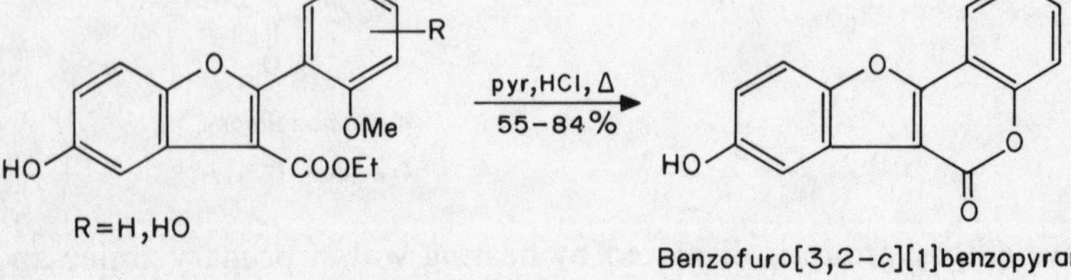

[1250]

Benzofuro[3,2-c][1]benzopyran-6-one

Other ring systems synthesized by one of the above methods:

[157, 172, 1341, 1637, 1898]

X=O,S
1-Benzopyran-2-one
1-Benzothiopyran-2-one
(review [1672])

[595]

Naphtho[2,3-b]pyran-2-one

[502]

Pyrano[3',2':5,6]benzofuro-
[3,2-c]pyridin-2-one

[395]

Furo[3,2-g][1]benzopyran-7-one

[750]

Pyrano[3,2-c]pyridin-4-one

[156]

Pyrano[3,2-c]quinoline-
2,5-dione

3. Pyran-4-one or Thiopyran-4-one

Many methods are available for the synthesis of a fused pyran-4-one ring system (review [B-12]). 2-Hydroxysalicyloyl chloride reacts with an enamine to give a moderate yield of a 4-oxopyran-3-carboxylic ester for which there are few good methods. A 2-acyloxy-acyl chloride reacts with a silylated alkene and phenyllithium to give good to high yields of a chromone. Several other derivatives of salicylic acid have been cyclized to a 4-benzopyranone.

[814]

1-Benzopyran-4-one

[1406]

R^1=H,Me; R^2=Me or
R^1R^2=(CH$_2$)$_n$
n = 3-5

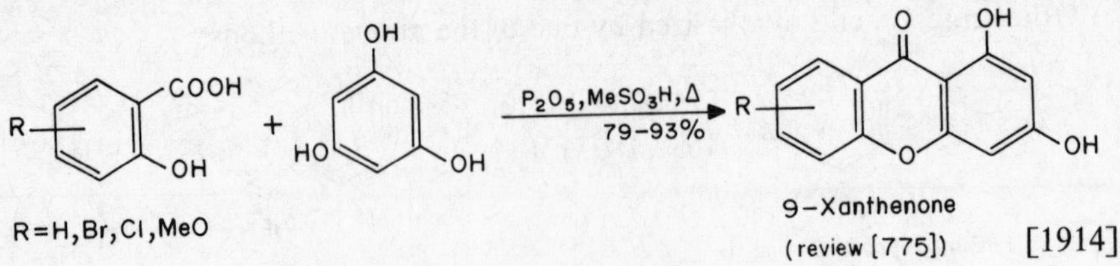

R=H,Br,Cl,MeO

9-Xanthenone
(review [775])   [1914]

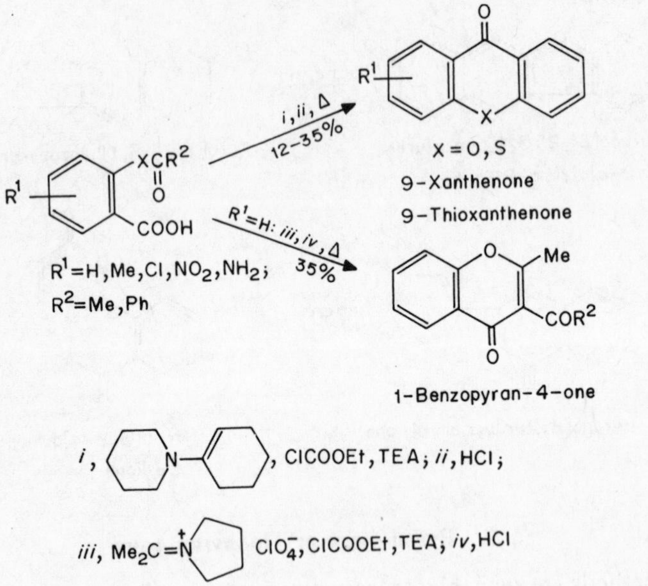

[1760]

X = O, S

9-Xanthenone
9-Thioxanthenone

R^1=H,Me,Cl,NO_2,NH_2;
R^2=Me,Ph

1-Benzopyran-4-one   [1760]

i, , ClCOOEt,TEA; ii, HCl;

iii, $Me_2C=\overset{+}{N}$ ClO_4^-,ClCOOEt,TEA; iv,HCl

[1]Benzopyrano[3,4-b][1]-
benzopyran-12-one
[594]

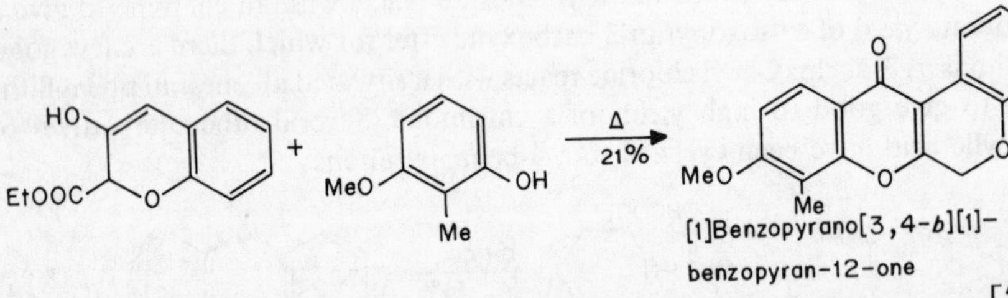

The anion from benzyl cyanide converts methyl 2-mercaptobenzoate into the thioisoflavone in high yield.

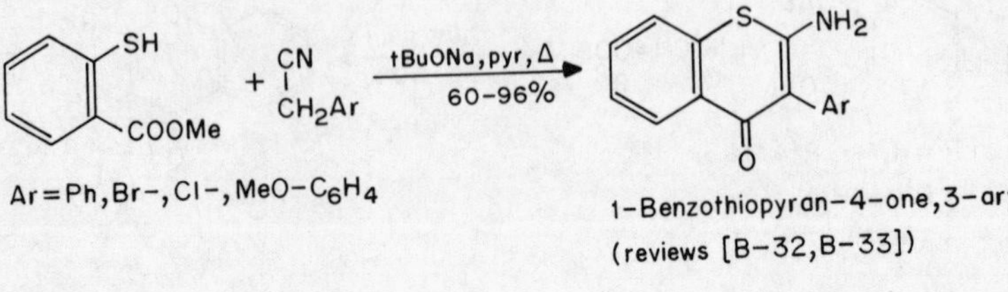

Ar=Ph,Br-,Cl-,MeO-C_6H_4

1-Benzothiopyran-4-one,3-aryl
(reviews [B-32,B-33])

[1592]

4.   1, 3-Oxazine-2, 4-dione

Sodium salicylate and cyanogen bromide react in DMF to form the benzoxazinedione

1,3-Benzoxazine-2,4-dione

[158]

5.   1, 3-Thiazin-4-one

Conversion of methyl 2-mercaptobenzoate into a benzothiopyranone is mentioned above (Section II.3); in contrast, the free acid in acetic acid or pyridine gives high yields of the 1, 3-thiazinone.

R = CN, COOEt, CONH₂,
CONHPh

1,3-Benzothiazin-4-one
(review [2010])

[1421]

III.   FORMATION OF A SEVEN-MEMBERED RING

1.   1, 4-Oxazepin-7-one

A fused lactone may be formed by treatment of the hydroxy-carboxylic acid with cyanogen bromide and TEA initially at ambient temperature and then under gentle heat for a short time.

R^1=H, Cl, Me; R^2=H, NO₂

Dibenz[be][1,4]oxazepin-11-one

[159]

2.   1, 4-Thiazepin-2-one or 1, 4-Thiazepine-2, 5-dione

These rings are formed by reaction of thiol and carboxy groups in the presence of DCC or DCC and DMAP.

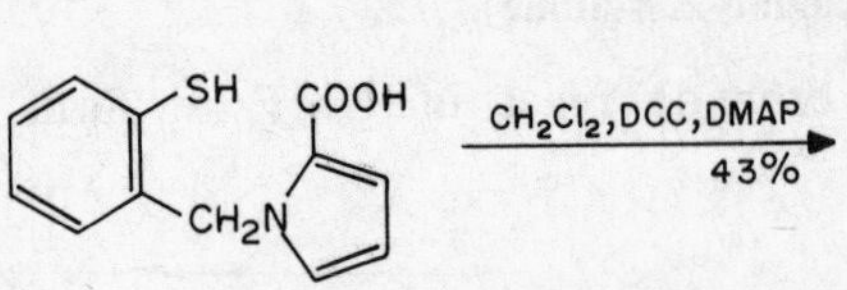

[1717]

Pyrrolo[2,1-c][1,4]benzothiazepin-
11-one

Other ring systems synthesized similarly:

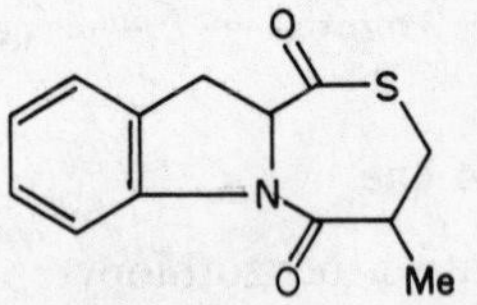

[165]

[1,4]Thiazepino[4,3-a]-
indole-1,5-dione

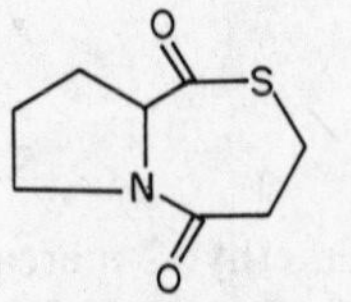

[481]

Pyrrolo[2,1-c][1,4]-
thiazepine-1,5-dione

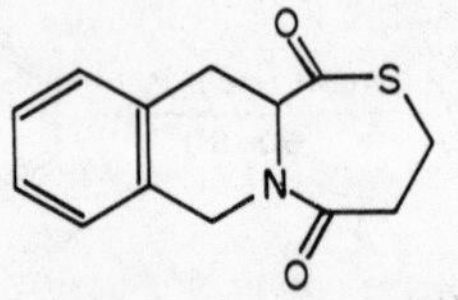

[593]

[1,4]Thiazepino[4,3-b]-
isoquinoline-1,5-dione

CHAPTER 69

Carboxylic Acid or its Derivative and Lactam Carbonyl or Isocyanate

The following derivatives of a carboxylic acid are mentioned in this chapter: carboxamide, acyl halide, ester and nitrile. An example of an *N*-carboxylic ester (a carbamate) is also included; the term lactam carbonyl covers thiocarbonyl.

I. FORMATION OF A FIVE-MEMBERED RING

1. Pyrrole

Decarboxylative cyclization of *N*-(ω-carboxypropyl)pyridin-2-one gives an indolizinium salt which is isolated as its perchlorate.

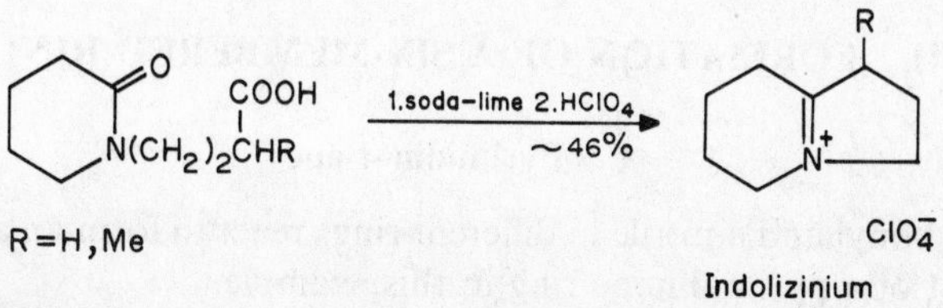

R = H, Me

Indolizinium

[164]

357

2. Thiophene

Cyano and lactam thiocarbonyl groups may be annulated by heating with methyl chloroacetate and an alkoxide.

Ar= 4-ClC$_6$H$_4$

Thieno[2,3-d]pyrimidine

[1164]

Another ring system synthesized similarly:

Thieno[2,3-d]pyrimidin-2-one

[1164]

3. Isothiazol-3-one

Oxidative cyclization of a thiolactam which has a neighbouring carboxamide proceeds at ambient temperature to give an isothiazolone ring.

R^1=H,Ph; R^2=aminoalkyl

Isothiazolo[5,4-b]pyridin-3-one

[154]

Other ring systems synthesized similarly:

X=N,Y=Z=CH

Isothiazolo[5,4-c]pyridin-3-one

X=Z=CH,Y=N

Isothiazolo[4,5-c]pyridin-3-one

X=Y=CH,Z=N

Isothiazolo[4,5-b]pyridin-3-one

[154]

II. FORMATION OF A SIX-MEMBERED RING

1. Pyrimidin-4-one

A lactam carbonyl and a nitrile in different rings react to form an amide linkage which is part of a pyrimidinone ring in this example.

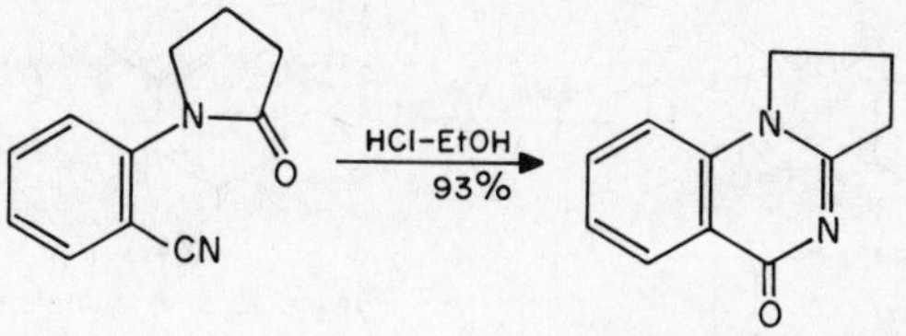

[18]

Pyrrolo[1,2-*a*]quinazolin-5-one

2. 1,2,4-Triazin-6-one or 1,2,4,5-Tetrazin-3-one

When a thiolactam carrying an ester or carbamate substituent is heated with hydrazine, a triazinone may be formed in good yield by reaction with the thioxo group.

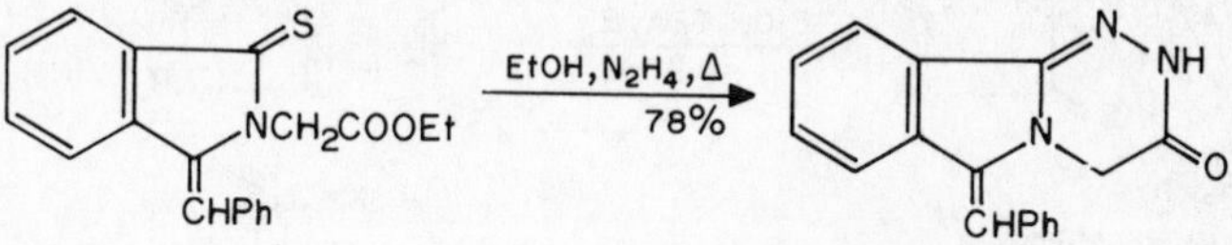

[4]

[1,2,4]Triazino[3,4-*a*]-
isoindol-3-one

[1052]

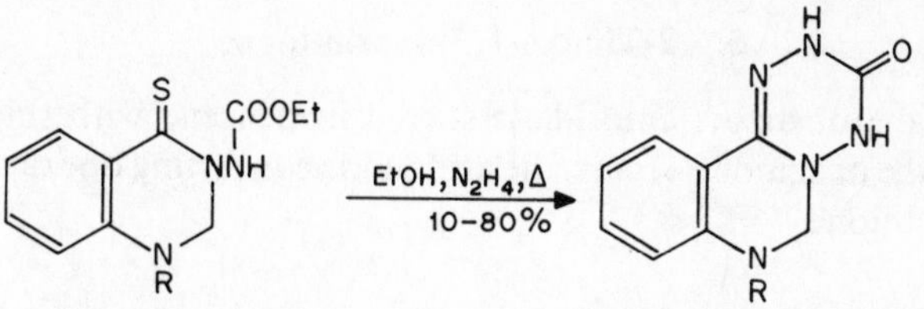

R=Ph,4-Me-,4-MeO-,
4-Cl-C$_6$H$_4$

1,2,4,5-Tetrazino[1,6-*c*]-
quinazolin-3-one

Another ring system synthesized similarly:

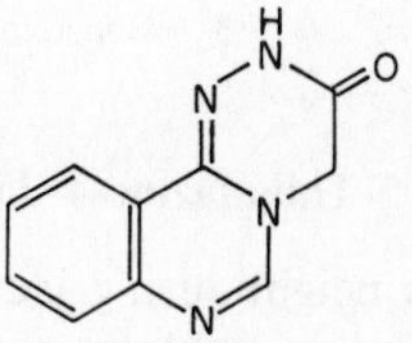

[1052]

[1,2,4]Triazino[4,3-*c*]-
quinazolin-3-one

3. Pyran-2-one

Photochemically assisted ring closure of a carboxylic ester on to a lactam carbonyl in the presence of PPA (review [B-21]) can give high yields of a fused pyranone ring.

[156]

Pyrano[2,3-b]quinolin-2-one

4. Thiopyran

A fused 4H-thiopyran ring is produced when a nitrile in a side-chain reacts with a neighbouring lactam thiocarbonyl in the presence of a tertiary amine.

[1098]

R=H,Ph
Ar=Ph,4-MeOC₆H₄

X=O,S

Thiopyrano[2,3-d]thiazol-2-one
Thiopyrano[2,3-d]thiazole-2-thione

5. 2-Thioxo-1,3-oxazin-6-one

When 2-isocyanatobenzoyl chloride is stirred in benzene with thioacetamide, cyclization to the oxazinone occurs, the acetamide converting one of the carbonyl groups into a thione.

[1201]

3,1-Benzoxazin-4-one, 2-thioxo

6. 1,3,5-Thiadiazine-4-thione

An N-thiocarboxamide and its neighbouring lactam thiocarbonyl react with cyanogen bromide to give a thiadiazinethione in high yield.

[1821]

R=Et,Bu

1,3,5-Thiadiazino[3,2-a]-
benzimidazole-4-thione

III.   FORMATION OF A SEVEN-MEMBERED RING

1.   1, 2, 4-Triazepine-3, 7-dione

An acyl chloride and an isocyanate group can be incorporated in a seven-membered ring by reaction with a hydrazine but the yield in this example is poor.

[1320]

1,3,4-Benzotriazepine-2,5-dione

CHAPTER 70

Carboxylic Acid, Ester, Nitrile or Isonitrile and Methylene

The term reactive methylene includes methyl, methylene and methine groups which are activated by a neighbouring group and/or a π-deficient ring.

I.   FORMATION OF A FIVE-MEMBERED RING

1.   Pyrrole or Pyrrol-3-one

The anion from a reactive N-methylene reacts with a nitrile to form a fused 3-aminopyrrole ring.

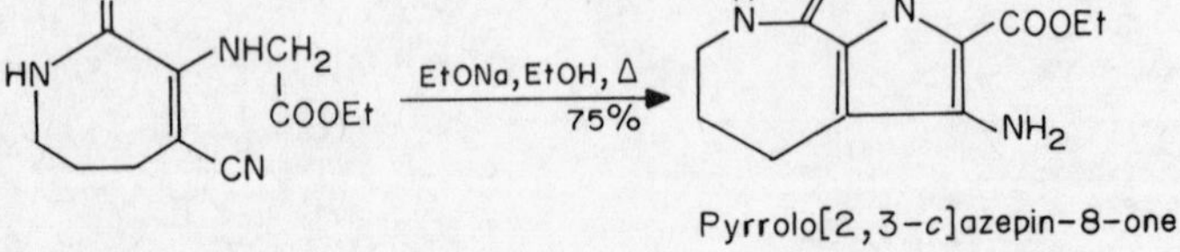

[1528]

Pyrrolo[2,3-c]azepin-8-one

Another ring system synthesized similarly:

Indole

[511]

Base-induced cyclization of a methylene-nitrile in which the methylene is attached to a ring-nitrogen gives an indolizine. A pyridinium-1-acetic ester (review [1870]) similarly yields an indolizinone.

Ar=Ph,4–Me–,4–Cl–C$_6$H$_4$

Indolizine
(review [1670])

[1686]

R^1,R^2=H,Me; R^3=H,Me,Ph   Indolizin–2–one

[1180]

An anion (from the methylene group) reacts with a neighbouring isonitrile to give a fused pyrrole ring in high yield; indoles are also formed either by heating with copper(I) oxide or with acid.

R^1=H,Me,Cl,MeO;
R^2=H,alkyl,COOMe

Indole   [97, 1067, 1132]

i,LDA or Li-diglyme or
Cu$_2$O, PhH or HCl

2. Furan

A 2-RCH$_2$O-carboxylic ester (R = electron-attractor) cyclizes to a fused furan in a Dieckmann-type condensation (review [1922]). A methyleneoxybenzonitrile yields a benzofuran [267].

R=Ac,CN,COOMe

X=S,NMe   [1261, 1350]
[1]Benzothieno[3,2–b]furan
Furo[3,2–b]indole

3. Thiophene or Thiophen-3-one

The thioether dioxide corresponding to the ether of the preceding section gives the tautomeric thiophen-3-one.

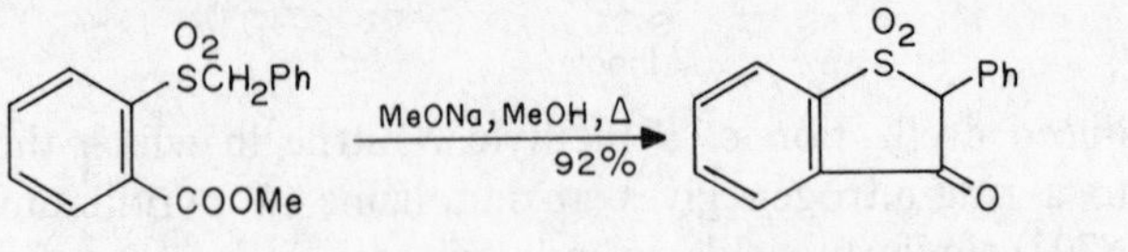

[1151]

Benzo[b]thiophen-3-one
1,1-dioxide

Replacement of the ester by a nitrile in the above reactions leads to a 3-aminothiophene in high yield.

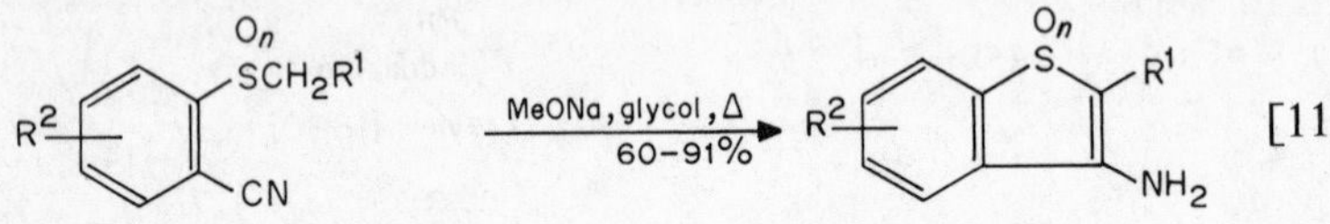

[1151, 1652]

R^1 = Ac,CN,Ph,PhCO,COOMe;
R^2= H,Cl,NO$_2$; n = 0-2

Benzo[b]thiophene(1,1-dioxide)
(review [1649])

Other ring systems synthesized similarly:

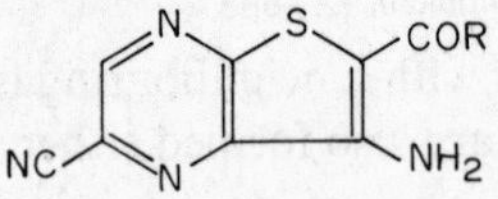

[1376]

Thieno[2,3-b]pyrazine

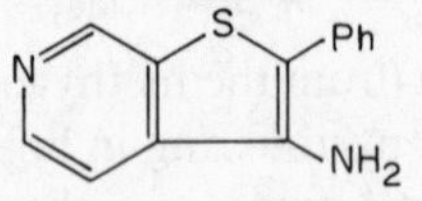

[1387]

Thieno[2,3-c]pyridine

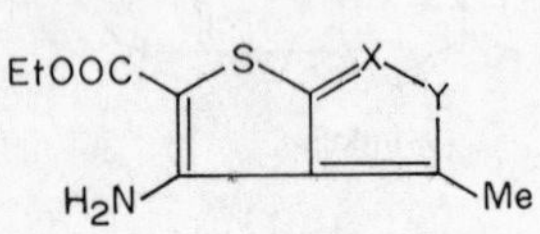

[1117, 1118]

X = N, Y = S
Thieno[2,3-c]isothiazole
X = S, Y = N
Thieno[3,2-d]isothiazole

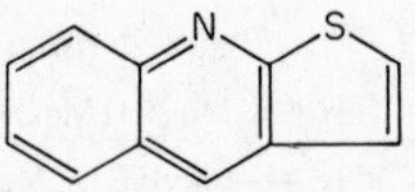

[1598]

Thieno[2,3-b]quinoline

Decarboxylative cyclization sometimes occurs on heating a dicarboxylic acid in a mild base.

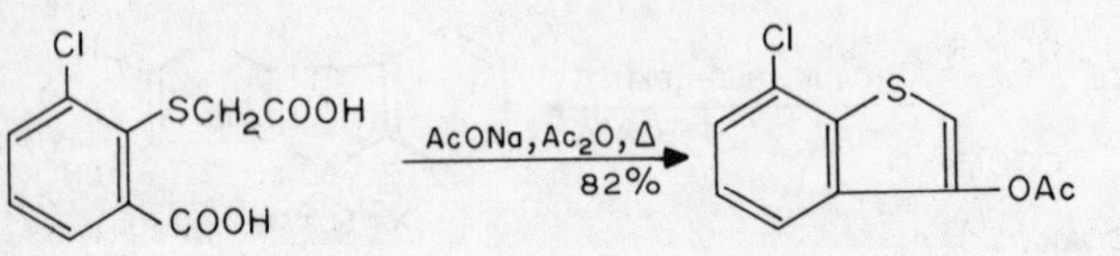

[896]

Benzo[b]thiophene

II.   FORMATION OF A SIX-MEMBERED RING

1.   Pyridine or Pyridin-2-one

Attack by an anion on a carboxylic ester—an intramolecular Dieckmann reaction (review [1922])—is a convenient way of synthesizing pyridin-2-one rings. The same result can sometimes be achieved thermally, for example, by heating in diphenyl ether [1024].

R = Ac, COOEt, CN

Thieno[3,2-*b*]pyridin-5-one

[752]

Other ring systems synthesized similarly:

Thieno[2,3-*b*]pyridin-6-one   [1024]

Benzo[*a*]quinolizin-2-one   [61]

Replacement of the ester by a nitrile group leads to high yields of a 4-aminopyridine ring.

R = H, Me, COOEt

Thieno[2,3-*b*]pyridin-6-one   [1094]

Other ring systems synthesized similarly:

2-Quinolinone   [1129]

Isoquinoline   [669]

When Vilsmeier–Haacke reaction conditions (review [1676]) are applied to a 2- or 4-methylpyridine containing an electron-withdrawing substituent, adjacent to the methyl, a fused pyridin-2-one ring is formed. 1,3,5-Triazine is another source of a C–N fragment [1983].

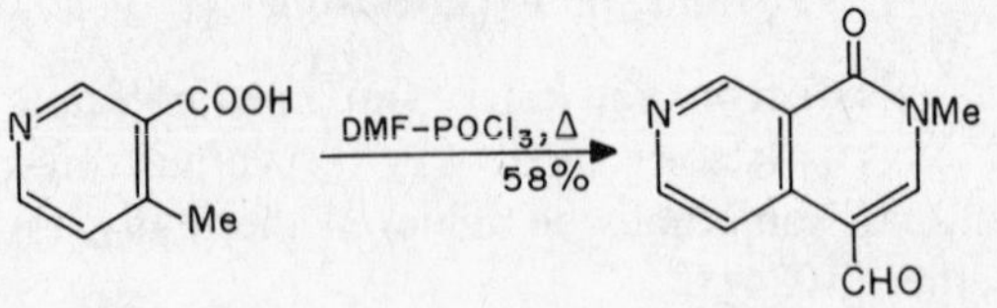

[1452]

2,7-Naphthyridin-1-one
(review [1648])

Other ring systems similarly synthesized:

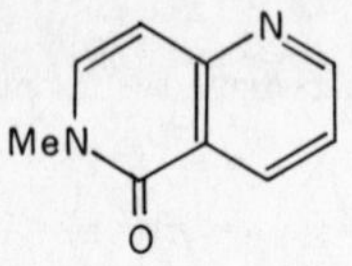

[1452,1983]

1,6-Naphthyridin-5-one

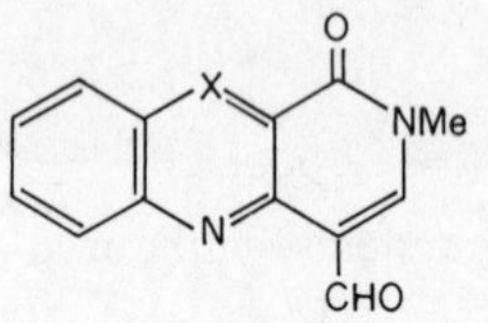

[1452]

X = CH, N
Pyrido[4,3-b]quinolin-1-one
Pyrido[3,4-b]quinoxalin-1-one

2. Pyridazine

When a 2-methylene-nitrile is coupled with an aryldiazonium salt (review [B-28]), a fused pyridazine ring is obtained.

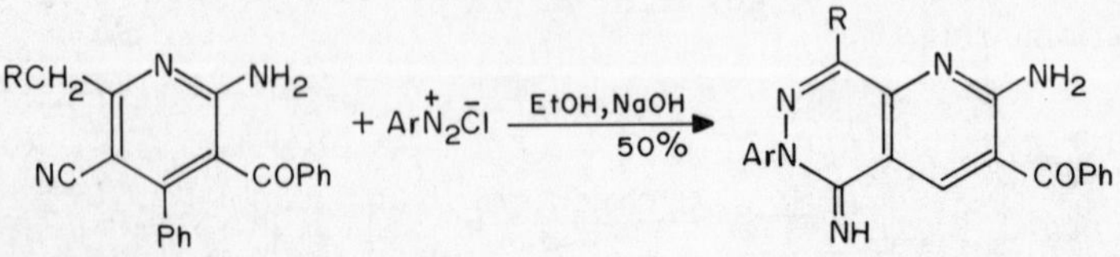

[1723]

R = CN, COOEt
Ar = 4-Cl-, 4-F-, 4-AcNH-,
4-COOEt-C_6H_4

Pyrido[2,3-d]pyridazine

3. Pyran or Pyran-2-one

Phenoxyacetic esters carrying a reactive methylene in the o-position give a pyran ring under basic conditions.

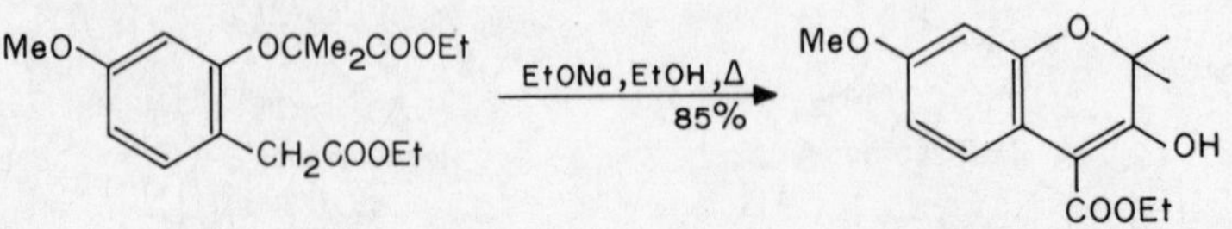

[671]

2H-1-Benzopyran
(review [1621])

When an *o*-toluic ester containing an activated methyl group is heated with DMF dimethylacetal in DMF, it gives a 2-benzopyranone.

[907]

2-Benzopyran-1-one

4. Thiopyran-3-one

An intramolecular Dieckmann reaction on a compound which has a thioether group leads to a thiopyranone. Simultaneous decarboxylation sometimes occurs [1276].

[1158]

Thiopyrano[3,4-*b*]pyridin-5-one

Other ring systems synthesized similarly:

[1158]

Thiopyrano[4,3-*b*]-
pyridin-8-one

[1158]

Thiopyrano[3,4-*c*]-
pyridin-4-one

[1158]

Thiopyrano[4,3-*c*]-
pyridin-4-one

[1276]

1-Benzothiopyran-3-one

5. 1,2-Thiazine 1,1-Dioxide

The chain containing the reactive methylene may contain a sulphonamide group and Dieckmann cyclization then yields a tautomeric 4-hydroxythiazine ring.

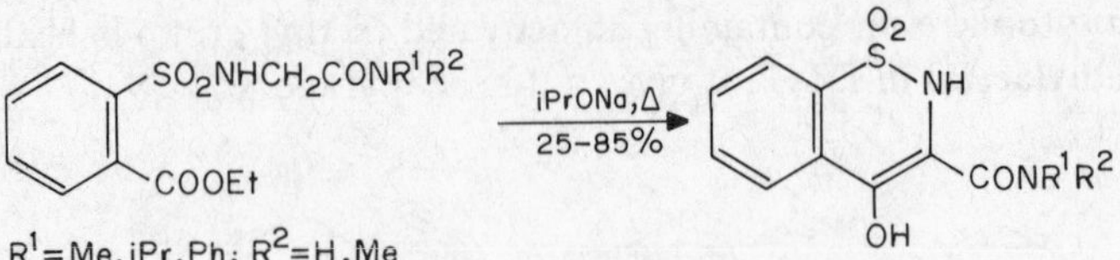

[465]

R^1=Me,iPr,Ph; R^2=H,Me

1,2-Benzothiazine 1,1-dioxide
(review [1773])

Other ring systems synthesized similarly:

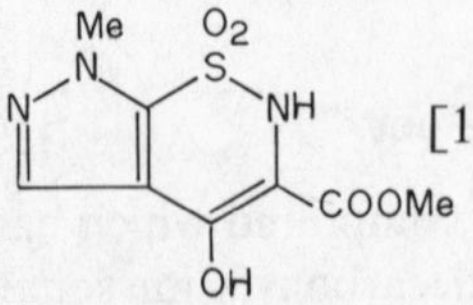

[1307]

Pyrazolo[4,3-*e*]-1,2-thiazine 1,1-dioxide

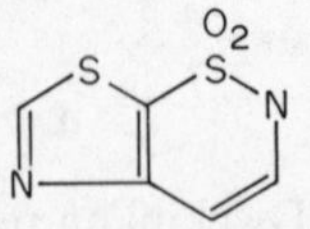

[1307]

Thiazolo[4,5-*e*]-1,2-thiazine 1,1-dioxide

III. FORMATION OF A SEVEN-MEMBERED RING

1. Azepine or Azepine-2, 5-dione

When the reactive methylene chain contains a nitrogen atom, the intramolecular Dieckmann reaction (review [1922]) can lead to an azepine ring.

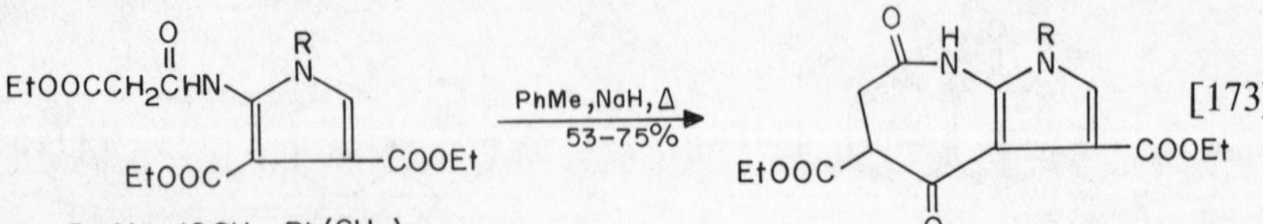

[173]

R=Me,$HOCH_2$,$Ph(CH_2)_2$

Pyrrolo[2,3-*b*]azepine-4,7-dione

Another ring system synthesized similarly:

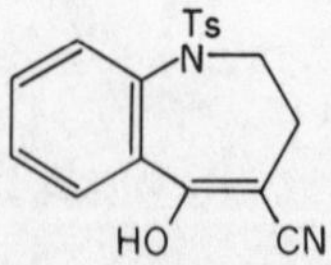

[1793]

1-Benzazepine
(review [B-19])

2. 1, 3-Oxazepine

In the presence of a strong base, a 2-alkylphenylisonitrile reacts with a ketone to form an oxazepine ring in high yield.

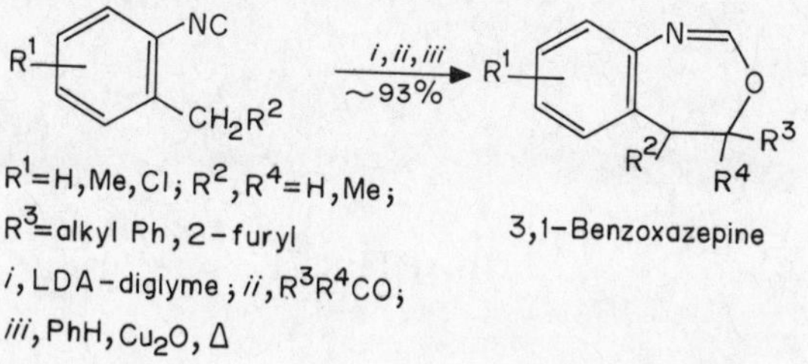
NC
R^1
CH_2R^2
i, ii, iii
~93%
R^1
N
O
R^2
R^3
R^4
R^1=H,Me,Cl; R^2,R^4=H,Me;
R^3=alkyl Ph,2-furyl
i, LDA-diglyme; ii, R^3R^4CO;
iii, PhH, Cu_2O, Δ
3,1-Benzoxazepine

[1884]

3. 1,3-Thiazepine or 1,3-Thiazepin-6-one

A sulphur and nitrogen in the side-chain or ring of a compound which can undergo an intramolecular Dieckmann or similar reaction leads to a fused thiazepine with simultaneous decarboxylation.

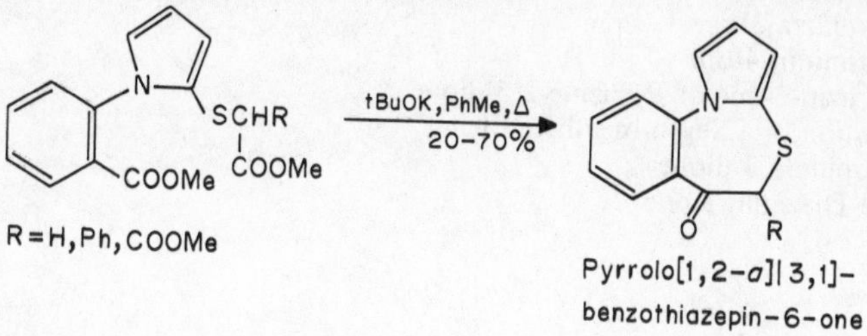
N
SCHR
COOMe
COOMe
tBuOK, PhMe, Δ
20-70%
N
S
O
R
R=H,Ph,COOMe
Pyrrolo[1,2-a][3,1]-
benzothiazepin-6-one

[512]

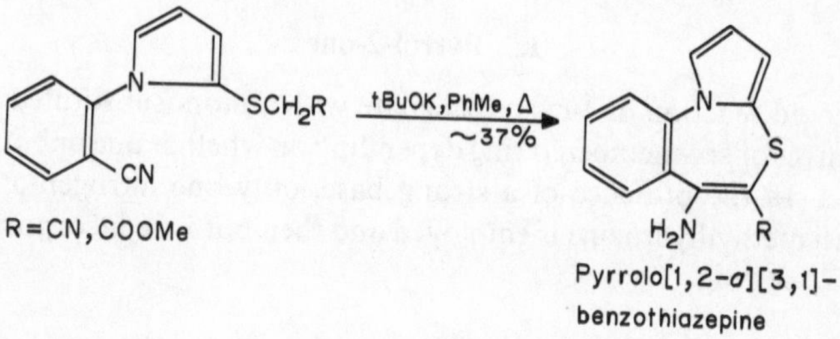
N
SCH_2R
CN
tBuOK, PhMe, Δ
~37%
N
S
H_2N
R
R=CN,COOMe
Pyrrolo[1,2-a][3,1]-
benzothiazepine

[512]

CHAPTER 71

Carboxylic Acid Halide or Ester and Nitrile

I. FORMATION OF A FIVE-MEMBERED RING

1. Pyrrol-2-one

Base-promoted reaction of 2-cyanobenzoate with a monosubstituted hydrazine may give a five- or six-membered ring depending on whether one or both nitrogen atoms react. In the presence of a strong base, only one nitrogen atom reacts except when methylhydrazine is employed and then both are incorporated into a new pyridazine ring.

$R = Me_2N, PhCH_2NH,$
$PhNH, PhCHMeNH$

1-Isoindolone
(review [1615])

[1855]

II. FORMATION OF A SIX-MEMBERED RING

1. Pyridine or Pyridin-2-one

An unusual and unexpected cyclization occurs when methyl 2-cyanomethyl-benzoate is heated with a mixture of phosphorus pentachloride and phosphorus oxychloride; trichloroisoquinoline is formed in high yield. In contrast, a similarly substituted acid chloride gives a moderate yield of the pyridin-2-one derivative.

Isoquinoline
(review [B—17])

[1892]

Furo[3,2-c]pyridin-4-one

[1188]

Nitrogenous reagents such as ammonia, hydrazine or hydroxylamine also react to form a new pyridin-2-one ring but the substituents are not as predictable.

Imidazo[4,5-c]pyridin-4-one

[175, 1131]

Imidazo[4,5-c]pyridin-4-one

[1313]

2. Pyridazin-3-one

When the two functions are attached directly to a ring, hydrazine hydrate in boiling ethanol gives a pyridazinone.

Pyrazolo[3,4-d]pyridazin-7-one

[1314]

3. Pyrimidin-4-one

An N-cyano is often more reactive than an ordinary nitrile and when it has an adjacent ester group, hydrazine cyclizes such a compound to a quinazolin-4-one.

4-Quinazolinone

[1856]

4. Pyrazin-2-one or Pyrazine-2,5-dione

Suitably separated nitrile and ester groups react to form a pyrazine ring on heating in an acidic medium but reaction with ethanolic ammonia gives a related imine (**71.1**, X = NH).

X = O, NH

Pyrazino[1,2-a]benzimidazole-1,3-dione
(**71.1**) Pyrazino[1,2-a]benzimidazol-3-one

[755]

III. FORMATION OF A SEVEN-MEMBERED RING

1. Azepine-2,7-dione

A compound containing an acyl chloride and nitrile groups separated by four carbon atoms is cyclized by hydrogen chloride to an azepinedione.

2-Benzazepine-1,3-dione

[1594]

2. 1,2-Diazepin-3-one

This ring is obtained from a 2-(cyanomethyl)-ester on heating with a two-molar excess of hydrazine hydrate.

[1,2]Diazepino[4,5-b]indol-1-one

[1081]

CHAPTER 72

Carboxylic Acid, its Derivative or Lactam Carbonyl and Nitro or Ureide

The carboxylic acid derivatives considered in this chapter are the carboxamide, ester and nitrile. As usual, the sulphur analogues of the lactam carbonyl and ureide are included.

I. FORMATION OF A FIVE-MEMBERED RING

1. Pyrrole

Partial reduction of a nitro group in the presence of a nitrile may give an *N*-hydroxypyrrole ring.

R = H, Me, Cl, AcNH

2. Thiophene

A nitro group may be displaced by a nucleophile especially when it is placed *o*- or *p*- to an electron-withdrawing substituent (for example, ester or nitrile). This type of reaction has been reviewed [1398, 1399, 1432]. In the presence of a suitable source of sulphur (such as sodium sulphide or a thiol), ring closure to the thiophene may occur.

R=Ac,CN,PhCO,CONH$_2$ Benzo[*b*]thiophene

(review[1649])

[1371]

R^1=COOMe,R^2=OH;   [284, 1385]
R^1=CN , R^2=NH$_2$

3. Oxazole

Cyclodesulphurization occurs when a lactam containing a thioureido side-chain is stirred at ambient temperature with DCC in DMF.

R = H, PhCH$_2$ Oxazolo[5,4-*d*]pyrimidine

[1317]

II.   FORMATION OF A SIX-MEMBERED RING

1.   Pyridin-2-one

Partial hydrogenation of a 2-nitrophenylpropanoic acid leads to a reduced quinolin-2-one; this reaction is accompanied by loss of bromine from the side-chain but chlorine is not affected.

R =H,Cl   2-Quinolinone

[1221]

2.   Pyrimidine, Pyrimidin-2- or -4-one or Pyrimidine-2, 4-dione

2-Cyano-*S*-methylisothioureides (obtained in high yield by alkylation of a thiourea) are ring-closed to pyrimidine by treatment with formamide.

$$NR_2 = $$ (furyl-CON-piperazine)

Cyclization of an alkoxycarbonyl- or cyano-ureide (or thioureide) may be promoted by either an acidic (mineral acid) or basic (aqueous alkali, sodium alkoxide, bicarbonate, ammonia or TEA) medium or by heating in DMSO. Some of the substrates are enamines and their use as synthons has been reviewed [1776].

R^1 = COOMe, COOEt, CN
R^2 = H, CH$_2$COOEt,
(CH$_2$)$_n$Cl (n = 2,3),
alkyl, aryl

2,4-Quinazolinedione [124, 286, 1146]

[73, 124, 1466]

2,4-Quinazolinedione   [1319]

Other ring systems synthesized similarly:

[401]

Pyrido[3,4-d]pyrimidin-4-one,
2-thioxo-

[1964]

Pyrrolo[3,2-d]pyrimidin-
4-one, 2-thioxo-

[705]

Pyrrolo[2,3-d]pyrimidine-
2,4-dione

[1056, 1751]

X = CH, S; n = 1,2
Cyclopentapyrimidine-2,4-dione
Thieno[3,2-d]pyrimidine-2,4-dione
Thiopyrano[3,2-d]pyrimidine-2,4-dione

[1751]

n = 1,2
Thieno[2,3-d]pyrimidine-2,4-dione
Thiopyrano[2,3-d]pyrimidine-2,4-dione

[287]

Pyrimido[4,5-b]quinolin-2-one

Conversion of an amino-carboxamide into a fused pyrimidinedione is usually possible by heating with urea [46] (see Chapter 46, Section II.3) but in one instance, this procedure failed. The corresponding ureido-carboxamide cyclized in high yield on heating with alkali.

[155]

2,4-Quinazolinedione

3.   1, 2, 4-Triazine-3, 5-dione or 1, 2, 4-Triazine-3-thione

Alkaline conditions (sodium hydroxide) cause cyclization of ureido-esters in which the ureido group is attached to an endocyclic nitrogen to give the triazine. Weaker bases (potassium carbonate) and ambient temperature suffice for cyclization of pyridinium ureides [1479].

R^1 = Me, Et;
R^2 = Me, COOH, COOEt, PhCO

[179, 941]

Pyrrolo[2,1-f][1,2,4]triazine-
2,4-dione

Another ring system synthesized similarly:

[1479]

Pyrido[2,1-*f*][1,2,4]triazin–
9–ium, 2,4–dioxo–,inner salt

Thermal cyclization of the lactam carbonyl-thiourea (**72.1**) gives a high yield of the triazinoindole.

[758]

(**72.1**)

1,2,4–Triazino[5,6–*b*]indole–3–thione

4. 1,3-Oxazin-4- or -6-one

Cyclization of a ureido-acid or -ester under acidic conditions at ambient temperature yields the oxazinone which can usually be converted into the pyrimidine by further reaction with an amine.

[73, 124, 1466]

$R^1=H,Me$; $R^2=H,alkyl,aryl$

3,1–Benzoxazin–4–one

5. 1,4-Thiazin-3-one

Neighbouring urea and lactam thiocarbonyl groups may be converted into a thiazinone ring by reaction with either α-bromoalkanoic acid or maleic acid.

[1818]

Pyrido[2,3–*b*][1,4]–
thiazin–2–one

R=H,Et,Bu

[1818]

CHAPTER 73

Carboxylic Acid or Ester and Ring-carbon

I. FORMATION OF A SIX-MEMBERED RING

1. Pyridine or Pyridin-4-one

Intermolecular cyclization of a side-chain ester or acid and a ring-carbon atom is accomplished either by heating with a Friedel–Crafts catalyst [178, 242, 1152] or by heating in a solvent such as methoxyethanol, diphenyl ether or triglyme [242, 751, 845, 1570]. Yields of the tautomeric products are usually good or very good. This type of reaction is related to the Gould–Jacobs reaction for the synthesis of quinolin-4-ones [1253, 1570].

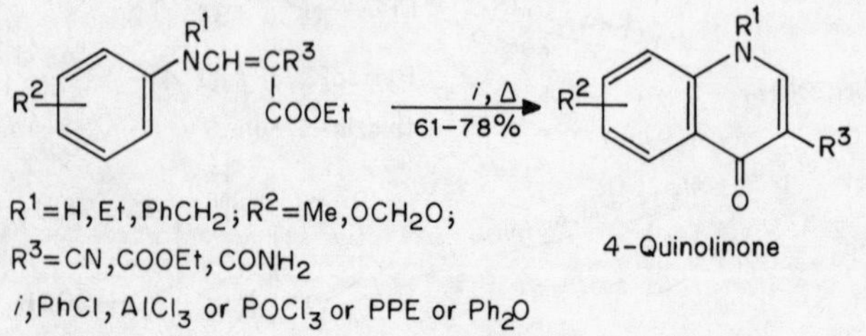

NCH=CR3
COOEt
i, Δ
61–78%
R1=H,Et,PhCH2; R2=Me,OCH2O;
R3=CN,COOEt,CONH2
i,PhCl, AlCl3 or POCl3 or PPE or Ph2O

4-Quinolinone

[178, 242, 1003]

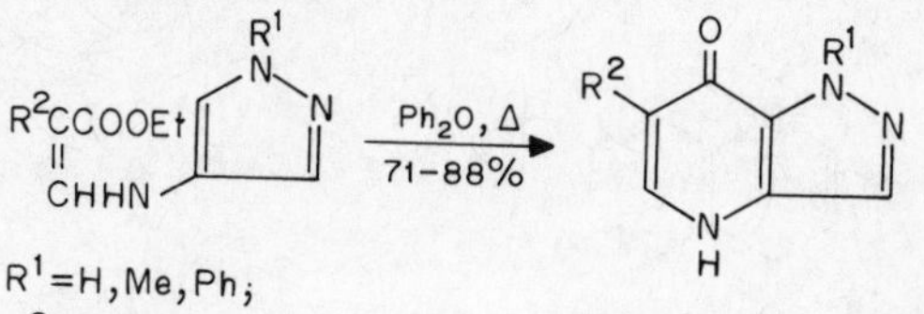

R^1 = H, Me, Ph;
R^2 = COOEt, CN

Ph$_2$O, Δ
71–88%

Pyrazolo[4,3-*b*]pyridin-7-one

[1281]

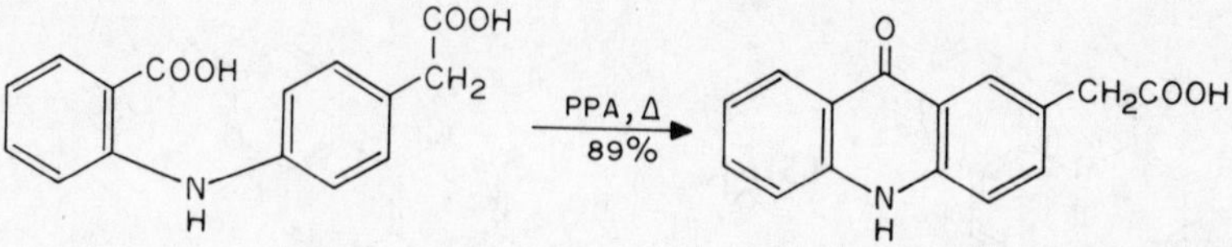

PPA, Δ
89%

9–Acridinone

[259]

Other ring systems synthesized from the ester:

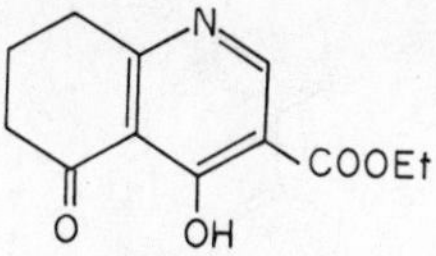

5–Quinolinone

[177]

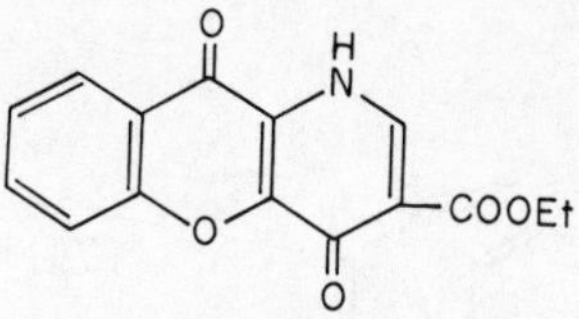

[1]Benzopyrano[3,2-*b*]-
pyridine–4,10–dione

[218]

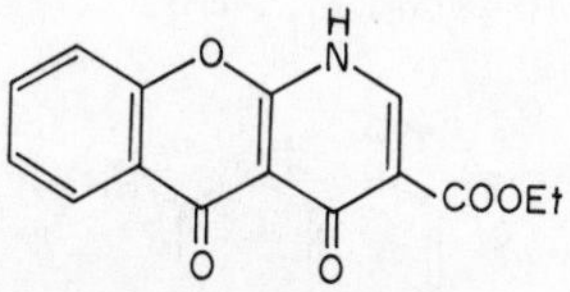

[1]Benzopyrano[2,3-*b*]-
pyridine–4,5–dione

[218]

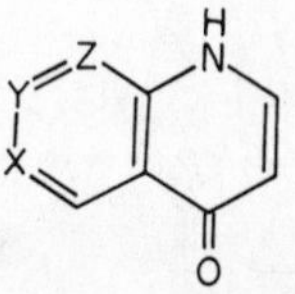

X, Y or Z=N, others CH
1,6–Naphthyridin–4–one
1,7–Naphthyridin–4–one
1,8–Naphthyridin–4–one
(review [1648])

[272, 349, 1254]

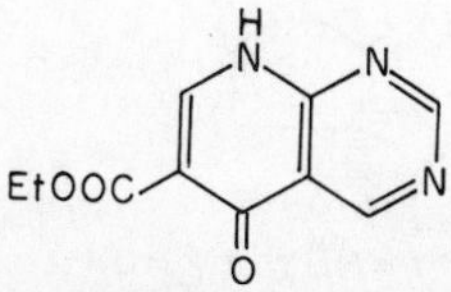

Pyrido[2,3-*d*]pyrimidin-5-one

[260]

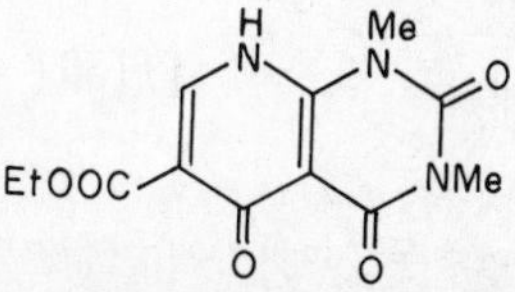

Pyrido[2,3-*d*]pyrimidine–
2,4,5–trione

[1256]

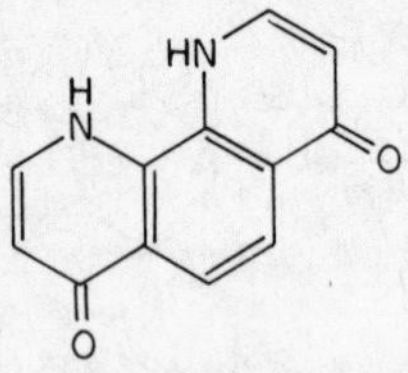

[515]

1,10-Phenanthroline-4,7-dione

(review [1888])

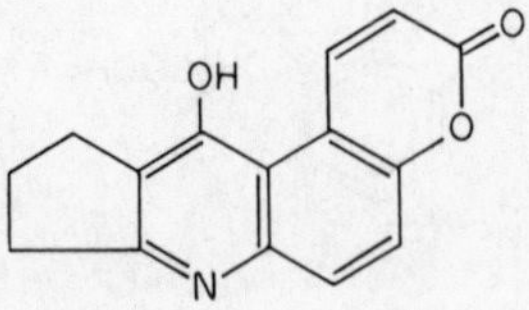

[516]

Cyclopenta[b]pyrano[3,2-f]-
quinolin-3-one

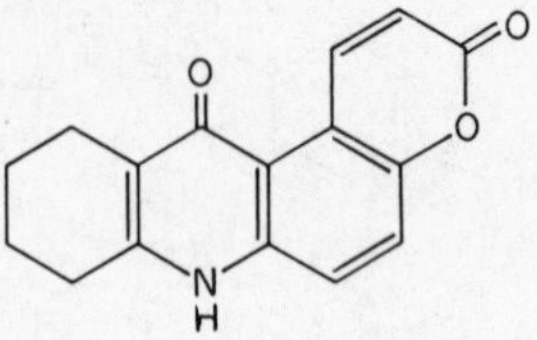

[516]

Pyrano[3,2-a]acridine-3,12-dione

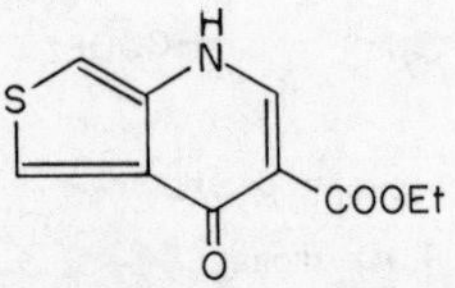

[751]

Thieno[3,4-b]pyridin-4-one

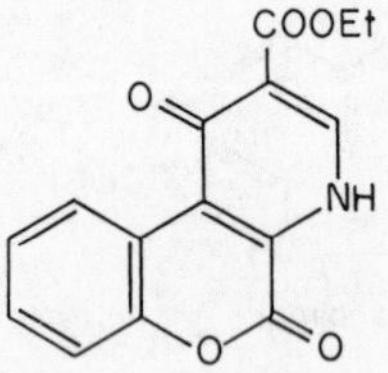

[218, 772]

[1]Benzopyrano[3,4-b]-
pyridine-1,5-dione

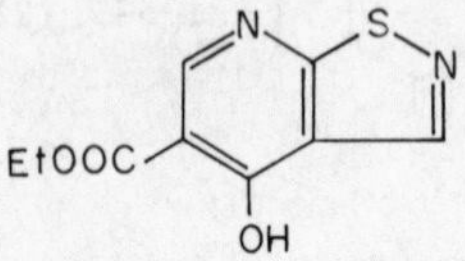

[815]

Isothiazolo[5,4-b]pyridine

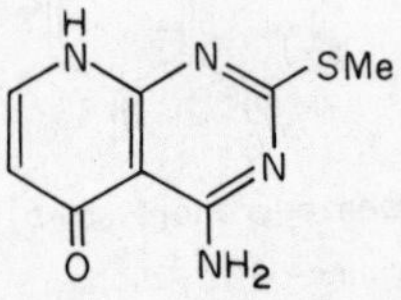

[991]

Pyrido[2,3-d]pyrimidin-5-one

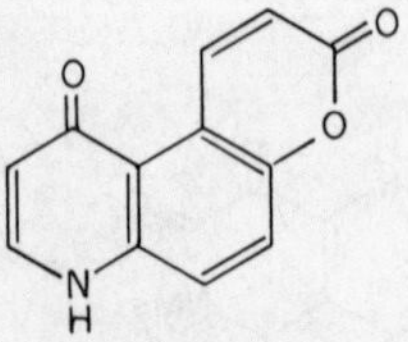

[1133]

Pyrano[3,2-f]quinoline-
3,10-dione

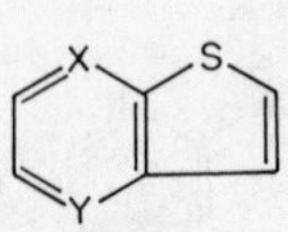

[1252]

X or Y = N, Y or X = CH
Thieno[2,3-b]pyridine
Thieno[3,2-b]pyridine

EtOOC — Pyrazolo[3,4-f]quinolin-9-one   [1253]

Pyrazolo[3,4-b]pyridin-4-one   [845, 1332]

Pyrido[2,3-a]indolizine-4,10-dione   [1792]

Thiazolo[5,4-b]pyridine (review [1865])   [1082]

The same type of cyclization can be effected under basic conditions.

$$\xrightarrow[70\%]{\text{NaH, THF, Me}_2\text{SO}_4, \Delta}$$

[1787]

Pyrrolo[3,4-b]pyridin-4-one

A doubly fused (non-tautomeric) pyridine is obtained when a diphenylamine containing a 2-carboxy group is treated with a Friedel–Crafts-type catalyst.

$$\xrightarrow[65\%]{\text{POCl}_3, \Delta}$$

[955]

Acridine (review [B-10])

Another ring system synthesized similarly:

[464]

Benz[c]acridine

2. Pyridin-2- or -3-one

A side-chain carboxylic acid under standard Friedel–Crafts conditions can cyclize to form a cycloalkanone ring. In order to form a heterocyclic ring the carboxyl group is activated (by either thionyl chloride or ethyl chloroformate), treated with sodium azide and the acyl azide on pyrolysis attacks the ring carbon atom to form a pyridin-2-one ring. 1-Isoquinolinone is synthesized similarly [262].

i,ClCOOEt,TEA;   Furo[2,3-c]pyridin-7-one
ii,NaN$_3$; iii, Δ [5]

Friedel–Crafts intramolecular acylation is a well-known method of forming rings; this may use the free carboxylic acid or its acyl chloride.

R=H,Me,Br,Cl   Benz[f]indolizine-3,10-dione   [1915]

3. Pyridazine or Pyridazin-4-one

Ester groups in side-chains which contain two nitrogen atoms cyclize into fused pyridazinol or the tautomeric pyridazinone. As for the pyridinone analogue discussed in the preceding section, either thermal or acidic conditions are effective.

4-Cinnolinone   [1507]

Another ring system synthesized similarly:

Pyridazino[2,3-a]indole   [1553]

4. Pyran-2-one

Using thallium(III) nitrate, a carboxyl in a side-chain cyclizes on to the adjacent carbon atom of a reactive ring in high yield under mild conditions.

R = phthalimido

Pyrano[2,3-*b*]indol-2-one

[343]

In the presence of an anhydride (or carboxylic acid) and a Lewis acid, a side-chain carboxylic acid is converted into a pyran-2-one ring.

2-Benzopyran-3-one

[923]

Other ring systems synthesized similarly:

X = NH, S
Pyrano[3,4-*b*]indol-3-one
[1]Benzothieno[2,3-*c*]pyran-
3-one

[756, 1897]

5. Pyran-4-one

Cyclization of a side-chain consisting of O—C—C—COOH is a useful route to chromanones and chromones. Substituents on this chain can be aryl (leading to flavones), carboxyl (which may or may not survive the cyclization conditions) or carboxymethyl. 2,3-Dichloro-6-methoxy-[1]benzopyran-4-one, for example, may be prepared in 82 per cent yield from the 3-aryloxypropanoic acid [266].

R = CH₂COOH, Ph

1-Benzopyran-4-one

[345, 1425]

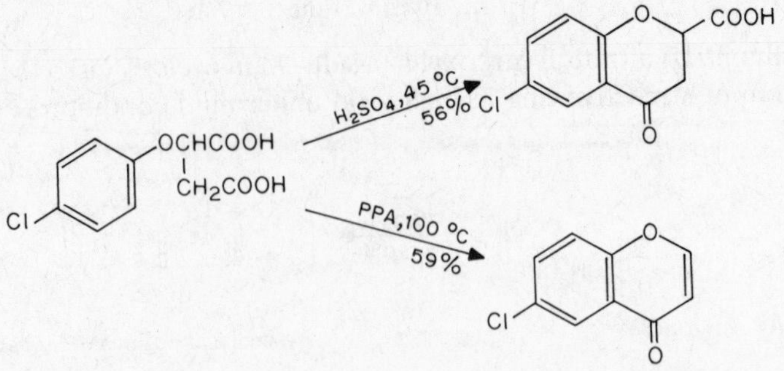

[712]

[712]

1-Benzopyran-4-one

A doubly fused Pyran-4-one is obtained from a diaryl ether containing a 2-carboxy group in one ring. Alternatively, one of the rings may be a quinone and the other may carry an ester group.

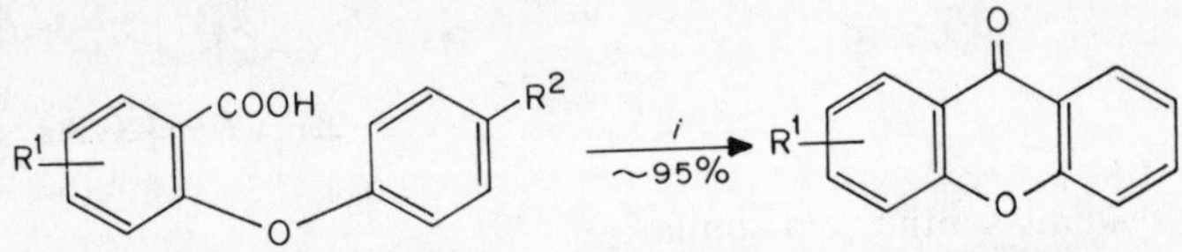

$R^1 = iPr, tBu; R^2 = H, COOH$

i, H_2SO_4 or $AlCl_3$–$NaCl, \Delta$

9-Xanthenone
(review [775])

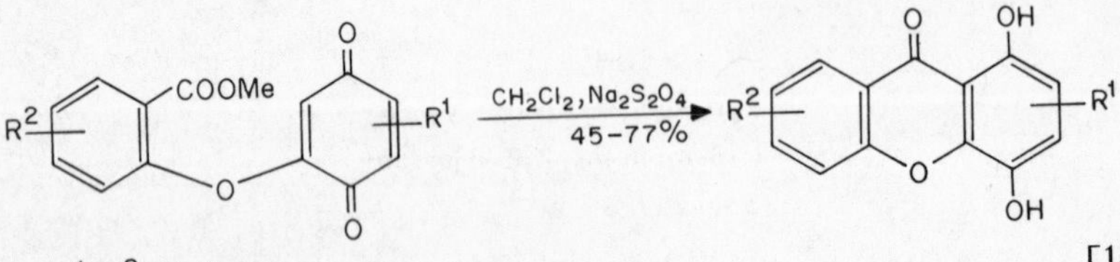

[243, 776]

$R^1, R^2 = H, MeO$

[1430]

Another ring system synthesized similarly:

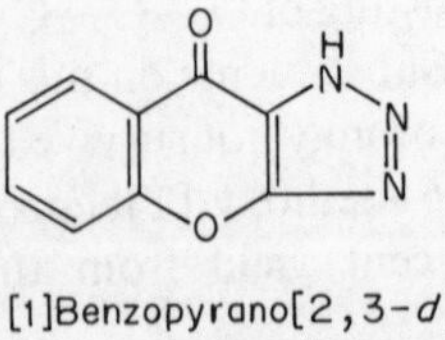

[497]

[1]Benzopyrano[2,3-d]-
1,2,3-triazol-9-one

6. Thiopyran-4-one

Subjecting the thioethers to reactions similar to those described in the preceding section gives the thiopyran-4-ones. The chemistry of 1-benzothiopyrans has been reviewed [1651].

Ring systems synthesized from the carboxylic acid and a catalyst by methods of the preceding section:

[417]

Thiopyrano[3,2-c][1,2]-
benzothiazin-4-one
6,6-dioxide

[891]

[1]Benzothiopyrano[2,3-d]-
1,2,3-triazol-9-one

[1645]

1-Benzothiopyran-4-one
(review [1651])

II. FORMATION OF A SEVEN-MEMBERED RING

1. Azepin-3-one or Azepine-2, 5-dione

Annulation of a side-chain carboxy group to the adjacent carbon atom in the ring is the usual method (cf. Sections I.4–I.6); several different catalysts are available, for example, aluminium chloride (with the acyl chloride) [344], tin(IV) chloride [548] or PPE (with the acid) [547]; it is sometimes necessary to raise the temperature by using diphenyl ether (with an acid chloride) [1040]. The following ring systems were synthesized by these methods:

[344]

3-Benzazepin-1-one
(review [1620])

[346]

Pyrazolo[3,4-c][2]benzazepine-
4,9-dione

[547, 548]

Pyrrolo[2,1-b][3]benzazepin-
11-one

[1040]

[1]Benzopyrano[2,3-c][1]-
benzazepine-6,12,13-trione

2. Oxepin-4-one

This ring, like that of azepinone, is formed by Friedel–Crafts acylation using one of the catalysts already mentioned.

Dibenz[*bf*]oxepin–10–one   [268]

Thieno[3,2–*c*][1]benzoxepin–
10–one   [899]

[1016]

[1]Benzoxepino[3,4–*c*]–
pyrazol–4–one

3. Thiepin-4-one, 1, 4-Thiazepin-5-one or 1, 3, 4-Thiadiazepin-6-one

The first and third of these rings may also be obtained by intramolecular acylation using PPA [270, 911] (review [B–21]) as catalyst-solvent.

[1]Benzothiepino[3,4–*d*]–   [270]
1,2,3–thiadiazol–10–one

Pyrrolo[1,2–*d*][1,2,4]triazolo–   [911]
[3,4–*b*][1,3,4]thiadiazepin–6–one

[270]

[1]Benzothiepino[4,3–*d*]–
1,2,3–thiadiazol–4–one

A synthesis of fused 1, 4-thiazepin-5-one was effected through amide formation between the carboxylic acid and the secondary amine at the 2-position.

1,5-Benzothiazepin–4–one   [735]

CHAPTER 74

Carboxylic Acid or Ester and Ring-nitrogen

I. FORMATION OF A FIVE-MEMBERED RING

1. Pyrrole or Pyrrol-2-one

A fused pyrrole ring is formed by heating a 2-carboxypiperidine or similar compound with DMAD and an anhydride.

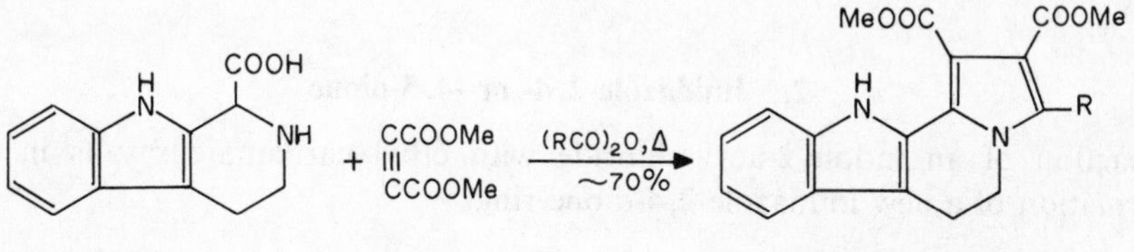

R = Me, Et

Indolizino[8,7-*b*]indole

[987]

Other ring systems synthesized similarly:

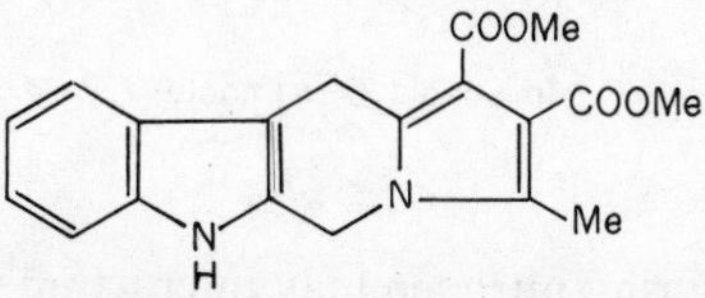

Indolizino[6,7-*b*]indole   [987]

Pyrrolo[1,2-*c*]thiazole   [1179]

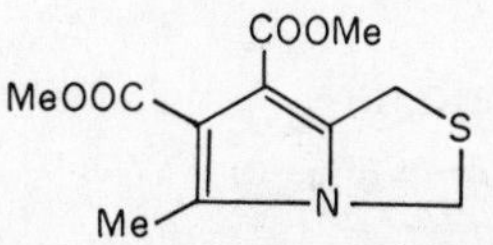

Annulation of a 2-(β-carboxyethyl)pyrimidin-4-one by hot acetic anhydride gives a fused pyrrolone ring. The two functions may also be in different rings.

[18]

Benzo[*f*]pyrrolo[2,1-*b*]-
quinazoline-10,12-dione

Another ring system synthesized similarly:

[703]

Pyridazino[4',5':3,4]pyrrolo-
[1,2-*a*]benzimidazol-11-one

Indolizinones are obtained in rather low yield by cyclization of a 2-pyrrolylacrylic acid in hot acetic anhydride. Simultaneous decarboxylation occurs when $R^2 = COOH$.

[1963]

$R^1 = H, Me$; $R^2 = H, COOH$

3-Pyrrolizinone
(review [1397])

2. Imidazole-2, 4- or -4, 5-dione

Reaction of an indole-2-acyl chloride with ethyl carbamate results in the formation of a new imidazole-2, 4-dione ring.

[592]

$R = HO, EtO, NH_2$;
$Ar = Ph, 2-FC_6H_4$

Imidazo[1,5-*a*]indole-1,3-dione

A side-chain ester and a neighbouring ring-nitrogen may interact on heating and in this example, an imidazole-4, 5-dione is formed.

$$R = Me, Et, PhCH_2$$

Imidazo[1,2-a]benzimidazole-
2,3-dione

[1321]

3. Thiazole or Thiazol-4-one

An unexpected heterocyclic system was obtained when *N*-arylpyrrol-2-ylthioacetic ester was treated with t-butoxide. An intramolecular rearrangement occurs to give the pyrrolothiazole. When the same side-chain is adjacent to an unsubstituted ring-nitrogen, it cyclizes on heating with acetic anhydride.

$$Ar = 2-NO_2C_6H_4$$

Pyrrolo[2,1-b]thiazol-3-one

[514]

Thiazolo[3,2-a]benzimidazol-
1-one

[1249]

Another ring system synthesized similarly:

Imidazo[2,1-b]thiazol-3-one

[1848]

II. FORMATION OF A SIX-MEMBERED RING

1. Pyridin-2-one

Lactam formation promoted by DCC and side-chain dehydrobromination occurs simultaneously in this example:

Pyrido[2,1-b][1,3]-
thiazin-6-one

[916]

2. Pyrimidin-4-one

A carboxylic acid or ester group reacts with a ring-nitrogen (whether N or NH) on heating the compound alone or with acid. The tendency to react with a ring-nitrogen is sometimes greater than it is with an amino group.

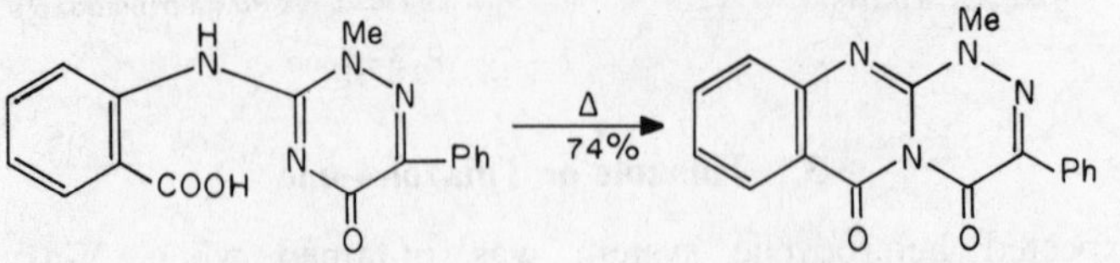

[1,2,4]Triazino[3,4-b]-
quinazoline-4,6-dione
[803]

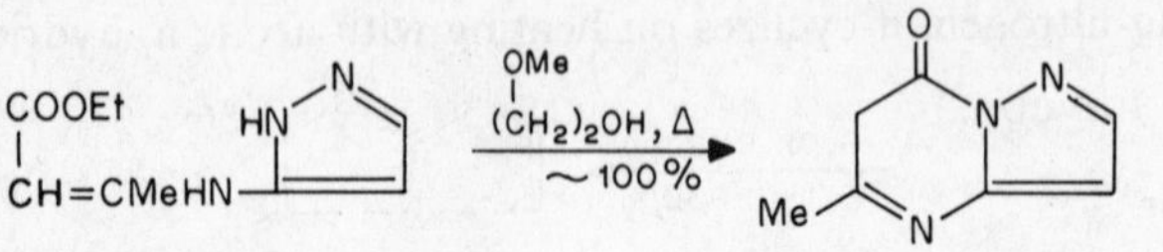

Pyrazolo[1,5-a]pyrimidin-7-one
[845]

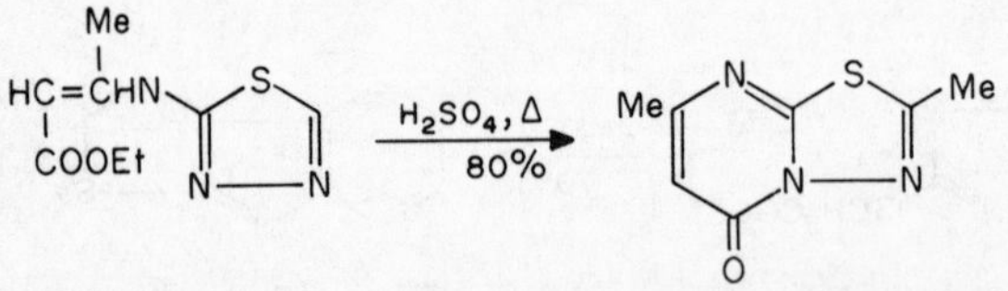

1,3,4-Thiadiazolo[3,2-a]-
pyrimidin-5-one
[1178]

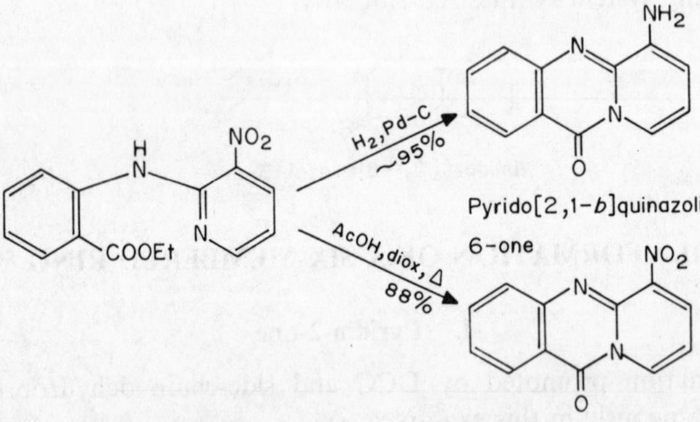

Pyrido[2,1-b]quinazolin-
6-one
[855]

[855]

Other ring systems synthesized by one of the above methods:

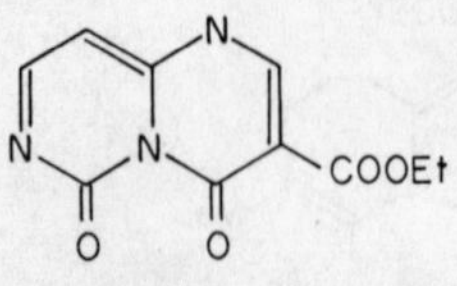

Pyrimido[1,6-a]pyrimidine-
4,6-dione
[260]

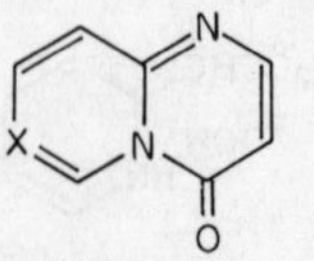

X = CH, N
Pyrido[1,2-a]pyrimidin-4-one
Pyrimido[1,6-a]pyrimidin-4-one
[991, 1155]

1,3,4-Thiadiazolo[3,2-*a*]-
pyrimidin-7-one

[1195]

Pyrimido[1,2-*a*][1,8]-
naphthyridin-10-one

[10]

Imidazo[1,5-*a*]pyrimidin-
4-one

[1852]

3. Pyrazin-2-one or Pyrazine-2,5-dione

Acylation of the ring NH with chloroacetyl chloride followed by treatment with a primary amine gives a new fused pyrazinedione in high yield.

i, ClCH$_2$COCl
ii, MeNH$_2$, NaHCO$_3$

Pyrazino[1,2-*a*]pyrrolo[2,1-*c*]-
[1,4]benzodiazepine-1,4-dione

[1238]

Dehydrative cyclization of the imidazol-4-carboxylic acid (**74.1**) gives a high yield of the pyrazinone.

(**74.1**)

Imidazo[1,5-*d*]pyrido-
[1,2-*a*]pyrazin-11-one

[1593]

4. 1,2,4-Triazin-5-one

A carboxylic acid or ester group in a side-chain containing a hydrazone function cyclizes on heating in an acid or a base to give this triazinone ring.

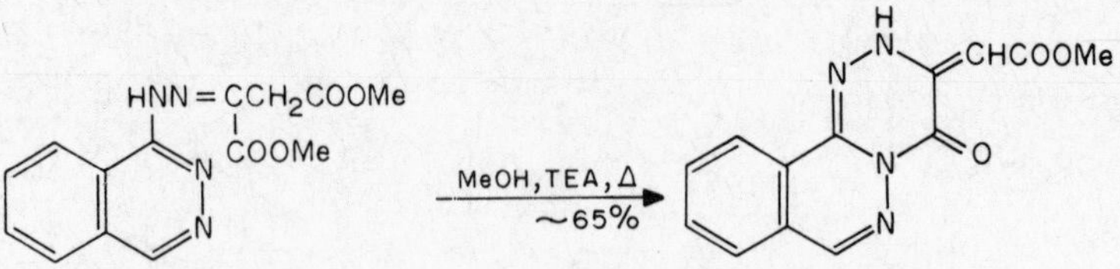

[701]

[1,2,4]Triazino[3,4-*a*]-
phthalazin-4-one

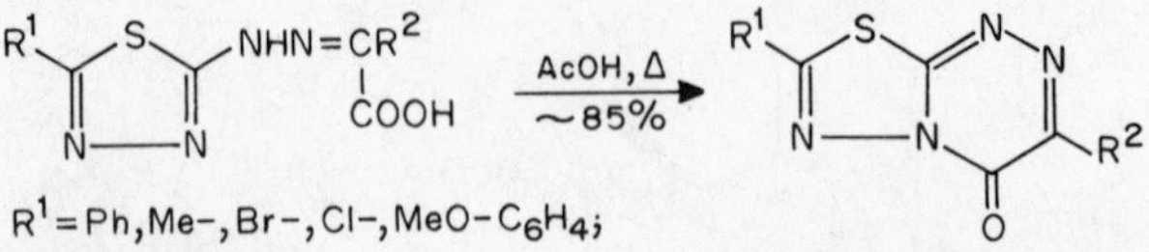

R¹ = Ph, Me-, Br-, Cl-, MeO-C₆H₄;
R² = Me, Ph

[1191]

[1,3,4]Thiadiazolo[2,3-*c*]-
[1,2,4]triazin-4-one

Another ring system synthesized similarly:

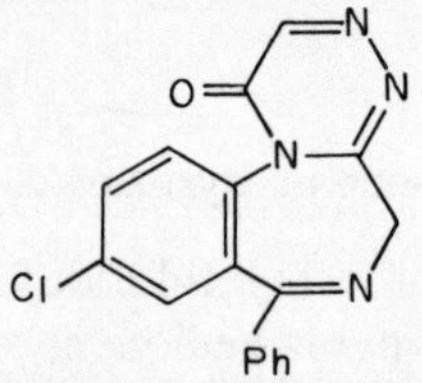

[1255]

[1,2,4]Triazino[4,3-*a*][1,4]-
benzodiazepin-1-one

5.   1,3-Thiazin-4-one

Thermal cyclization of a 2-SCH=CHCOOEt side-chain of an imidazole yields a
tricyclic thiazinone. A saturated carboxylic acid side-chain cyclizes in high yield
on heating in acetic anhydride.

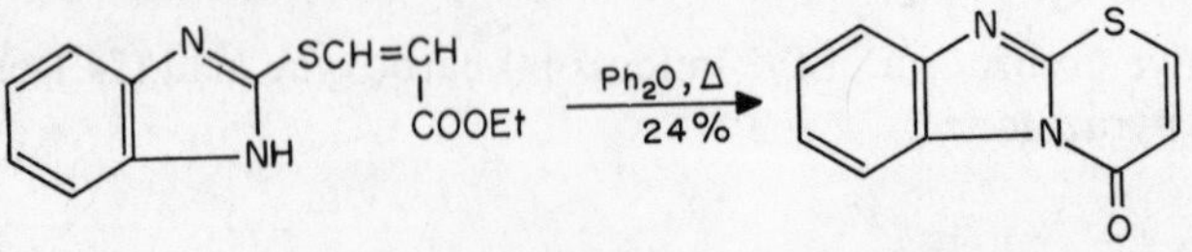

[912]

Ph

[1,3]Thiazino[3,2-*a*]-
benzimidazol-4-one

Another ring system synthesized similarly:

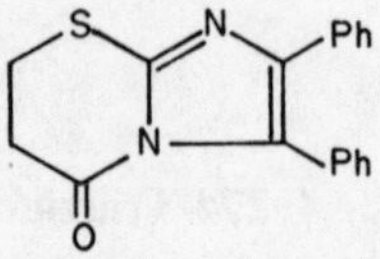

[1848]

Imidazo[2,1-*b*][1,3]-
thiazin-5-one

6. 1,3,5-Thiadiazine-2,4-dione or -2,4-dithione

A heterocyclic 2-trithiocarbonate (and its oxo analogue) is cyclized to this ring by reaction with an isocyanate or isothiocyanate in the presence of a base.

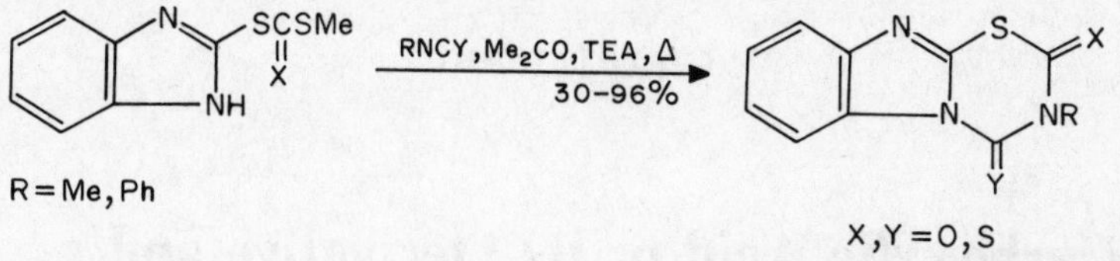

R = Me, Ph

X, Y = O, S
1,3,5-Thiadiazino[3,2-a]-
benzimidazole-2,4-dione
1,3,5-Thiadiazino[3,2-a]-
benzimidazole-2,4-dithione

[733]

CHAPTER 75

Carboxylic Acid or its Derivative and a Sulphinic or Sulphonic Acid Derivative

I. FORMATION OF A FIVE-MEMBERED RING

1. Pyrrol-3-one

Strictly, the acyl chloride is accompanied in this example by a reversed sulphonamide function, that is, a sulphonylamino group. Cyclization to a pyrrol-3-one is brought about by triethylamine at ambient temperature. The chemistry of indol-3-ones has been reviewed [1662].

$Ar = 4\text{-}MeC_6H_4$ 3-Indolone   [1299]

2. Isothiazol-3-one or its 1,1-Dioxide

Heating an N-t-butylaminosulphonylbenzoic acid with PPA (review [B-21]) causes simultaneous cyclization and dealkylation to an isothiazol-3-one 1,1-dioxide; the same ring is obtained from the sulphonamide-ester. 1,2-Benzisothiazoles have been reviewed [1617].

R = H, Me, MeO, Cl, F

1,2-Benzisothiazol-3-one
1,1-dioxide

[180]

Thieno[3,4-*d*]isothiazol-
3-one 1,1-dioxide

[993]

A fused isothiazol-3-one ring is formed from sulphoxide (or methylsulphinyl) and a carboxamide by heating with thionyl chloride.

R = alkyl, PhCH₂, Ph,
4-MeC₆H₄

1,2-Benzisothiazol-
3-one

[1102]

II. FORMATION OF A SIX-MEMBERED RING

1. Pyridin-2-one

A phenylsulphonyl group in a side-chain is displaced (cf. Chapter 63, Section I.1) when its *o*-carboxylic ester derivative is treated with a primary alkylamine at ambient temperature.

Pyrido[3,4-*b*]quinoxalin-
1-one 5,10-dioxide

[2025]

2. 1,2-Thiazin-3-one 1,1-Dioxide

When either the carboxyl or sulphonamide group is separated from the ring by a carbon atom, a thiazin-3-one is formed by heating under Dean and Stark conditions or with acetic anhydride or by conversion into the acyl chloride. 1, 2-Benzothiazine 1, 1-dioxides have been reviewed [1773].

i or ii, Δ
48–61%

[1748, 1993]

R = Me, 4-H_2NSO$_2$C$_6$H$_4$

i, xylene, TsOH ;

ii, AcOH, AcONa, Ac$_2$O

1,2–Benzothiazin–3–one
1,1–dioxide

SOCl$_2$
58–85%

[1748]

R = alkyl, PhCH$_2$, ClC$_6$H$_4$

2,3–Benzothiazin–4–one
2,2–dioxide
(review [942])

3. 1, 2, 6-Thiadiazin-3-one 1, 1-Dioxide

Aqueous alkali causes ring closure of a sulphonamide and an ester or carboxamide group.

aq. NaOH, Δ
68%

[1751]

Pyrazolo[3,4-c][1,2,6]-
thiadiazin–4–one 2,2–dioxide

aq. NaOH, Δ
55%

[67]

2,1,3–Benzothiadiazin–
4–one 2,2–dioxide

III. FORMATION OF A SEVEN-MEMBERED RING

1. 1, 2-Thiazepin-3-one 1, 1-Dioxide

2-(Sulphonamido)arylpropanoyl chlorides undergo thermal cyclization to a seven-membered ring.

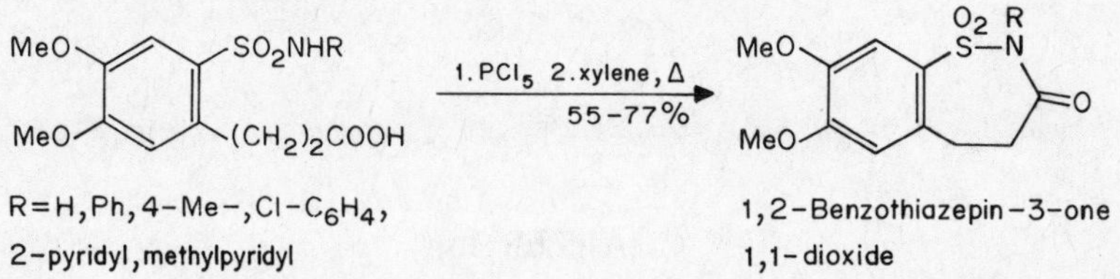

MeO—⟨benzene⟩—SO₂NHR + (CH₂)₂COOH →[1.PCl₅ 2.xylene,Δ / 55-77%] MeO—⟨benzothiazepinone⟩

R=H,Ph,4-Me-,Cl-C₆H₄,
2-pyridyl,methylpyridyl

1,2-Benzothiazepin-3-one
1,1-dioxide

[1217]

2. 1, 2, 4-Thiadiazepin-5-one 1, 1-Dioxide

Reaction of an *o*-alkoxycarbonyl sulphonyl chloride with an aminopyrazole yields a thiadiazepin-5-one ring. The reactions of 3(5)-aminopyrazoles have been reviewed [1123].

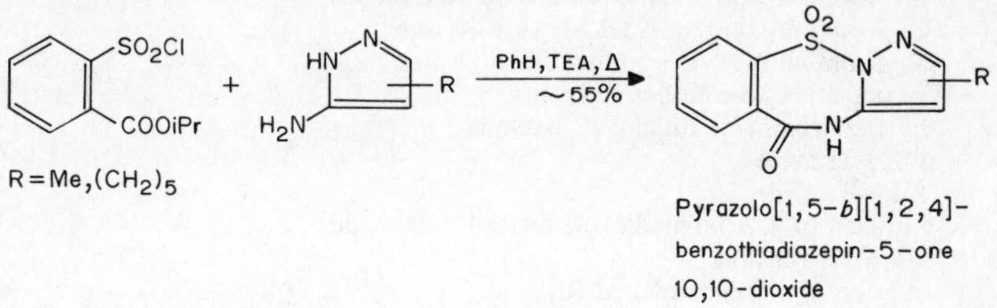

⟨C₆H₄⟩—SO₂Cl / COOiPr + HN–N / H₂N—⟨pyrazole⟩—R →[PhH,TEA,Δ / ~55%] ⟨product⟩—R

R=Me,(CH₂)₅

Pyrazolo[1,5-*b*][1,2,4]-
benzothiadiazepin-5-one
10,10-dioxide

[240]

CHAPTER 76

1, 2-Diamine

Cyclizations of 1, 2-diamines, their monoacyl and diacyl derivatives are closely related; the reaction conditions often (but not always) promote hydrolysis of the acyl group(s). Another difficulty in classifying such reactions is that both diamines and their monoacyl derivatives are often produced *in situ* from the nitroamine or nitroacylamine. In view of these problems, and the number of examples of each type, cyclization of monoacyldiamines is discussed separately (in Chapter 5). Cyclizations of nitro-amines in which the nitro is not reduced to the amine are included in Chapter 14. Cyclization of diamines, o-nitroamines and NN'-diacylamines are discussed in the present chapter together with one or two examples of reactions where a monoacyldiamine is clearly hydrolysed to the diamine before cyclization. Other examples of cyclizations of 1, 2-diamines will be found in Chapters 30 (Section III.2), 51 (Section I.2), 65 (Section III.2), 66 (Sections III.2, IV.1), 80 (Section III.1), 96 (Section II.1), 102 (Section II.2).

I. FORMATION OF A FIVE-MEMBERED RING

1. Imidazole

Many examples of the conversion of a 1, 2-diamine into a fused imidazole ring are described in the literature. This section is therefore subdivided according to the type of cyclizing reagent. The sulphur analogue of each reagent where appropriate is mentioned alongside the oxygen-containing compound.

A. *Reaction with a Carboxylic Acid*

The well-known Phillips synthesis of benzimidazoles has been applied to many heterocyclic and other diamines under a variety of conditions. A comprehensive survey of the synthesis of benzimidazoles up to 1977 [B-18] and a more recent treatise [B-4] provide detailed accounts. Reaction of a diamine with a carboxylic acid is the most commonly used method; this may proceed without a catalyst [1456] but usually requires the presence of a mineral acid, PPA, PPE (see below), the trimethylsilyl ester of polyphosphoric acid (PPSE) [1559] or phosphorus pentoxide-methanesulphonic acid [1730]. One of the amine groups may be substituted by an alkyl group but when one of the amine groups is tertiary (see [1618] for a review of this type of cyclization), the presence of hydrogen peroxide is necessary. Alkaline hydrolysis of a monoacyldiamine may be followed by acidification and reaction with a carboxylic acid [1896].

$$\text{HCOOH, H}_2, \text{Pd-C} \quad 29\text{-}98\%$$

R = H, Me

Imidazo[4,5-*f*]quinazoline-7,9-dione

[608]

$$\text{R}^2\text{COOH, PPA or PPE, } \Delta \quad 9\text{-}95\%$$

R^1 = H, Me, MeO, Cl, CF$_3$, NH$_2$;
R^2 = H, Ph, Cl-, Me-, NO$_2$-C$_6$H$_4$,
pyridyl

X = Y = CH
Benzimidazole
X = N, Y = CH
Imidazo[4,5-*b*]pyridine
X = CH, Y = N
Imidazo[4,5-*c*]pyridine

[199, 334, 420, 483, 924, 1458]

Another ring system synthesized similarly:

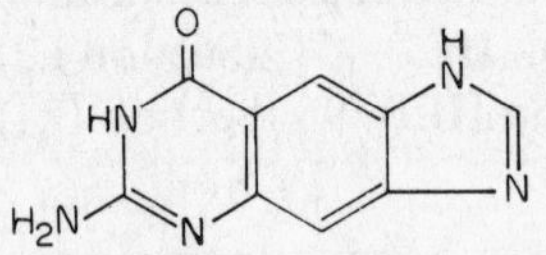

Imidazo[4,5-g]quinazolin-8-one

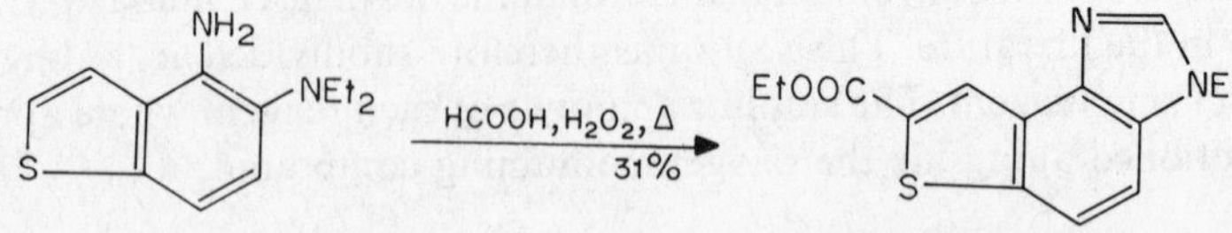

[75]

$\xrightarrow[31\%]{HCOOH, H_2O_2, \Delta}$

[1005]

Thieno[3,2-e]benzimidazole

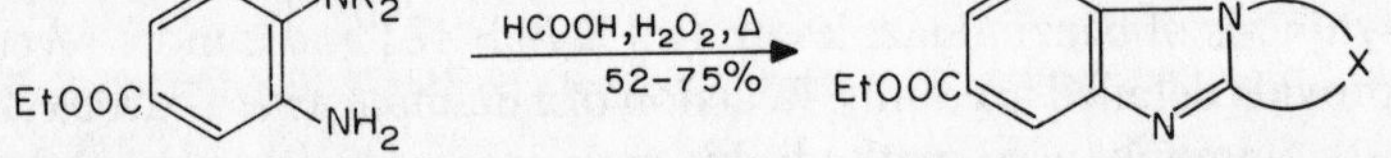

$\xrightarrow[52-75\%]{HCOOH, H_2O_2, \Delta}$

[217, 1491]

$NR_2 = N(CH_2)_n (n = 4,5),$
$N[(CH_2)_2]_2O$

$X = (CH_2)_4, (CH_2)_5$ or $N[(CH_2)_2]_2O$
Pyrrolo[1,2-a]benzimidazole
Pyrido[1,2-a]benzimidazole
[1,4]Oxazino[4,3-a]benzimidazole

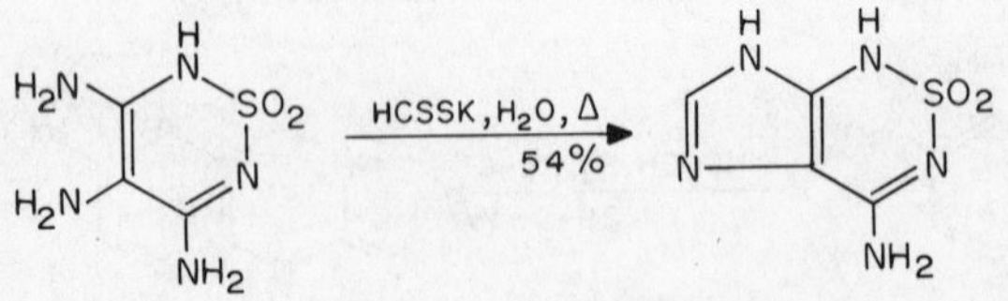

$\xrightarrow[54\%]{HCSSK, H_2O, \Delta}$

[1199]

Imidazo[4,5-c][1,2,6]-
thiadiazine 2,2-dioxide

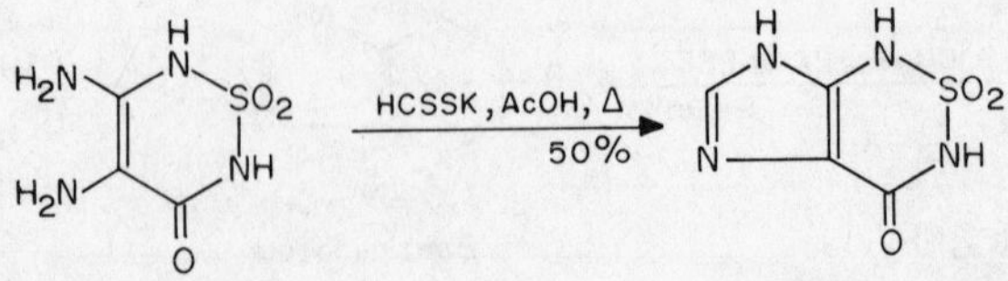

$\xrightarrow[50\%]{HCSSK, AcOH, \Delta}$

[1144]

Imidazo[4,5-c][1,2,6]-
thiadiazin-4-one 2,2-
dioxide

B. Reaction with a Carboxylic Acid Derivative

Esters, orthoesters, selenoesters, dithiocarbamates, anhydrides, acyl halides, nitriles, amidines, and imidate salts react with diamines to give a fused imidazole ring. In general, these reagents offer little advantage over the carboxylic acid but in particular compounds their use is recommended, for example, where the acid is thermally unstable or unstable to mineral acid. A few reagents will cyclize the diamine without the need to heat the mixture, for example, the seleno-ester (**76.1**, R^3 = EtO) [1231] and cyanogen bromide [295]. Synthesis of 2-aminobenzimidazoles has been reviewed [1429].

R^1 = alkyl; R^2 = H,
4-Cl-3-COOEt-
2-Me; R^3 = H, Me

Imidazo[4,5-*b*]pyridine   [1042, 1259]

R^1, R^2 = H, Me

Imidazo[4',5':4,5]thieno-
[2,3-*b*]pyridine   [1086]

Other ring systems synthesized similarly:

Imidazo[4,5-*b*][1,8]-
naphthyridin-8-one   [1994]

Imidazo[4,5-*d*]pyridazin-
4-one   [421]

Imidazo[4,5-*d*]-
pyridazine   [468]

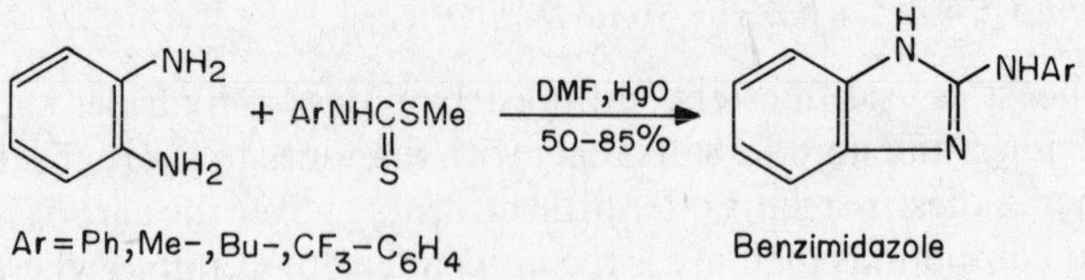

$$\text{Ar} = \text{Ph}, \text{Me}-, \text{Bu}-, \text{CF}_3-\text{C}_6\text{H}_4 \qquad \text{Benzimidazole}$$

[819]

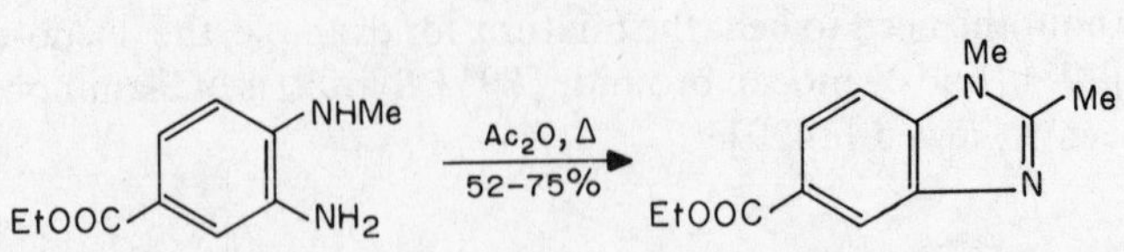

[217]

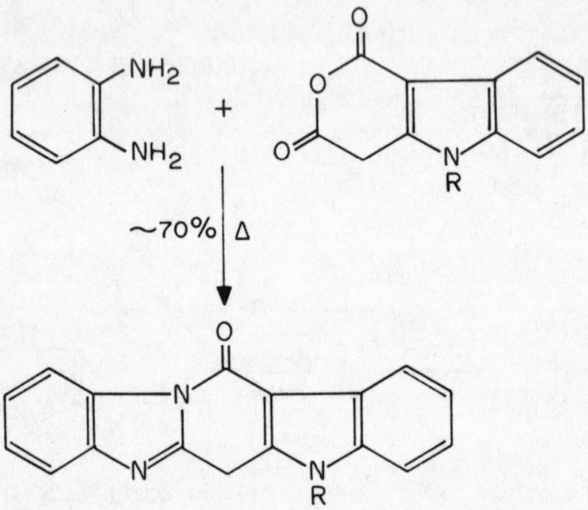

[198]

Indolo[2′,3′:4,5]pyrido[1,2-a]–
benzimidazol–13–one
R = Me, Et

Other ring systems synthesized similarly:

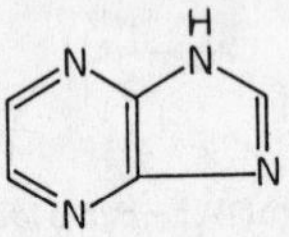

[1126]

Imidazo[4,5-b]–
pyrazine

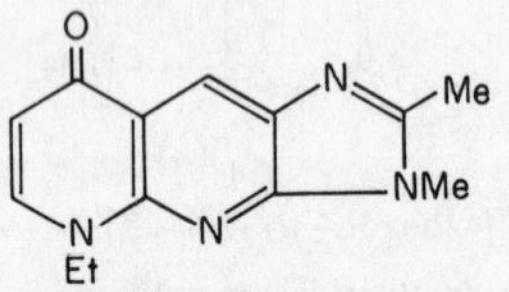

[1994]

Imidazo[4,5-b][1,8]–
naphthyridin–8–one

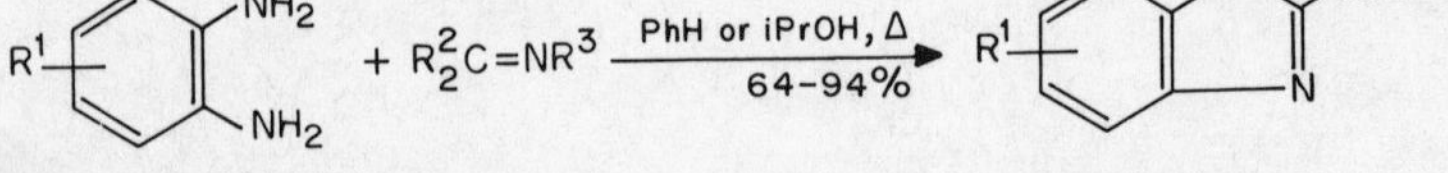

[99, 291]

$R^1 = H, Me, Cl; R^2 = Cl, PhO;$
$R^3 = CN, SO_2Me, SO_2Ph$ Benzimidazole

(76.1)

[403, 1231]

$R^1 = H, Me, Cl$; $R^2 =$ alkyl, aryl, heteroaryl;
$R^3 = EtO, NH_2$
$i, R^3 = EtO : EtOH$; $ii, R^3 = NH_2 : PhMe, \Delta$

$R = H, Me$

Imidazo[4,5-f]quinoline

[295]

[1447, 1996]

$R^1 = Bu, Cl, NO_2, Ph$; $R^2 = H, Me$
$R^3 = H$, pyrimidinyl derivative

[57, 1042]

$R^1 = H, CH_2CMeCOOEt$; $R^2 = H, NH_2$
      |
    NHAc

$X = CH, NR^2$
Benzimidazole
Imidazo[4,5-b]pyridine

[1997]

Ar = 2-ClC$_6$H$_4$

C. Reaction with an Aldehyde or a Ketone

Formaldehyde reacts with diamines to give 1-methylbenzimidazoles while other aldehydes may yield either 2- or 1,2-substituted benzimidazoles. Useful selectivity is shown for the formation of the 6-nitro (rather than the readily available 5-substituted) isomer in the formaldehyde cyclization.

[358]

[358]

Benzimidazole

Aromatic aldehydes in the presence of mild oxidizing agents (e.g., nitrobenzene, copper(II) acetate or iron(III) chloride) give a 2-arylbenzimidazole but, under non-oxidizing conditions, a 1,2-disubstituted benzimidazole is obtained. When $R^1 \neq H$, the oxidative method gives a 1-R^1-2-arylbenzimidazole. Pyridine diamines react with dimethylformamide dimethylacetal in hot DMF to form imidazopyridines.

[1388, 1396, 1742]

$R^1 = H, Me; R^2 = H, Me, Br, Cl, NO_2;$
$R^3 = 2$-furyl-, 5-Me-2-furyl,
2-aminopyridin-3-yl

[1388, 1396, 1742]

[1814]

$R =$ alkyl; $Ar = 4$-$HOOCCH_2OC_6H_4$

2,6-Purinedione

[1683]

X = Y = CH
Benzimidazole

X = N, Y = CH
Imidazo[4,5-b]pyridine

X = CH, Y = N
Imidazo[4,5-c]pyridine

Unsaturated and heteroaryl ketones react to form benzimidazoles with disproportionation of part of the molecule.

R^1 = 2-thienyl;
R^2 = 3- or 4-pyridyl;
R^3 = 3- or 4-pyridyl, 2-thienyl

Benzimidazole

[786]

[786]

D. From an NN'-Diacyldiamine

Cyclization of monoacyldiamines is discussed in Chapter 5. One of the acyl groups of the diacyldiamines is lost in this acid-catalysed cyclization which gives good yields of the fused imidazoles.

R^1=H,NH$_2$,AcNH; R^2=H,Me

Imidazo[5,4-c]carbazole

[641]

R^1, R^2=H,Me

Imidazo[4',5':4,5]thieno-
[2,3-b]pyridine

[1086]

Thieno[3,4-d]imidazole

[805]

E. Reaction with Other Reagents

An imidazole ring is formed by reaction of a diamine with Vilsmeier reagents, DMF-sodium mercaptide, or tetraethoxymethane. The mono-Schiff base of a diamine may be oxidatively cyclized at ambient temperature. Vilsmeier reagents are less likely than orthoesters to give a mixture containing the undesired uncyclized byproducts [687].

$R^1 = H, Me, Et; R^2 = Me, Et$ → Purine (DMF, $POCl_3$, Δ, ~40%) [687]

$R = Cl, MeS$ → Imidazo[4,5-d]pyridazine (DMF, MeSNa, Δ, ~93%) [667]

$R = CH_2OHCH, aryl, glycosyl$ → 2,6-Purinedione ($HgCl_2$, ~100%) [609]
       |
       OH

$(EtO)_4C$, Δ, 37–93% [698]

X = Y = CH — Benzimidazole
X = CH, Y = N — Imidazo[4,5-b]pyridine
X = Y = N — Purine

Isothiocyanates in the presence of DCC convert diamines into fused 2-aminoimidazoles. Substituted 2-aminobenzimidazoles are of considerable biological importance (as antiparasitic compounds) and may also be prepared from several imidic compounds such as N-cyano-, N-alkyl-, N-aryl-, or N-ethoxycarbonyl-derivatives as illustrated in the examples.

R^1, X, NH$_2$, NH$_2$ + R^{2}NCS $\xrightarrow{\text{PhMe or MeCN,DCC,}\Delta}$ benzimidazole/purine ring with NHR^2

[186, 850, 1808]

R^1=H,NH$_2$,PhCH$_2$O,PhCO;
R^2=Ph,COOMe

X=CH,N
Benzimidazole
Purine

R^1, NH$_2$, NH$_2$ + R$_2^2$C=NR^3 $\xrightarrow{i,\Delta}$ benzimidazole with NHR^3

R^1	R^2	R^3	%	Ref.
H, Cl	PhO	CN	~90	99
H, Me, Cl	Cl	SO$_2$Me	64–94	291
H, Me, Cl, NO$_2$	MeS	Me, Ar	50–80	684
H	MeS, EtO	COOEt	72	1589

Ar = Ph, Me-, MeO-, Cl-C$_6$H$_4$

Iron(II) oxalate has reducing and other properties and converts a 2-nitro-*N*-arylamine into a 1-aryl-2-methylbenzimidazole [1176]. The *N*-piperidinyldiamine (76.2) requires the presence of a mercury salt while the reactive halide (76.3) cyclizes the diamine on prolonged heating in dioxan.

structure with NHR, NHC=NCN, SMe $\xrightarrow[\sim85\%]{\text{MeOH,Hg(OAc)}_2}$ benzimidazole with R, NHCN

[422]

(76.2)

Benzimidazole

R = piperidine-NCOOEt

R, NH$_2$, NH$_2$ + chloropyridine with NO$_2$, Cl (76.3) $\xrightarrow[35-80\%]{\text{HCl,diox,}\Delta}$ pyridobenzimidazole with R, NO$_2$

[220]

(76.3)

Pyrido[1,2-*a*]benzimidazole

R=H,Me,Cl,NO$_2$

Deoxygenative cyclization of a 2-nitro-imine with triethyl phosphite in boiling t-butylbenzene gives moderate yields of 2-arylbenzimidazoles. An unusual reaction is that in which a methyl group becomes the 2-C atom of an imidazole on heating a 2-picoline with a diamine and sulphur.

[1451]

R^1 = H, Me, MeO, Cl, NO_2, NH_2;
R^2 = H, alkyl, COOEt, $CONH_2$, HO, Cl

[1458]

Palladium-mediated alkyl transfer gives a mixture of two products.

37%   25%   [6]

2. Imidazol-2-one or Imidazole-2-thione

In this section these tautomeric compounds are regarded as existing in the keto or thione form unless specific evidence for the enol form is available. A diamine is converted into the imidazolone by reaction with phosgene or urea and into the corresponding thione with carbon disulphide in alkali. Synthetic uses of carbon disulphide have been reviewed [2017]. Isothiocyanates when heated with diamines yield imidazolethiones; di-isothiocyanates react in a strongly basic medium with 1, 3-diketones to give a similar ring system.

[449, 950]

R^1 = H, Ac; R^2 = MeO,
$CH_2CMeCOOEt$
         |
        NHAc

2-Benzimidazolone

[195, 1072]

R = H, Cl

i, $CO(NH_2)_2$;
ii, CS_2, pyr, NaOH

X = O, S

Imidazo[4, 5-b]pyrazin-2-one
Imidazo[4, 5-b]pyrazine-2-thione

$R^1 = Et_2N(CH_2)_2;$
$R^2 = H, Cl, CF_3, NO_2$

2-Benzimidazolethione   [292]

Other ring systems synthesized using carbon disulphide:

[1994]

Imidazo[4,5-*b*][1,8]naphthyridin-
8-one, 2-thioxo

[204]

Imidazo[4,5-*d*]pyridazin-
4-one, 2-thioxo

R = Me, Ph

[1696]

2-Benzimidazolethione

When 2,3-diaminopyridine is treated with ethyl 2-oxocyclohexanecarboxylate, cyclization occurs only when the temperature is sufficiently high. In refluxing xylene, the product is an imidazopyridine and it may be produced by thermal breakdown of a diazepine intermediate.

[1990]

Imidazo[4,5-*b*]pyridin-2-one

3. 1, 2, 3-Triazole

This ring is commonly formed from a 1,2-diamine by the action of nitrous acid but nitrosodiphenylamine has also given good results and under weakly acidic conditions. The nitrous acid method is effective when both groups are primary or when one carries an aryl or heteroaliphatic ring.

[468, 835]

1,2,3-Triazolo[4,5-*d*]-
pyridazine

Other ring systems synthesized similarly:

X = CH, N

Benzotriazole

1,2,3-Triazolo[4,5-c]-
pyridine

[199, 290, 430, 1306, 1965, 1998]

1,2,3-Triazolo[4,5-c]-
pyridazine   [181]

[918]

1,2,3-Triazolo[4,5-d]-
pyrimidin-7-one
(review[2023])

1,2,3-Triazolo[4,5-b][1,8]-
naphthyridin-8-one   [1994]

1,2,3-Triazolo[4,5-b]-
quinoline   [174]

$\xrightarrow[71-98\%]{AcOH, \Delta}$

X = CH, N

Benzotriazole

1,2,3-Triazolo[4,5-b]-
pyridine   [51]

4. 1,2,4-Triazole

A carboxylic acid, a carboxylic ester, an acyl chloride, an anhydride or an orthoester cyclizes a heterocyclic diamine in which one of the amine groups is part of a hydrazine; a fused 1,2,4-triazole ring is thus formed.

$\xrightarrow[90\%]{HCOOH, \Delta}$

[1,2,4]Triazolo[5,1-c]-
[1,2,4]triazin-4-one   [748]

Another ring system synthesized similarly:

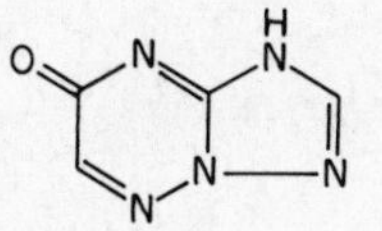

[832]

[1,2,4]Triazolo[1,5-b]-
[1,2,4]triazin-7-one

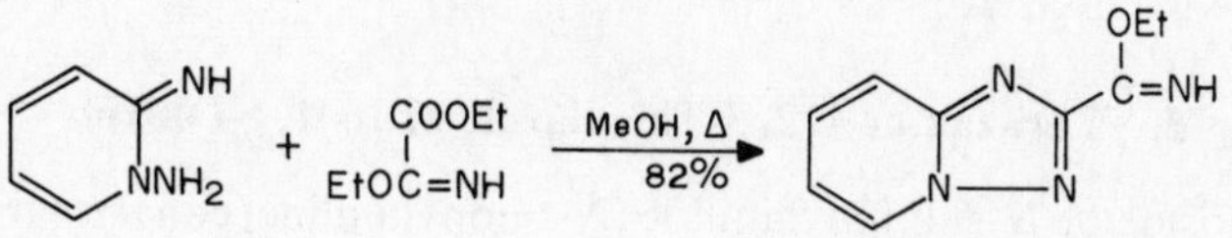

[1419]

[1,2,4]Triazolo[1,5-a]-
pyridine

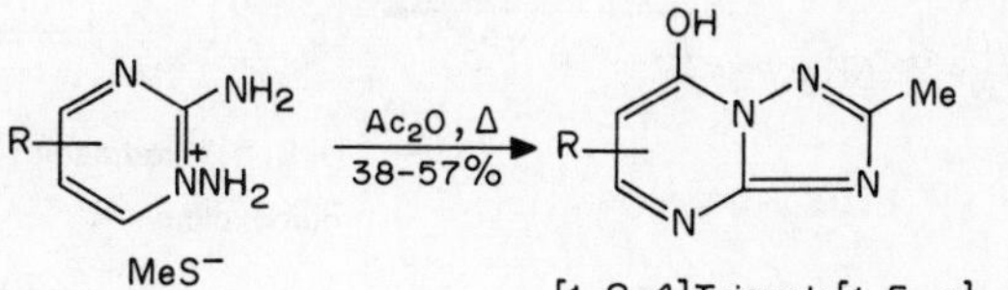

[1338]

MeS⁻

R=H,Me,MeO,Cl

[1,2,4]Triazolo[1,5-a]-
pyrimidine

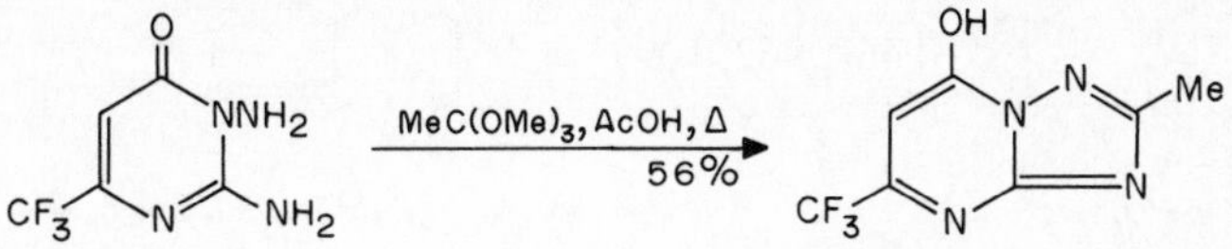

[1457]

[1,2,4]Triazolo[1,5-a]-
pyrimidine

Other ring systems synthesized similarly:

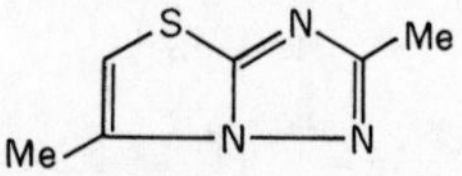

[1257]

Thiazolo[3,2-b][1,2,4]-
triazole

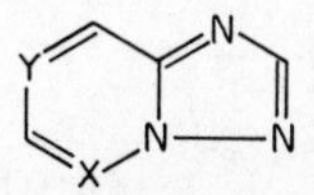

[1338]

X=CH,Y=N
[1,2,4]Triazolo[1,5-a]pyrazine
X=N,Y=CH
[1,2,4]Triazolo[1,5-b]pyridazine

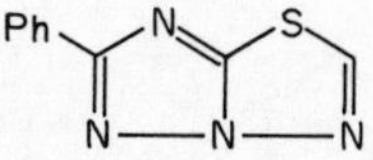

[1436]

[1,2,4]Triazolo[5,1-b]-
1,3,4-thiadiazole

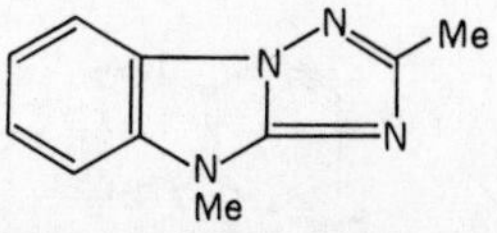

[1436]

[1,2,4]Triazolo[1,5-a]-
benzimidazole

[1856]

[1,2,4]Triazolo[5,1-b]-
quinazolin-5-one

[557]

[1]Benzothieno[3,2-e][1,2,4]-
triazolo[1,5-c]pyrimidine

5. Furazan or 1, 2, 5-Thiadiazole or its 1, 1-Dioxide

Thionyl chloride or N-sulphinylaniline (N-thionylaniline) converts diamines (or
their N-alkyl or N-acyl derivatives) into 1, 2, 5-thiadiazole derivatives.

$\xrightarrow[\sim 95\%]{SOCl_2,\,pyr,\,(CH_2Cl)_2}$

[1185]

[1,2,5]Thiadiazolo[3,4-b]-
quinoxaline

Other ring systems synthesized similarly:

[467, 962, 1099]

n = 0,1
2,1,3-Benzothiadiazole
2,1,3-Benzothiadiazole-
2-oxide

[1339]

[1,2,5]Thiadiazolo[3,4-b]-
pyrazine

[810]

Isothiazolo[3,4-e]-2,1,3-
benzothiadiazole

[1538]

[1,2,5]Thiadiazolo[3,4-c]-
quinoline

$\xrightarrow[51-98\%]{PhMe,\,PhNSO,\,\Delta}$

[187, 962]

R = Me, Et, Br, MeO

2,1,3-Benzothiadiazole
(review [1078])

Other ring systems synthesized similarly:

[1,2,5]Thiadiazolo[3,4-d]-
pyrimidine

[197]

[1,2,5]Thiadiazolo[3,4-c][1,2,6]-
thiadiazine 5,5-dioxide

[109]

A fused 1,2,5-thiadiazole or its 1,1-dioxide may also be formed by reacting *o*-phenylenediamine with one of three other reagents: piperidinesulphenyl chloride, sulphur dioxide or sulphamide. An example of the synthesis of a fused furazan (1,2,5-oxadiazole) ring is mentioned in Section II.1.F.

[1177]

[193]

2,1,3-Benzothiadiazole

[1492]

2,1,3-Benzothiadiazole
2,2-dioxide

6. 1,3,4-Thiadiazole

A 1,3,4-thiadiazole ring may be obtained from a diamine and carbon disulphide. The use of this reagent in organic synthesis has been reviewed [2017].

R=Ph,4-Me-,
4-MeO-C$_6$H$_4$

[1,3,4]Thiadiazolo[2,3-c]-
[1,2,4]triazine-7-thione

[196]

II.   FORMATION OF A SIX-MEMBERED RING

1.   Pyrimidine

Imine (C=NH) and amino groups may be annulated by either an anhydride or a ketone.

[1706]

[1706]

Purine

$R^1 = Me, Et, CF_3$; $R^2 = Me, Ph$;
$R^3 = H, alkyl, Ac, PhCO,$
CH_2COOEt, Ph

2. Pyrazine, 2-Pyrazinone or 2,3-Pyrazinedione

Annulation of a 1,2-diamine into a pyrazine ring is a common type of reaction
and the amino groups may be attached to a variety of rings, as the following
examples show. These are subdivided according to the type of reagent used.

A. Reaction with Glyoxal

A diamine usually reacts with glyoxal under mild conditions; the fully aromatic
pyrazine ring is not formed when one amine group is substituted, for example, by
a methyl group. Sodium hydrogen sulphite is sometimes added to the aldehyde
prior to cyclization or during the reaction [187, 293, 1995].

[187]

R = MeO, Cl

Quinoxaline
(review [1669])

[607]

$R^1 = H, Me$; $R^2 = H, Me$

Pyrazino[2,3-e]-1,2,4-triazine

Other ring systems synthesized similarly:

[608]

Pyrazino[2,3-f]quinazoline-
8,10-dione

[190]

Pyrazino[2,3-f]quinoxaline

Pteridine   [293]

Pyrazino[2,3-c]-
[1,2,6]thiadiazine
2,2-dioxide   [1260]

Isoxazolo[4,5-b]pyrazine   [1995]

B.   Reaction with a 2-Oxoaldehyde

This cyclization may be effected in an alkaline, acidic or neutral medium; 1,1-dichloroacetone probably behaves as an oxoaldehyde when it is stirred with a diamine in aqueous sodium acetate. Control of the pH at 5.5–6.0 is important in this cyclization [1601].

R = Me, Ph

X = NPh
1,2,3-Triazolo[4,5-b]pyrazine
X = O
[1,2,5]Oxadiazolo[3,4-b]pyrazine   [1173]

Other ring systems synthesized similarly:

Thieno[3,4-b]pyrazine   [805]

X = CH, N
Quinoxaline
Pteridine   [977, 1601, 1921]

C.   Reaction with a 1,2-Diketone

Cyclic and acyclic diketones react to give a pyrazine ring but benzocyclobutane-1,2-dione can give unexpected products, depending on the substituents and reaction conditions (see Section IV). Unsymmetrical ketones give mixtures of isomers whose ratio varies with the acidity of the medium in some examples [1853].

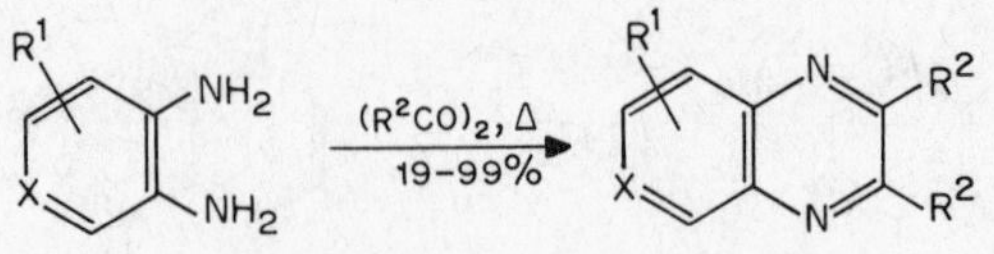

[199, 1403, 1999]

R^1=H,Ph,COOH,COOEt;
R^2=Me,Ph

X=CH,N
Quinoxaline
Pyrido[3,4-b]pyrazine

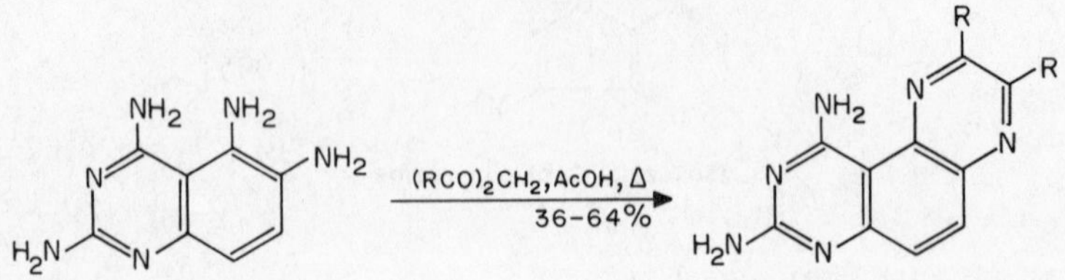

[945]

2HCl

R=Me,Ph

Pyrazino[2,3-f]quinazoline

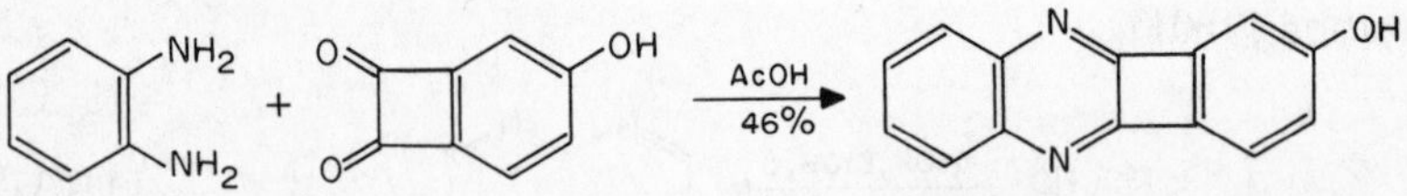

[219]

Benzo[3,4]cyclobuta[1,2-b]-
quinoxaline

Other ring systems synthesized similarly:

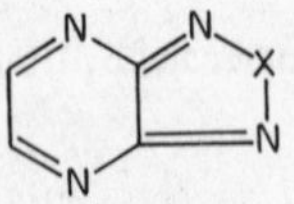

[1173]

X=NPh
1,2,3-Triazolo[4,5-b]-
pyrazine
X=O
[1,2,5]Oxadiazolo[3,4-b]-
pyrazine

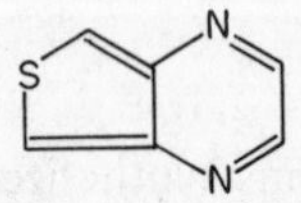

[805]

Thieno[3,4-b]pyrazine

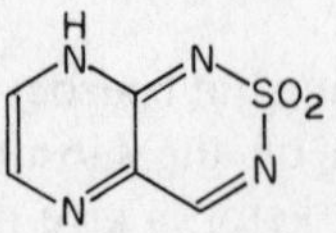

[1260]

Pyrazino[2,3-c][1,2,6]-
thiadiazine 2,2-dioxide

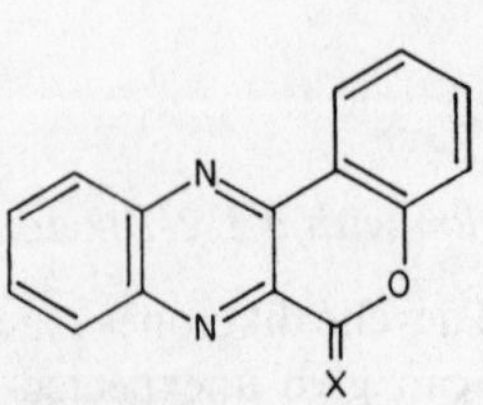

[189]

X=H₂ or O
[1]Benzopyrano[3,4-b]-
quinoxalin-6-one

Pyrazino[2,3-*b*]quinoline [174]

4-Pteridinone [1967, 2000]

2,3-Quinoxalinedione [1999]

X=CH
Quinoxaline
X=N
Pyrido[2,3-*b*]pyrazine [2000]

When the amine groups are aliphatic, reaction with a cyclic 1,2-diketone proceeds under mild conditions.

Isoquino[2,1-*a*]quinoxaline [426]

D. Reaction with a 2- or 3-Oxoester, a Diacid or a Diester

Oxalic acid or its esters condense with a diamine to give 2,3-quinoxalinediones [1741] while derivatives of mesoxalic acid yield the 2-oxo-3-carboxylic derivative. The chloromethyleniminium salt (76.4) derived from oxalic acid reacts with a diamine by displacement of the chlorine, condensation of the ester and retention of the amine group.

4-Pteridinone [383]

X or Y = OH
Pyrazino[2,3-*f*]quinoxaline [384]

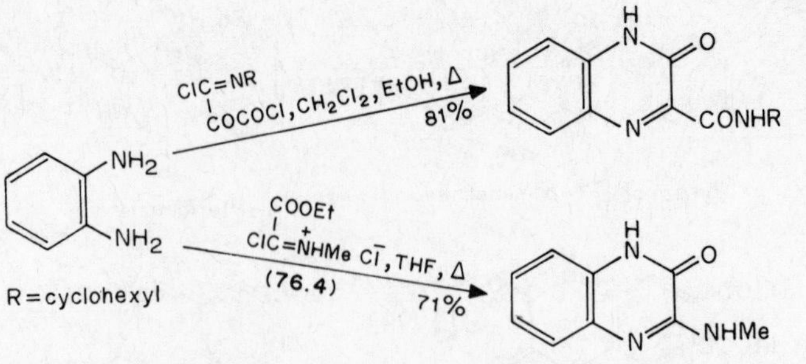

[1828]

[451]

R=cyclohexyl

2-Quinoxalinone

Enolizable 3-oxo-esters or -acids on heating with a diamine in ethanol give good yields of 2-quinoxalinones.

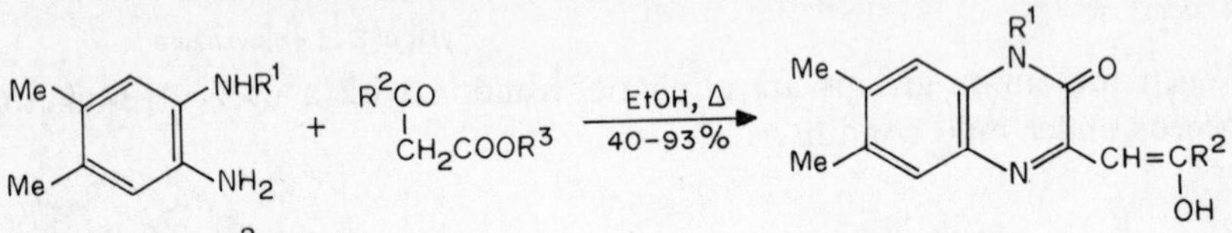

$R^1, R^3 = H, Me$; $R^2 = Me, Ph$

2-Quinoxalinone

[1927]

Other ring systems synthesized similarly:

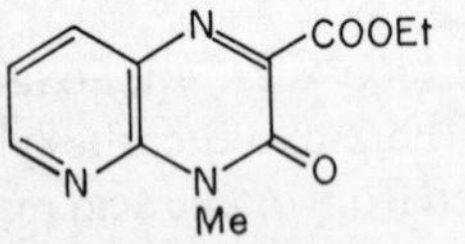

[246]

Pyrido[2,3-*b*]pyrazin-3-one

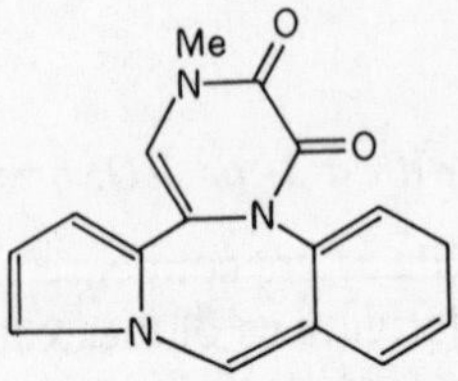

[828]

Pyrazino[1,2-*a*]pyrrolo-[2,1-*c*][1,4]benzodiazepine-3,4-dione

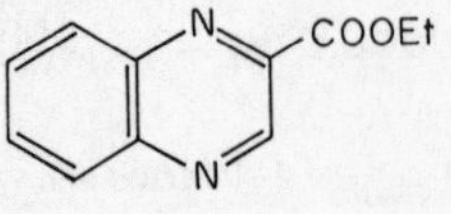

[1921]

Quinoxaline

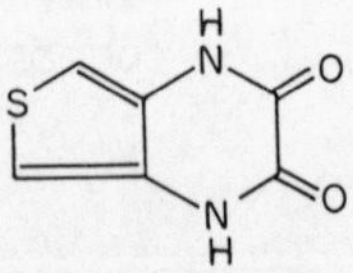

[805]

Thieno[3,4-*b*]pyrazine-2,3-dione

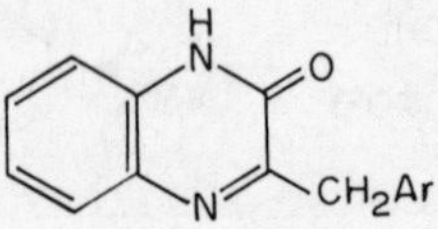

[7]

2-Quinoxalinone

E. Reaction with an Oxosulphoxide or a Sulphone

The sulphur function is displaced when these compounds are heated with a diamine in weakly acid or basic solution (cf. Chapter 63, Section I.1).

R = H, Me, MeO, Cl, NO₂
i, PhCOCH₂SOMe, AcOH, PhH
ii, PhCH—CHSO₂Ts, DMF
        \O/

Quinoxaline
(review [B-15])

[188, 482]

Other ring systems synthesized similarly:

[545]

4-Pteridinone

[606]

Pyrido[3,4-b]pyrazine

F. Reaction with Other Reagents

A compound containing two reactive halogen atoms or a reactive halogen adjacent to a keto group or a functionalized keto group reacts with a 1,2-diamine to form a pyrazine ring. In contrast to the benzimidazole formed by the action of iron(II) oxalate with N-methyldiphenylamines (Section I.1.E), the demethyl homologue yields phenazine. High yields of pyrazinone derivatives are obtained from diamines and ethyl ethoxyiminoacetate.

Bis[1,2,5]thiadiazolo-
[3,4-b:3',4'-e]pyrazine

[1185]

2-Quinoxalinone

[1085]

Quinoxaline

[1085]

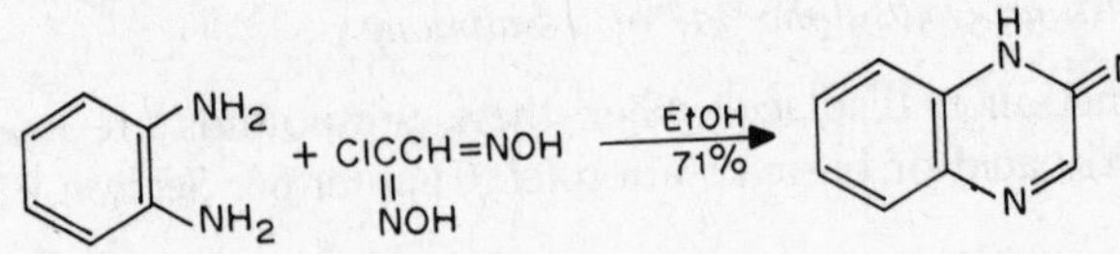

[1337]

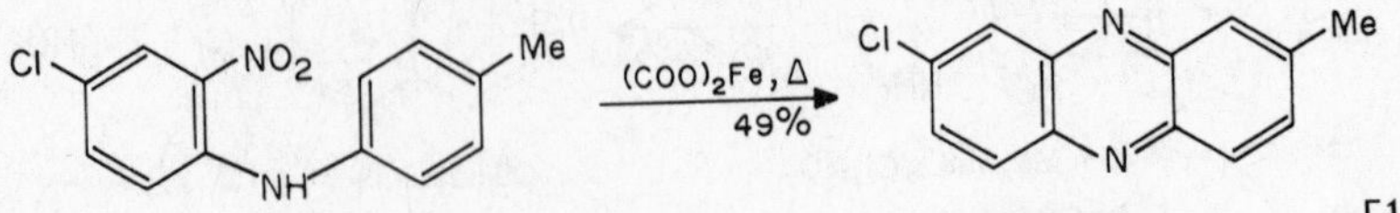

Phenazine   [1176]

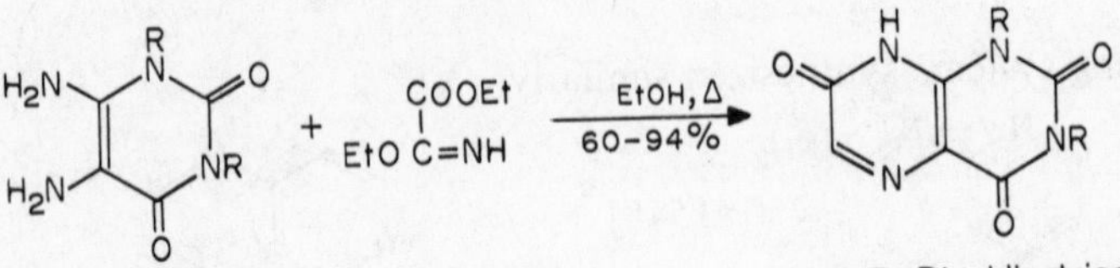

R = H, Me   2,4,7-Pteridinetrione

Other ring systems synthesized similarly:

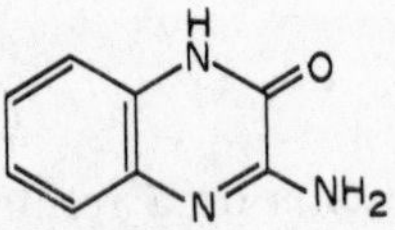

[1419]

2-Quinoxalinone

[1419]

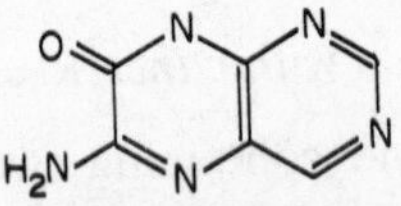

7-Pteridinone

The dithioic ester, NCCSSMe, annulates a diamine and forms the 3-amino-pyrazine-2-thione in moderate yield. When dichloroglyoxime is treated with alkali, the dinitrile oxide is obtained and is so reactive that at 0 °C it converts o-phenylenediamine into the quinoxaline 'dioxime'. When this is treated with iodosobenzene bis(trifluoroacetate), the di(hydroxyimine) forms a furazan 1-oxide (furoxan) ring in high yield.

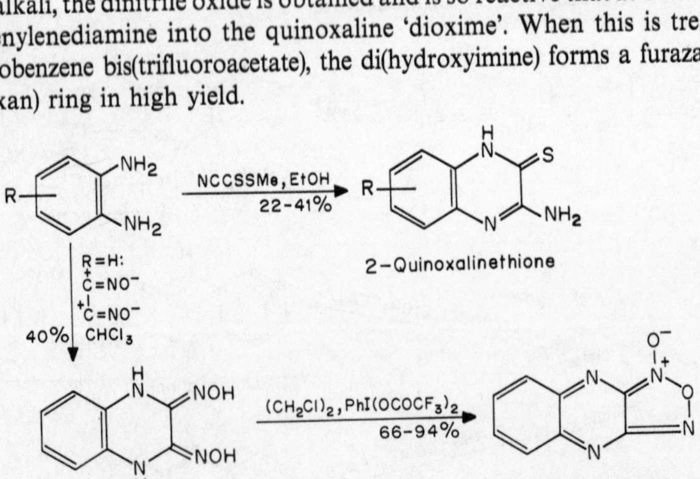

[1630]

2-Quinoxalinethione

R = H, Me   [1,2,5]Oxadiazolo[3,4-b]-quinoxaline 1-oxide

[1090, 1843]

3. 1, 2, 4-Triazine or 1, 2, 4-Triazin-5-one

Reaction of a 1, 2-diamine (one being an N-amino group) with a 2-oxo-acid or -ester or a 1, 2-dicarbonyl compound gives high yields of a fused triazine or triazinone depending on the oxidation level of the carbonyls. Cyclization can occur in either an acidic or basic medium.

MeHN— / H₂NN— HCl + COOEt / COR → (AcONa, AcOH, Δ, ~85%) → 1,2,4-Triazolo[4,3-b]-[1,2,4]triazin-7-one   [748]

R = H, Me, COOEt

Ph— —NNH₂ / —NH₂ + PhCCOOH → (HCl, Δ, 90%) → Imidazo[1,2-b][1,2,4]-triazin-3-one   [201]

H₂N— / H₂NN— —R¹ + COR² / COR³ → (AcOH, 38–65%) → 1,2,4-Triazolo[4,3-b]-[1,2,4]triazine   [1367]

R¹–R³ = H, Me

Other ring systems synthesized similarly:

Pyrazolo[1,5-b][1,2,4]-triazine   [1874]

1,2,4-Triazolo[4,3-b]-[1,2,4]triazine   [385]

III. FORMATION OF A SEVEN-MEMBERED RING

1. 1, 4-Diazepine or 1, 4-Diazepin-5-one

The three carbon atoms needed to complete the diazepine ring may be provided by a 1,3-dione, a 2,3-unsaturated ketone, an allenic group, an alkyne or by carbon atoms of another ring. In (76.5), when R³ ≠ H or when R² = Cl or 2,4-dinitrophenyl and R² = H, no diazepines are formed. From the allenic diester (**76.6**), a single product is obtained when $R^3 = H$ but substitution of this hydrogen leads to a mixture of benzodiazepines in which either R^3 or R^4 is a substituent. Ketene dithiols or dithioacetals condense with a diamine to give good yields of sulphur-containing diazepines. The chemistry [1619] and medicinal chemistry [1895, B-24] of benzodiazepines have been reviewed.

$R^1 = H, COOEt$; $R^2 = Me, PhCH_2,$
$HC\equiv CCH_2$

(**76.5**)

1,5–Benzodiazepine
(review [1619])

[1127, 1999]

Other ring systems synthesized similarly:

Thieno[3,4-*b*][1,4]–
diazepine

[805]

[1,2,5]Oxadiazolo[3,4-*b*]–
[1,4]diazepine

[958]

Thieno[3,4-*b*][1,4]diazepin–
2–one

[191]

Isoxazolo[4,5-*b*][1,4]–
diazepine

[1995]

1,5–Benzodiazepine

[786]

$R^1 = 2$-thienyl;
$R^2 = 3$- or 4-pyridinyl
i, CHCOOMe
‖
C=CR^{3}COOMe

(**76.6**)

1,5–Benzodiazepin–2–one

[1100]

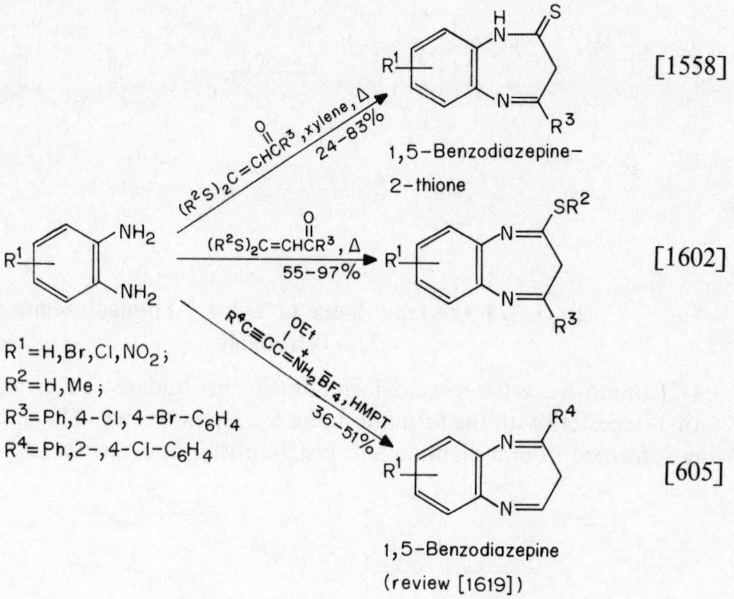

$(R^2S)_2C{=}CH\overset{O}{\overset{\|}{C}}R^3$, xylene, Δ
24–83%
1,5-Benzodiazepine-2-thione
[1558]
$(R^2S)_2C{=}CH\overset{O}{\overset{\|}{C}}R^3$, Δ
55–97%
[1602]
$R^4C{\equiv}C\overset{OEt}{C}{=}\overset{+}{N}H_2\ \overset{-}{B}F_4$, HMPT
36–51%
[605]
1,5-Benzodiazepine
(review [1619])
R^1=H,Br,Cl,NO_2;
R^2=H,Me;
R^3=Ph,4-Cl,4-Br-C_6H_4
R^4=Ph,2-,4-Cl-C_6H_4

3-Oxoesters, ethoxymethylenemalonic esters or a reactive 2-chlorobenzoate annulate diamines and form a fused diazepine in good yield.

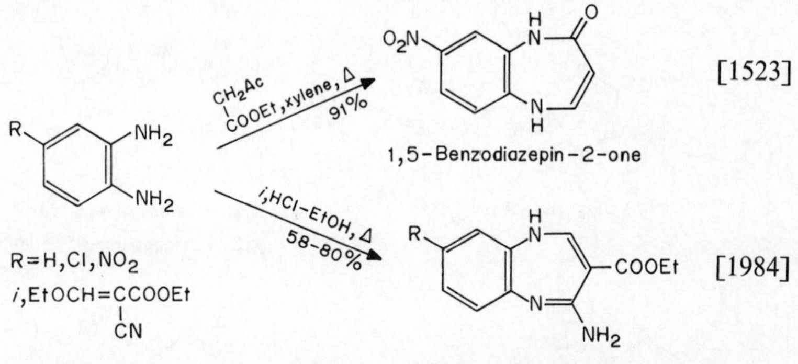

CH_2Ac / $COOEt$, xylene, Δ
91%
1,5-Benzodiazepin-2-one
[1523]
i, HCl–EtOH, Δ
58–80%
1,5-Benzodiazepine
[1984]
R=H,Cl,NO_2
i, EtOCH=C(CN)COOEt

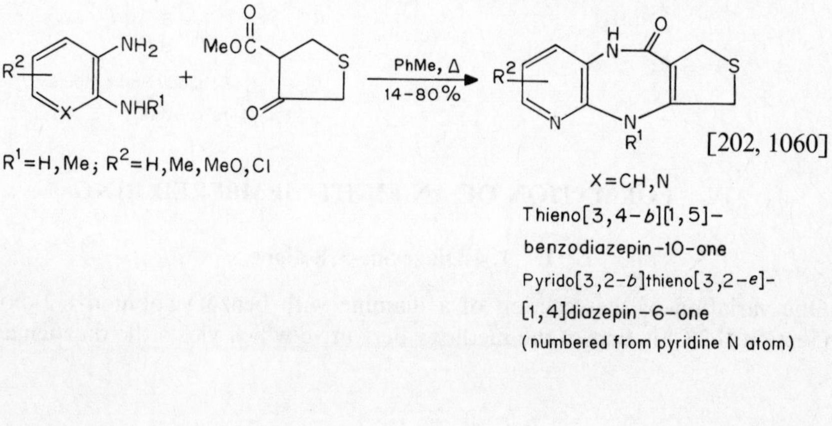

PhMe, Δ
14–80%
[202, 1060]
R^1=H,Me; R^2=H,Me,MeO,Cl
X=CH,N
Thieno[3,4-*b*][1,5]-benzodiazepin-10-one
Pyrido[3,2-*b*]thieno[3,2-*e*]-[1,4]diazepin-6-one
(numbered from pyridine N atom)

Dibenzo[*be*][1,4]diazepin-
11-one

[302]

2. 1,2,4-Triazepin-5-one or 1,3,4,7-Dithiadiazepine
1,3-Tetraoxide

3,4-Diamino-1,2,4-triazoles and similar triazines undergo a thermal cyclization with 3-oxoesters with the formation of a triazepinone ring. The dithiadiazepine ring is formed from a diamine and bis(chlorosulphonyl)methane.

R = H, Me

1,2,4-Triazolo[4,3-*b*]-
[1,2,4]triazepin-8-one

[1358]

R = Me, Ph

[1,2,4]Triazino[4,3-*b*]-
[1,2,4]triazepine-4,9-dione

[1359]

R = H, Cl

2,4,1,5-Benzodithiadiazepine
2,2,4,4-tetraoxide

[1258]

IV. FORMATION OF AN EIGHT-MEMBERED RING

1. 1,4-Diazocine-5,8-dione

One variation of the reaction of a diamine with benzocyclobuten-1,2-dione (Section II.2.C) is that of the methoxy derivative which yields the diazocine.

Dibenzo[bf][1,4]diazocine-6,11-dione

[219]

CHAPTER 77

1, 3-, 1, 4- and 1, 5-Diamine

I. FORMATION OF A FIVE-MEMBERED RING

1. Pyrrole

1, 4-Diamines such as 2-aminovinylaniline cyclize with loss of one amine group to give a pyrrole ring. The diamine is rarely isolated but is formed *in situ* by catalytic [227, 550] or chemical [540, 812] reduction of the dinitro compound; the choice of solvent is important for the best results [1777] and the presence of silica minimizes side-reactions [540]. The two nitro groups are often attached to different rings.

R = H, Ac   Indole

[227, 550, 1777]

Other ring systems synthesized similarly:

Furo[2,3-f]indole   [531]

Pyrrolo[3,2-b]indole
(review [921])   [1456]

426

Two nuclear amino groups attached to different rings may be converted by Tauber's synthesis into a doubly fused pyrrole ring by heating in diethylene glycol, formic acid [1771] or *NN*-dimethylaniline [423]. Orthophosphoric acid in the absence of a solvent removes t-butyl groups simultaneously. One diamine was isolated as an intermediate in a Fischer indole synthesis and was then separately cyclized to the fused pyrrole [423]. Sulphuryl chloride in carbon tetrachloride effects cyclization but this may be accompanied by nuclear chlorination.

Carbazole
(review [1695])

[206]

[206]

[718]

Other ring systems synthesized similarly:

[423, 1771]

X=O,S
Pyrimido[4,5-*b*]indole–
2,4-dione
Pyrimido[4,5-*b*]indol–
4-one,2-thioxo–

II. FORMATION OF A SIX-MEMBERED RING

1. Pyridine

Oxidative cyclization of a 1,5-diamine in sulpholane produces a doubly fused pyridine ring in good yield. In this example, cyclization failed when $R^2 = H$.

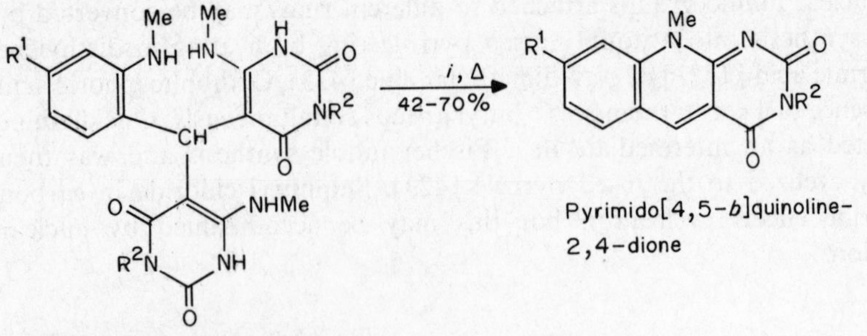

R^1=Cl,F,NO$_2$,4-NO$_2$C$_6$H$_4$;
R^2= CH$_2$CONHMe,
i,(=NCOOEt)$_2$, sulpholane

Pyrimido[4,5-b]quinoline-
2,4-dione

[1805]

Another ring system synthesized similarly:

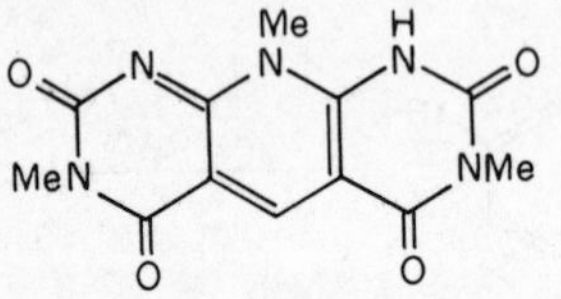

[738]

Pyrido[2,3-d:6,5-d']-
dipyrimidine-2,4,6,8-
tetraone

2. Pyrimidine or Pyrimidin-2-one

An efficient method of annulating two *peri*-positioned amine groups is to heat the compound with an isothiocyanate.

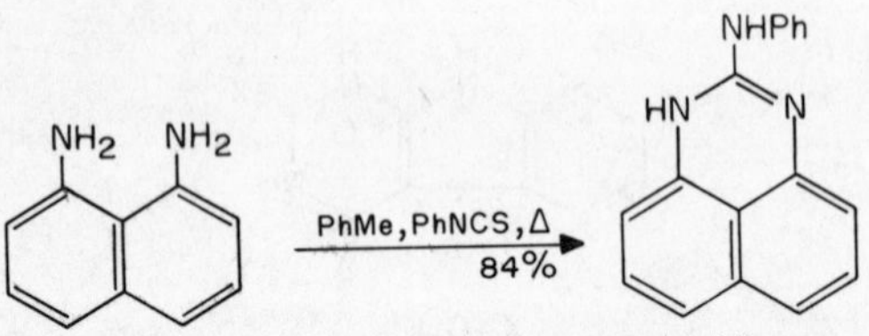

Perimidine

[292]

A 2-aminobenzylamine needs one more carbon atom in order to form a fused pyrimidine ring. This may be provided by formamidine or a bis(methylthio) cyanamide.

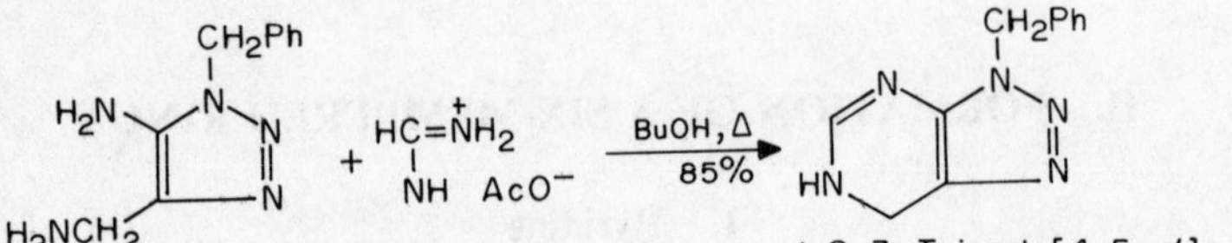

1,2,3-Triazolo[4,5-d]-
pyrimidine
(review [2023])

[221]

[422]

Quinazoline

A 1,3-diamine is converted into a pyrimidin-2-one ring by reaction with diethyl carbonate and a base. The chemistry of polyazaphenanthrenes has been reviewed [1888].

[294]

Pyrimido[6,1-*a*]isoquinolin-4-one

3. 1,2,3- or 1,2,4-Triazine

Two *peri*-positioned primary amine groups undergo diazotization with the formation of a doubly fused 1,2,3-triazine ring [192, 1838]. Pentyl nitrite sometimes gives better results than the conventional sodium nitrite [1838]. High yields have also been obtained by using nitrosodiphenylamine [51].

[900]

Naphtho[1,8-*de*]-
1,2,3-triazine

Another ring system synthesized similarly:

[192]

Acenaphtho[5,6-*de*]-
1,2,3-triazine

Reduction of a 1,5-dinitro compound which has two nitrogen atoms between the two functions spontaneously gives a 1,2,4-triazine ring. In this example, the nuclear bromine atom is lost but this does not always happen in this reaction.

[427]

4. 1,2,6-Thiadiazine, 1,2,4- or 1,2,6-Thiadiazine 1,1-Dioxide

Reaction of a *peri*-diamine with a sulphinyl chloride provides a sulphur atom to form a 1,2,6-thiadiazine ring.

[1177]

Naphtho[1,8-*cd*]-
[1,2,6]thiadiazine

Two primary aromatic amine groups separated by five atoms react on heating in acid (cf. Section I.1) to give a 1,2,4-thiadiazine ring.

[1200]

R = Me, Ph   Pyrazolo[1,5-*b*][1,2,4]-
benzothiadiazine 9,9-
dioxide

A doubly fused 1,2,6-thiadiazine ring is formed from a *peri*-disubstituted diamine and sulphamide.

[1492]

Naphtho[1,8-*cd*][1,2,6]-
thiadiazine 2,2-dioxide

III.   FORMATION OF A SEVEN-MEMBERED RING

1.   1,3-Diazepine, 1,3-Diazepin-5-one or 1,3-Diazepine-2-thione

This diazepine ring is completed by treating a 1,4-diamine with a source of one carbon atom, for example, di(methylthio)cyanoimidocarbonate, a dichloromethine derivative of chloroacetamide, carbon disulphide or an orthoformate ester.

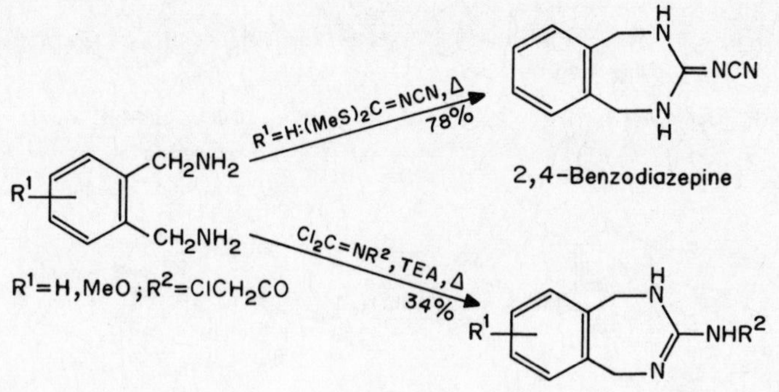

2,4-Benzodiazepine   [205]

$R^1=H$; $(MeS)_2C=NCN, \Delta$; 78%

$Cl_2C=NR^2, TEA, \Delta$; 34%

$R^1=H,MeO$; $R^2=ClCH_2CO$

[842]

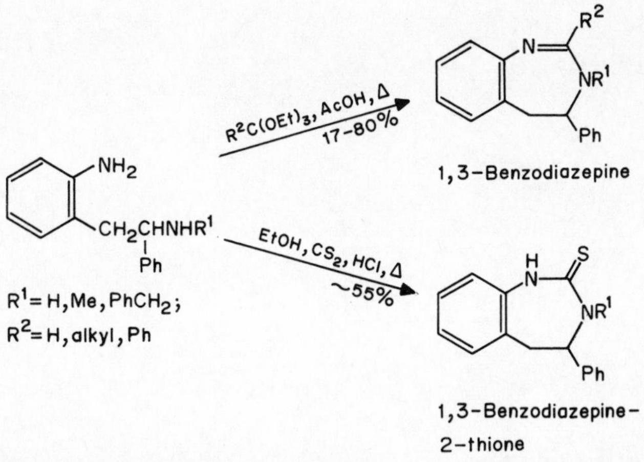

$R^2C(OEt)_3, AcOH, \Delta$; 17-80%

1,3-Benzodiazepine   [8]

$R^1=H,Me,PhCH_2$;
$R^2=H,alkyl,Ph$

$EtOH, CS_2, HCl, \Delta$; ~55%

1,3-Benzodiazepine-
2-thione   [1043]

Another ring system synthesized similarly:

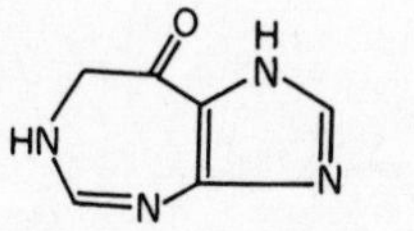

Imidazo[4,5-d][1,3]-
diazepin-8-one   [1309]

2. 1,2,3- or 1,3,5-Triazepine

Two primary amine groups attached to linked rings may on diazotization cyclize to form a 1,2,3-triazepine ring. The 1,3,5-triazepine ring may be obtained by a variation on the orthoester synthesis mentioned in the previous section.

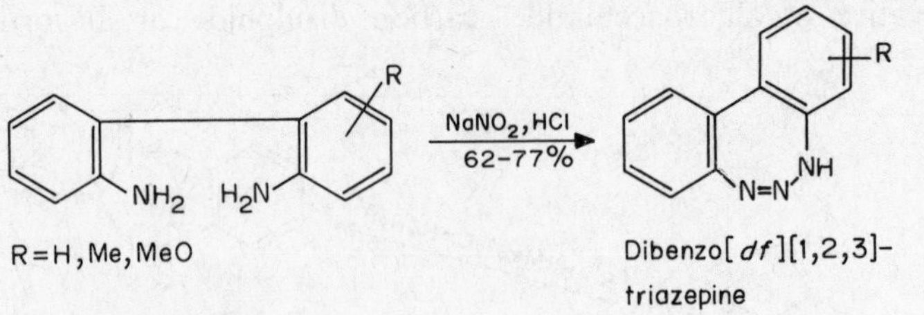

R = H, Me, MeO

$\xrightarrow[\text{62-77\%}]{\text{NaNO}_2, \text{HCl}}$

[1838]

Dibenzo[df][1,2,3]-
triazepine

$\xrightarrow[\text{43\%}]{\text{HC(OEt)}_3, \Delta}$

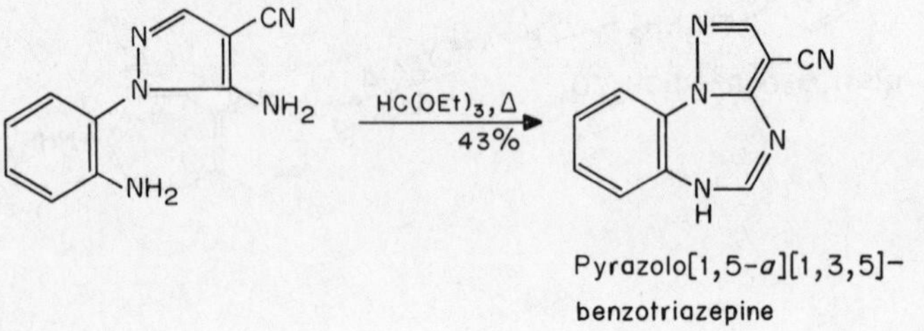

[1767]

Pyrazolo[1,5-a][1,3,5]-
benzotriazepine

Diazo or Diazonium Salt and Methylene

The chemistry of heterocyclic diazo compounds and diazonium salts has been reviewed [1326, B-28].

I. FORMATION OF A FIVE-MEMBERED RING

1. Pyrazole

The coupling of a diazonium salt with a reactive methyl or methylene group proceeds at low temperatures and usually gives good yields of a fused pyrazole. Diazotization may be effected with either sodium nitrite-acetic acid or pentyl nitrite. Variations include treatment of a diazonium tetrafluoroborate with potassium acetate in chloroform [1315], replacement of the amino by isocyanate [1643], and Skraup reaction conditions [163].

$$R^1 = H, Me, MeO, Br, Cl, NO_2, CN; \quad R^2 = H, 1\text{-isoquinolyl}$$

Indazole

[163, 454, 568, 938, 1315, 1643]

Another ring system synthesized similarly:

Pyrazolo[4,3-c]pyrazole

[1909]

Reductive cyclization of a 2-acetyl-diazo compound gives a high yield of a fused pyrazole.

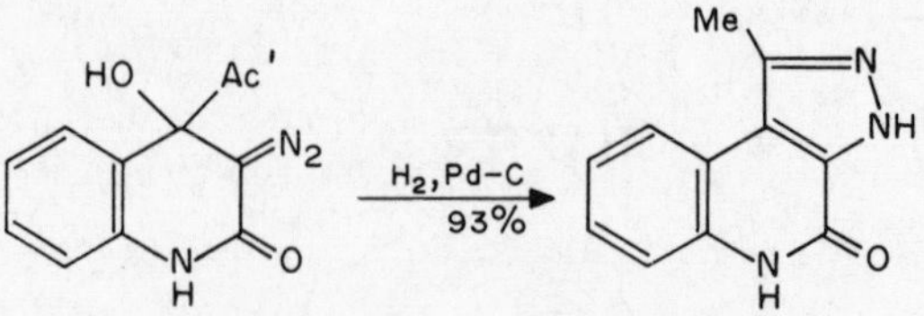

Pyrazolo[3,4-c]quinolin-
4-one

[965]

II. FORMATION OF A SIX-MEMBERED RING

1. Pyridazin-4-one

In an extension of the cyclization mentioned in the preceding section, the methylene may be separated from the ring by a carbonyl; a pyridazinone ring is thus formed.

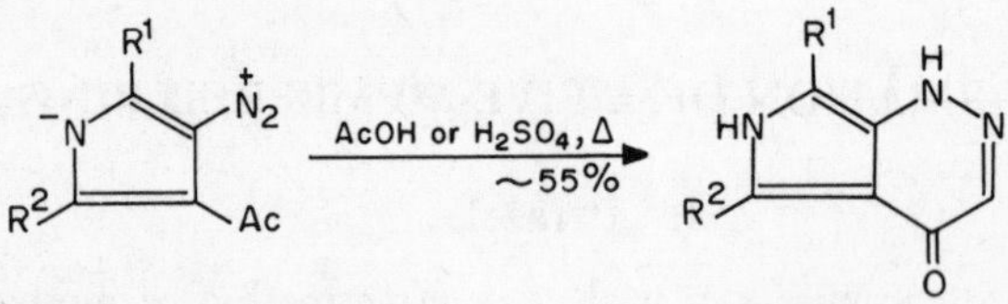

R^1=H, 3-MeOC$_6$H$_4$;
R^2=Me, Ph

Pyrrolo[3,4-c]pyridazin-
4-one

[1076, 1291]

Other ring systems synthesized similarly:

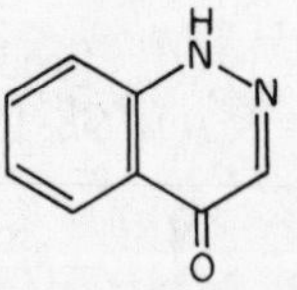

[277]

4-Cinnolinone

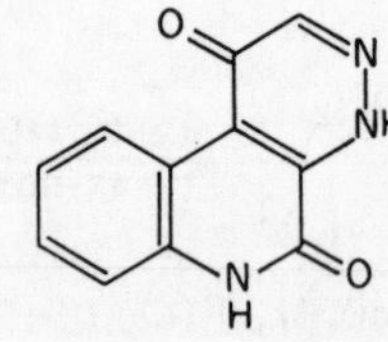

[965]

Pyridazino[3,4-c]quinoline-
1,5-dione

2. 1, 2, 3-Triazine 1-Oxide

The coupling of a diazonium salt with an active methyl or methylene sometimes leads to the formation of a triazine N-oxide.

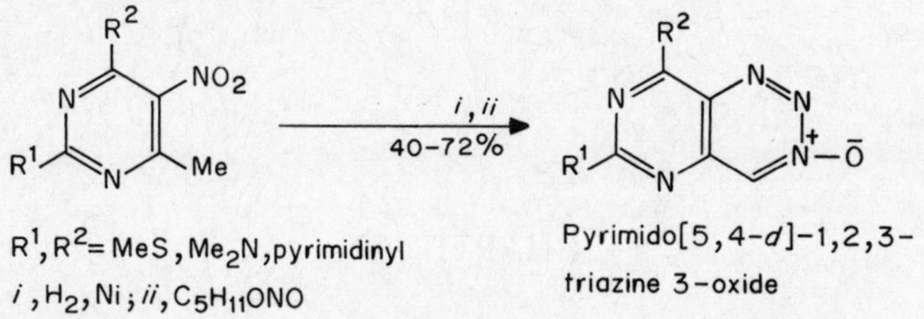

$R^1, R^2 = MeS, Me_2N, pyrimidinyl$

$i, H_2, Ni; ii, C_5H_{11}ONO$

[1154]

Pyrimido[5,4-*d*]-1,2,3-triazine 3-oxide

Another ring system synthesized similarly:

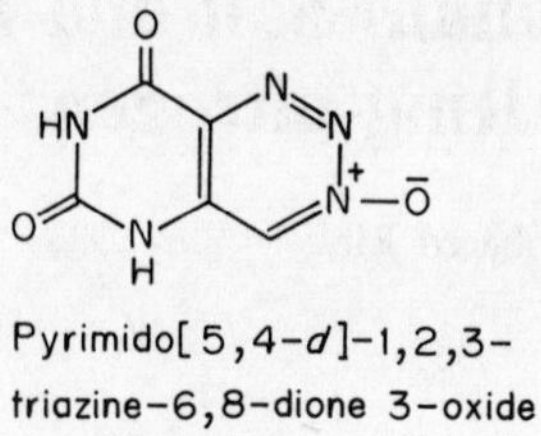

[953]

Pyrimido[5,4-*d*]-1,2,3-triazine-6,8-dione 3-oxide

CHAPTER 79

Diazo or Diazonium Salt and Ring-carbon or Ring-nitrogen

Reactions of diazo compounds and diazonium salts relevant to this chapter have been reviewed [1326, B-28].

I. FORMATION OF A FIVE-MEMBERED RING

1. Pyrazole

When an aliphatic diazo compound is heated in boiling toluene, a pyrazole ring is formed.

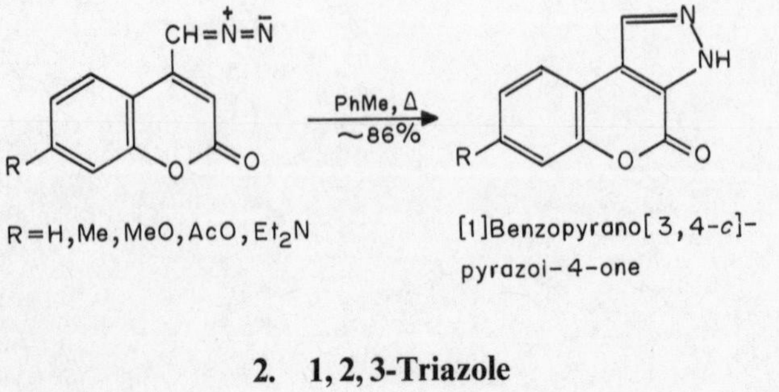

CH=N̄=N̄
PhMe, Δ
~86%
R
O
O
N
NH
R=H,Me,MeO,AcO,Et2N
[1]Benzopyrano[3,4-c]-
pyrazol-4-one

[1049]

2. 1, 2, 3-Triazole

A diazo group, generated *in situ* by careful thermolysis of a tetrazole in mesitylene, cyclizes on to an adjacent ring-nitrogen atom to give a fused triazole.

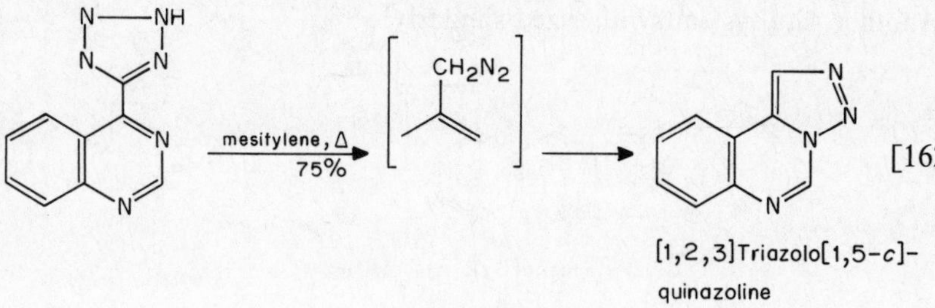
mesitylene, Δ
75%
CH_2N_2
[1,2,3]Triazolo[1,5-c]-quinazoline

[1624]

3. Tetrazole

A fused tetrazole ring is formed when a diazonium salt is treated at a low temperature with sodium azide.

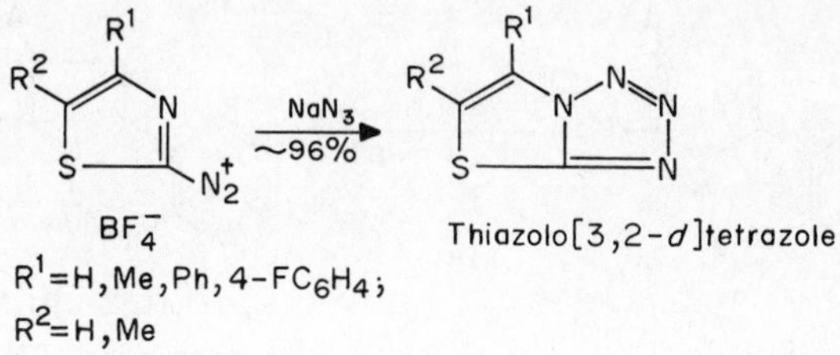
NaN_3
~96%
BF_4^-
R^1=H,Me,Ph,4-FC_6H_4;
R^2=H,Me
Thiazolo[3,2-d]tetrazole

[223]

II. FORMATION OF A SIX-MEMBERED RING

1. Pyridazine or Pyridazin-4-one

Coupling of a diazonium group attached to one ring to the carbon atom of another (more electron-rich) ring is a frequently used route to fused pyridazines.

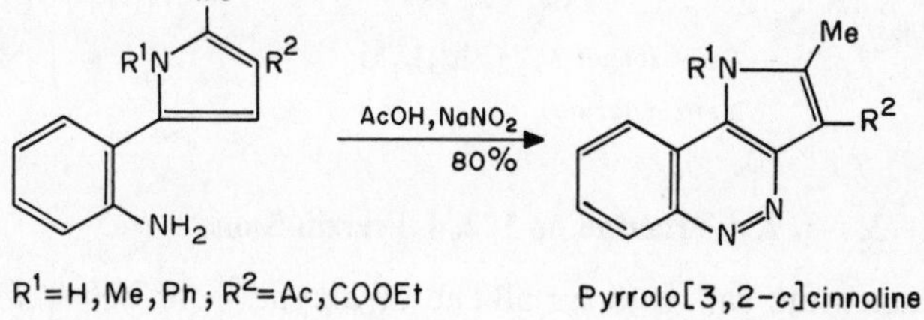
$AcOH, NaNO_2$
80%
R^1=H,Me,Ph; R^2=Ac,COOEt
Pyrrolo[3,2-c]cinnoline

[1845, 1857]

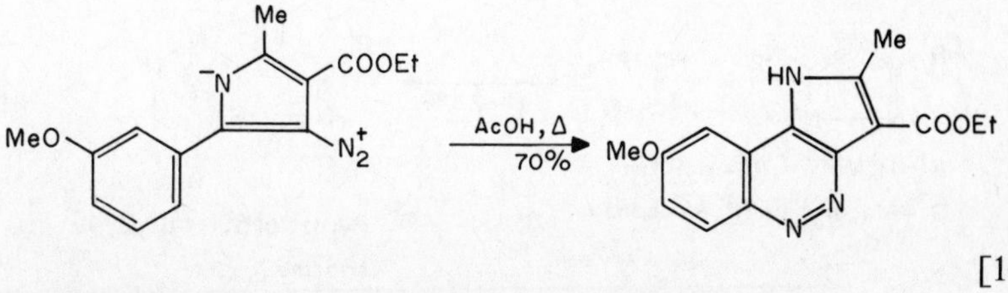
AcOH, Δ
70%

[1291]

Another ring system synthesized similarly:

[1606]

Thieno[3,2-c]cinnoline

2. 1, 2, 3-Triazine

When 2-(2-pyrrolyl) anilines are diazotized, coupling may occur at either N-1 or C-3 of the pyrrole ring. In practice, it occurs preferentially at N-1 when this is unsubstituted but at C-3 when the nitrogen is substituted (see preceding section).

[1857]

Pyrrolo[1,2-c][1,2,3]-
benzotriazine

Another ring system synthesized similarly:

[1007]

Quinazolino[3,2-c][1,2,3]-
benzotriazine

3. 1, 2, 4-Triazine or 1, 2, 4-Triazin-5-one

Coupling of a diazonium salt with an alkene or alkyne at or below ambient temperature can lead to the formation of a fused triazine ring.

[222]

$R^1 = H, Me, Ph; R^2 = H, Ph;$
$R^3 = Me, aryl$ or $R^1, R^2 = benzo$

Pyrazolo[5,1-c][1,2,4]-
triazine

[780]

$R^1 = Me, Ph; R^2 = Me, Et;$
$R^3 = Et, Ph$

X = CH, N
Pyrazolo[5,1-c][1,2,4]triazine

[1,2,3]Triazolo[5,1-c][1,2,4]triazine

Another ring system synthesized similarly:

[780]

Pyridazino[3',4':3,4]pyrazolo-
[5,1-c][1,2,4]triazine

Coupling of a diazonium salt with a malonic acid derivative can lead to triazine ring formation when the diazonium group is positioned adjacent to a ring nitrogen.

1.NaNO$_2$,HCl 2.R^{2}CH$_2$CN
68–88%

[1305]

$R^1 = Br, SH; R^2 = CN, PhCO;$
$R^3 = NH_2, Ph$

Pyrazolo[5,1-c][1,2,4]triazine

Other ring systems synthesized similarly:

[1293]

[1,2,4]Triazolo[5,1-c][1,2,4]triazine

[145]

X = NH, O
[1,2,4]Triazino[4,3-a]-
benzimidazole
[1,2,4]Triazino[4,3-a]-
benzimidazol-3-one

X = CH, N
Pyrazolo[5,1-c][1,2,4]triazin-4-one
[1,2,4]Triazolo[5,1-c][1,2,4]triazin-4-one

Reaction of a Wittig reagent with a diazotized 3-aminopyrazole leads to the formation of a triazine ring under mild conditions.

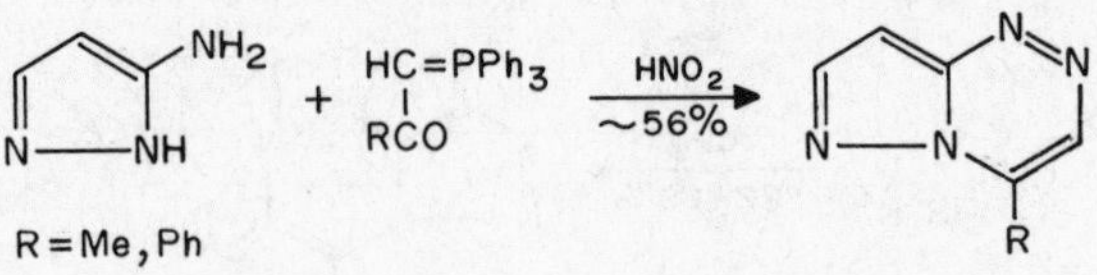

[222]

$$R = Me, Ph$$

Pyrazolo[5,1-c][1,2,4]-
triazine

Attack by a diazonium group on another ring-carbon can lead to a new fused triazine ring (cf. preceding section).

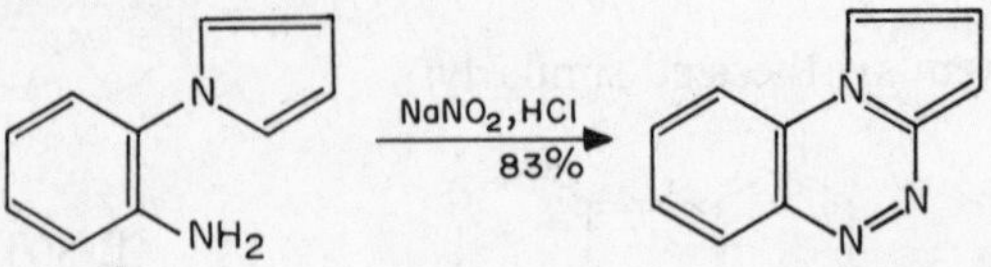

[494]

Pyrrolo[2,1-c][1,2,4]-
benzotriazine

Other ring systems synthesized similarly:

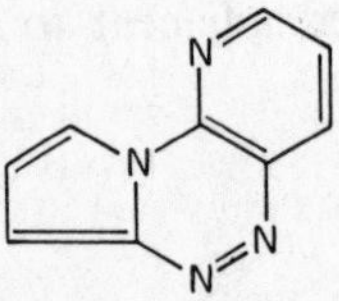

[1885]

Pyrido[2,3-e]pyrrolo-
[2,1-c][1,2,4]triazine

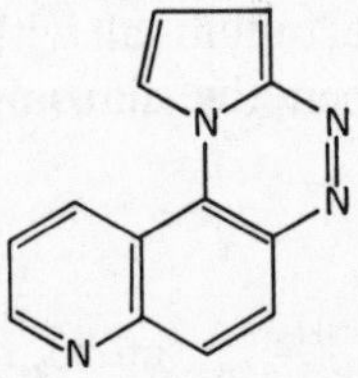

[711]

Pyrido[3,2-f]pyrrolo[2,1-c]-
[1,2,4]benzotriazine

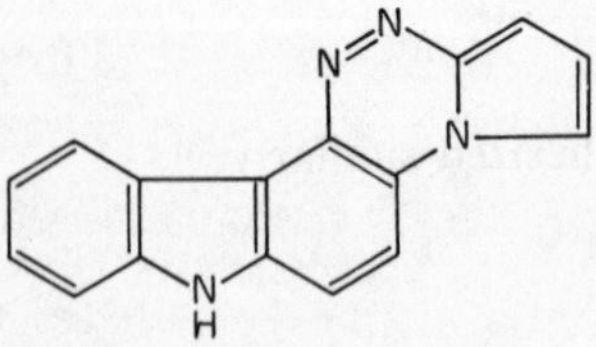

[332]

Pyrrolo[2',1':3,4][1,2,4]-
triazino[5,6-c]carbazole

4. 1, 2, 3, 5-Tetrazin-4-one

Treatment of an N-heterocycle-α-diazonium salt with an isocyanate produces a fused tetrazinone in high yield.

[781]

$$R^1 = H, Me, Ph, COOEt;$$
$$R^2 = alkyl, aryl$$

Pyrazolo[5,1-d]-1,2,3,5-
tetrazin-4-one

Another ring system synthesized similarly:

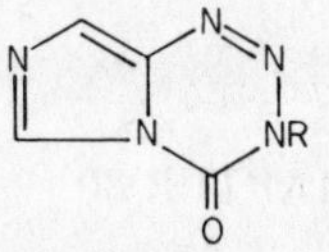

[890]

Imidazo[5,1-*d*]-1,2,3,5-
tetrazin-4-one

III. FORMATION OF AN EIGHT-MEMBERED RING

1. 1, 2, 3, 5-Tetrazocin-6-one

Suitably separated diazonium and ring-nitrogen functions couple to form this
large ring in good yield.

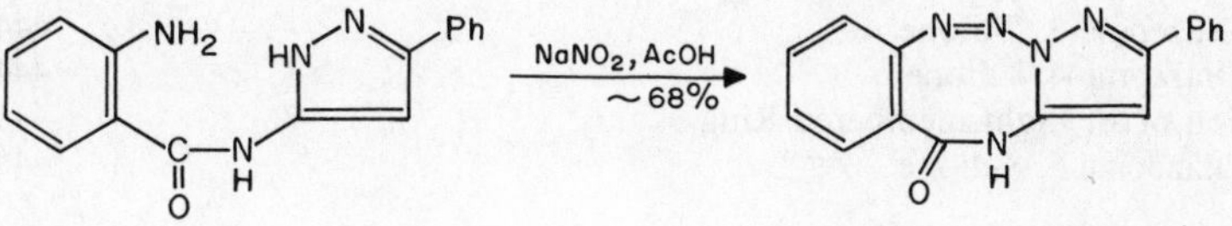

[1352]

Pyrazolo[1,5-*c*][1,2,3,5]benzo-
tetrazocin-5-one

CHAPTER 80

Dicarboxylic Acid, its Anhydride, Diacyl Halide, Diester or Dihydrazide

I. FORMATION OF A SIX-MEMBERED RING

1. Pyridin-4-one

An unusual reaction between a carboxyl and a carboxylic ester in the presence of PPE leads to the formation of a pyridinone ring.

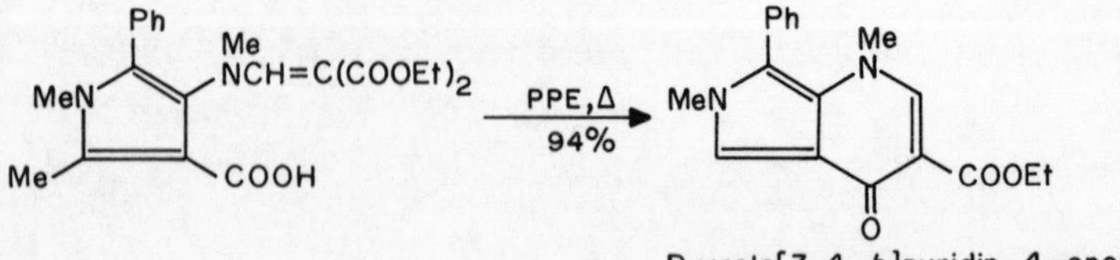

Pyrrolo[3,4-*b*]pyridin-4-one

[1787]

2. Pyridazine or Pyridazine-3, 6-dione

Reaction between a dicarboxylic anhydride [996] or ester and hydrazine hydrate gives a fused pyridazine; this ring is tautomeric and in this context assumed to be in the diketo form.

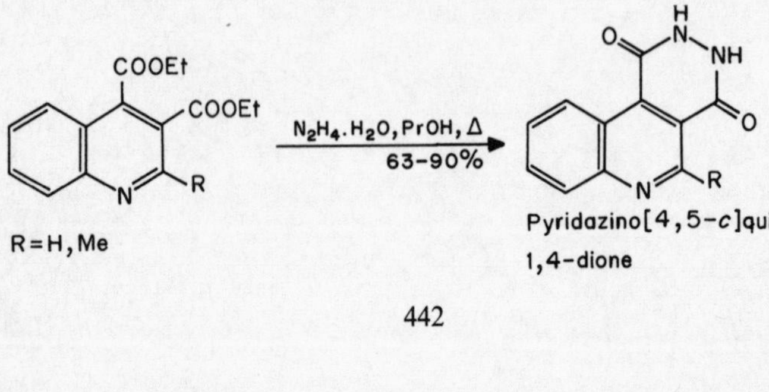

R = H, Me

Pyridazino[4,5-*c*]quinoline-1,4-dione

[996, 1699]

Other ring systems synthesized similarly:

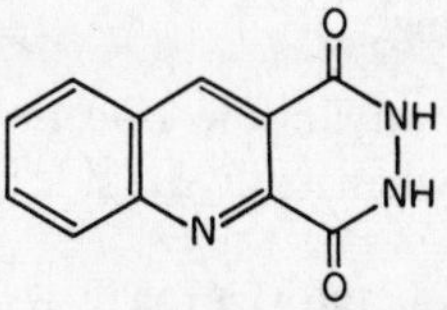

[1488, 1699]

Pyridazino[4,5-b]quinoline-
1,4-dione

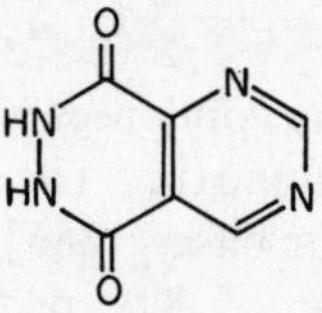

[655]

Pyrimido[4,5-d]pyridazine-
5,8-dione

When a 1,2-dihydrazide is heated neat, in hydrochloric acid or propanol, the pyridazinedione (or its tautomer) is formed in high yield.

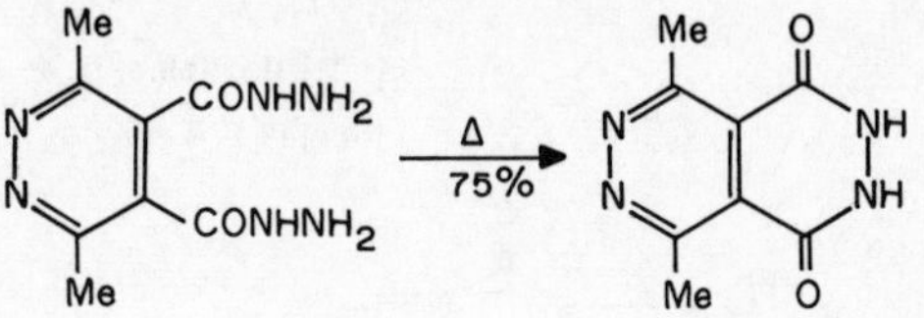

[1340]

Pyridazino[4,5-d]pyridazine-
1,4-dione

Other ring systems synthesized similarly:

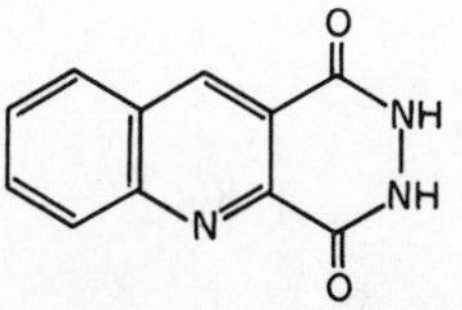

[1866]

Pyridazino[4,5-b]quinoline-
1,4-dione

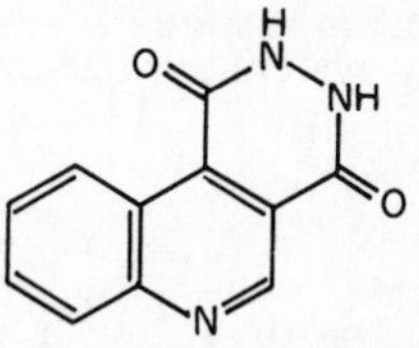

[1866]

Pyridazino[4,5-c]quinoline-
1,4-dione

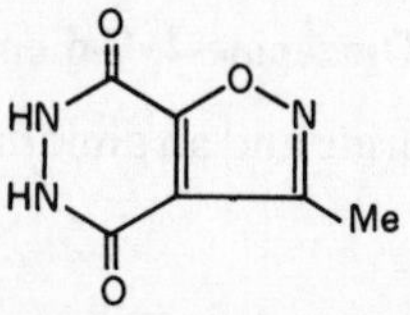

[1939]

Isoxazolo[4,5-d]pyridazine-
4,7-dione

A comparatively rare annulation of neighbouring carboxamide and hydrazide groups leads to the same pyridazinedione on heating without solvent [1939].

II. FORMATION OF A SEVEN-MEMBERED RING

1. 1, 2-Diazepine-3, 7-dione

Hydrazine hydrate condenses with a 3-methoxycarbonylindole-2-acetic ester to form a diazepinedione ring. The same ring is formed when a dihydrazide is heated with either acid or base [1866].

Note added in proof: The product from the diester (80.1) [1081] has a six-membered ring as in (80.2) according to a recent paper [1679].

(80.1)

$N_2H_4 \cdot H_2O, \Delta$
58%

[1081]

[1,2]Diazepino[5,4-b]-
indole-1,4-dione

[1679]

(80.2)Pyrido[4,3-b]-
indole-1,3-dione

Another ring system synthesized similarly:

[1866]

[1,2]Diazepino[5,4-b]-
quinoline-1,4-dione

2. 1, 3-Diazepine-4, 7-dione

Reaction between phthaloyl dichloride and an amidine in a basic medium gives a high yield of this diazepinedione.

$R = Me, PhCH_2; Ar = Ph,$
$4-Br-, 4-EtO-C_6H_4$

RC=NAr
|
NHAr

Et$_2$O, TEA
~95%

[469]

2,4-Benzodiazepine-
1,5-dione

III.   FORMATION OF AN EIGHT-MEMBERED RING

1.   1, 4-Diazocine-5, 8-dione

Condensation of an o-dicarboxylic ester with o-phenylenediamine in a basic medium gives an eight-membered ring.

[703]

Pyridazino[4,5-c][1,6]-
benzodiazocine-5,12-dione

CHAPTER 81

Dihalogen

I. FORMATION OF A FIVE-MEMBERED RING

1. Pyrrole

A reactive 1,2-dihalide reacts with an enamine to form a fused pyrrole but a mixture of isomers is sometimes formed [596]. Pyrido[2,3-b]indoles have been reviewed [1752]. A pyrrole is obtained from the reaction of a 1,4-dihalide and an acylhydrazine or a tosylhydrazine in the presence of a tertiary amine or sodium hydride.

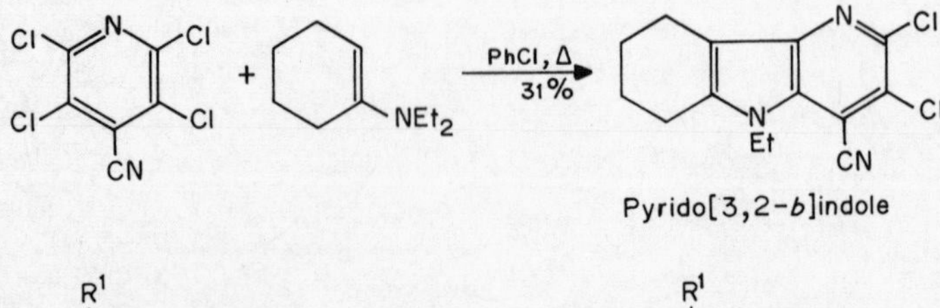

Pyrido[3,2-b]indole   [596]

R^1=Me,COOMe; R^2=H,COOMe

Isoindole
(review [1615])   [471, 587]

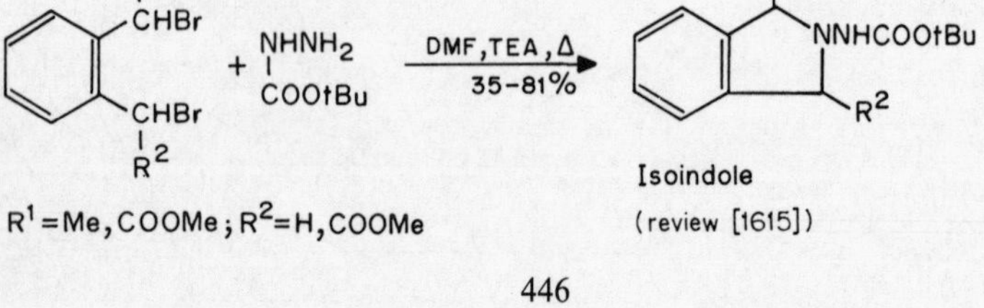

446

Another ring system synthesized similarly:

Benzo[1,2-*c* : 4,5-*c'*]dipyrrole

[2002]

Halogens attached to two different rings may react with a primary aliphatic amine to form a doubly fused pyrrole. The chemistry of phenanthrolines has been reviewed [1920].

R = Me₂N(CH₂)₃

Pyrrolo[*lmn*][4,7]-phenanthroline

[515]

2. Furan

1,2-Dichlorides react with an anion from methyl ketones to give a fused furan. Butyl-lithium causes ring closure of *o*-(2-bromoethoxy)bromobenzenes to the benzofuran in good yields.

R = EtO, Et₂N

Furo[2,3-*b*]pyrazine

[470]

R = H, Me, MeO, Br, Cl

Benzofuran

[424]

The copper-catalysed reaction of a 2,2'-dibromobiphenyl ether yields dibenzo-furans in a Ullmann-type reaction.

R = alkyl, MeO

Dibenzofuran
(review [2027])

[1146]

3. Thiophene

When two suitably positioned reactive halogen atoms are treated with sodium sulphide, a thiophene ring is formed. Thienopyridines have been reviewed [1661].

X=N, Y=CH
Thieno[3,4-*b*]pyridine
X=CH, Y=N
Thieno[3,4-*c*]pyridine

[951]

II. FORMATION OF A SIX-MEMBERED RING

1. Pyrazine or 1,4-Dioxin

When a 1,2-dihalide is heated with a 1,2-diamino- or 1,2-dihydroxy-benzene, a pyrazine or dioxin ring is formed in good or high yield.

X = NH   [1174]
Pyrido[2,3-*b*]quinoxaline
X = O
[1,4]Benzodioxino[2,3-*b*]-
pyridine

2. Pyran

Under very mild conditions, butyl-lithium causes ring closure of *o*-(3-bromopropoxy)bromobenzenes and the formation of a benzopyran.

1-Benzopyran
(review [B-16])

[424]

3. 1,4-Thiazine

Sodium sulphide displaces halogen atoms and with a neighbouring 2-chloro-substitute amine, this method leads to the formation of a thiazine ring. The same ring is formed from the reaction of a 1,2-dichloride with a 2-mercaptoaniline (see also Chapter 10).

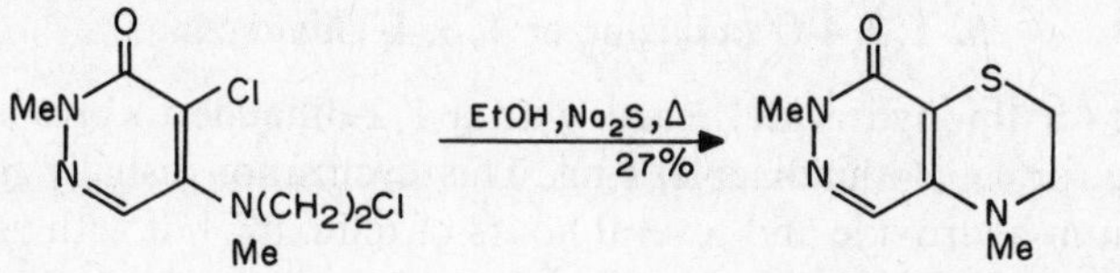

[207]

Pyridazino[4,5-*b*]-
1,4-thiazin-8-one

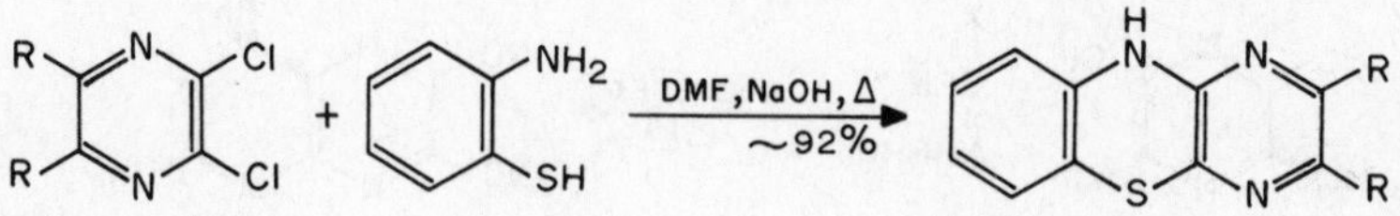

[598]

R=H, benzo

Pyrazino[2,3-*b*][1,4]-
benzothiazine
Quinoxalino[2,3-*b*][1,4]-
benzothiazine

Other ring systems synthesized similarly:

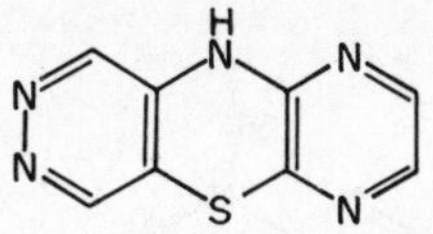

[582]

Pyrazino[2,3-*b*]pyridazino-
[4,5-*e*][1,4]thiazine

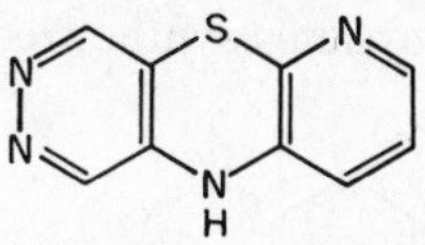

[582]

Pyridazino[4,5-*b*]pyrido-
[3,2-*e*][1,4]thiazine

4. 1,4-Oxathiin

When an unsymmetrical 1,2-dihalide reacts with 2-mercaptophenol, two positional isomers are possible and a mixture of these is often obtained. A slightly smaller amount (21 per cent) of the 4-one is isolated in this reaction.

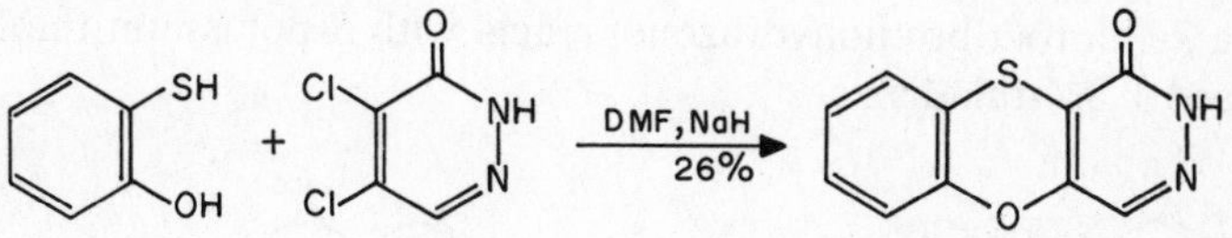

[224]

[1,4]Benzoxathiino[2,3-*d*]-
pyridazin-1-one

Another ring system synthesized similarly:

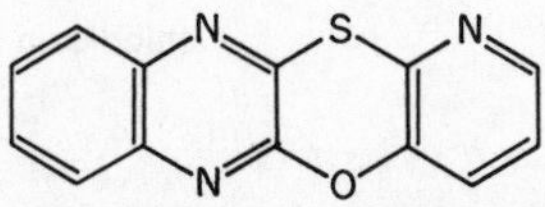

[597]

Pyrido[2',3':5,6][1,4]oxathiino-
[2,3-*b*]quinoxaline

5. 1, 3, 4-Oxadiazine or 1, 3, 4-Thiadiazine

A hydrazide (or thiohydrazide) reacts with a 1, 2-dihalide to give a fused 1, 3, 4-oxadiazole (or 1, 3, 4-thiadiazole) ring. This cyclization usually needs either TEA or sodium hydroxide and several hours of refluxing but with very reactive halides such as 2, 3-dichloropyrazine-5, 6-dicarbonitrile, ambient temperature is sufficient.

[361]

X = O
Pyrazino[2,3-e][1,3,4]-
oxadiazine

X = S
Pyrazino[2,3-e][1,3,4]-
thiadiazine

Other ring systems synthesized similarly:

[599]

[1,3,4]Thiadiazino[5,6-b]-
quinoxaline

[361, 599]

X = O
Pyrimido[4,5-e][1,3,4]-
oxadiazine

X = S
Pyrimido[4,5-e][1,3,4]-
thiadiazine

A reactive 2-chloro(α-bromohydrazone) reacts with S-potassium thioacetate to give a fused 1, 3, 4-thiadiazine.

[1577]

R = H, Me, PhCH$_2$, Ph

Pyridazino[4,5-e][1,3,4]-
thiadiazin-5-one

CHAPTER 82

Diol or Dithiol

A few examples of the annulation of 2-mercaptophenol are included in this chapter.

I. FORMATION OF A FIVE-MEMBERED RING

1. Furan

Two hydroxy groups (generated by demethylation of the diether) separated by four carbon atoms belonging to two benzene rings, react on heating in acetic anhydride to form a doubly fused furan.

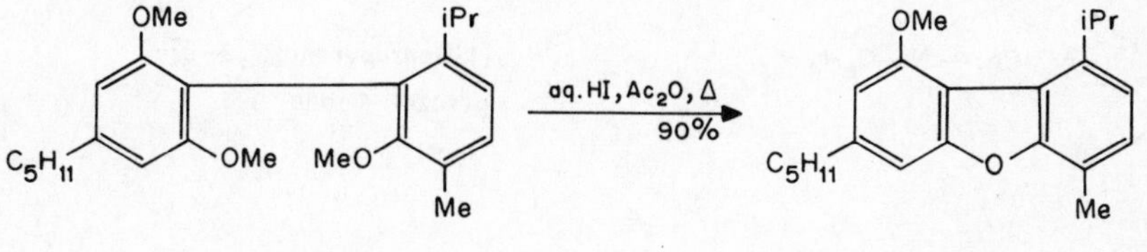

[720]

Dibenzofuran
(review [1974])

2. 1, 3-Oxathiole or 1, 3-Oxathiol-2-one

Reaction of 2-mercaptophenol with an alkyne yields 1, 3-benzoxathiols in moderate yields. A stronger base did not improve the yield and the reaction failed when mono- or di-bromoalkenes replaced the alkyne.

R^1 = Ac, COOMe;
R^2 = H, Me, Ph, COOMe

1,3-Benzoxathiole

[1269]

When a 2-methoxythiophenol, protected as its *S*-arylthio, is heated with a dealkylating agent, good to high yields of the benzoxathiolone is obtained.

R = H, Me, Cl, HO, MeO

1,3-Benzoxathiol-2-one

[460]

II. FORMATION OF A SIX-MEMBERED RING

1. Pyran or Pyran-4-one

A doubly fused pyran ring may be formed from a 1,5-diol containing two rings separated by a carbon atom (cf. preceding section). Several reagents are effective in this cyclization, for example, phosphorus oxychloride, PPA (review [B-21]), boron trifluoride-etherate, hydrogen chloride, or trifluoroacetic acid; refluxing in mesitylene containing silica gel is also effective.

Ar = Ph, 4-NO$_2$C$_6$H$_4$

[1]Benzopyrano[2,3-*c*]-pyrazol-4-one

[351]

R^1, R^2 = H, alkyl;
R^3 = H, Et, PhCH$_2$;
R^4 = H, Me

2*H*-1-Benzopyran

[668]

Other ring systems synthesized by one of the above-mentioned methods:

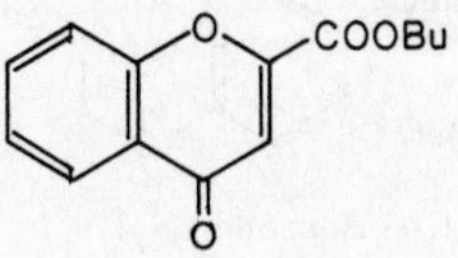

1-Benzopyran-4-one [171]

Pyrano[3,2-c:5,6-c']diquinoline [425]

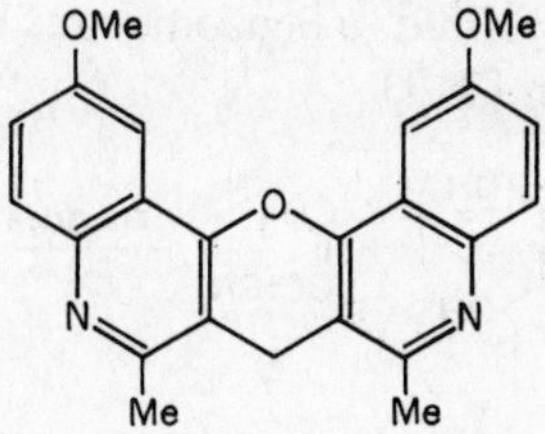

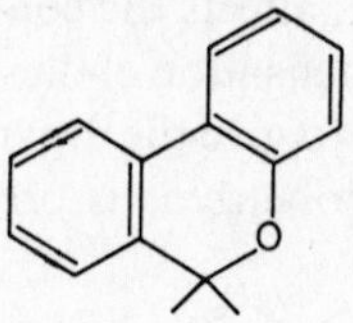

Dibenzo[bd]pyran [760]

X=O
Pyrano[3,2-b]benzofuran-4-one
X=S
[1]Benzothieno[3,2-b]pyran-4-one [1482]

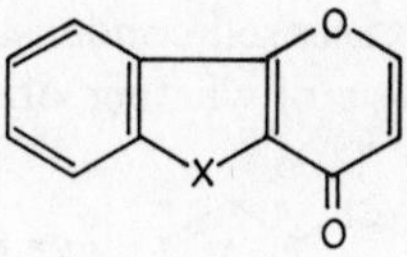

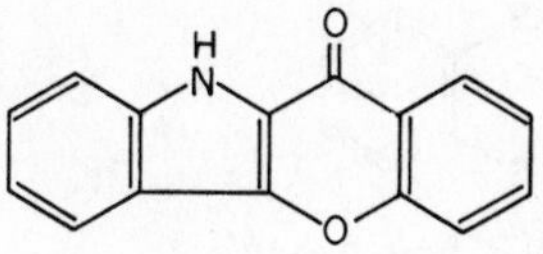

[1]Benzopyrano[3,2-b]indol-11-one [1769]

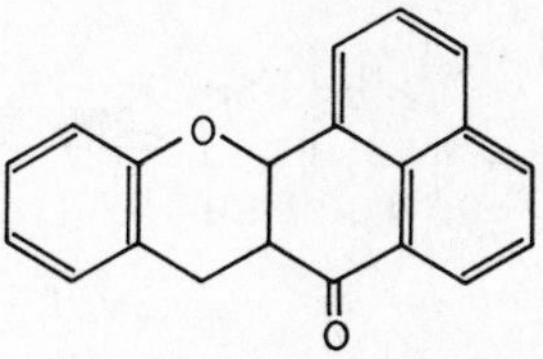

Naphtho[1,8-bc]xanthen-7-one [149]

A tautomeric 2-hydroxyquinone cyclizes regiospecifically and in high yield according to the acidic catalyst present.

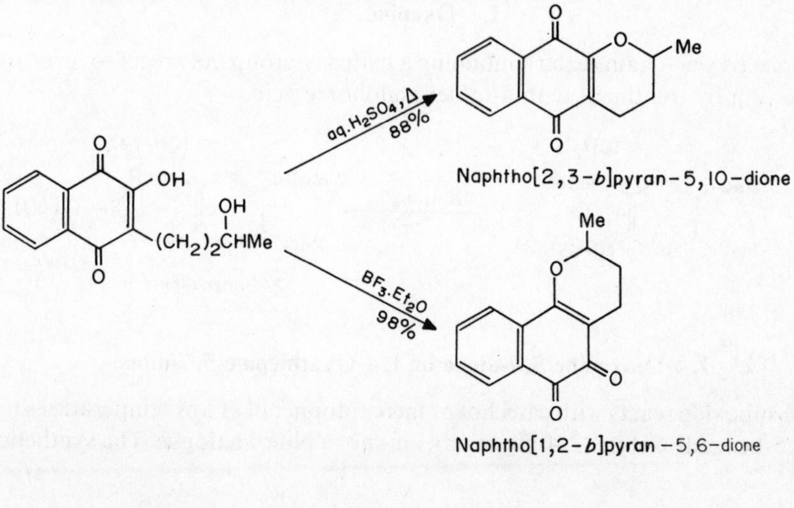

Naphtho[2,3-b]pyran-5,10-dione [209]

Naphtho[1,2-b]pyran-5,6-dione [209]

2.   1, 4-Dioxin or 1, 4-Dithiin

Catechols are cyclized to benzodioxans by reaction with 1-chloroacrylonitrile or epichlorohydrin [719].

1,4-Benzodioxin

A basic medium promotes interaction between a 1, 2-dithiol and 2-chloronitro-benzene to form a 1, 4-dithiin ring. The base used, however, affects the course of the reaction. Sodium hydride promotes intermolecular condensation of these two reactants while TEA causes self-condensation of the dithiol to give the dipyridine in high yield irrespective of whether or not the chloronitrobenzene is present.

[1,4]Benzodithiino[2,3-*b*]-
pyridine

[1,4]Dithiino[2,3-*b*:5,6-*b*']-
dipyridine

III.   FORMATION OF A SEVEN-MEMBERED RING

1.   Oxepine

Two *o*-placed side-chains each containing a hydroxy group may react to form an oxepine ring by treatment with 4-toluenesulphonic acid.

2-Benzoxepine

2.   1, 4-Dioxepine-5, 7-dione or 1, 4-Oxathiepine-5, 7-dione

Carbon suboxide reacts with catechol or mercaptophenol at low temperatures to form 1, 5-benzodioxepine-2, 4-dione or its monosulphur analogue. The synthetic uses of this reagent have been reviewed [100].

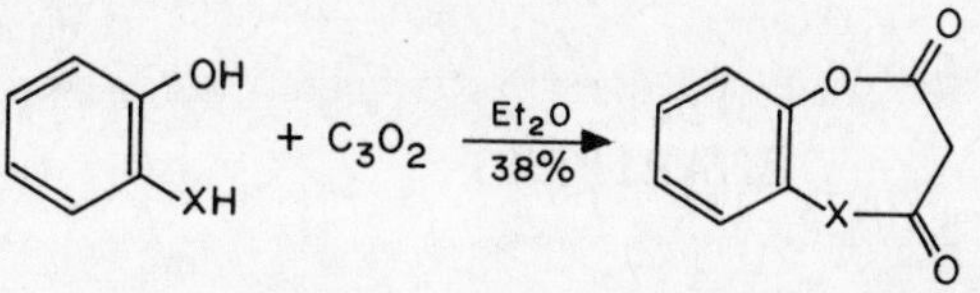

[208]

$X = O$
1,5-Benzodioxepine-2,4-dione
$X = S$
1,5-Benzoxathiepine-2,4-dione

3.   1,4-Oxazepine or 1,4-Oxazepin-5-one

Two hydroxy groups separated by five carbon and one nitrogen atoms may be dehydratively cyclized to form a fused oxazepine or oxazepinone by the use of PPA or phosphorus pentoxide.

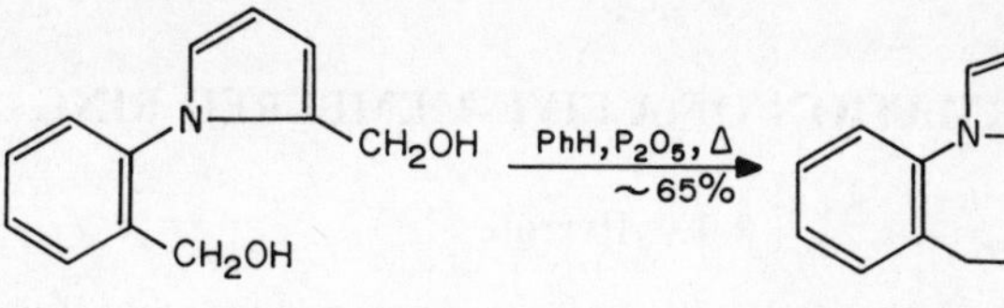

[602]

Pyrrolo[1,2-*a*][4,1]-
benzoxazepine

Another ring system synthesized similarly:

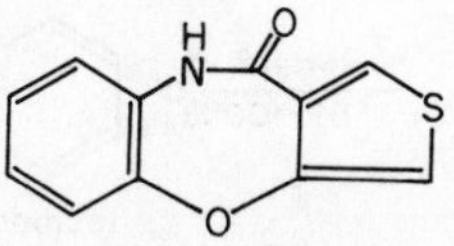

[600]

Thieno[3,4-*b*][1,5]-
benzoxazepin-10-one

Dinitrile

I. FORMATION OF A FIVE-MEMBERED RING

1. Pyrrole

When a 1, 2-dinitrile is heated with ammonia or a primary alkylamine and sulphur, an isoindole is formed in high yield. Sulphur and amine appear to add across the C≡N initially.

R = H, alkyl   Isoindole (review[1615])   [1513]

II. FORMATION OF A SIX-MEMBERED RING

1. Pyridine

Hydrogen bromide at ambient temperature converts 1, 3-dinitriles into a 2-aminopyridine ring. A 7-amino-1, 6-naphthyridine is similarly synthesized [48].

Pyrazolo[3, 4-c]pyridine   [1901]

2. Pyridazine

1,2-Dinitriles react readily with hydrazine to give a fused 1,4-diaminopyridazine ring.

$$N_2H_4, MeOH \quad 74\%$$

[1340]

Pyridazino[4,5-d]-
pyridazine

3. Pyrimidine

The reaction of hydrogen halides with o-cyano-cyanamides can give one of several products depending on the temperature of reaction and the halide used. For example, hydrogen iodide in dioxan at 35 °C gives a 2-amino-4-iodopyrido[2,3-d]pyrimidin-7-one but at 110 °C, the main products were the deiodinated amine (**83.1**, R = NH$_2$) and the iodide (**83.1**, R = I).

diox, HI, 35 °C
88%

[1904]

Pyrido[2,3-d]pyrimidin-
7-one

diox, HI, Δ
48%

[1904]

(**83.1**)

III. FORMATION OF A SEVEN-MEMBERED RING

1. 1,4-Diazepine-2,7-dione

A 1,4-dinitrile may be cyclized by heating with a mixture of sulphuric and acetic acids.

AcOH, H$_2$SO$_4$, Δ
28%

[755]

[1,4]Diazepino[1,2-a]benzimidazole-
1,3-dione

Di-Ring-carbon or Di-Ring-nitrogen

An example of the formation of a heterocyclic ring by annulation of two methylene groups is included.

I. FORMATION OF A FIVE-MEMBERED RING

1. Pyrrole

Indolizine derivatives may be prepared by reacting an alkyne with an *N*-oxo- or *N*-carboxy-methylpyridinium salt (review of reactivity of these salts [1870]) in the presence of a phase transfer catalyst (an ammonium salt). In this way, the use of high temperature and a strong base such as sodium hydroxide can be avoided in favour of ambient temperature and a weaker base. Ultrasound improved the yields of some indolizines prepared in this way.

R^1 = MeO, Ph, 4-MeO-, 4-Cl-,

4-NO_2-C_6H_4;

R^2 = Ph, COOMe; R^3 = Me, Et

i, CH_2Cl_2, K_2CO_3, $R_4\overset{+}{N}\overset{-}{Cl}$, 18 °C

[438]

Indolizine

(review[1670])

A doubly fused pyrrole is formed during a palladium-mediated cyclization of a diphenylamine.

$R = H, Me, MeO, Br, Cl, NO_2, COOH$ Carbazole
(review [1695])

2. Pyrazole or Pyrazol-3-one

When 2-diazopropane is stirred with a pyrimidopyridazine, a moderate yield of the pyrazolo-derivative is obtained. Cyclization of the pyridazine, $R^2 = N_3$, gave a tricyclic product in which $R^2 = NH_2$; this is believed to be brought about by azo transfer.

$R^1 = H, Br, Ph, COOEt; R^2 = H, Cl$ Pyrazolo[4,3-d]pyrimido-
[1,2-b]pyridazin-4-one

Another ring system synthesized similarly:

Pyrazolo[4,3-d]-1,2,4-
triazolo[4,3-b]pyridazine

Formation of a new fused ring by joining two endocyclic nitrogen atoms is not a commonly met reaction but the two nitrogen atoms of 3-pyrazolone undergo this type of reaction on heating with a 3-oxocarboxylic ester (neat or with aqueous sodium carbonate). Small amounts of other isomers are sometimes formed [1262].

$R^1 = Me, Ph; R^2 = Me, PhCH_2;$ Pyrazolo[1,2-a]pyrazole-
$R^3 = Ac, COOEt$ 1,5-dione

3. Furan or Thiophene

A doubly fused furan may be constructed by reaction of an aryl ether with either PPA (review [B-21]) or a palladium salt (cf. Section I.1).

R = H, Me, MeO, NH₂

Benzofuro[2,3-*b*]quinoxaline

[688]

Another ring system synthesized similarly:

Dibenzofuran
(review [1974])

[689, 903]

A less common cyclization (which strictly does not fall under the title of this chapter) is the annulation of two methylene groups by sulphur.

R = H, PhCO

Benzo[*c*]thiophene
(review [1650])

[1636]

4. Isoxazole

Hydroxamoyl chlorides (which generate nitrile oxides) add across cycloalkenes in boiling toluene or in the presence of iron(III) chloride to form a fused isoxazole ring. For unsymmetrical alkenes, regiospecificity is dependent on the substituent attached to the hydroxamoyl and alkene groups.

R = H, Me

Naphtho[2,1-*d*]isoxazole

[1831]

Other ring systems synthesized similarly:

Naphtho[1,2-*d*]isoxazole

[1831]

[1,3]Dioxolo[4,5-*c*]-
isoxazol-5-one

[1831]

[1839]

Benzo[3,4]cyclobut-
[1,2-*d*]isoxazole

II. FORMATION OF A SIX-MEMBERED RING

1. Pyridine or Pyridin-2-one

Vilsmeier reagents are useful in annulating two ortho positions of diarylamines [793, 2001] and a similar result is obtained by employing acetic acid-zinc chloride [535], acetic acid-palladium acetate [903] or a ketone in acid solution [1514].

R^1=H,Me; R^2=H,Me,Cl

Pyrido[4,3-*b*]quinolin-1-one

[793]

Other ring systems synthesized using the reagents mentioned:

[2001]

Pyrimido[4,5-*b*]quinoline-
2,4-dione

Acridine
(review [B-10])

[535, 1514]

[903]

6-Phenanthridinone
(review[1616])

2. Pyridazine or Pyridazin-4-one

Some heterocycles which contain an azo group (—N=N—) undergo a Diels–Alder reaction which yields a fused pyridazine or pyridazinone ring. The use of azodicarbonyl compounds in syntheses has been reviewed [1774].

[1607]

[1,2,4]Triazolo[1,2-a]-
pyridazine-1,3,6-trione

i, AcOCH=CHCH=CH₂

Pyrazolo[1,2-a]
pyridazin-1-one

[1607]

3. Pyrazine N-Oxide

An aryl-heteroarylamine undergoes annulation (probably through a hydroxyl-amine intermediate) on treatment with nitrous acid. Potassium nitrate in acetic acid gives a similar result.

[1869]

$R^1, R^3 = H, Me$; $R^2 = alkyl$

Pyrimido[4,5-b]quinoxaline-
2,4-dione 5-oxide

4. 1,3,5-Triazin-2-one

Phosgene reacts with nitrogen atoms of different rings and, in suitable substrates, can form a doubly fused 1,3,5-triazinone.

Pyrido[2′,1′:4,5][1,3,5]triazino-
[1,2-a]benzimidazol-12-one

[1590]

5. Pyran-4-one

Annulation of a diphenyl ether with a carbonyl group is achieved in good yield by treatment with oxalyl chloride and a Friedel–Crafts catalyst.

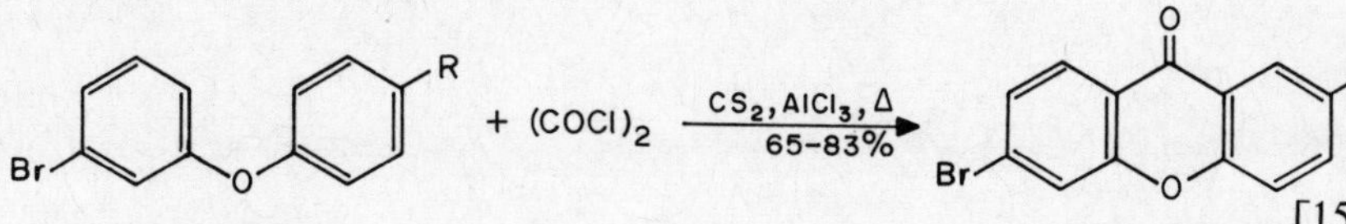

R = Br, F

9-Xanthenone
(review [775])

[1536]

6. 1,2-Oxazine

Cycloaddition of a halogenoisonitroso ketone or ester to the 2,3-bond of an indole gives an oxazinoindole in high yield.

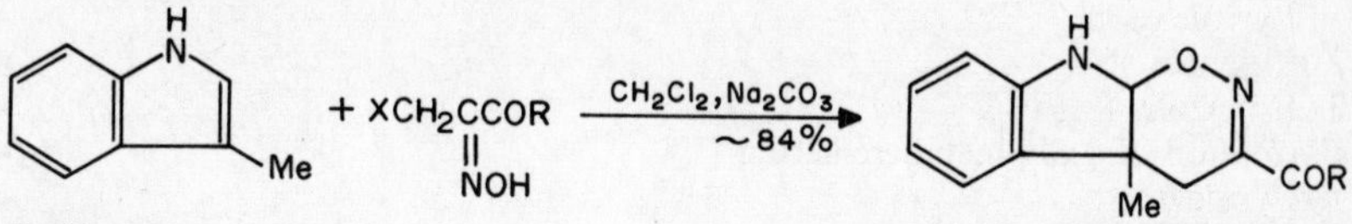

R = Me, EtO
X = Br, Cl

1,2-Oxazino[6,5-b]indole

[659]

CHAPTER 85

Enamine and Ester or Ketone

I. FORMATION OF A FIVE-MEMBERED RING

1. Pyrrole

A 2-keto-enamine reacts with glycine in alkali to form a pyrrole ring.

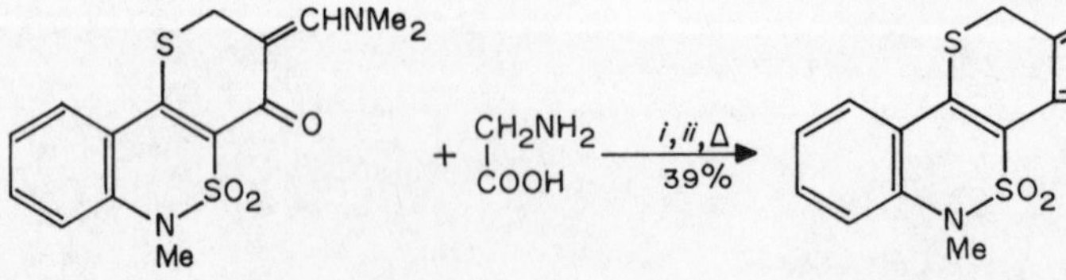

[229]

i, KOH, MeOH, Δ; ii, Ac_2O

Pyrrolo[3′,4′:4,5]thiopyrano-[3,2-*c*][2,1]benzothiazine 4,4-dioxide

2. Pyrazole

A hydrazine cyclizes a 2-keto-enamine to a fused pyrazole in high yield.

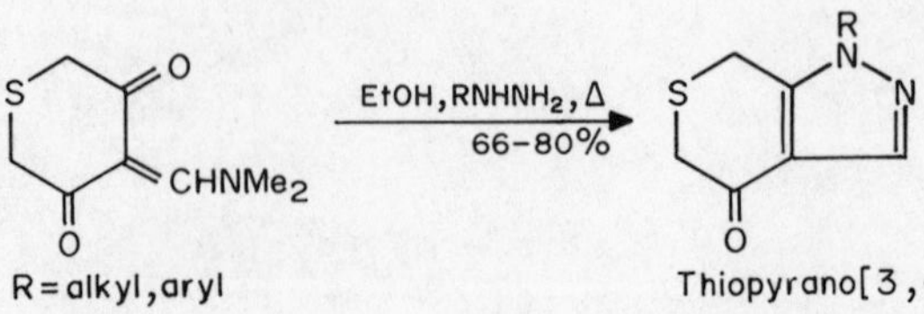

[1468]

R = alkyl, aryl

Thiopyrano[3,4-*c*]-pyrazol-4-one

Other ring systems synthesized similarly:

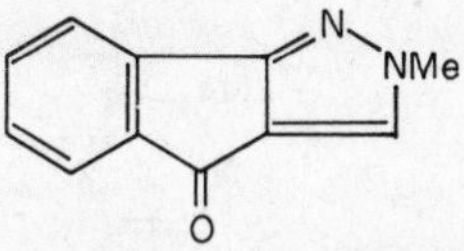

[230]

Indeno[1,2-c]pyrazol-4-one

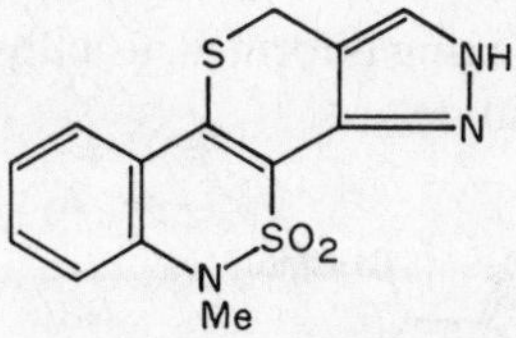

[229]

Pyrazolo[3',4':4,5]thiopyrano
[3,2-c][2,1]benzothiazine
4,4-dioxide

3. Isoxazole

Replacing the hydrazine in the preceding section with hydroxylamine leads to the formation of a fused isoxazole ring in good yield.

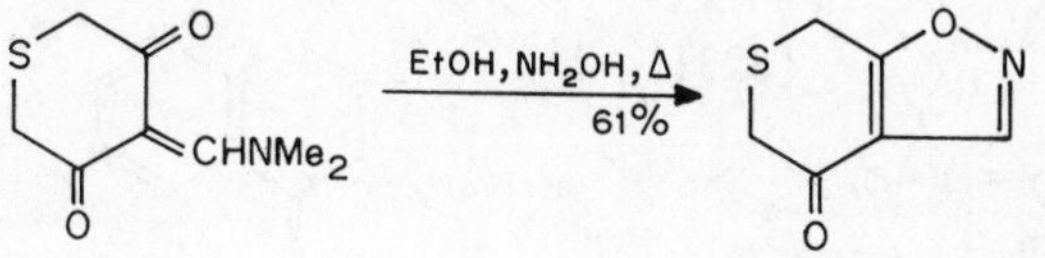

[1468]

Thiopyrano[4,3-d]-
isoxazol-4-one

Another ring system synthesized similarly:

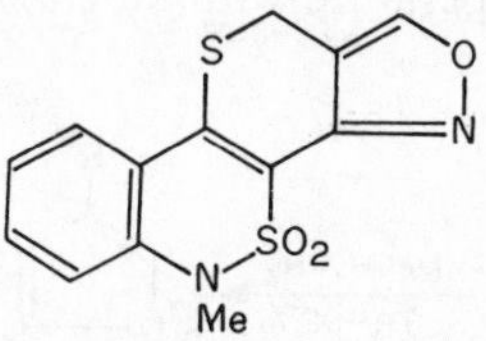

[229]

Isoxazolo[3',4':4,5]thiopyrano-
[3,2-c][2,1]benzothiazine
4,4-dioxide

II. FORMATION OF A SIX-MEMBERED RING

1. Pyridazine

When the enamine and carbonyl functions are separated by a carbon, reaction with a hydrazine gives a fused pyridazine ring.

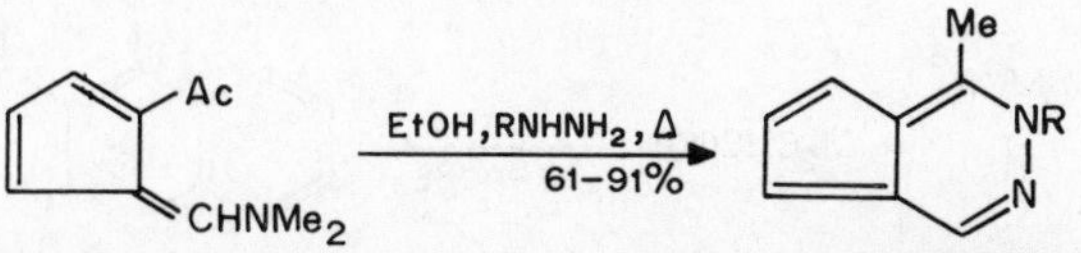

[1656]

R=H,Me,Ph

Cyclopenta[d]-
pyridazine

2. Pyrimidine or Pyrimidin-4-one

A pyrimidine ring is formed, usually in high yield, when a keto-enamine is heated with an amidine.

Pyrimido[5,4-*d*][1]-
benzazepin-6-one

Other ring systems synthesized similarly:

[823]

Pyrimido[5,4-*d*][2]-
benzazepine
(6-oxide)

[1468]

Thiopyrano[3,4-*d*]-
pyrimidin-5-one

Ammonia reacts at room temperature with an enamine-ester to give a high yield of a fused pyrimidinone.

R^1=H, PhCH$_2$; R^2=Me, Ph

Pyrrolo[3,2-*d*]pyrimidin-
4-one

3. Pyran-2-one

Chloroketenes react with enaminones with the formation of a pyranone ring but low yields may be obtained unless the enamine has an *N*-aryl ring.

R^1=Me, R^2=Ph; R^1=R^2=Ph or
NR^1R^2=N(CH$_2$)$_5$

Pyrano[3,2-*c*][1]benzopyran-2-one

Another ring system synthesized similarly:

[1925]

1-Benzopyran-2-one

Under basic conditions, enaminones are converted into pyran-2-ones by reaction with activated nitriles.

[801, 1803]

R^1=H,Me,MeO,HO,Me$_2$N; R^2=H,R^3=Ph or
R^2R^3=(CH$_2$)$_4$; R^4=O,H$_2$, Me$_2$; R^5=Ph,CN,
PhSO$_2$,COOMe,COOEt
i,DMF,tBuOK or EtONa,PhMe or Triton B.

X=CH$_2$,NH,NMe,NCHO,O
Naphtho[1,2-b]pyran-2-one
Pyrano[3,2-c]quinolin-2-one
Pyrano[3,2-c][1]benzopyran-2-one
Pyrano[3,2-c][1]benzopyran-2,5-dione
Pyrano[3,2-c]quinoline-2,5-dione

4. 1,2-Oxathiin 2,2-Dioxide

A mixture of methanesulphonyl (or phenylmethylsulphonyl)chloride and TEA produces a sulphene, RCH=SO$_2$ (R = H, Ph), which cyclizes enaminones *in situ* at low temperatures to an oxathiin ring in yields which vary considerably with the structure of the substrate.

[359, 524]

NR$_2$=NMe$_2$, NEt$_2$, N(CH$_2$)$_n$,
n=4,5

X=O,S
Furo[2,3-h]-1,2-benzoxathiin-
2,2-dioxide
Thieno[2,3-h]-1,2-benzoxathiin-
2,2-dioxide

Other ring systems synthesized similarly:

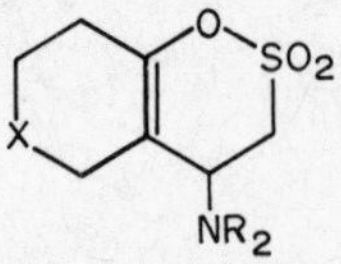

[611, 1469]

X = CH$_2$, O
1,2-Benzoxathiin 2,2-dioxide
Pyrano[3,4-e]-1,2-oxathiin
2,2-dioxide

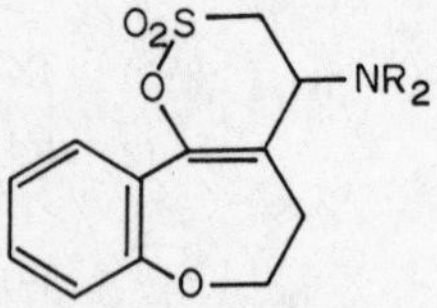

[521]

1,2-Oxathiino[5,6-d][1]-
benzoxepine 2,2-dioxide

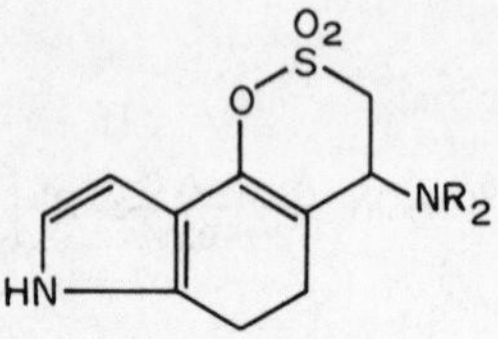

[28, 29]

1,2-Oxathiino[6,5-e]-
indole 2,2-dioxide

CHAPTER 86

Enamine and a Non-carbonyl Group

The following non-carbonyl functions are included in this chapter: hydroxy, nitrile, ring-carbon and ring-nitrogen.

I. FORMATION OF A SIX-MEMBERED RING

1. Pyridine or Pyridin-2- or -4-one

Treatment of an enamine with hydrogen chloride-chloroform at ambient temperature in a Pictet–Spengler-type cyclization sometimes results in elimination of one side of the double bond—in this example, as methoxyacetone.

R=H,MeO

Pyrido[3,4-b]indole

[485]

Chloromethylenemalonitrile (more effective than methoxymethylenemalonitrile) reacts with an enamine to form a 2-aminopyridine-3-carbonitrile ring in high yield. Intramolecular reaction between enamine and nitrile group also results in a pyridine ring being formed under acidic conditions.

Ar=4-COOHC$_6$H$_4$

1,6-Naphthyridine
(review [1648])   [1608]

[522]

1,6-Naphthyridine

Inter-ring C–C bond formation involving an enamide can produce different products according to the method used; photochemical or acylative cyclization may yield products which are different from those obtained by heating.

[249]

Isoquino[2,1-*b*][2,7]-
naphthyridin-8-one

Intramolecular cyclization of an ester-carrying enamine to a pyridin-4-one needs heating in either 1,2-dichlorobenzene (when $R^1 = H$) or with PPA (review [B-21]) (when $R^1 = Me$).

[1425]

$R^2 = Me, HO, Ac$
i, $R^1 = H$: 1,2-Cl$_2$C$_6$H$_4$
ii, $R^1 = Me$: PPA

4-Quinolinone

Enamines may cyclize on to ring-nitrogen atoms when treated with anhydride or diketene (reviews [116, 2018]); a fused pyridin-2-one ring is formed.

[926]

Pyrido[1,2-*a*]benzimidazol-
1-one

i, [diketene], CHCl$_3$, R=Ac

ii, (RCH$_2$CO)$_2$O, R=H, Me

Another ring system synthesized similarly:

4-Quinolizinone

(review [1692])

[1634]

2. Pyrazine

An enamine-nitrile may react with a primary alkylamine at ambient temperature to form a pyrazine ring (with loss of the enamine's t-amino group). Cyclizations of this kind have been reviewed [1876]. A similar cyclization of the malononitrile derivative (86.1) proceeds in high yield.

NR_2^1 = morpholinyl;
R^2 = Me, MeOCH$_2$CH$_2$

Quinoxaline

[1875]

(86.1)

Pyrido[3,4-b]pyrazine

[1778]

3. Pyran-4-one

Enamines react in boiling xylene with 3-oxocarboxylic esters to yield reduced chromones.

NR_2^1 = morpholinyl;
R^2 = H, R^3 = Me, Ph
or R^2R^3 = (CH$_2$)$_4$

1-Benzopyran-4-one

[1263]

A phenolic hydroxy reacts intramolecularly with an enamine and forms a fused pyran-4-one ring.

Ether or Thioether and Ring-carbon or Ring-nitrogen

I.   Formation of a Five-membered Ring   473
 1. 1,2,4-Triazole or 1,2,4-Triazol-3-one   473
 2. Isothiazole   474
II. Formation of a Six-membered Ring   474
 1. Pyridin-4-one   474
 2. Pyrimidin-4-one or Pyrimidine-2,4-dione   474
 3. 1,2,4-Triazin-6-one   475
 4. Pyran   476

A comparatively rare example of ring closure onto a ring-sulphur atom is cited in Section I.2.

I.   FORMATION OF A FIVE-MEMBERED RING

1.   1,2,4-Triazole or 1,2,4-Triazol-3-one

S-Methylthiolactims are cyclized by reaction with acid hydrazides giving the triazole while ethyl carbazate yields the triazolone.

[1023]

1,2,4-Triazolo[4,3-d]-[1,4]benzodiazepine

[1023]

1,2,4-Triazolo[4,3-d]-[1,4]benzodiazepin-3-one

2. Isothiazole

The valency of sulphur increases from two to four during the formation of the isothiazole ring by the reaction of a side-chain methylthioether with the isothiazolium salt.

$$R^3 \underset{R^2}{\overset{\text{Me}}{\underset{}{\text{N-S}}}} C=CR^1 \quad \xrightarrow[\text{31-53\%}]{\text{MeCN,MeNH}_2} \quad R^3 \overset{\text{Me} \quad \text{Me}}{\underset{R^2}{\text{N-S-N}}} \overset{R^1}{}$$

[693]

$$R^1-R^4=H,Me$$

Isothiazolo[5,1-e]isothiazole

II. FORMATION OF A SIX-MEMBERED RING

1. Pyridin-4-one

A dianion generated *in situ* from 5-methylisoxazole and LDA displaces a lactim ethoxy group at low temperature and forms a fused pyridinone ring.

$$(H_2C)_n \overset{\text{OEt}}{\underset{N}{}} + \overset{\bar{C}HCN}{\underset{CO\bar{C}H_2}{}} \quad \xrightarrow[\sim 35\%]{\text{THF}} \quad (H_2C)_n \overset{O}{\underset{NH_2}{}}$$

[1911]

Pyrrolo[1,2-a]pyridin-2-one
Pyrido[1,2-a]pyridin-2-one
Pyrido[1,2-a]azepin-2-one

2. Pyrimidin-4-one or Pyrimidine-2,4-dione

Displacement of a methylthio group of a cyclic enamine by reaction with a two-fold molar proportion of aryl isocyanate or isothiocyanate in refluxing toluene leads to the formation of a fused pyrimidine-2,4-dione or -2,4-dithione.

$$(H_2C)_n \overset{\text{Me}}{\underset{}{\text{N}}} SMe \quad + 2 ArNCX \quad \xrightarrow[\text{11-78\%}]{\text{PhMe},\Delta} \quad (H_2C)_n \overset{\text{Me} \quad \text{Me}}{\underset{X}{\text{N-N}}} \overset{X}{\underset{NAr}{}}$$

[1531, 1886]

$$n=1,2,3$$
$$Ar=Ph,4-Me-,4-Cl-,$$
$$4-Br-C_6H_4$$

X=O,S

Pyrrolo[2,3-d]pyrimidine-2,4-dione
Pyrrolo[2,3-d]pyrimidine-2,4-dithione
Pyrido[2,3-d]pyrimidine-2,4-dione
Pyrido[2,3-d]pyrimidine-2,4-dithione
Pyrimido[4,5-b]azepine-2,4-dione

An ethoxy group attached to a double-bonded carbon is easily displaced and when a ring-nitrogen atom is adjacent, ketene reacts to form a fused pyrimidinone ring. The *S*-methylthiolactim mentioned in Section I.1 forms a pyrimidin-4-one ring on reaction with methyl anthranilate [1023] and a similar cyclization occurs when the 3-methylthio-1, 2, 4-triazin-5-ones react with anthranilic acid.

R = H, Me

Pyrido[1,2-*a*]pyrimidin-4-one

[1097]

R = Me, PhCH$_2$, Ph, PhCH=CH,
4–ClC$_6$H$_4$CH=CH

[1,2,4]Triazino[3,2-*b*]-
quinazoline-2,6-dione
(review [1542])

[803]

Another ring system synthesized similarly:

Quinazolino[3,2-*d*][1,4]-
benzodiazepin-9-one

[1023]

3. 1, 2, 4-Triazin-6-one

When the reactive ethoxymethylidene (EtOCH=) group in a side-chain adjacent to a ring-nitrogen is treated with hot ethanolic alkali, a triazinone ring is readily formed.

R = Me, Et

[1,2,4]Triazino[4,5-*b*]-
indazol-1-one

[999]

Another ring system synthesized similarly:

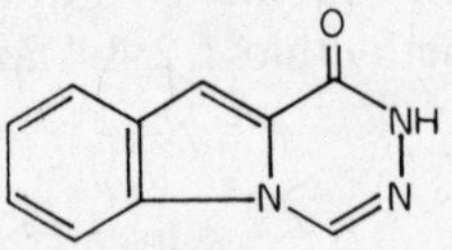

[1265]

[1,2,4]Triazino[4,5-*a*]-
indol−1−one

4.   Pyran

Lithiation at one carbon of a benzene ring followed by reaction with citral is a convenient but not high yielding method of synthesizing 2*H*-1-benzopyrans of biological interest.

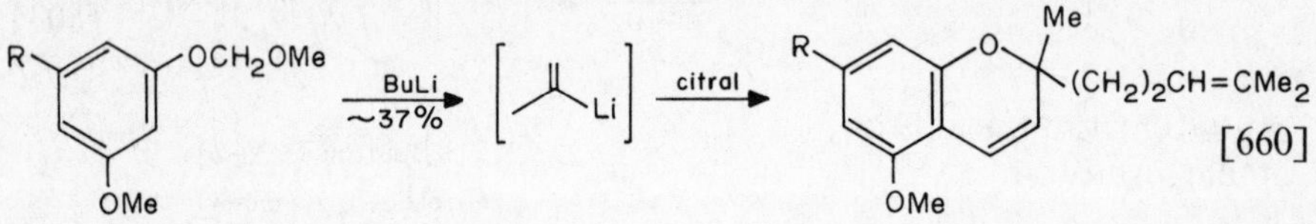

R = H, alkyl

2*H*−1−Benzopyran

[660]

Halogen and Hydroxy or Thiol

Some annulations of ether and halogen or acyloxy and halogen are described in this chapter; in many of these, the reaction conditions produce the free hydroxy group prior to cyclization. A few cyclizations which involve interaction between alkoxy and hydroxy groups are included because an alkoxy (like a halogen) is often displaced during the reaction.

I. FORMATION OF A FIVE-MEMBERED RING

1. Furan

Claisen rearrangement of a 2-bromoallyloxyarene gives an o-(2-bromo-allyl)phenol which cyclizes *in situ* to a new furan ring. Copper(I) alkynes react with an o-iodophenol (or its acetate) in pyridine to give a similar ring.

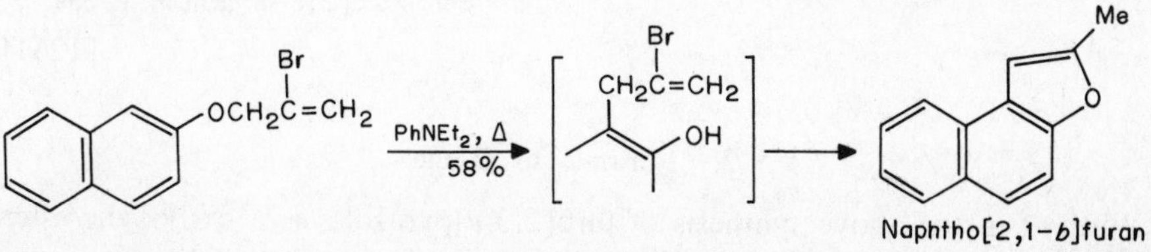

Naphtho[2,1−b]furan

[1181]

$$R^1 = HO, Ac, 4,5\text{-}OCH_2O;$$
$$R^2 = H, Ac; R^3 = MeC=CH_2$$

Benzofuran
(review [1622])

[486, 487, 1493]

Another ring system synthesized by the copper alkyne method:

[1626]

Furo[3,2-c]quinolin-4-one

o-Halogenophenols condense with nitriles in the presence of piperidine to give a 2-aminofuran ring.

R = CN, PhCO

Furo[2,3-c]pyrazole

[1586]

As a consequence of the tautomeric character of quinoline-2,4-diones, two isomeric products are formed when one or other of the enol forms (containing a 3-(2-methoxyphenyl) group) simultaneously undergoes demethylation and cyclodehydration.

pyr, HCl, Δ   47%   Benzofuro[3,2-c]quinolin-5-one

[1264]

19%

Benzofuro[2,3-b]quinolin-11-one

[1264]

2. Furan-2- or -3-one

When R in the above synthesis of furo[2,3-c]pyrazoles is a carboxylic ester group, the reaction takes a different course, the nitrile group remaining unchanged and a furan-2-one ring being formed.

Furo[2,3-c]pyrazol-5-one [1586]

A reactive halogeno-side-chain of a phenol is cyclized in high yield to give a furan-2-one ring on heating with a mild base.

3-Benzofuranone [473]

Another ring system synthesized similarly:

[1413]

Furo[3,2-c]pyridin-3-one

3. Thiophene

O-Arylthiocarbamates on heating rearrange to their S-aryl isomers which can react with a nuclear halogen on heating with alkali. A doubly fused thiophene is formed when the two functions are attached to different rings. A nuclear chlorine is also displaced by a side-chain thiol and a fused thiophene is obtained in rather low yield.

[1]Benzothieno[3,2-b]pyridine

[836]

Benzo[b]thiophene
(review [1649])

[896]

4. Oxazole

Reaction between a nuclear halogen and a 2-hydroxyethyl group is brought about at ambient temperature by a strong base.

[232]

[1]Benzothieno[2,3-*d*]oxazolo-
[3,2-*a*]pyrimidine

II. FORMATION OF A SIX-MEMBERED RING

1. Pyran, Pyran-2- or -4-one or Thiopyran-4-one

Lithiation of 2-bromophenols and treatment with an $\alpha\beta$-unsaturated ketone gives good yields of 2*H*-benzopyrans. A fused pyran is formed from a reactive halogen and a phenolic group.

R^1=H, Me, MeO, Cl; R^2=H, Me

[854]

2*H*-1-Benzopyran

[898]

Pyrano[3,2-*c*][1]-
benzopyran-4-one

Other ring systems synthesized from a halogen and a phenolic group:

[1409]

1-Benzopyran-4-one

[1687]

[1]Benzopyrano[2,3-*d*]-
1,2,3-triazol-9-one

2-(α-Fluoroacetyl)phenols reacts with ammonia or primary amines to give 2*H*-benzopyrans but secondary amines give chromones.

2H-1-Benzopyran   [1532]

1-Benzopyran-4-one   [1532]

A fused pyran-2-one ring is obtained when a 2-cyano- or 2-benzoyl-cinnamonitrile reacts with an *o*-bromophenol.

R = COOEt, CN   [1586]

X = O, NH
Pyrano[2,3-*c*]pyrazol-6-one
Pyrano[2,3-*c*]pyrazole

Experiments to demethylate a phenolic or thiophenolic ether which also contained either reactive bromine atoms or an enolic hydroxy group [1489] in a side-chain gave rise to a fused pyran- or thiopyran-4-one ring.

                                                                     [1145]

X = O, S
[1]Benzothieno[2,3-*b*]pyran-4-one
Thiopyrano[2,3-*b*][1]benzothiophen-4-one

Other ring systems synthesized similarly:

[1145]

X = O, S
[1]Benzothieno[3,2-*b*]pyran-4-one
Thiopyrano[3,2-*b*][1]benzothiophen-4-one

[1489]

Pyrano[3,2-*c*]pyridin-4-one

Interaction between a methoxy on one ring and a hydroxy on another in strongly basic media probably involve displacement of the alkoxy group so that the two rings become linked at those positions [774].

R^1=H,Me; R^2=H,Cl

i, NaH–DMSO or aq. NaOH

9–Xanthenone
(review [775])

[777]

2. 1,4-Dioxin or 1,4-Dithiin

Self-condensation of two molecules of a 2-chloro-3-pyridinol under basic conditions gives rise to a 1,4-dioxin ring. A thiophenol when subjected to the action of HMPT and a tertiary amine produces a thianthrene.

[1,4]Dioxino[2,3-*b*:5,6-*b*']dipyridine

[1192]

Thianthrene

[1053]

3. 1,3-Oxazine

This ring is formed under reductive conditions during the synthesis of the ring system of cephalosporin antibiotics in which sulphur is replaced by oxygen. More recent work in this and the penicillin field is covered by reviews of this specialized field [B-26, B-27] which is outside the scope of this book but an illustrative example is given (see also Chapter 90, Section I.2). The abnormal numbering of this ring system is shown.

5–Oxa–1–azabicyclo[4.2.0]–
oct–2–en–8–one

[1929]

4. 1,4-Oxazine or 1,4-Oxazin-3-one

Under strongly basic conditions, a side-chain alcohol displaces a nuclear chlorine to form an oxazine ring. When the functional groups are interchanged and the halogen is activated by a carbonyl, an oxazin-3-one ring is formed.

R^1 = alkyl, Ph; R^2 = alkyl
R^1, $R^2 \neq$ H

EtONa, EtOH, Δ
63–90%

Pyridazino[4,5-b]-
1,4-oxazin-8-one

[231]

AcOK, EtOH, Δ
84%

Pyrano[3,2-g]-1,4-
benzoxazine-3,6-dione

[360]

NaH, DMF, Δ
33%

[1,4]Oxazino[2,3,4-kl]-
phenothiazin-1-one

[153]

5. 1,4-Oxathiin

Phenolic and halogen functions on different rings which are joined together through a sulphur atom react under basic conditions to give a doubly fused oxathiin.

R = H, Me

DMF, NaH, Δ
27–95%

[1,4]Benzoxathiino[2,3-d]-1,2,3-
triazole

[1687]

III. FORMATION OF A SEVEN-MEMBERED RING

1. 1,4-Oxazepine or 1,4-Oxazepin-5-one

Base-induced reaction of suitably placed halogen and alcohol or phenol functions leads to the formation of an oxazepine ring.

Pyrrolo[2,1-c][1,4]benzoxazepine   [1816]

Pyrazolo[4',3':5,6]pyrido[2,3-b]-
[1,5]benzoxazepine

[523]

Other ring systems synthesized similarly;

[1037]

Dipyrido[2,3-b:2',3'-f]-
[1,4]oxazepine

[1941]

1,2,3-Triazolo[4,5-b][1,5]-
benzoxazepin-10-one

An alkoxy group probably behaves like the halogen in the above cyclization.

[984]

Thieno[3,4-b][1,5]benzoxazepin-10-one

IV. FORMATION OF AN EIGHT-MEMBERED RING

1. 1,3,6-Thiadiazocin-7-one

Reductive cyclization of a nuclear thiocyanate and a side-chain chlorine (during which the thiocyanate is converted into a thiol which displaces the halogen) gives good yields of this large ring.

R^1=H,Me,Cl; R^2= H,Me

Pyrrolo[1,2-a][3,1,6]-
benzothiadiazocin-6-one

[1719]

CHAPTER 89

Halogen and Lactam Carbonyl or a Sulphur-containing Group

The sulphur-containing functions included in this chapter are chlorosulphonyl, isothiocyanate, lactam thiocarbonyl, thiocyanate or thiourea.

I. FORMATION OF A FIVE-MEMBERED RING

1. 1,2,4-Triazole

Neighbouring chloromethyl and lactam thione groups are annulated into a triazole ring by heating with hydrazine. The many uses of this reagent in heterocyclic synthesis have been reviewed [1437].

$$\text{Ph} \quad \overset{O}{\diagup}\overset{S}{\diagdown} \quad \xrightarrow[38\%]{N_2H_4,\,\Delta} \quad \text{Ph} \quad \overset{O}{\diagup}\overset{N}{\diagdown}\!\!-\!\text{NH}$$

N——NCH₂Cl   N——N

[911]

1,2,4-Triazolo[3,4-b]-
[1,3,4]oxadiazole

2. Furan

Interaction between a 2-iodo-lactam carbonyl and phenylacetylene is catalysed by a palladium-triphenylphosphine complex in the presence of a tertiary amine; a fused furan ring is formed in high yield.

485

[317]

Furo[2,3-*d*]pyrimidine

i, CuI, TEA, Pd(PPh$_3$)$_2$Cl$_2$

3. Oxazole

A reduced oxazole ring may be formed by stirring an *N*-2-chloroethyl lactam in aqueous alkali at ambient temperature.

[286]

Oxazolo[2,3-*b*]quinazolin-5-one

4. Isothiazole 1,1-Dioxide

Annulation of neighbouring chloromethyl and chlorosulphonyl groups may be effected by warming the compound with a primary amine.

[767]

R = 4-HOOCC$_6$H$_4$CHMe

1,2-Benzisothiazole
1,1-dioxide
(review[1775])

5. Thiazole

A thiazolopyridine may be obtained from either a 2-chloropyridin-3-yl-isothiocyanate and an amine, or a 2-chloropyridin-3-ylurea and bromine. The latter method was expected to yield the [4,5-*c*] isomer [1780] and deserves further study.

[1288]

R = cyclohexyl, EtOOCNH,
PhNH, PhCONH

Thiazolo[5,4-*b*]pyridine

[1780]

Thiazolo[4,5-*b*]pyridine

II. FORMATION OF A SIX-MEMBERED RING

1. 2-Thioxopyrimidin-4-one

When a reactive chloro-thiourea is stirred at ambient temperature with a base in DMF, this pyrimidine ring is formed.

[1928]

R=H,alkyl,PhCH$_2$,Ph

Pyrido[2,3-*d*]pyrimidin-
4-one,2-thioxo-

2. 1,4-Oxathiin

2-Chloropyridin-3-yl thiocyanate may be cyclized to a doubly fused oxathiin by heating it in alkaline solution with an epoxyketone.

[289]

[1,4]Benzoxathiino[2,3-*b*]-
pyridine

3. 1,3-Thiazin-4-one or 2-Thioxo-1,3-thiazin-4-one

Photochemical cyclization of a 2-chloro-thioureidobenzothiophene occurs in good yield but the more labile 2-chloropyridine analogue cyclizes on heating in ethanol. A fused thiazinone ring is also formed when a 2-chloro-isothiocyanate is annulated by warming with a secondary amine.

[1544]

R^1,R^2=Me,Et,Ph,(CH$_2$)$_5$

[1]Benzothieno[2,3-*e*]-
1,3-thiazin-4-one

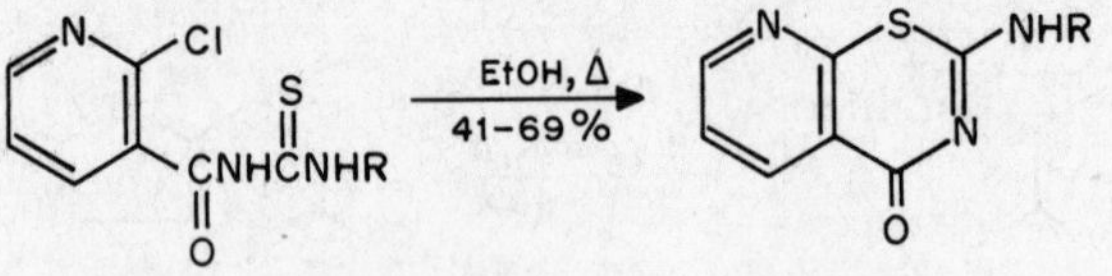

[1928]

R=H,alkyl,PhCH$_2$,Me-,
MeO-,Br-,NO$_2$-,Me$_2$N-C$_6$H$_4$

Pyrido[3,2-e]-1,3-
thiazin-4-one

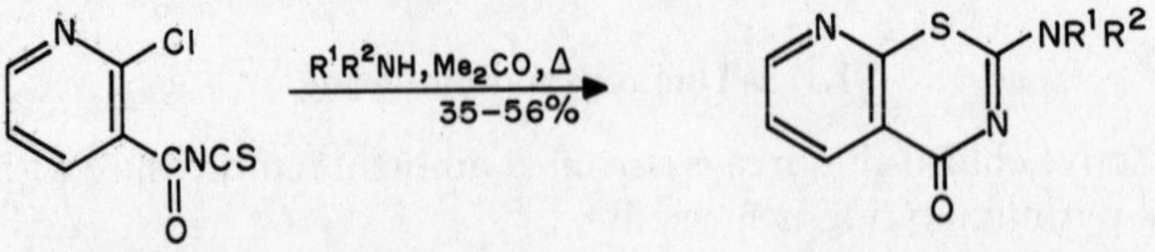

[1928]

R^1=Me,R^2=Ph or R^1=R^2=Ph
or R^{1}R^2=(CH$_2$)$_5$ or NR^{1}R^2=N[(CH$_2$)$_2$]O

Pyrido[3,2-e]-1,3-
thiazin-4-one

CHAPTER 90

Halogen and Methylene or Ring-carbon

I. FORMATION OF A FIVE-MEMBERED RING

1. Pyrrole

An anion in a side-chain displaces a nuclear chlorine in liquid ammonia with the formation of an isoindole; in this reaction a cyano group is also lost, possibly after cyclization.

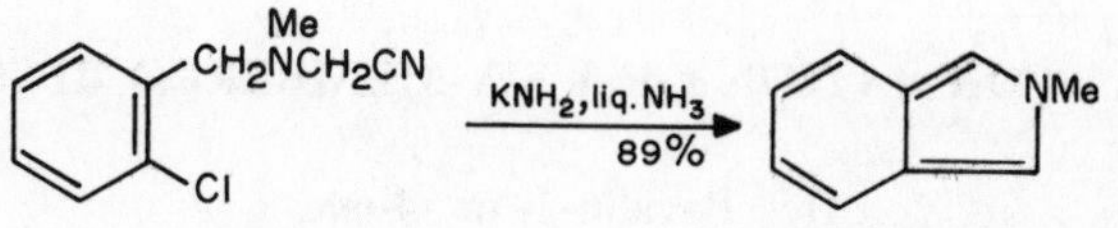

[1597]

Isoindole
(review [1615])

Photochemically induced bond-formation between two carbon atoms in different rings is effected in high yield when one of the carbons has a reactive chlorine atom.

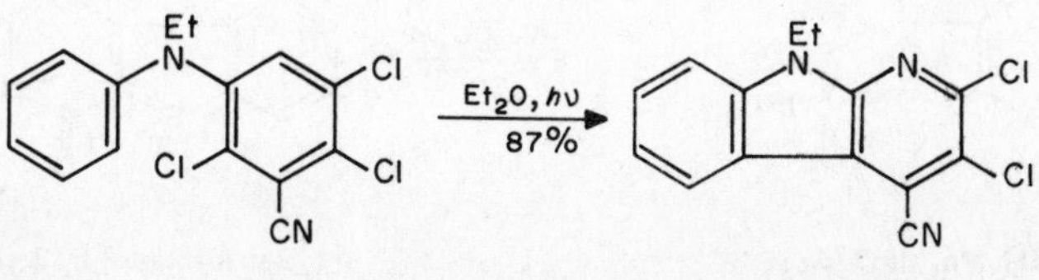

[596]

Pyrido[2,3-*b*]indole

2. Imidazole-4-thione

An important step in the synthesis of 1-azapenem is the formation of this ring fused to an azetidinone. The chloro-methylene is cyclized by treatment with sodium thiocyanate and lithium hexamethyldisilazide.

[1285]

$R^1 = tBu\,Me_2SiO$; $R^2 = 4-NO_2C_6H_4CH_2$
i, NaSCN, Me_2CO, Δ
ii, LiN$(SiMe_3)_2$, THF

1,4-Diazabicyclo[3.2.0]-
heptan-7-one, 3-thioxo-

3. Furan

Cyclization of a type similar to that described in Section I.1 can also be achieved by reaction of the halide with palladium(II) acetate-sodium carbonate.

R = H, Me, CH_2OH, NO_2, CN

Dibenzofuran
(review [1974])

[1527]

II. FORMATION OF A SIX-MEMBERED RING

1. Pyridin-2- or -4-one

Photocyclization is effective in forming a bond from a benzene ring to a chlorine-bearing thiophene ring.

Me_2CO, hʋ
~70%

[1882]

R = H, Me, MeO, Ac, COOEt

[1]Benzothieno[2,3-c]-
quinolin-6-one

Another ring system synthesized similarly:

[1882]

[1]Benzothieno[2,3-c]-
isoquinolin-5-one

Cyclic imino ethers react with halides which also contain a reactive methylene group.

[499]

Benzo[c]quinolizin-6-one

2. Pyridazin-4-one

The fluoro-ester used in the preceding section couples with a diazonium salt and the NH of the resulting 'hydrazone' displaces the fluorine and forms a fused pyridazinone ring.

[1529]

4-Cinnolinone

Ar = Ph, Me-, MeO-, Cl-,
NO₂-C₆H₄

3. 1, 3, 4-Thiadiazine

Reaction between an α-chlorohydrazone and a neighbouring methylene group proceeds by a [2,3] sigmatropic course in which a cyclic five-membered transition state is postulated.

[1653]

R = Ph, CH₂=CH, CH₂=CMe

4,1,2-Benzothiadiazine

CHAPTER 91

Halogen and Nitro

I. FORMATION OF A FIVE-MEMBERED RING

1. Pyrrole

The combination of an N-phenacylpyridinium function (review [1870]) and a reactive halogen favours the formation of the indolizine from the chloronitrobenzene at ambient temperature.

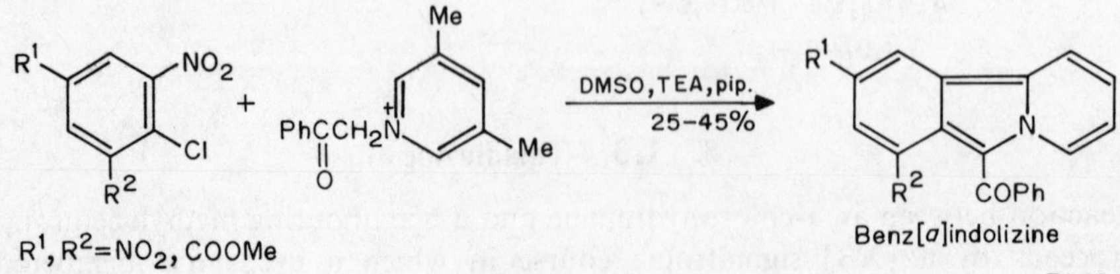

$R^1, R^2 = NO_2, COOMe$

Benz[a]indolizine

[1434]

2. Pyrazole or Pyrazole 1-Oxide

Two molar proportions of a primary arylamine react unexpectedly with 4-bromomethyl-3-nitropyridine to produce a 2-aryl-3-arylaminopyrazolopyridine in good yield.

Ar = Ph, Me-, Br-, Cl-C$_6$H$_4$

[1431]

X = N, Y = CH
Pyrazolo[3,4-c]pyridine
(review[1600])
X = CH, Y = N
Pyrazolo[4,3-b]pyridine
(review[1123])

Another ring system synthesized similarly:

[252]

Pyrazolo[4,3-d]pyrimidine-
5,7-dione 1-oxide

3. Thiazole or Thiazole 3-Oxide

Halogens may be displaced by sulphide groups and the presence of the neighbouring nitro group enhances their reactivity. Reaction with carbon disulphide and sodium sulphide cyclizes these two groups to form a thiazole ring. The chemistry of thiazolopyridines has been reviewed [1865]. A 2-mercapto-N-heterocycle reacts to form a doubly fused thiazole, the nitro group being displaced (reviews of such reactions [1398, 1399, 1432]).

[1202]

Thiazolo[5,4-b]pyridine

[1833]

[1]Benzopyrano[4',3':4,5]thiazolo-
[2,3-c][1,2,4]triazol-6-one

One of the less frequently used reactions of nitro groups is that with reactive methylene groups in a basic environment; thus, ethyl mercaptoacetate and triethylamine converts 2-chloronitrobenzenes into benzothiazole 3-oxides.

R = H, Cl, NO$_2$, CF$_3$, COOMe,
    SO$_2$NH$_2$

Benzothiazole 3-oxide

[1400]

Another ring system synthesized similarly:

[1718]

Thiazolo[5,4-*f*]quinazoline−
7,9−dione 3−oxide

4. Furazan 2-Oxide

Sodium azide converts chloro-nitro compounds into a furazan 2-oxide ring. The chemistry of benzofurazan 2-oxides (benzofuroxans) has been reviewed [1690, 1726].

[1609]

[1,2,5]Thiadiazolo[3,4-*e*]−
2,1,3−benzoxadiazole 1−oxide

II. FORMATION OF A SIX-MEMBERED RING

1. Pyrazine 1-Oxide

The nitro group is converted into the *N*-oxide when a chloronitro compound reacts with an amine or a 2-nitroso-amine.

R = H, NO

Pyrimido[5,4-*g*]pteridine−
2,4,6,8−tetraone 5−oxide

[1433]

2. 1,4-Oxazine

2-Hydroxyaniline condenses with reactive chloronitro compounds to form a fused oxazine ring. The synthesis of heterocycles by displacement of nitro groups has been reviewed [1432].

Pyridazino[3,4-b][1,4]-
benzoxazin-3-one

[231]

3. 1,4-Thiazine

A parallel reaction to that just mentioned (in which 2-hydroxyaniline is used) leads to a fused thiazine, the nitro group being displaced (reviews [1398, 1399, 1432]). A Smiles rearrangement accounts for the isomer formed. Examples of thiazine formation from 2-chloronitrobenzene may also be found in Chapter 82, Section II.3.

Phenothiazine   [148]

$R^1 = H, Br, Cl, NO_2$; $R^2 = Cl, Me$;
$R^3 = Br, Cl$

Another ring system synthesized similarly:

[1]Benzopyrano[3,4-b][1,4]-
benzothiazin-6-one

[1833]

4. 1,4-Oxathiin

Treatment of a chloronitro compound with 2-mercaptophenol in alkali is another cyclization in which the nitro group is displaced [1398, 1399, 1432].

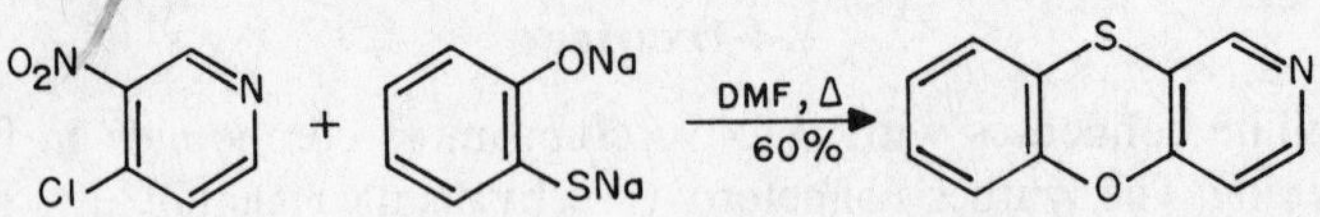

[1,4]Benzoxathiino[3,2−c]−
pyridine

[238]

III.   FORMATION OF A SEVEN-MEMBERED RING

1.   1,4-Thiazepine

Mildly alkaline conditions suffice for the double condensation of a chloronitrobenzene with 2-(2-mercaptophenyl)imidazole.

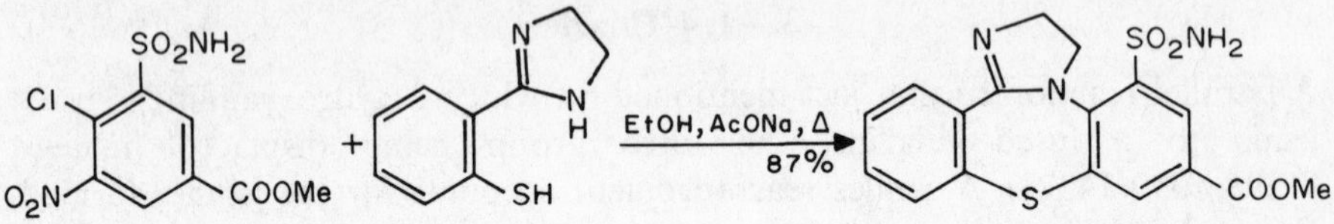

Dibenz[bf]imidazo[1,2−d]−
[1,4]thiazepine

[1347]

CHAPTER 92

Halogen and Ring-nitrogen

I. FORMATION OF A FIVE-MEMBERED RING

1. Pyrrole or Pyrrol-2-one

A reduced pyrrole ring may be formed by treatment of a 2-(3-halogenopropyl)heterocycle with either sodamide or alkoxide. When the halogenoalkyl is placed between two nitrogen atoms (as in a pyrimidine or triazine), cyclization on to either is possible.

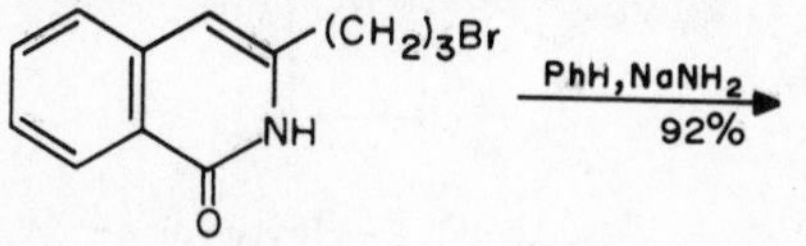

[722]

Pyrrolo[1,2-*b*]isoquinolin-5-one

Other ring systems synthesized similarly:

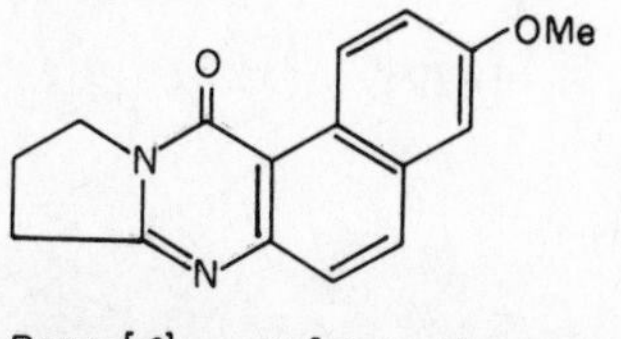

[18]

Benzo[*f*]pyrrolo[2,1-*b*]-quinazolin-12-one

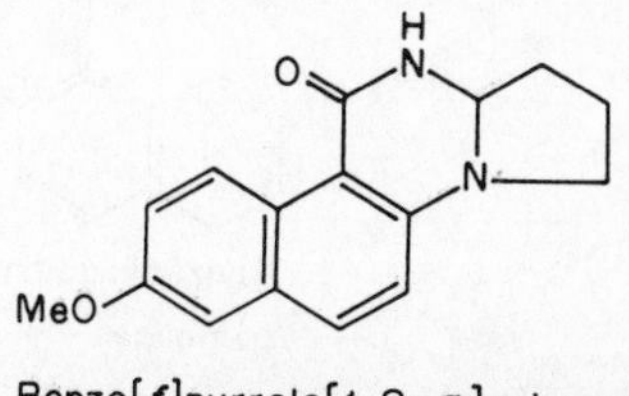

[18]

Benzo[*f*]pyrrolo[1,2-*a*]quinazolin-11-one

Copper-assisted bond formation between an anionic nitrogen and a bromo-benzene can result in the formation of a pyrrolone ring.

R = Me, MeO

Pyrrolo[1,2-a]indol-4-one
(review [909])

[708]

2. Imidazole or Imidazol-4-one

A 2-(chloroethyl) or chloroacetyl group cyclizes on to a neighbouring ring nitrogen on heating in a basic or an acidic medium.

R = H, Me

Imidazo[1,2-a]pyridine

[1083, 1509]

Imidazo[1,2-a]pyrazine

[89]

Other ring systems synthesized similarly:

Imidazo[1,2-a]benzimidazol-2-one

[842]

Imidazo[1,2-c]pyrimidine-3,7-dione

[350]

Benz[f]imidazo[1,2-a]-quinoline

[429]

Ethyl α-halogenoalkanoates react with 2-halogeno-N-heterocycles to form an N-ethoxycarbonylmethyl intermediate which is cyclized by ethanolic ammonia.

$R = H$, alkyl

i, MeCOEt, K_2CO_3, Δ;

ii, NH_3–EtOH

Imidazo[1,2-a]thieno[2,3-d]-

pyrimidin–2–one

[721]

3. 1, 2, 4-Triazole

A π-deficient 2-chloroheterocycle reacts with an acid hydrazide to form a fused triazole.

$R = H$, alkyl, aryl;

$Ar = Ph, Me-, F-, Cl-,$

$CF_3-, MeO-, NO_2-C_6H_4$

1,2,4–Triazolo[4,3-b]–

pyridazine

[70]

4. 1, 2, 4-Thiadiazole or 1, 2, 4-Thiadiazol-3-one

A 2-trichloromethylsulphenamido-N-heterocycle cyclizes at ambient temperature to form either of these two rings according to the base used.

[1,2,4]Thiadiazolo[4,3-a]-

pyridin–3–one

[278]

[1,2,4]Thiadiazolo[4,3-a]-

pyridine

[278]

II. FORMATION OF A SIX-MEMBERED RING

1. Pyridin-2-one

When 1-chloroisoquinolone is heated with glutaconic anhydride, a benzo-quinolizinone is formed in moderate yield. Quinolizinones may be prepared by heating 2-(4, 4-dichloro-1, 3-butadienyl)pyridines in aqueous dioxan.

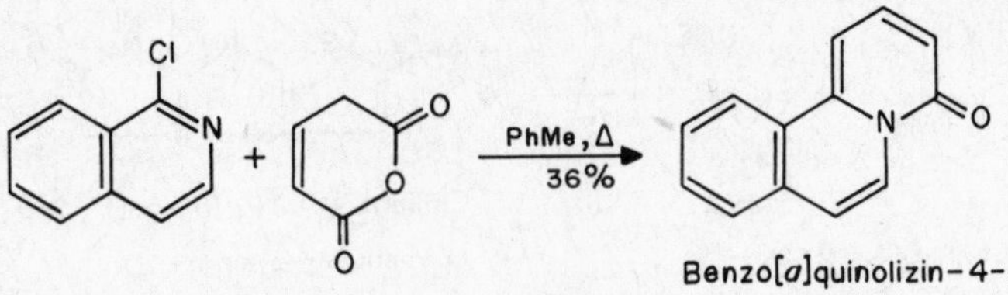

Benzo[a]quinolizin-4-one

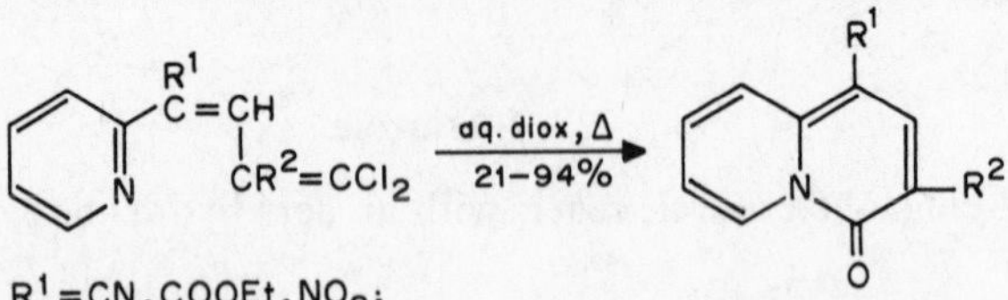

[804]

R¹ = CN, COOEt, NO₂;
R² = H, Me, Et

4-Quinolizinone

[1969]

2. Pyrimidin-4-one

Displacement of a halogen attached to one ring and formation of a bond to a nitrogen in another ring is a method which can be used to form one of several new rings depending on the structure of the substrate; a fused pyrimidin-4-one ring provides an example.

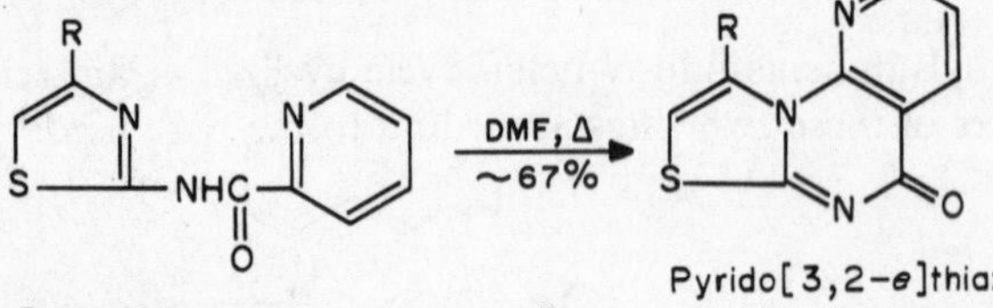

R = H, Me

Pyrido[3,2-e]thiazolo-
[3,2-a]pyrimidin-5-one

[617]

Another ring system synthesized similarly:

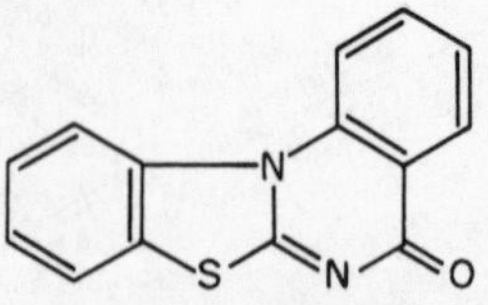

[65]

Benzothiazolo[3,2-a]-
quinazolin-5-one

A terminal chlorine in a side-chain adjacent to a ring-nitrogen can be cyclized with mild base.

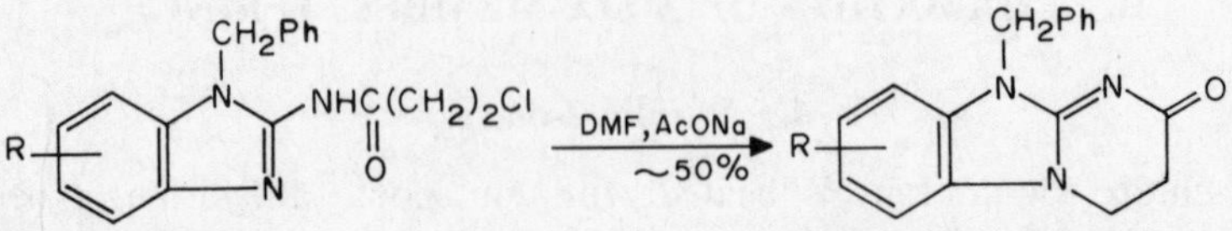

R = MeO, CF₃

Pyrimido[1,2-a]benzimidazol-
2-one

[842]

2-Chloropyridine undergoes a thermal cyclization with 2-amino-carboxylic esters to form a doubly fused pyrimidinone.

R^1=H, alkyl, Ac, COOH;
R^2=H, Me, iPr, Br

Pyrido[2,1-*b*]quinazolin-11-one

[1902]

Another ring system synthesized similarly:

[1]Benzothieno[2,3-*d*]pyrido-
[1,2-*a*]pyrimidin-12-one

[1223]

3. Pyrazin-2-one

Terminal halogen atoms of side-chains containing appropriate atoms cyclize by a base-assisted reaction with a ring-nitrogen to form a pyrazinone ring.

Ar = 2-ClC$_6$H$_4$

DMF, TEA, Δ
74%

Pyrazino[1,2-*a*][1,4]-
benzodiazepin-2-one

[563]

Another ring system synthesized similarly:

[1]Benzothieno[2,3-*d*]pyrazino-
[1,2-*a*]pyrimidine-3,6-dione

[1002]

4. 1,4-Oxazin-2-one

A base-catalysed cyclization similar to that described in the previous section but in which the acylamine is replaced by an acyloxy group leads to an oxazinone.

[1,4]Oxazino[2,3,4-*kl*]-
phenothiazin-2-one   [153]

III. FORMATION OF A SEVEN-MEMBERED RING

1. 1,4-Diazepin-2- or -5-one

Interaction between a ring-nitrogen atom and a side-chain halogen is assisted by
a basic medium and sometimes copper powder is added when the halide is not
very reactive. On the other hand, an α-chlorocarbonyl may react on heating in
propanol [340].

Pyrrolo[1,2-*d*][1,4]-
benzodiazepin-6-one   [87]

Other ring systems synthesized similarly:

[1,4]Benzodiazepino[4,5-*d*]-   [340]
[1,4]benzoxazepin-10-one

[1,4]Diazepino[6,7,1-*kl*]-   [616]
phenothiazin-4-one

CHAPTER 93

Hydrazide or Hydrazine and Nitro or Ring-nitrogen

In addition to the functions given in the title, isolated examples of the following pairs are mentioned: hydrazine and nitroso; hydroxamic acid and ring-N-oxide; hydrazone and nitro.

I. FORMATION OF A FIVE-MEMBERED RING

1. Imidazole

Diazotization of a hydrazide followed by gentle heating results in annulation (probably through the isocyanate) to the neighbouring ring-nitrogen.

Imidazo[5,1-*f*][1,2,4]-triazine-4,7-dione

[307]

2. 1, 2, 3-Triazole or its 1-Oxide

Annulation of hydrazine and nitroso groups may be achieved by treatment with Vilsmeier reagents. This reaction should be allowed to proceed without cooling, otherwise a triazine ring is obtained (see Section II.2). Replacement of the nitroso by nitro leads to a 1-hydroxytriazole.

1,2,3-Triazolo[4,5-d]-
pyrimidine-5,7-dione
(review [2023])

[904]

Benzotriazole

[1147]

A triazole N-oxide is obtained by treatment of a 2-nitro-N^2-alkyl- or -aryl-hydrazine with hydrogen chloride-alkanol or by methylation of the unsub-stituted nitro-hydrazine.

Benzotriazole 1-oxide

[986]

Another ring system synthesized similarly:

1,2,3-Triazolo[4,5-d]-
pyrimidin-7-one

[166]

3. 1, 2, 4-Oxadiazol-5-one

When a hydroxamic acid group is adjacent to an N-oxide function, heating with DCC produces a fused oxadiazole, probably by way of a Lossen rearrangement to the isocyanate.

CONHOH ... $\xrightarrow[60\%]{\text{diox,DCC,}\Delta}$ ... [1768]

[1,2,4]Oxadiazolo[2,3-c]-
quinazolin-2-one

II.   FORMATION OF A SIX-MEMBERED RING

1.   Pyridazin-4-one

Displacement of a nitro group by a nucleophile (reviews [1398, 1399, 1432]) is demonstrated by the cyclization of 2-nitro-hydrazones to give a fused pyridazinone ring.

$\xrightarrow[37-98\%]{\text{EtOH,Na}_2\text{CO}_3,\Delta}$ ... [1380]

R = Ac, PhCO, CONH$_2$,
COOEt, CN, NO$_2$

4-Cinnolinone

2.   1, 2, 4-Triazin-6-one or 1, 2, 4-Triazine 4-Oxide

The additional carbon atom required to cyclize a hydrazide, such as imidazole 2-carbohydrazide, may be provided by an orthoester; alternatively, the pre-formed ethoxymethylene hydrazide may be cyclized by heating with alkali.

H$_2$NHNC ... + MeC(OEt)$_3$ $\xrightarrow[51\%]{\text{EtOH,}\Delta}$ ... [380]

Imidazo[1,2-d][1,2,4]-
triazin-8-one

CNHN=CHOEt $\xrightarrow[51\%]{\text{EtOH,KOH,}\Delta}$ ... [387]

Pyrazolo[1,5-d][1,2,4]-
triazin-4-one

Other ring systems synthesized similarly:

[380]

Imidazo[1,5-*d*][1,2,4]-
triazin-1-one

[799]

[1,2,4]Triazino[4,5-*a*]-
indol-1-one

A 2-nitroso-hydrazine when treated at 0 °C with Vilsmeier reagents cyclizes to a fused triazine oxide in high yield (review [1676]); at higher temperatures, a triazole is formed (see Section I.2).

[999]

[1,2,4]Triazino[4,5-*b*]-
indazol-1-one

[1424]

Furo[2',3':4,5]pyrrolo[1,2-*d*]-
[1,2,4]triazin-5-one

3. 1,3,5-Triazine-2,4-dione

Two molecules of the substrate are incorporated in the product when an oxazole- or thiazole-2-carbohydrazide is treated with nitrous acid. An isocyanate is probably produced and dimerizes to the triazinedione.

$\xrightarrow[\sim 67\%]{HCl, NaNO_2}$

R = Me, Ph

[946]

X = O, S

Oxazolo[3,2-*a*][1,3,5]triazine-5,7-dione
Thiazolo[3,2-*a*][1,3,5]triazine-5,7-dione

4. 1,4-Thiazine

Diphenyl sulphide which has a 2-nitro in one ring and a 2-hydrazino in the other cyclizes in basic media to give a doubly fused thiazine ring. In this example, the terminal nitrogen of the hydrazino group is acylated.

$\xrightarrow[55-70\%]{DMF, K_2CO_3, \Delta}$

[1028]

Phenothiazine

R = R$_2^1$NCH$_2$CNH, R^1 = Me, Et or
                  ||
                  O

NR$_2^1$ = N(CH$_2$)$_5$

5. 1, 3, 4-Thiadiazine

Displacement of a nitro group by an adjacent hydrazone function is a well-known reaction (Section II.1) and requires heating of a 2-nitrophenylthiopyruvaldehyde hydrazone in ethanol and a mild base.

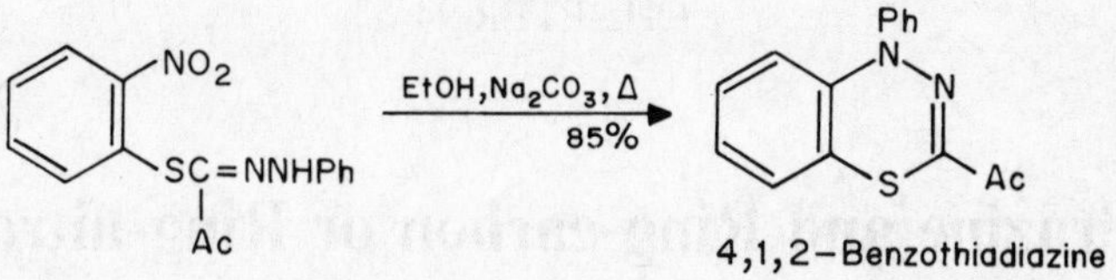

[710]

4,1,2−Benzothiadiazine

III. FORMATION OF A SEVEN-MEMBERED RING

1. 1, 2, 4-Triazepin-7-one

When the hydrazide and ring-nitrogen are on different rings a seven-membered ring product is formed in the orthoester reaction (see Section II.1).

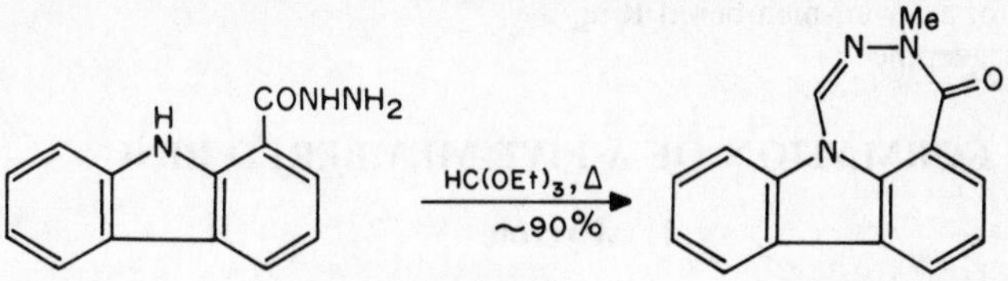

[1268]

[1,2,4]Triazepino[6,5,4−*jk*]−
carbazol−4−one

CHAPTER 94

Hydrazine and Ring-carbon or Ring-nitrogen

I. FORMATION OF A FIVE-MEMBERED RING

1. Pyrrole

One of the best known reactions which lead to the formation of a new fused heterocyclic ring is the Fischer indole synthesis. Since its discovery in 1883, it has been extensively and intensively studied but continues to interest chemists in extending its mechanism, scope and utility. It was comprehensively reviewed in 1982 [B-29]; its application to the synthesis of indoles was surveyed in 1970 [B-30] and 1979 [B-15a] and again in 1984 [B-4]. In this section, Fischer cyclizations of a pre-formed arylhydrazone and of a mixture of an arylhydrazine and a carbonyl compound are discussed together.

Many acidic catalysts promote Fischer's reaction, for example, zinc chloride [532, 1976], acetic or formic acid [1033], PPA (review [B-21]) [944, 1021, 1033, 1189, 1930], mineral acid [534, 536, 1189, 1203, 1975]; examples of the use of these in the synthesis of a pyrrole ring fused to benzene or a heterocycle continue to appear.

Most of the variations require heating but using phosphorus trichloride in benzene, high yields of many indoles are obtained from the hydrazones at ambient temperature and within a short reaction time [1101].

[1021]

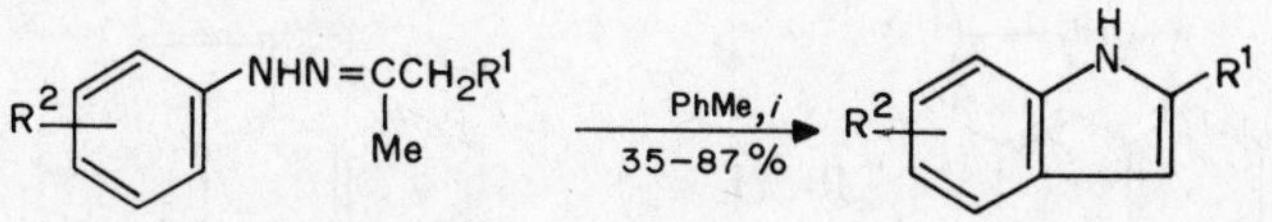

$R^1 = COOEt, Ac; R^2 = H, Me, MeO, Cl$
i, TsOH or Amberlyst-15

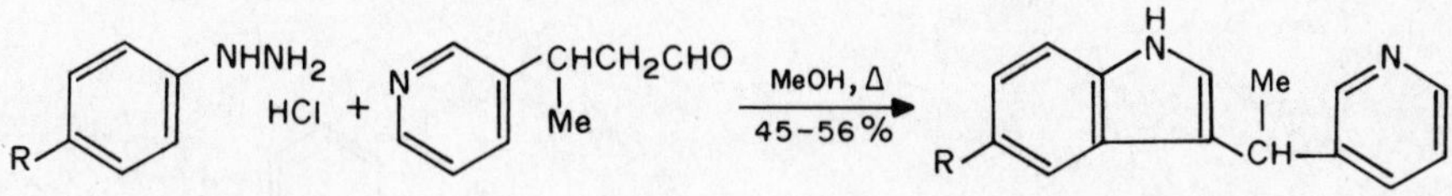

[1051]

$R = H, MeO$

[534]

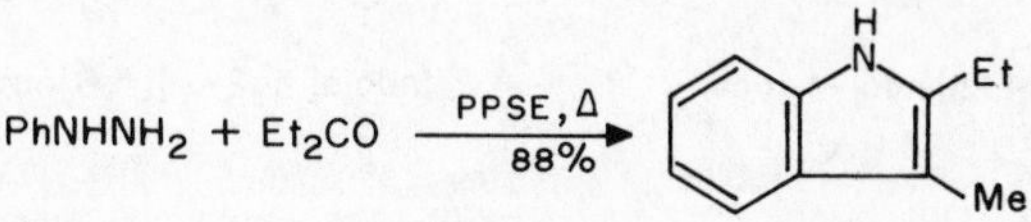

$PhNHNH_2 + Et_2CO \xrightarrow[88\%]{PPSE, \Delta}$

[1559]

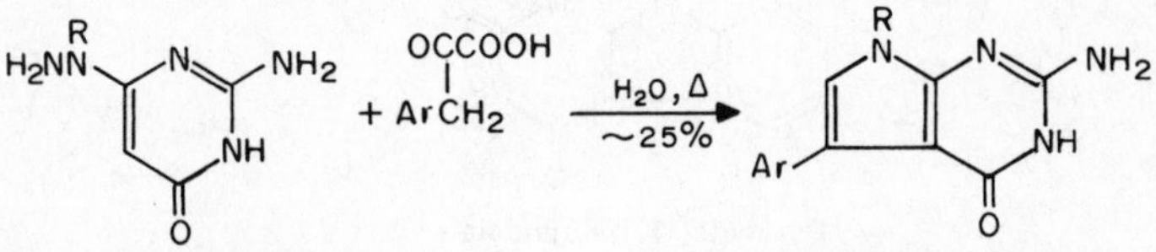

$R = H, Me, HO(CH_2)_2;$
$Ar = Ph, 4-HOC_6H_4$

Pyrrolo[2,3-d]pyrimidin-4-one

[1369]

Other ring systems synthesized by one of the above methods.

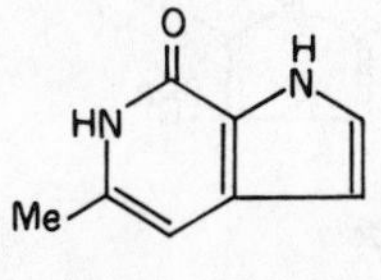

[532]

Pyrrolo[2,3-c]pyridin-7-one

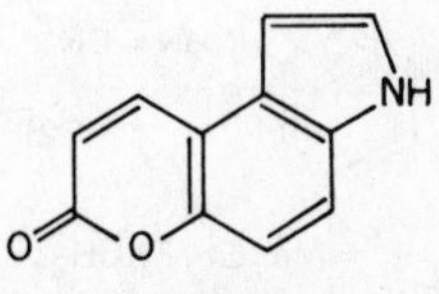

[944]

Pyrano[3,2-e]indol-7-one

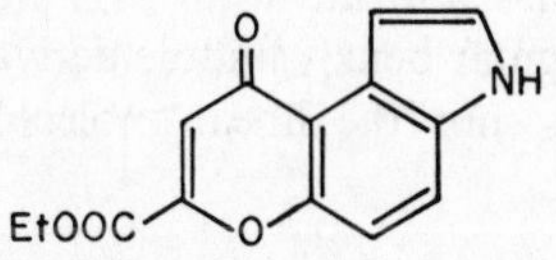

[944]

Pyrano[3,2-e]indol-9-one

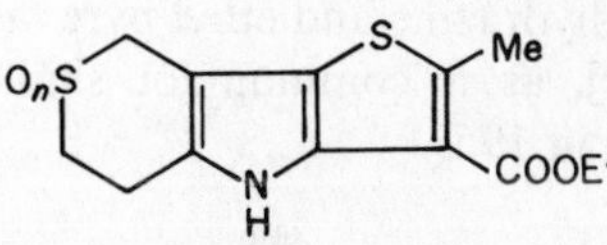

[536]

$n = 0, 2$
Thieno[3,2-b]thiopyrano[3,4-d]pyrrole
Thieno[3,2-b]thiopyrano[3,4-d]pyrrole
7,7-dioxide

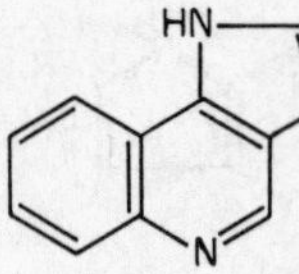

[1022]

Pyrrolo[3,2-c]quinoline

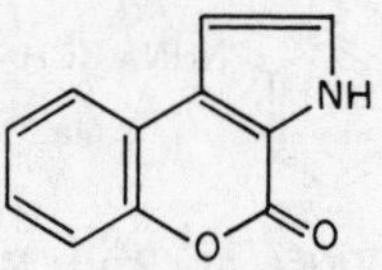

[1033]

[1]Benzopyrano[3,4-b]-
pyrrol-4-one

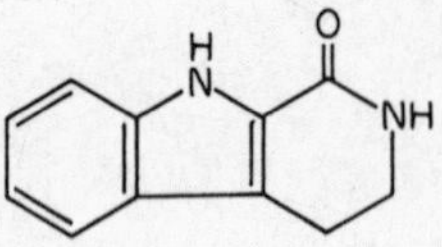

[1277, 1292]

Pyrido[3,4-b]indol-1-one

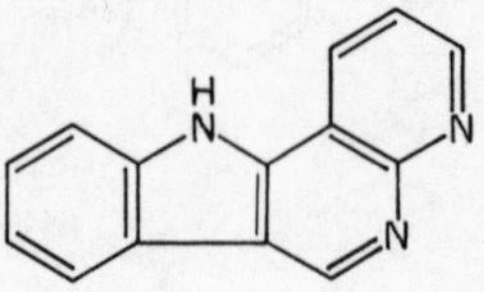

[1189]

Indolo[3,2-c][1,8]naphthyridine

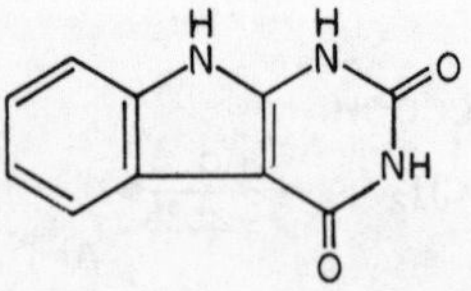

[1203]

Pyrimido[4,5-b]indole-
2,4-dione

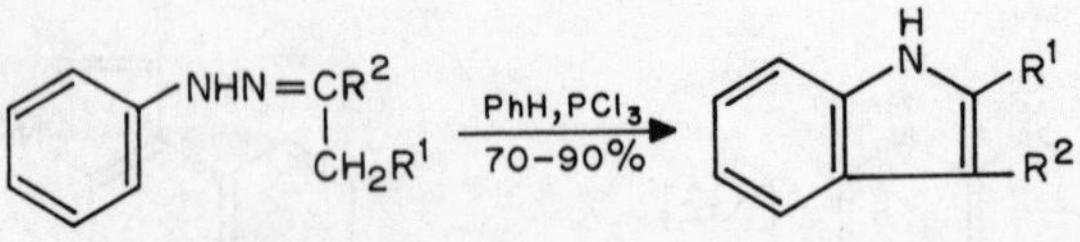

[1101]

$R^1 = Me, Et, PhCH_2, Ph$;
$R^2 = Me, Ph, Me_2C = CHCH_2$

Indole

Fischer indolization is sometimes accompanied by an additional reaction such as chlorination or loss of a methoxy group [237] or a fluorine atom [679]. 2-Naphthylhydrazine and ethyl pyruvate yield the angular benz[e]indole derivative [237], as is common for similar compounds, not the linear molecule described in 1953.

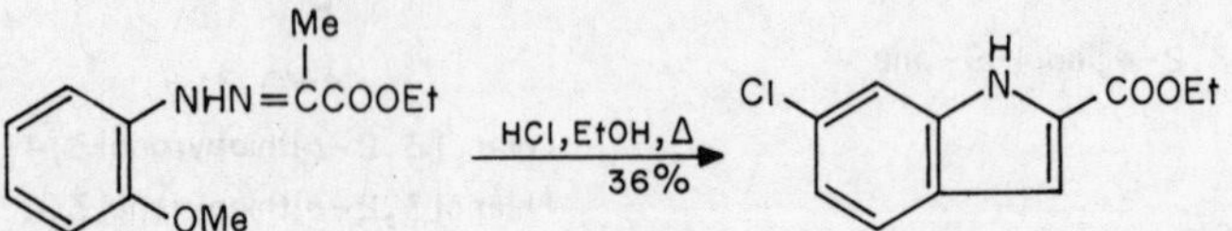

[237, 273]

Benz[e]indole

[679]

Replacement of NH of the hydrazone by an oxygen atom to give an *O*-phenylketoxime allows the Fischer cyclization to be used in the synthesis of a benzofuran (review [B-11] but aldoximes fail to cyclize.

R^1 = TsO; R^2 = H, R^3 = Me or
R^2R^3 = (CH$_2$)$_4$

Benzofuran
Dibenzofuran

[1389]

Partially reduced carbazoles are readily obtained by using a cyclohexanone in the Fischer cyclization which is the first stage in the Borsch carbazole synthesis, the cyclization being followed by dehydrogenation.

R^1, R^2 = H, Me

Carbazole
(review[1695])

[575]

Regiospecificity in the Fischer indolization has been studied, for example, the cyclohexenone (**94.1**) gives a mixture of about equal amounts of the two isomers.

(**94.1**)

24%

23%   [1975]

1-Carbazolone   [1930]

Fischer's reaction occasionally takes an unexpected course as in this formation of a pyridobenzimidazole.

R^1 = H, Me; R^2 = Me, Ph,
4-Br-, 4-Cl-C_6H_4

Pyrido[1,2-a]benzimidazole

[1976]

2. Pyrazole

Formation of a pyrazole ring by reaction of a hydrazine group with its neighbouring ring carbon atom requires a reagent which supplies one carbon atom. Vilsmeier reagent, an aryl isothiocyanate, DMFDMA and phosgenimmonium chloride are effective. When the hydrazine has an $NHNH_2$ grouping, DMFDMA or DMFDEA simultaneously effects cyclization and alkylation of the terminal nitrogen but a substrate containing NHNHPh is simply cyclized [1141]. The reactions of hydrazinopyrimidines have been reviewed [1977].

R = H, Me, PhCH$_2$, Ph

Pyrazolo[3,4-d]pyridazin-4-one

[1722]

Ar = 4-MeC$_6$H$_4$

Pyrazolo[3,4-d]pyrimidine-4,6-dione

[619]

Pyrazolo[3,4-d]pyrimidine-4,6-dione

[1141]

R^1 = Me, Et; R^2 = alkyl
i, Me$_2$NCH(OMe)$_2$;
ii, Me$_2$N$\overset{+}{=}$CCl$_2$, CH$_2$Cl$_2$

[927]

3. 1, 2, 4-Triazole

Annulation of a hydrazine group to the neighbouring ring-nitrogen is a well-exemplified cyclization. The commonest reagents used for this are carboxylic acids, acyl chlorides, anhydrides, carboxylic esters, orthoesters, isothiocyanates, S-methylisothiourea, aldehydes and cyanogen bromide.

$$R^1 \xrightarrow[30-80\%]{R^2COOH, \Delta} R^1$$ [972]

$R^1 = H, NH_2, NO_2;$
$R^2 = H, Me, Ph$

[1,2,4]Triazolo[1,5-a]pyridine
(review[1694])

Other ring systems synthesized similarly:

[621]

1,2,4-Triazolo[3,4-f]-
[1,2,4]triazin-8-one

[279]

[1,2,4]Triazolo[4,3-a]-
quinoxaline 5-oxide

[585]

Pyrrolo[1,2-a]thieno[3,2-e]-
1,2,4-triazolo[3,4-c]pyrazine

[385, 1367, 1465]

X = CH, N
1,2,4-Triazolo[4,3-b]-
pyridazine
1,2,4-Triazolo[4,3-b]-
[1,2,4]triazine

[748]

1,2,4-Triazolo[3,4-c]-
[1,2,4]triazin-5-one

[799]

1,2,4-Triazolo[4',3':2,3]-
pyridazino[4,5-b]indole

[799]

1,2,4-Triazolo[4',3':1,6]-
[1,2,4]triazino[4,5-a]indole

[963]

1,2,4-Triazolo[4,3-a]-
pyridine

Benzofuro[3,2-d]-1,2,4-
triazolo[4,3-b]pyridazine   [1025]

Thieno[3,2-e]-1,2,4-
triazolo[4,3-c]pyrimidine   [1343]

1,2,4-Triazolo[3,4-c]-
[1,2,4]triazine   [1367]

Pyrazolo[3,4-e][1,2,4]triazolo-
[3,4-c][1,2,4]triazine   [1474]

R = H, Me, Ph

1,2,4-Triazolo[4′,3′:1,6]-
[1,2,4]triazino[4,5-a]indole   [799]

Other ring systems synthesized similarly:

1,2,4-Triazolo[4,3-b]-
pyridazine   [70]

[1,2,4]Triazolo[4,3-a]-
[1,4]benzodiazepine   [1879]

1,2,4-Triazolo[4′,3′:1,6]-
pyridazino[4,5-b]quinoxaline   [1460]

R¹ = pyrrol-1-yl;
R² = alkyl, Ph

1,2,4-Triazolo[4,3-a]-
pyridine   [1891]

Other ring systems synthesized similarly:

[585]

Pyrrolo[1,2-a]thieno-
[3,2-e]1,2,4-triazolo-
[3,4-c]pyrazine

[622]

[1,2,4]Triazolo[4,3-a]-
quinoxaline

R^1=H,Et,Pr;
R^2=H,Me

HCOOMe or R^2C(OEt)$_3$,BuOH,Δ
44-97%

(94.1)   [622, 1444]

[1,2,4]Triazolo[4,3-a]-
quinoxaline

[681]

1,2,4-Triazolo[4,3-b]-
pyridazine

Replacing the malonic ester with trialkyl orthoacetate or orthoformate gives high yields of the 3-methyl derivative or the parent respectively of this ring system [137, 618]. The following are other examples of ring systems which have been synthesized using an orthoester.

[557]

Thieno[3,2-e]-1,2,4-
triazolo[4,3-c]pyrimidine

[1782]

1,2,4-Triazolo[4,3-a]-
purin-5-one

[1970]

Pyrazolo[1,5-a][1,2,4]-
triazolo[4,3-c]pyrimidine

[1851]

(94.2)
[1,2,4]Triazolo[4,3-a]-
quinoxalin-4-one

Pyrazolo[4,3-e][1,2,4]triazolo-
[4,3-a]pyrimidin-4-one

[972, 1815]

X=CH
1,2,4-Triazolo-
[4,3-a]pyridine
X = N
1,2,4-Triazolo-
[4,3-b]pyridazine

[623]

R=Bu, PhCH₂,
Me-, F-C₆H₄

(94.3)
1,2,4-Triazolo[3,4-a]-
phthalazine

When the isothiocyanate and DCC were replaced by S-methylisothiourea, the same product (**94.3** R = H) was obtained in 51 per cent yield [1139]. The ring systems (**94.1**, R^1 = Me, R^2 = NH₂) and (**94.2**) were thus obtained. The latter (with the 3-NHR group replaced by 3-Ph) may also be synthesized from the hydrazine, benzaldehyde and acetic acid [1084]. Heating the hydrazine with cyanogen bromide gives good yields of the 3-aminotriazole derivative.

[1148]

1,2,4-Triazolo[4,3-a]-
pyridine

Another ring system synthesized similarly:

[1294]

1,2,4-Triazolo[4,3-b]-
[1,2,4]triazine

4.   1,2,4-Triazol-3-one or 1,2,4-Triazole-3-thione

Amongst the reagents which are used to convert an α-hydrazino-N-heterocycle into the fused triazolone are ethyl chloroformate, urea and phosgene, any one of which can give good yields.

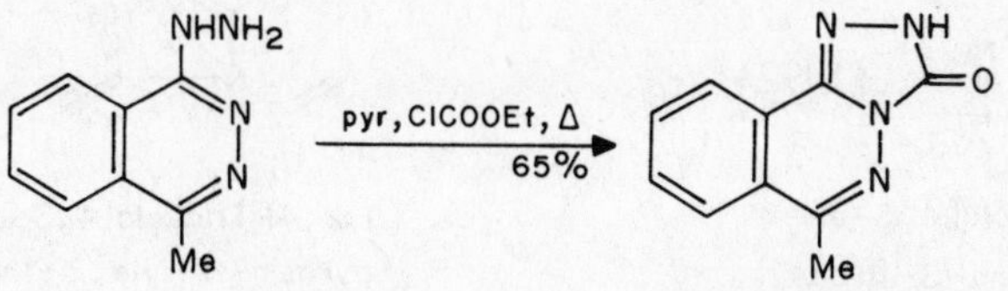

[1084]

1,2,4-Triazolo[3,4-a]-
phthalazin-3-one

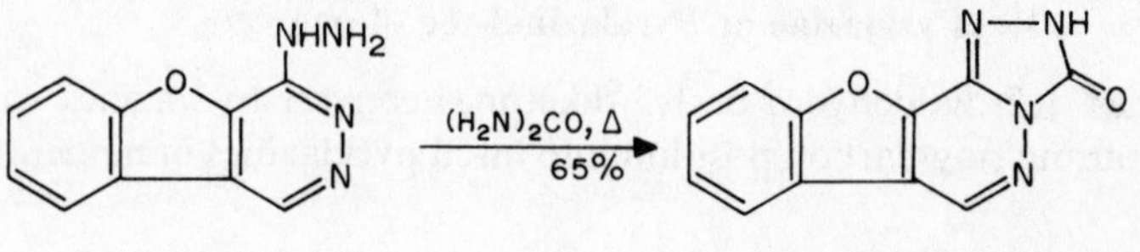

[1032]

Benzofuro[2,3-d]-1,2,4-triazolo-
[4,3-b]pyridazin-3-one

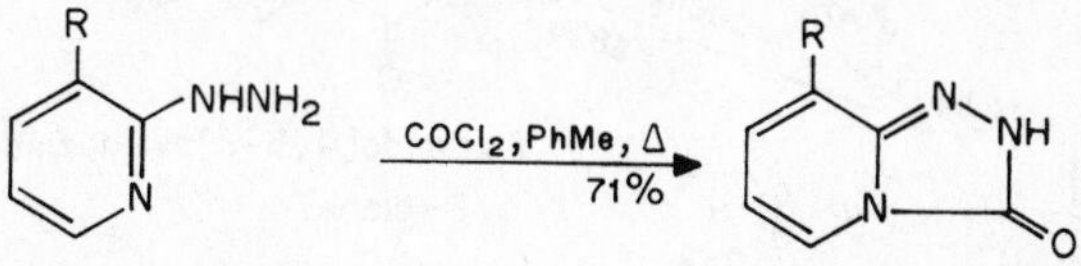

[1891]

R = i-pyrrolinyl

1,2,4-Triazolo[4,3-a]-
pyridin-3-one

The thione is prepared in high yield by using carbon disulphide [1084, 1971] or thiocarbonyldi-imidazole [1992] as reagents. The reactions of hydrazinopyrimidines [1977] and carbon disulphide [2017] have been reviewed.

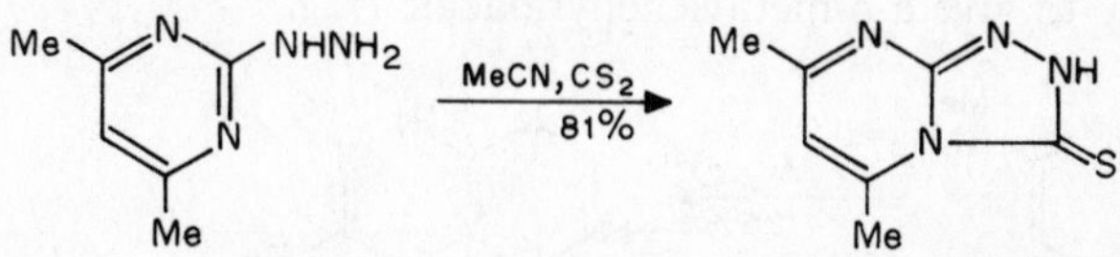

[133]

1,2,4-Triazolo[4,3-a]-
pyrimidine-3-thione

Other ring systems synthesized similarly:

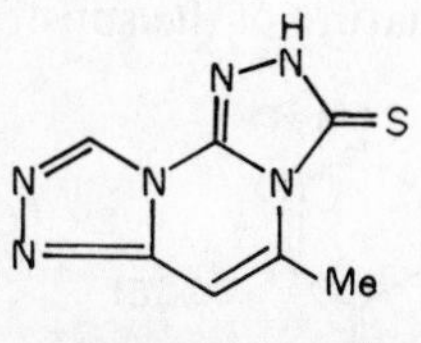

[1344]

Bis[1,2,4]triazolo[4,3-a:4',3'-c]-
pyrimidine-3-thione

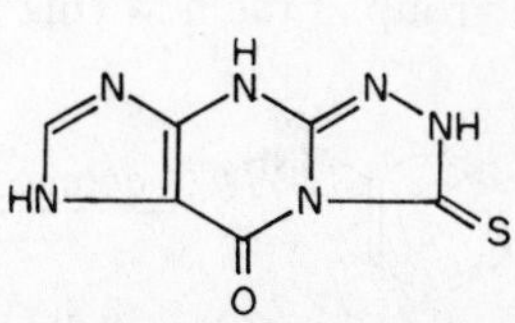

[1782]

1,2,4-Triazolo[4,3-a]purin-
5-one, 3-thioxo

1,2,4-Triazolo[4,3-*a*]-
[1,3,5]triazine-3-thione

[1992]

1,2,4-Triazolo[4,3-*a*]-
pyrazin-8-one, 3-thioxo-

[1971]

II. FORMATION OF A SIX-MEMBERED RING

1. Pyridazine or Pyridazin-3- or -4-one

α-Bromoketones, 1,2-dialdehydes or 1,2-diketones convert hydrazines with a
vacant neighbouring ring-carbon position into fused pyridazines in moderate to
good yields.

Pyrimido[4,5-*c*]pyridazine-
5,7-dione

[1017]

[1363]

R^1 = H, alkyl; Ar = Ph,
4-Cl-, 4-Me-, 4-MeO-C_6H_4;
R^2 = Me, PhCH$_2$; R^3 = H, Me, Ph

In a variation of this cyclization, 4-oxopyrimidin-6-ylhydrazine reacts with
2,3-butanedione to give a 4-methylenepyridazine ring.

[1614]

Pyrimido[4,5-*c*]pyridazin-5-one

α-Ketoesters react to form fused pyridazinone rings, the position of the
carbonyl group in the new ring being dependent on the nature of the substrate.

[620]

Pyrimido[4,5-*c*]pyridazin-4-one

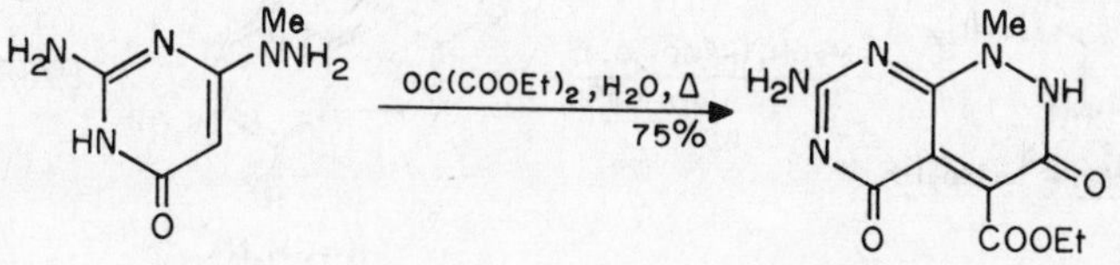

[620]

Pyrimido[4,5-c]pyridazine-
3,5-dione

Another ring system synthesized similarly:

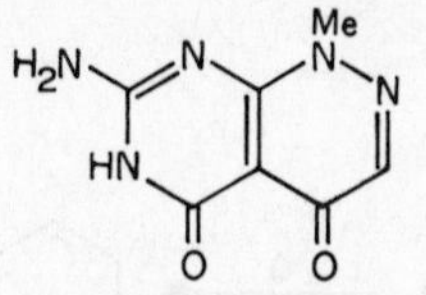

[1369]

Pyrimido[4,5-c]pyridazine-
4,5-dione

2. 1, 2, 4-Triazine or 1, 2, 4-Triazin-5-one

6-Hydrazinopyrimidine-2, 4-diones which are unsubstituted at the nitrogens may react with phenacyl bromides to give a product of the kind shown above for their *N*-substituted analogues or may react with the neighbouring ring-nitrogen. A mixture of these isomers is often obtained, the ratio depending on the reagent and the conditions.

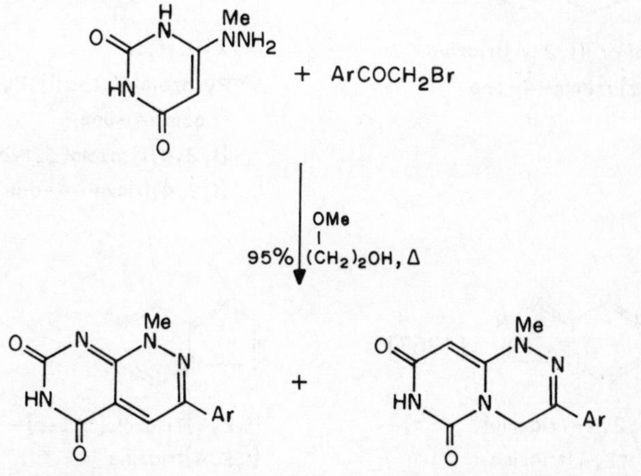

[184]

Pyrimido[4,5-c]pyridazine-
5,7-dione

Pyrimido[4,3-c][1,2,4]-
triazine-6,8-dione

Hydrazino and a neighbouring ring-nitrogen may be linked to form a triazine ring by reaction with a 1, 2-diketone or a 2-oxoaldehyde, and into a triazinone with a 2-oxoester or its enolate.

$R^1 = H, NH_2; R^2 = Me, Ph$

MeOH, $(R^2CO)_2O, \Delta$
40-94%

[968, 1294]

X = CH, N
Pyrazolo[5,1-c][1,2,4]triazine
[1,2,4]Triazolo[5,1-c][1,2,4]-
triazine

$R = tBu, EtO, Ph, 2,5-Me_2C_6H_3,$
$3,4-(MeO)_2C_6H_3$

HCCOR
CCOOEt
ONa

Δ
32-92%

[943]

[1,2,4]Triazino[3,4-a]-
phthalazin-4-one

Other ring systems synthesized similarly:

[798]

Dibenzo[cf][1,2,4]triazino-
[4,3-a]azepin-4-one

[968]

X = CH, N
Pyrazolo[5,1-c][1,2,4]-
triazin-4-one
[1,2,4]Triazolo[5,1-c]-
[1,2,4]triazin-4-one

[1367]

1,2,4-Triazolo[3,4-c]-
[1,2,4]triazine

[1367]

[1,2,4]Triazolo[5,1-c]-
[1,2,4]triazine

III.   FORMATION OF A SEVEN-MEMBERED RING

1.   1,2,4-Triazepine

Replacing the 1,2-diketone used in the preceding section in the formation of a new triazine by a 1,3-diketone yields a triazepine in good yield.

$R = Me, Ph; Ar = 4-NO_2C_6H_4$

Thiazolo[2,3-c][1,2,4]-triazepine

[787]

CHAPTER 95

Hydrazone or Oxime and Ring-carbon or Ring-nitrogen

I.   FORMATION OF A FIVE-MEMBERED RING

1.   Pyrazole

Some phenylhydrazones are thermally cyclized to give a fused pyrazole ring but when the hydrazone function is reversed, treatment with N-bromosuccinimide-acetic acid has the same overall effect.

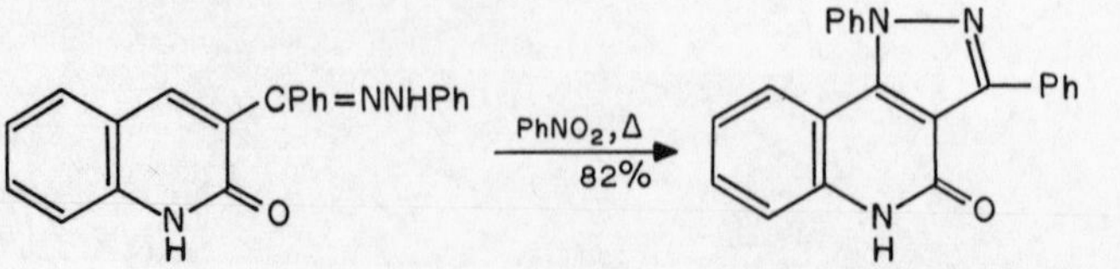

Pyrazolo[4,3-c]quinolin-4-one

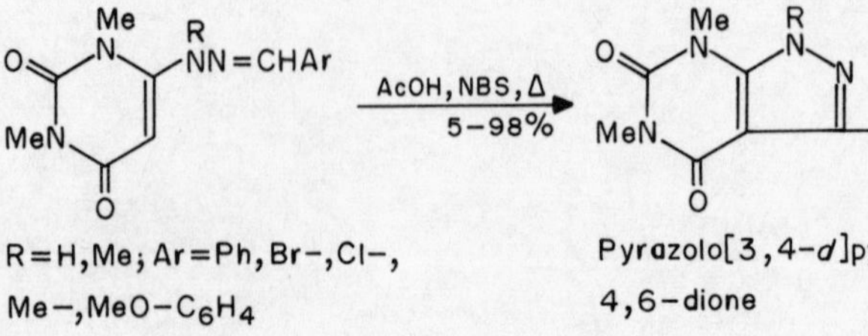

[827]

$R = H, Me;$ $Ar = Ph, Br-, Cl-,$
$Me-, MeO-C_6H_4$

Pyrazolo[3,4-d]pyrimidine-4,6-dione

[1310]

2. 1, 2, 3-Triazole

Oxidation of the phenylhydrazone of a 2-acylpyridine by lead tetra-acetate (LTA) (review [B-36]) gives good yields of a triazolopyridinium salt but an unsubstituted hydrazone is cyclized to the triazole base.

R=Me,Ph; Ar=Ph,
4-NO$_2$C$_6$H$_4$

[1,2,3]Triazolo[1,5-a]-
pyridin-8-ium

[1518]

[1,2,3]Triazolo[1,5-c]-
pyrimidine

[1026]

Oxidative cyclization of a hydrazone may also be achieved using NBS and perchloric acid.

X=S,NH
R=Ph,benzothiazol-2-yl
i, AcOEt, NBS, HClO$_4$

[1,2,3]Triazolo[1,5-a]benzimidazole

[1149]

Treatment of a tosylhydrazone group placed adjacent to a ring nitrogen with alkoxide results in cyclization to a triazole ring.

R^1,R^2=H,Me

[1,2,3]Triazolo[1,5-a]-
quinoxaline 5-oxide

[279]

3. 1, 2, 4-Triazole

Hydrazones with a neighbouring ring-nitrogen atom are oxidatively cyclized by lead tetra-acetate [985, 1518, B-36] or bromine [560, 905].

$$R = Me_2N, N[(CH_2)_2]_2O;$$
$$Ar = Ph, Cl-, NO_2-, NH_2-, N_3-C_6H_4$$

1,2,4-Triazolo[4,3-a]-
[1,3,5]triazine   [985]

$$\xrightarrow[\sim 80\%]{CH_2Cl_2, LTA}$$

Other ring systems synthesized similarly:

[1]Benzothieno[3,2-e]-1,2,4-
triazolo[4,3-c]pyrimidine   [560]

1,2,4-Triazolo[4,3-a]-
pyridine   [1518]

Pyrazolo[5,1-c]-1,2,4-triazole   [905]

5-Phenylhydrazinylideneimidazole is cyclized by heating with methanal.

$$\xrightarrow[50\%]{EtOH, HCHO, \Delta}$$

Imidazo[5,1-c]-1,2,4-
triazol-5-one   [1766]

Oximes are annulated to a ring-carbon or -nitrogen by heating with PPA (review [B-21]) and to a ring-nitrogen by phosphorus oxychloride [1367]. Where there is a choice of sites, the PPA reaction appears to favour ring-carbon [1370].

$$\xrightarrow[\sim 81\%]{PPA, \Delta}$$

R = H, Me, Cl

[1,2,4]Triazolo[1,5-a]-
pyrimidine   [1370]

Other ring systems synthesized similarly:

[1367, 1370]

X = Y = CH
[1,2,4]Triazolo[1,5-*a*]pyridine
X = Y = N
[1,2,4]Triazolo[1,5-*b*][1,2,4]triazine
X = N, Y = CH
[1,2,4]Triazolo[1,5-*b*]pyridazine

II. FORMATION OF A SIX-MEMBERED RING

1. Pyridazine or Pyridazin-3-one

A heteroarylhydrazone may be cyclized by heating with triethyl orthoformate in DMF.

R = Ph, Br–, Cl–, MeO–C$_6$H$_4$

Pyrimido[4,5-*c*]pyridazine-5,7-dione

[908]

A hydrazide which has been condensed again with a carbonyl compound to give the side-chain, —CONHN=CR'R" can react with an acyl chloride-TEA to give a pyridazinone ring. The isomeric indol-1-one was similarly prepared [794].

+ R^{5}COCl

R^1= H, Me, Ac, PhCO; R^2= H, Me, Et;
R^3= Et, Ph; R^4= Me, Ph, Me–, Cl–,
AcO–C$_6$H$_4$; R^5= Me, Ph

Pyridazino[4,5-*b*]indol-4-one

[794]

2. 1,2,4-Triazine

Neighbouring hydrazone and ring-nitrogen functions are annulated by reaction with an orthoester which supplies C-3 of the triazine ring.

$R^1 = H, Ph; R^2 = H, Me$

[1,2,4]Triazino[4,5-a]-
benzimidazole

[544]

Another ring system synthesized similarly:

[674]

2,2a,4,5-Tetraazabenz[cd]azulene

α-Carbonylhydrazones in which the carbonyl may be a ketone or carboxylic ester, are converted into a triazine by heating with acid. Alternatively, an α-halogeno-α-carbonylhydrazone may be cyclized by heating with sodium benzenesulphinate [1765].

Pyrazolo[5,1-c][1,2,4]triazine

[1765]

$R = Me, EtO, Ph; X = Br, Cl$

[1765]

Other ring systems synthesized similarly:

[1030]

Pyrido[2',3':3,4]pyrazolo-
[5,1-c][1,2,4]triazine

[969]

Imidazo[5,1-c][1,2,4]triazine

3. 1,2,4-Triazin-3- or -5-one

A triazin-3-one is obtained from a hydrazone by heating with ethyl chloroformate-pyridine but the 5-one with pyridine alone if an ester group is already incorporated in the hydrazone.

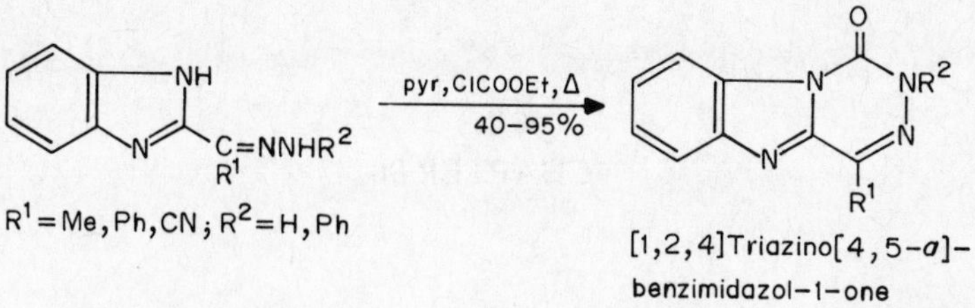
NH
N
C=NNHR2
R1
R1=Me,Ph,CN; R2=H,Ph
pyr,ClCOOEt,Δ
40–95%
O
N
NR2
N
N
R1
[1,2,4]Triazino[4,5-a]-
benzimidazol-1-one

[1000, 1171]

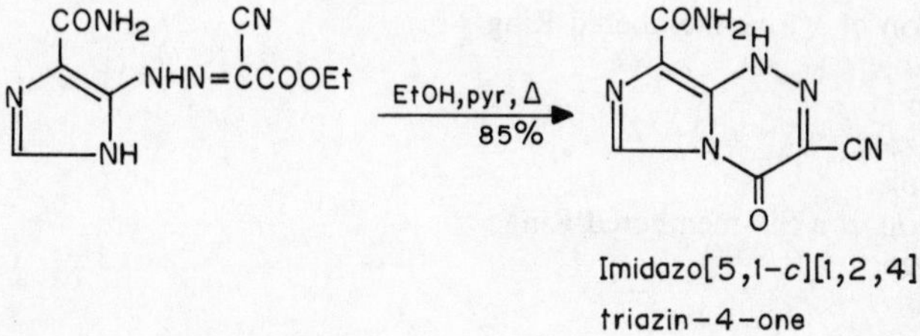
CONH2
CN
NHN=CCOOEt
N
NH
EtOH,pyr,Δ
85%
CONH2
H
N
N
N
N
CN
O
Imidazo[5,1-c][1,2,4]-
triazin-4-one

[969]

III. FORMATION OF A SEVEN-MEMBERED RING

1. 1,2-Diazepine

The sodium salts of tosylhydrazones undergo 1,7-cyclization on heating in DME.

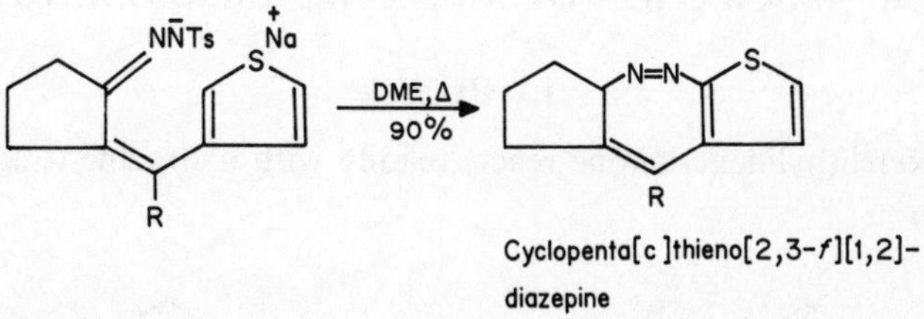
NN̄Ts
Na+
S
R
DME,Δ
90%
N=N
S
R
Cyclopenta[c]thieno[2,3-f][1,2]-
diazepine

[791]

CHAPTER 96

Hydroxy and Ketone or Lactam Carbonyl

I. FORMATION OF A FIVE-MEMBERED RING

1. Pyrazole

A 2-hydroxymethylidene-ketone reacts readily with hydrazine to form a fused pyrazole ring.

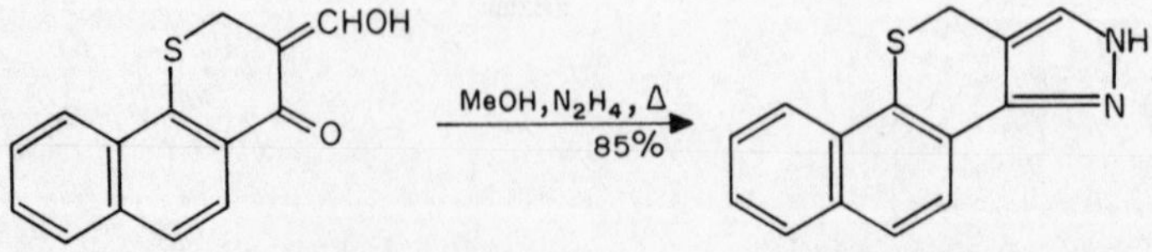

Naphtho[2',1':5,6]thiopyrano–
[4,3-c]pyrazole

[800]

2. Furan

Annulation of hydroxymethyl and ketone or lactam carbonyl [431] groups occurs readily by heating in an organic solvent for a few minutes.

528

Furo[3,4-*b*]indole

[1390]

Another ring system synthesized similarly:

[22]

Furo[3,2-*b*]pyridine

[431]

Benzofuro[2,3-*b*]quinoline

3. Isoxazole

The oxime of a ketone reacts with a neighbouring phenolic function (sometimes in a basic medium) to form an isoxazole in high yield.

[241]

1,2-Benzisoxazole

Another ring system synthesized similarly:

[1168]

Furo[3,2-*f*]-1,2-benzisoxazole

4. Oxazole

Treatment of a phenol-oxime with PPA (cf. Section I.3) (review [B-21]) causes a Beckmann rearrangement and the formation of a benzoxazole.

[427]

Benzoxazole

II. FORMATION OF A SIX-MEMBERED RING

1. Pyrazine

The tautomeric benzotropolone reacts with an aliphatic or benzenoid 1,2-diamine and a pyrazine ring is formed.

[642]

Benzo[4,5]cyclohepta–
[1,2-*b*]pyrazine

[642]

Benzo[4,5]cyclohepta–
[1,2-*b*]quinoxaline

2. Pyran

2-Hydroxyacetophenone is the 'raw material' for many syntheses of benzopyrans; when stirred for some hours with dimethyl acetylenedicarboxylate it gives the 4*H*-benzopyran but with the allenedioic ester, a 2*H*-benzopyran is formed.

[30]

4*H*-1–Benzopyran

[1425]

3. Pyran-2-one

2-Hydroxyacetophenone and its analogues also make good substrates from which pyran-2-ones may be synthesized by reactions which supply C-2 and sometimes C-3 also of the benzopyran. Examples of the latter type are those with an alkanoic acid-phenyl dichlorophosphate, malonic acid derivatives, a ketene or a Wittig reagent.

[895, 1092]

R^1=Me, PhCH$_2$, Ph; R^2=Ph,
COOEt, PhCH$_2$; R^3=OH, Cl
i, PhOPOCl$_2$–TEA when R^3=OH;
Me$_2$CO, K$_2$CO$_3$ when R^3=Cl

1–Benzopyran–2–one

[1]Benzothieno[3,2-b]-
pyran-2-one [1004]

[1394, 1809]

R^1 = H, Me, Ph; R^2 = H, MeO, HO, NO_2 1-Benzopyran-2-one

i, $Me_3SiCH=C=O$, DMF, NaH

ii, $Ph_3P=CHCOOEt$, PhH, Δ

An attempt to nitrate a 2-hydroxyalkanophenone in acetic acid gave instead moderate yields of the dinitrocoumarin.

[1807]

R^1 = Me, Et; R^2 = H, Me, Cl

A regioselective synthesis of coumarins in which no chromones are produced (cf. the Pechman synthesis) involves the reaction of a 2-methoxymethoxyaceto-phenone with a cerium(III) dichloro-ester (formed *in situ* under anhydrous conditions and low temperature from cerium(III) chloride and the lithio derivative of the ester). Presumably the methoxymethyl group is removed and a pyran-2-one is formed in high yield.

[2021]

R = Me, Cl, $MeOCH_2O$

1-Benzopyran-2-one

4. Pyran-4-one

One of the most frequently used methods of obtaining this ring is acid treatment of a 2'-hydroxyphenyl-1,3-diketone. The latter may be prepared *in situ* by a base-catalysed reaction of an acetophenone with an ester [577] or by a Baker–Venkataraman rearrangement of an O-acyl derivative of the phenolic ketone [B-4, B-12, 1454]. Another variation is the halogenation of the 1,3-diketone which causes simultaneous cyclization to the 3-halogenochromone [250].

Diethoxymethyl acetate reacts with the pyrazole (96.1) to give high yields of 5-acylpyranopyrazolones [1368].

Pyrano[2,3-c]pyrazol-4-one

(96.1)

$R^1 = Me, Ph; R^2 = Me, Et, Ph$

[1368]

$R = Me, (MeO)_4$
$Ar = 4-PhCH_2O-3-MeOC_6H_3,$
$2-NO_2C_6H_4$

1-Benzopyran-4-one, 2-aryl
(reviews [B-32, B-33])

[697, 1454]

5. 1, 2, 3-Oxathiazine 2, 2-Dioxide

Chlorosulphonyl isocyanate converts 2-hydroxyphenyl ketones into benzoxathiazines in good yields.

$R^1 = Me, Ph; R^2 = H, Cl$

1,2,3-Benzoxathiazine
2,2-dioxide

[875]

III.   FORMATION OF AN EIGHT-MEMBERED RING

1.   1, 5-Oxazocine

This large ring is formed in moderate yield by reaction of the 2-hydroxy-ketone with epichlorohydrin and methanolic ammonia. The chemistry of benzoxazocines has been reviewed [1872].

$R^1 = alkyl, PhCH_2, Ph, 4-MeC_6H_4;$
$R^2 = H, Me, Cl, NO_2, Ph$

1,5-Benzoxazocine

[1871]

CHAPTER 97

Hydroxy and Methylene

A few examples of the cyclization of acyloxy and methylene are included.

I. FORMATION OF A FIVE-MEMBERED RING

1. Furan-3-one

Diacetyloxyiodobenzene, a dehydrogenating agent [B-31], converts 2-hydroxy-propiophenone into a benzofuran-3-one and a similar product is obtained from the sulphoxide (97.1), pyridine and phosgene with simultaneous deoxygenation of the sulphoxide.

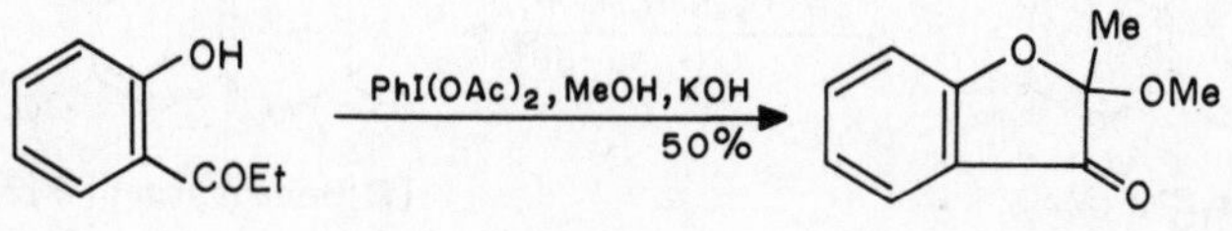

3-Benzofuranone

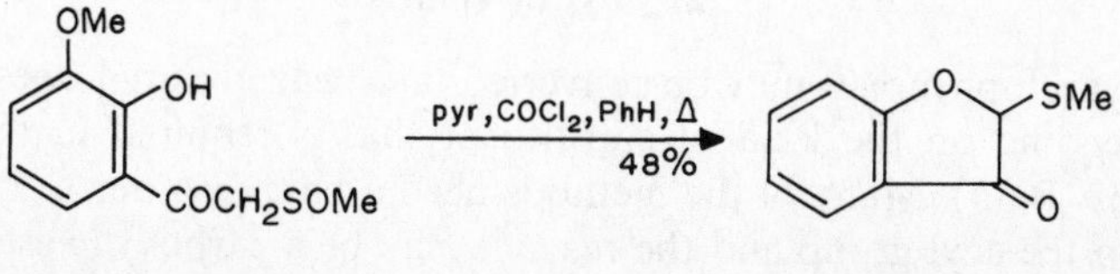

[1519]

(97.1)

[595]

533

II. FORMATION OF A SIX-MEMBERED RING

1. Pyran or Pyran-2-one or -2-thione

A doubly fused $4H$-pyran is formed in good yield by joining a hydroxy in one ring to a methylene in another; the linking carbon atom is provided by diethoxymethyl acetate which gives a product in which $R^2 = H$ or triethyl orthoacetate and R^2 is then methyl.

$R^1 = H, Me, MeO, HO$

i, MeCOOCH(OEt)$_2$

ii, MeC(OEt)$_3$

[1]Benzopyrano[4,3-c]–
pyrazol–3–one

[1736]

The additional carbon atom required to complete the pyran-2-one or -2-thione ring can be supplied by either carbon disulphide (review of its use in synthesis [2017]) or diethyl carbonate.

1–Benzopyran–2–thione

[643]

R=H,Br,Cl,F,MeO

[2]Benzopyrano[4,3-g][1]–
benzopyran–9–one

[1183]

2. Pyran-4-one

2-Hydroxyacetophenones may be converted into benzopyran-4-ones in several ways depending on the kind of substituent that is required in the product [B-12, B-32, B-33]. Many of the methods need a base in order to generate an anion from the acyl group and the reagent may be a carboxylic ester, an acid chloride [1079, 1411, 1840] or an anhydride [388, 1711, 1980].

Pyrano[2,3-c]pyrazol-4-one [363]

R=HO,(HO)₂,AcO,

1-Benzopyran-4-one
[354, 1779, 2020]

R¹=H,MeO ; R²=cycloalkyl

[1411]

[1711]

R¹=Ac,PhSO,MeSO,MeSO₂,NO₂
R²= MeO,NO₂

[364]

Among other reagents which give good results for particular types of pyran-4-ones are Vilsmeier reagents [176, 1490, 1497] (a review of this cyclization [1676] is available), orthoesters [364], carbon disulphide-dimethyl sulphate [245, 2013], DMF dimethyl acetal [762, 1666, 1978, 1979], and an aldehyde [1135, 1481]. Examples of some of these are given, substituent R² is often critical for the success of the particular method [105] and each reference should be examined.

Reaction with methyl dimethoxyacetate, and enamine (formed *in situ*) or formaldehyde gives high yields of a chromanone but when other aldehydes and a higher boiling solvent are used, the MeSO group is eliminated and a chromone is obtained; when R¹ = Ar, heating in benzene yields isoflavones [762].

R^2=H;(MeO)$_2$CHCOOMe,MeONa,Δ
83%   [853]

R^2=H;R^3COR^4,pip,PhMe,Δ
66–99%   [1182]

R^2=MeSO: HCHO,pip,MeOH,Δ
75%   [1135]

R^2=MeSO: R^{3}CHO,pip,PhMe,Δ
68–89%   [1931]

R^1=H,Me,MeO,Br,Cl,HO;
R^2,R^3=alkyl,(CH$_2$)$_2$COOH
or (CH$_2$)$_n$ (n=4,5)

1-Benzopyran-4-one

The well-known base-catalysed Baker–Venkataraman rearrangement of a 2-acyloxy-ketone is a versatile method, especially for the synthesis of flavones (2-arylchromones) [B-32, B-33, 1454]. When base-sensitive substituents are present, it is preferable to apply a thermal method as hydrolysis of a chlorine results in cyclization on to a different benzene ring. Chromanones are obtained when an enamine (prepared *in situ* is heated with a 2-hydroxyacetophenone [325, 1182].

glycerol,Δ
50%   [27]

Me$_2$CO,
90% K$_2$CO$_3$,Δ

1-Benzopyran-4-one, 2-aryl   [27]

CHAPTER 98

Hydroxy and Nitro, Nitroso or Ring-nitrogen

I. FORMATION OF A FIVE-MEMBERED RING

1. Imidazole

Under the influence of PPA (review [B-21]) a 2-hydroxyethylamino side-chain forms an imidazole ring.

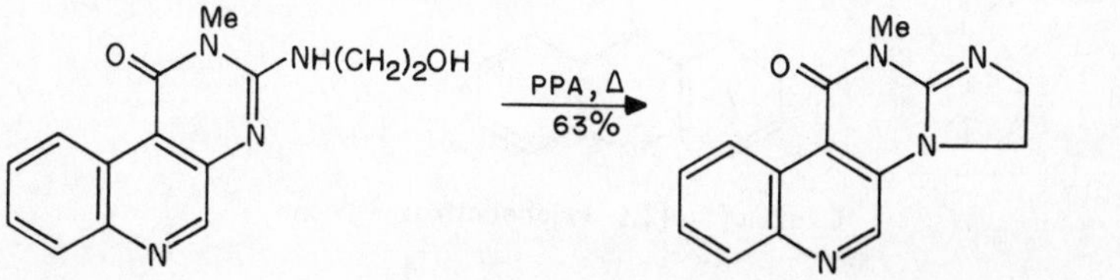

[394]

Imidazo[2′,1′:2,3]pyrimido-[4,5-*c*]quinolin-11-one

2. Furan-2, 3-dione

An unexpected product was isolated in good yield when the hydroxy-nitroacetophenone (**98.1**) was treated with ethyloxalyl chloride in pyridine in an attempt to prepare a chromone.

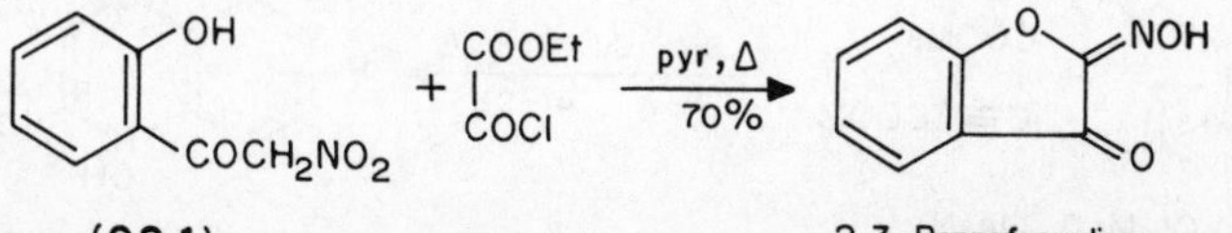

[364]

(**98.1**) 2,3-Benzofurardione (2-oxime)

3. Oxazole or Oxazol-2-one

Nucleophilic displacement of a nitro group is a well-known reaction (reviews [1398, 1399, 1432]) and in strongly alkaline solution a side-chain hydroxy group reacts in this way to form an oxazole ring.

$$\xrightarrow[\text{52\%}]{\text{diox, NaOH}}$$

[1402]

Oxazolo[2,3-c]-1,2,4-
triazol-3-one

Annulation of a hydroxy group to a ring-nitrogen atom to form a fused oxazol-2-one ring is easily effected by reaction at ambient temperature with phosgene. The same product may be prepared from the 1-aroylisoquinoline by successive reduction and cyclization without isolating the alcohol (79 percent yield of the 1-(4-methoxyphenyl)oxazole).

$$\xrightarrow[\text{91\%}]{\text{COCl}_2,\text{TEA, NaHCO}_3}$$

[647]

Ar = Ph, MeO-,
PhCO-C$_6$H$_4$

Oxazolo[4,3-a]isoquinolin-3-one

Another ring system synthesized similarly:

[153]

Oxazolo[5,4,3-kl]phenothiazin-1-one

II. FORMATION OF A SIX-MEMBERED RING

1. 1,4-Oxazine or 1,4-Oxazine-2,3-dione

Copper-assisted reaction of a nitroso-phenol with an activated acetylene gives good yields of 1,4-benzoxazines. The reduced oxazine ring is formed by the base-induced displacement of a nitro group (see Section I.3) at ambient temperature by a 2-hydroxyethylamino side-chain.

$$\xrightarrow[\text{61-98\%}]{\text{DME, Cu(OAc)}_2, \Delta}$$

R = Me, Br, Cl, MeO, Me$_2$N

1,4-Benzoxazine

[1417]

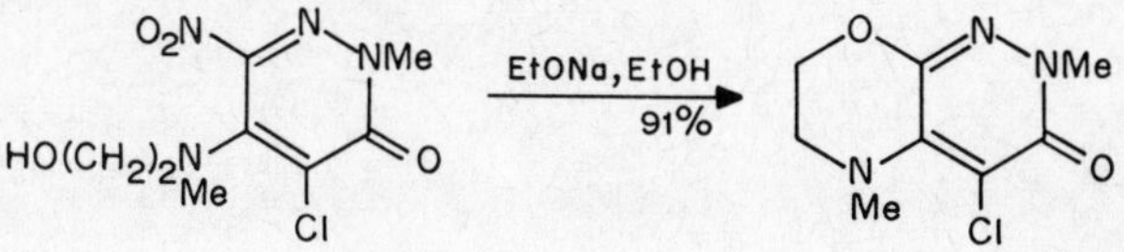

[231]

Pyridazino[3,4-b][1,4]-
oxazin-3-one

Replacing phosgene in the annulation of 1-hydroxyphenothiazine (Section I.3) by oxalyl chloride in refluxing benzene produces a fused 1,4-oxazine-2,3-dione in good yield.

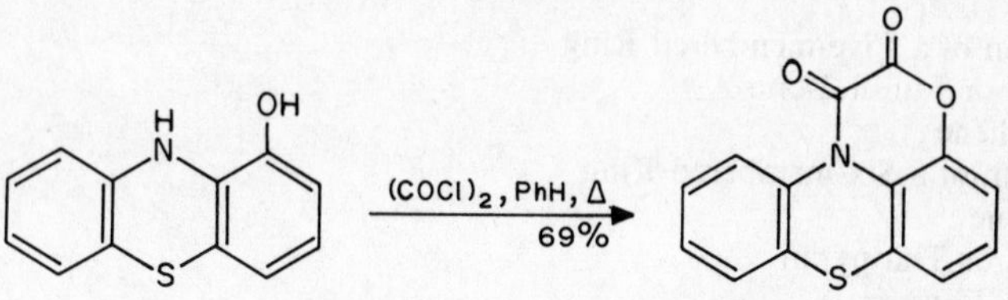

[153]

[1,4]Oxazino[2,3,4-kl]-
phenothiazine-1,2-dione

CHAPTER 99

Hydroxy or Thiol and Ring-carbon

I. FORMATION OF A FIVE-MEMBERED RING

1. Furan or Furan-2-one

Cyclodehydration (by PPE or PPA) can occur between an alcohol group and an unsubstituted ring-carbon to form a C—C bond. Closely related is the unusual formation of a C—O bond and simultaneous deoxygenation of a neighbouring *N*-oxide.

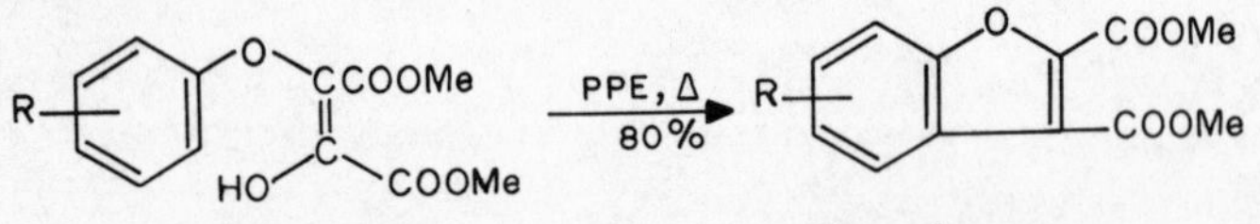

Benzofuran

[1467]

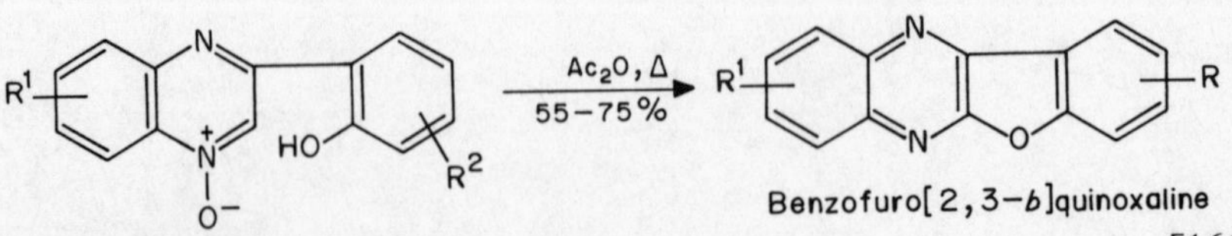

Benzofuro[2,3-*b*]quinoxaline

R^1=H, Me, Cl; R^2=H, Me, Br [1667]

Chloroacetaldehyde annulates a hydroxypyridinone in a weakly basic medium at ambient temperature to give the 2,3-dihydro-3-hydroxyfuran which is dehydrated by acid. An α-hydroxycarboxylic acid converts a phenol into a benzofuran-2-one at ambient temperature.

R = H, Me   Furo[3,2-c]pyridin-4-one

[1342]

2-Benzofuranone   [255]

2. Thiophene

A side-chain thiol under dehydrogenative conditions may be converted into a thiophene ring.

Benzo[b]thiophene

[640]

II. FORMATION OF A SIX-MEMBERED RING

1. Pyridine

Hydroxyalkylaminomethylbenzenes are cyclized by Lewis acids to reduced isoquinolines in moderate to high yield depending on the N-substituent.

R = alkyl; Ar = 3,4-$(MeO)_2C_6H_4$ Isoquinoline
                                           (review [2003])

[638]

Other ring systems synthesized similarly:

X = N, Y = CH
Thieno[2,3-c]pyridine
X = CH, Y = N
Thieno[3,2-c]pyridine
(review [1661])

[637]

2. Pyran or Thiopyran

Reaction of a phenol (or thiophenol) with an $\alpha\beta$-unsaturated aldehyde or an α-chloroacetylene in a basic medium gives 2H-benzopyran (or 2H-benzothiopyran). The best yields are obtained from reactive phenols such as resorcinols. An alternative method is to treat the phenol successively with a Grignard reagent and an $\alpha\beta$-unsaturated acetal [680] or a diene [645].

R^1 = H, Me, MeO, HO, Cl, Ac, C_5H_{11};
R^2 = H, Me, Ph, Me_2C=CH$(CH_2)_2$
i, PhMe, Ti(OEt)$_4$, or pyr or
PhMe, tBuNH$_2$

2H-1-Benzopyran

[625, 670, 716]

[636]

[1044]

R = Me_2C=CH$(CH_2)_2$

2H-1-Benzopyran
(review [1621])   [1664]

Other ring systems synthesized similarly:

Naphtho[2,1-b]pyran   [680]

1-Benzothiopyran   [1044]

In order to incorporate potentially useful functions at C-2 and C-3 of a chromene, a phenol may be heated with arylidenemalononitrile and a mild base.

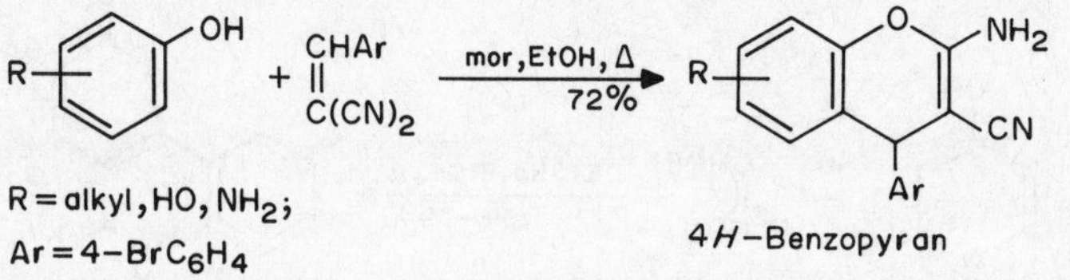
OH
R
+
CHAr
C(CN)$_2$
mor, EtOH, Δ
72%
O
NH$_2$
CN
Ar
R = alkyl, HO, NH$_2$;
Ar = 4-BrC$_6$H$_4$
4H-Benzopyran

[1511]

A hydroxy may be linked through carbon to the ring-carbon in another (π-excessive, preferably) ring by reaction with acetone and boron trifluoride.

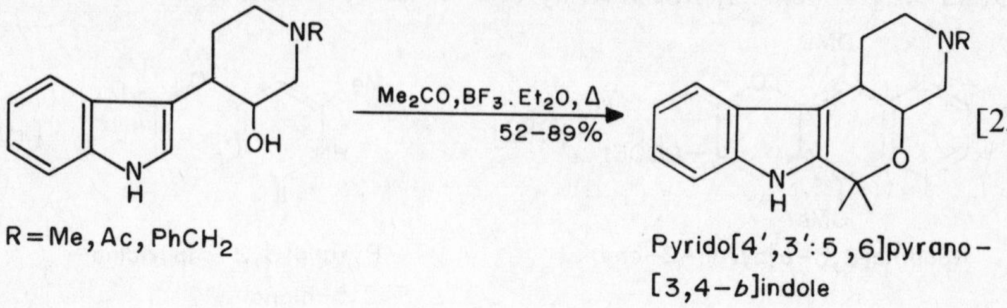
NR
OH
N
H
Me$_2$CO, BF$_3$.Et$_2$O, Δ
52–89%
O
R = Me, Ac, PhCH$_2$
Pyrido[4′,3′:5,6]pyrano-[3,4-b]indole

[256]

3. Pyran-2-one

Phenols have traditionally been converted into coumarins by the Pechmann reaction (reviews [B-4, B-35, B-39]); considerable modifications to this method have been described, for example, by the replacement of the phenol by its acetate [172], diaryl benzyl- and ethoxymethylene-malonates [282, 889, 1001], ethyl 2-aminobut-2-enoate [1197] or malonic acid-zinc chloride-phosphorus oxychloride [595] as well as different catalysts such as hydrogen fluoride [817], butyllithium [282], ammonium acetate [1088], sodium ethoxide [1001] or trifluoroacetic acid [1665]. Some of these methods are illustrated below.

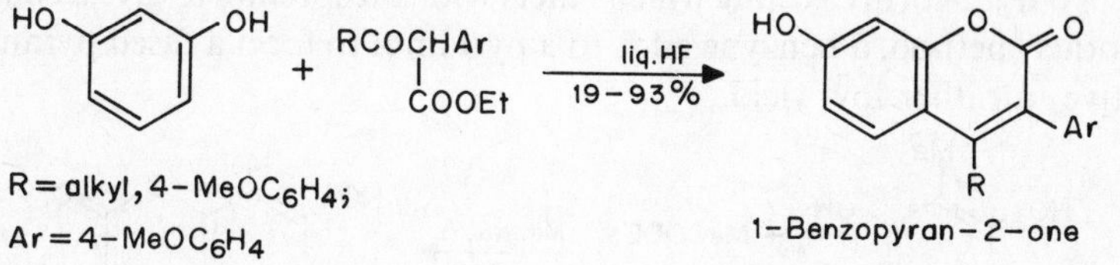
HO
OH
+
RCOCHAr
COOEt
liq. HF
19–93%
O
Ar
R
R = alkyl, 4-MeOC$_6$H$_4$;
Ar = 4-MeOC$_6$H$_4$
1-Benzopyran-2-one

[817]

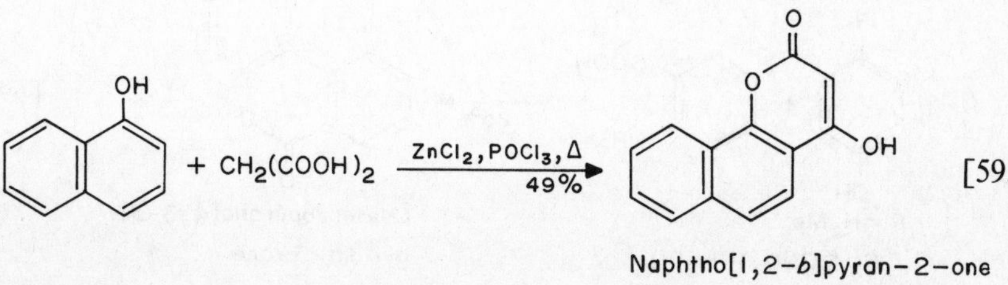
OH
+ CH$_2$(COOH)$_2$
ZnCl$_2$, POCl$_3$, Δ
49%
O
OH
Naphtho[1,2-b]pyran-2-one

[595]

OH + CHOEt / R²CCN — EtONa, EtOH, Δ 69–79% → Pyrano[3,2-c]quinoline-2,5-dione   [1001]

R¹ = H, Me ; R² = CN, COOEt

Other ring systems synthesized by one of the above methods:

OMe ... COOEt [282]
Naphtho[2,3-b]pyran-2-one

Me ... [1001]
Pyrano[3,2-c]pyridine-2,5-dione

Ph ... CH₂Ph OH [889]
Pyrano[2,3-c]pyrrole-2,5-dione

X = O, NMe [1088]
Pyrano[3,2-c][1]benzopyran-2,5-dione
Pyrano[3,2-c]quinoline-2,5-dione

Two other methods in which an intermediate is generated and reacted *in situ* are of interest: a mixture of ethyl formate, ethyl acetate and sodium methoxide produces ethyl 2-formylacetate which reacts with a resorcinol to give a coumarin. In the other method, a benzyne adds to a pyridinol to form a fused pyran-2-one derivative in rather low yield.

Me; HO—OH + MeCOOEt + HCOOEt — MeONa, Δ 76% → [644]
1-Benzopyran-2-one

R ... + COOH / NH₂ — i, Δ 25% → [696]
[2]Benzopyrano[4,3-b]pyridin-6-one

R = H, Me
i, $C_5H_{11}ONO$, $(CH_2Cl)_2$

Treatment of 2,4-dimethylphenol with 4-methoxycinnamic acid gives the dihydro-4-arylcoumarin, but other dimethylphenols give poor results.

Ar = 2- or 4-MeOC₆H₄

$Ar = 2$- or 4-$MeOC_6H_4$

[646]

1-Benzopyran-2-one

4. Pyran-4-one or Thiopyran-4-one

Methods of converting phenols into chromones are less well-developed than those into coumarins. Several kinds of malonic acid derivatives, however, react with phenols in this way; an ester-amide in the presence of phosphorus oxychloride yields 2-t-aminochromones [1410, 1972, 1973], and a chromanone is formed in moderate yield when a phenol is heated with isopropylidenemalonic acid monoester.

$R^1 = H, Me, MeO, iPr, Me_2N$;
$R^2 = Me, Et$ or $NR_2^2 = N(CH_2)_n$
$n = 4, 5$.

[1410]

[639]

1-Benzopyran-4-one

A Friedel–Crafts-type acylation of a phenol with 3-chlorocrotonic (3-chloroprop-2-enoic) acid deserves more thorough exploration of its scope.

[1409]

When a thiophenol is heated with a 3-oxoester, moderate to good yields of a thiochromone are obtained.

$R^1 = H, MeO$; $R^2 = Me, Ph$

[998]

1-Benzothiopyran-4-one

5. 1,4-Oxazin-2-one

Phosgene forms a link between a hydroxy group and the carbon of another ring.

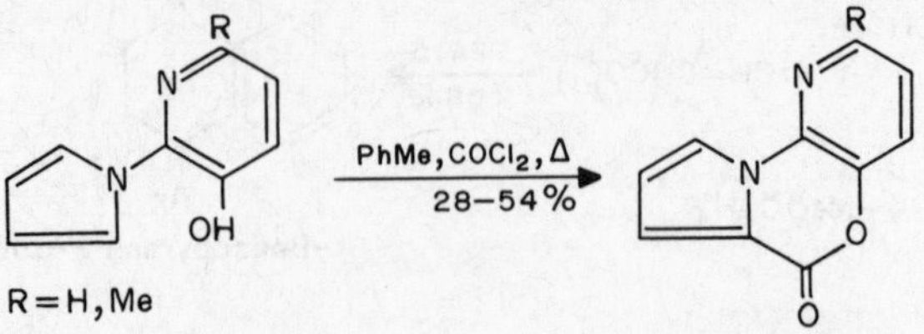

[1781, 1885]

Pyrido[3,2-*b*]pyrrolo[1,2-*d*]-
[1,4]oxazin-6-one

III. FORMATION OF A SEVEN-MEMBERED RING

1. 1,3-Diazepine

Cyclodehydration by sulphuric acid results in the formation of a C—C bond between two rings; yields are very good.

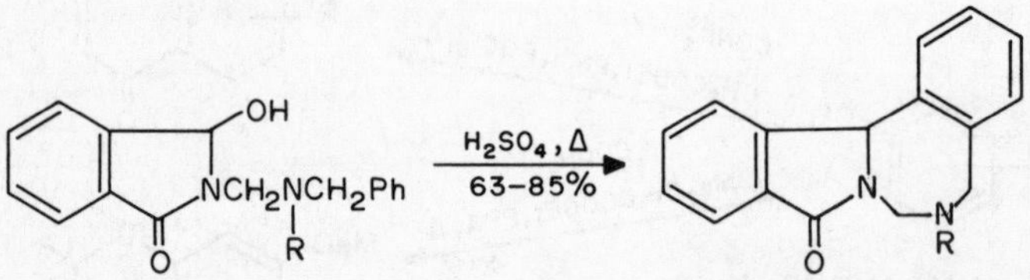

[768]

R = Me, PhCH₂, Ph

Isoindolo[1,2-*a*][2,4]-
benzodiazepin-9-one

2. Oxepine

Boron trifluoride etherate promotes the cyclization of a 3-hydroxyalkylbenzene with an acetal which provides one of the ring carbon atoms.

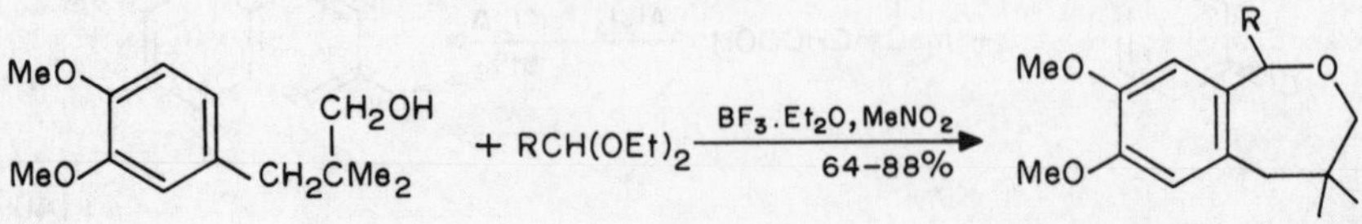

R = Br(CH₂)ₙ , n = 1,2

2 – Benzoxepine

[601]

Imine and Hydroxy, Ring-carbon or Ring-nitrogen

Compounds which contain a —CH=N—R function are variously known as anils, azomethines, imines or Schiff's bases. In this chapter, the word 'imine' is used for the —CH=N—R grouping where R ≠ H and is usually a ring; the CH part is usually attached to a benzene ring and such a group would be called benzylideneamino. The use of imines in synthetic heterocyclic chemistry has been reviewed [1573].

I. FORMATION OF A FIVE-MEMBERED RING

1. Oxazole

Several oxidizing agents convert a 2-hydroxy-imine into a fused oxazole in good yield, for example, lead tetra-acetate, N-iodosuccinimide or complex plumbo-phosphates [1849].

$Ar = 3-, 4-NO_2C_6H_4$

Pyrano[3,2-f]benzoxazol-5-one

$Ar = Ph, HO-, MeO-, NO_2-,$
$-OCH_2O-, Me_2N-C_6H_4$

Benzoxazole

Another ring system synthesized similarly:

[354]

Pyrano[2,3-e]benzoxazol-6-one

II. FORMATION OF A SIX-MEMBERED RING

1. Pyridine or Pyridin-2-one

The cyclization of an imine in which the nitrogen atom is separated from the ring by two carbon atoms is catalysed by benzoic acid; a good yield of product is obtained in refluxing xylene.

[690]

Pyrido[3,4-b]indole

Another example of annulation of an imine and a ring-carbon requires a chloro- or dichloro-ketene and a base.

$R^1 = H, Me; R^2 = Cl, Ph$
$Ar = Me-, MeO-, Cl-C_6H_4$

[1]Benzopyrano[3,2-c]-
pyridine-3,10-dione

[1560]

2. Pyrimidine or Pyrimidin-4-one

An imine and a ring-nitrogen may form a pyrimidine ring by reaction with diketene (review [2018]) or a chloroketene. The imine carbon must carry a hydrogen for this reaction to succeed.

[1083, 1553]

i, , R = Ac

ii, Cl$_2$CHCOCl — TEA, DME, R = Cl

Pyrido[1,2-a]pyrimidine
(review[1693])

Other ring systems synthesized using a chloroketene:

[1445]

X = S , Y = CH
Thiazolo[3,2-*a*]pyrimidin-5-one
X = S , Y = N
1,3,4-Thiadiazolo[3,2-*a*]-
pyrimidin-5-one

[1445]

X = NH , O , S
Pyrimido[1,2-*a*]benzimidazol-4-one
Pyrimido[2,1-*b*]benzoxazol-4-one
Pyrimido[2,1-*b*]benzothiazol-4-one

3. Pyran-2-one or Thiopyran-2-one

2-Hydroxy-imines react with malononitrile and perchloric acid to give a benzopyrylium salt which is converted into a coumarin-3-carbonitrile in good overall yields. Replacing the nitrile by malonyl chloride in xylene gives the corresponding carboxamide.

(100.1)
R = Me, Et, Ph,
PhCH₂, 4-Me-,
4-MeO-, 4-Cl-
C₆H₄

X = O: CH₂(CN)₂, HClO₄
50-90%

95% Ac₂O

[415]
1-Benzopyran-2-one

X = O: xylene, CH₂(COCl)₂

[1715]

X = S: Et₂O, C₃O₂
21-51%

[1678]

(100.2)
1-Benzothiopyran-2-one

The thiopyran (100.2) and its bz-substituted derivatives are obtained in moderately good yield by treating the thiophenol (100.1, X = S) with carbon suboxide (review of its reactions [100]).

CHAPTER 101

Isocyanate or Isothiocyanate and Methylene, Ring-carbon or Ring-nitrogen

I. FORMATION OF A FIVE-MEMBERED RING

1. Pyrazole

o-Tolylisocyanate is cyclized in moderate yield to indazole by treatment with pentyl nitrite.

Indazole   [1643]

2. Imidazol-2-one

Although some isocyanates are isolated and purified before being converted into another functional group, their high reactivity (review [251]) dictates that they may be prepared and treated *in situ* with further reagents. For example, diazotization of a hydrazide converts it into an acylazide which on warming can lose nitrogen and react with a ring-nitrogen. The acylazide may be prepared separately and purified before cyclization through the isocyanate.

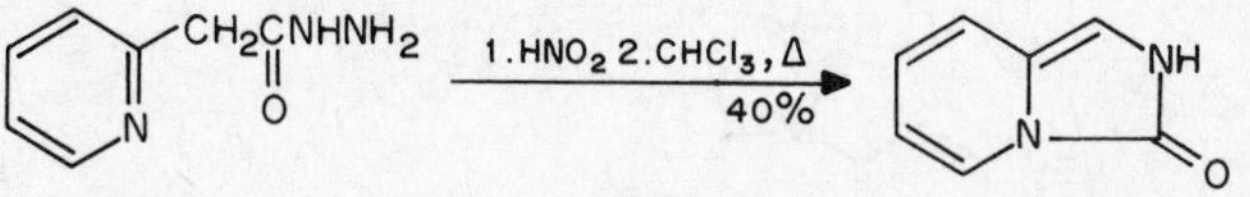

[1270]

Imidazo[1,5-a]pyridin-3-one

Other ring systems synthesized similarly:

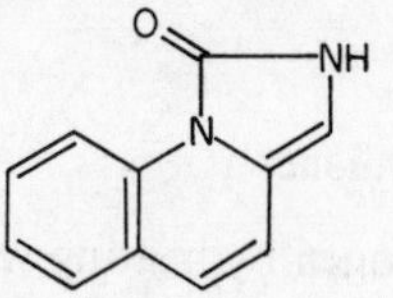

[414]

Imidazo[1,5-a]quinolin-1-one

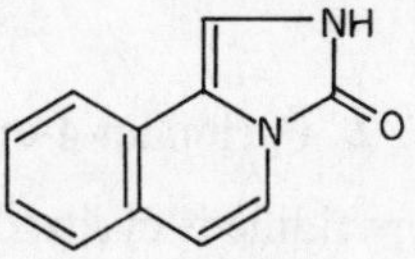

[414]

Imidazo[5,1-a]isoquinolin-3-one

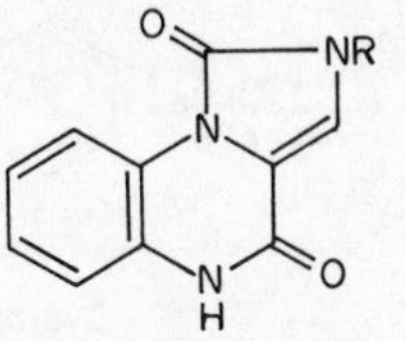

[1075]

Imidazo[1,5-a]quinoxaline-
1,4-dione

II. FORMATION OF A SIX-MEMBERED RING

1. Pyridin-2-one

Heating a compound containing a side-chain isocyanate (or isothiocyanate) alone or with PPA (review [B-21]) causes ring-closure on to a ring-carbon atom. An acylazide behaves similarly and may attack a carbon in the same or in a different ring.

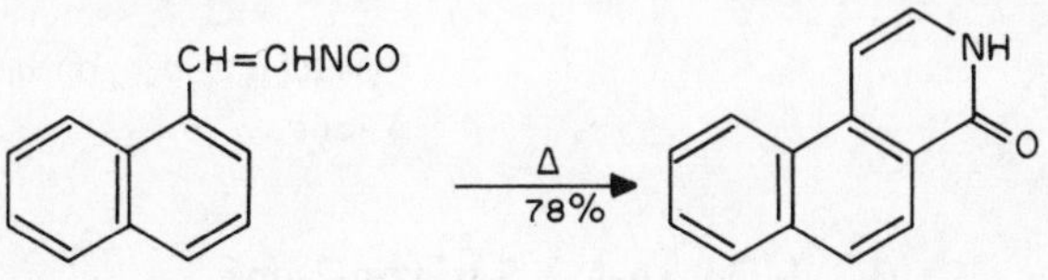

[390]

Benz[f]isoquinolin-4-one

Other ring systems synthesized similarly:

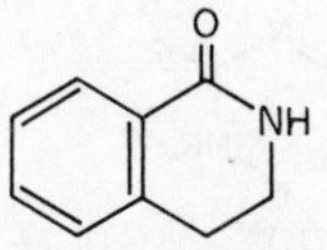

[707]

1-Isoquinolinone

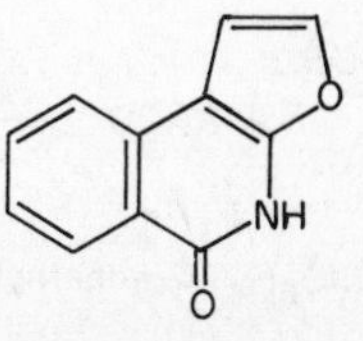

[1140]

Furo[2,3-c]isoquinolin-5-one

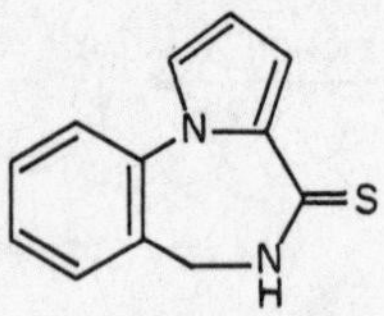

[1571]

Pyrrolo[1,2-a][1,4]-
benzodiazepine-4-thione

2. Pyrimidin-4-one or Pyrimidine-4-thione

2-Isothiocyanatopyridine is cyclized on to the ring-nitrogen by heating with an enamine, ketene or diketene.

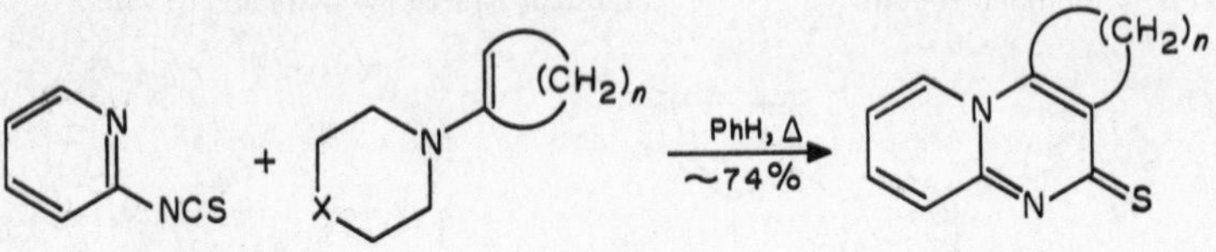

[852]

$$n = 3, 4$$

Cyclopenta[e]pyrido[1,2-a]-
pyrimidine-4-thione
Pyrido[1,2-a]quinazoline-5-thione

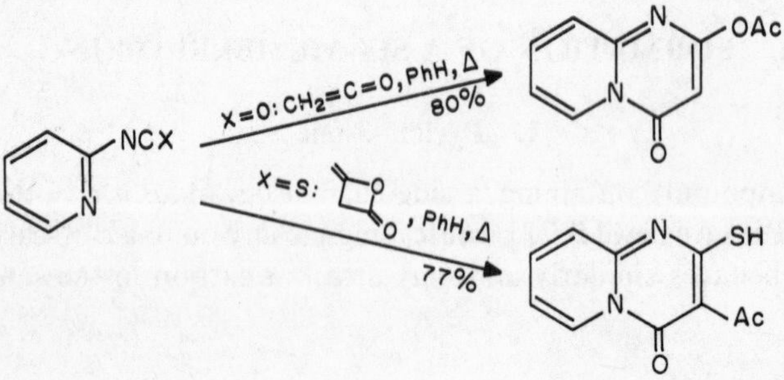

[1551]

Pyrido[1,2-a]pyrimidin-
4-one

3.   4-Thioxo-1, 3, 5-triazin-2-one

This ring is obtained by warming a 2-isothiocyanatopyridine with an isocyanate in benzene.

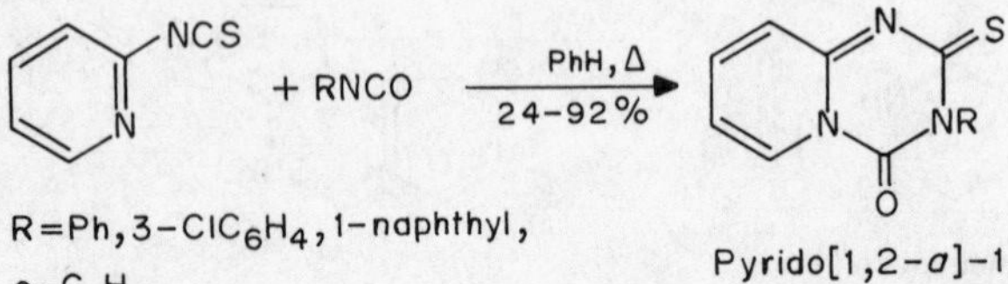

[851]

R=Ph, 3-ClC$_6$H$_4$, 1-naphthyl,
c-C$_5$H$_{11}$

Pyrido[1,2-a]-1,3,5-
triazin-4-one, 2-thioxo

III.   FORMATION OF A SEVEN-MEMBERED RING

1.   Azepin-2-one

An isocyanate behaves as a *C*-acylating agent and this is catalysed by a Friedel–Crayts catalyst.

Pyrazolo[3,4-*c*][2]–
benzazepin–9–one

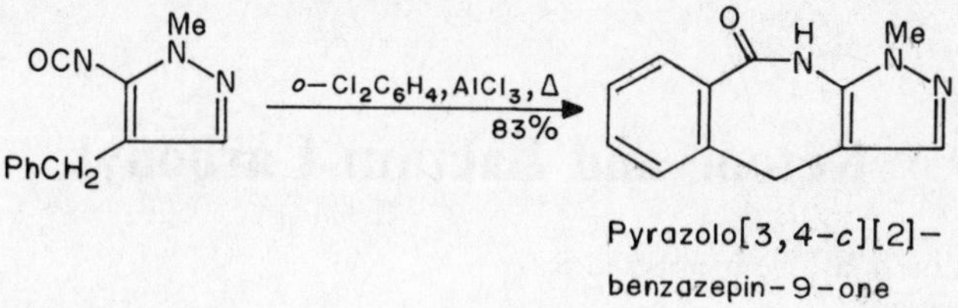

[369]

IV.   FORMATION OF AN EIGHT-MEMBERED RING

1.   Azocin-2-one

A homologous extension of the reaction described in the previous section gives an azocinone.

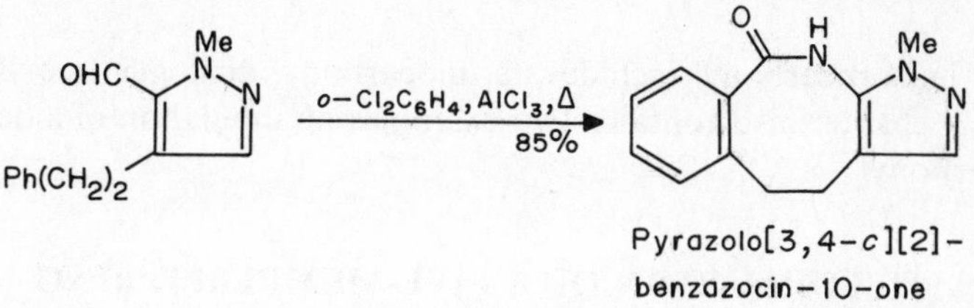

Pyrazolo[3,4-*c*][2]–
benzazocin–10–one

[369]

CHAPTER 102

Ketone and Lactam Carbonyl

The term lactam carbonyl includes its thiocarbonyl analogue (see also Chapter 69); this chapter also contains two examples of annulation of aldehyde and lactam carbonyl.

I. FORMATION OF A FIVE-MEMBERED RING

1. Pyrazole

A ketone hydrazone cyclizes on heating in acetic acid.

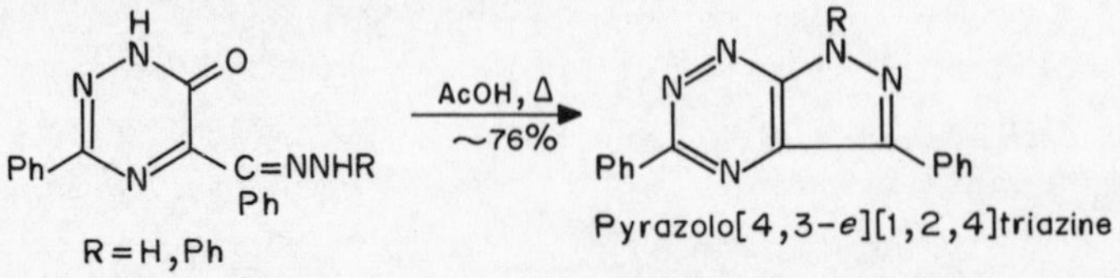

AcOH, Δ
~76%
C=NNHR
R=H,Ph
Pyrazolo[4,3-e][1,2,4]triazine

[1517]

2. Thiophene

A side-chain ketone and a lactam carbonyl react with phosphorus pentasulphide in pyridine to form a fused thiophene ring in high yields.

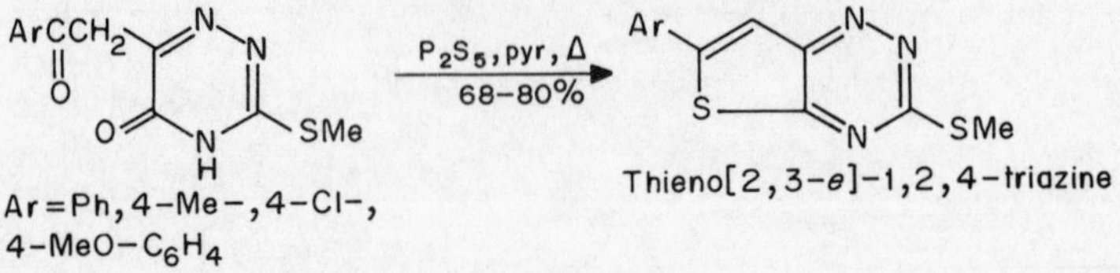

P_2S_5, pyr, Δ
68–80%
Ar=Ph, 4-Me-, 4-Cl-, 4-MeO-C_6H_4
Thieno[2,3-e]-1,2,4-triazine

[320]

Other ring systems synthesized similarly:

[226]

Thieno[2,3-*b*]quinoxaline

[320]

Thieno[2,3-*e*]-1,2,4-
triazin-3-thione

A fused thiophene may also be obtained from a 2-formyl-lactam thiocarbonyl by first reacting the aldehyde with an α-halogenoketone; ketone and thione groups then react to form a thiophene ring.

$$+ ArCCH_2Br \xrightarrow[64-76\%]{EtOH, \Delta}$$

[1899]

R=alkyl,Ph; Ar=Ph,4-Br-,
4-NO$_2$-,4-Ph-C$_6$H$_4$

Thieno[2,3-*b*]pyridinium

Another ring system synthesized similarly:

[1598]

Thieno[2,3-*b*]quinoline

3. Oxazole

Although the oximido (or isonitroso) pyrazole used in this reaction was not prepared from a dicarbonyl compound, it is regarded as an oxime of a dicarbonyl compound for classification purposes. When this is heated with benzylamine (or benzyl cyanide-sodium ethoxide), a new oxazole ring is formed.

$$+ PhCH_2NH_2 \xrightarrow[54\%]{EtOH, \Delta}$$

[1059]

Ar = 2,4-(NO$_2$)$_2$C$_6$H$_3$

Pyrazolo[4,3-*d*]oxazole

4. Thiazole

When a compound which has a side-chain ketone and a thiolactam is heated with acetic anhydride and sulphuric acid, a fused thiazole ring is obtained.

$$\xrightarrow[81\%]{Ac_2O, H_2SO_4, \Delta}$$

[1948]

Ar= 3-HOC$_6$H$_4$

Thiazolo[3,2-*a*]benzimidazole

II.   FORMATION OF A SIX-MEMBERED RING

1.   Pyridazine

A carbonyl group in a side-chain and the lactam carbonyl react with hydrazine
hydrate to form a pyridazine ring.

$R^1 = Me, Ph; R^2 = H, Me$

N$_2$H$_4$·H$_2$O, AcOH, Δ
30–90%

Pyridazino[3,4-b]quinoxaline

[1927]

2.   Pyrazine

This ring is formed in high yield by reaction of a 2-oxo-lactam carbonyl with o-
phenylenediamine.

R = H, halogen, NO$_2$

diox, Δ
83–100%

Indolo[2,3-b]quinoxaline

[275]

3.   1, 2, 4-Triazine or 1, 2, 4-Triazine-3-thione

A benzimidic acid hydrazide reacts in refluxing acetic acid with the two carbonyl
groups of a keto-lactam to form a new triazine ring but thiosemicarbazide in the
presence of a weak base condenses to give the triazinethione, both in high yield. A
preformed monothiosemicarbazone may be cyclized in aqueous alkali to give a
similar result [1420].

R = H, 6,7-Me$_2$

NH
‖
ArCNHNH$_2$, AcOH, Δ
87%

[210]

1,2,4-Triazino[5,6-b]indole

aq. K$_2$CO$_3$, CSNH$_2$, Δ
NHNH$_2$
81%

[758]

1,2,4-Triazino[5,6-b]-
indole-3-thione

Another ring system synthesized similarly:

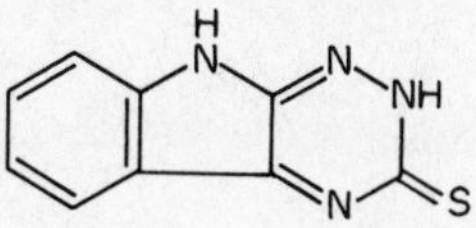

[1420]

1,2,4-Triazino[6,5-*b*]-
indole-3-thione

4. Pyran-4-one

A fused pyran is readily built by passing hydrogen chloride into a methanolic solution of the keto-lactam.

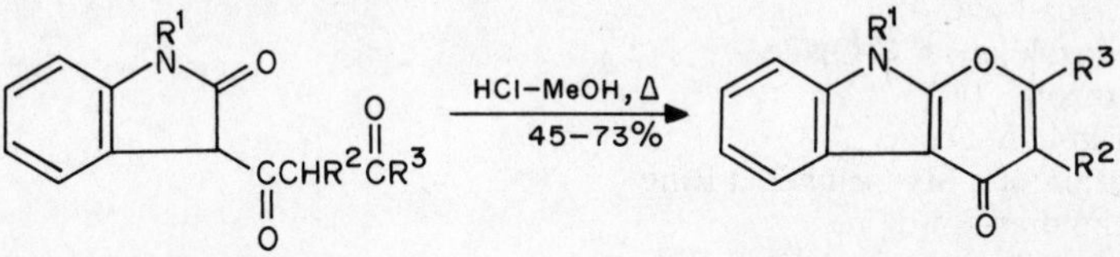

R^1=H,Me; R^2=Me,Et;
R^3=Me,Ph,4-MeOC$_6$H$_4$

HCl–MeOH, Δ
45–73%

Pyrano[2,3-*b*]indole-4-one

[879, 1710]

CHAPTER 103

Methylene and Nitro or Nitroso

I. FORMATION OF A FIVE-MEMBERED RING

1. Pyrrole

The acidic character of the hydrogens of a methyl placed near a nitro group allows reaction with several electrophilic reagents, for example, carbonyl groups of aldehydes or esters.

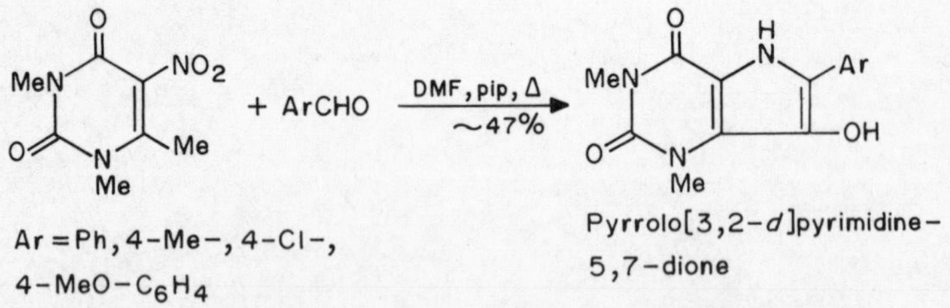

$$\text{(reaction scheme)} \qquad [373]$$

Ar = Ph, 4-Me−, 4-Cl−,
4-MeO−C$_6$H$_4$

Pyrrolo[3,2-*d*]pyrimidine−
5,7−dione

$$\text{(reaction scheme)}$$

Indole

$$[964]$$

558

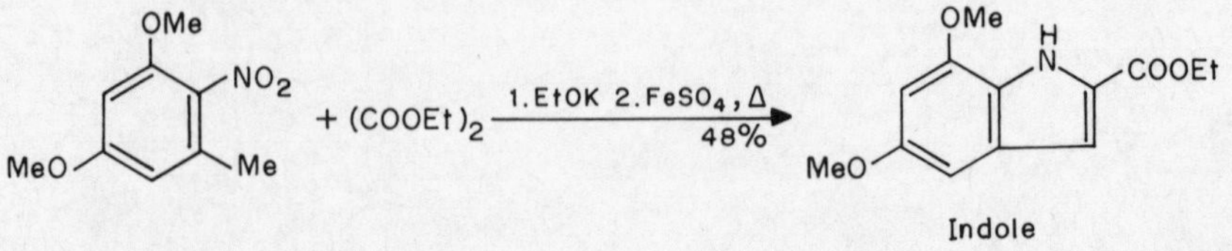

[331]

R = H, PhCH$_2$

N-Methoxyindoles are more stable than their hydroxy homologues and are useful intermediates in the synthesis of other indoles. The parent compound may be synthesized in good yield from *o*-nitrotoluene in a three-step reaction but the intermediates are not isolated.

[2012]

i, DMF, DMFDMA, DBU, Δ;
ii, Zn–NH$_4$Cl; *iii*, MeI–NaH

2. Pyrazol-3-one

An anion from a side-chain containing a —NPh—CH$_2$—CO— group was expected to attack the nitro group but the pyrazolone isolated may have resulted from the transient formation of 2-hydroxyimino-*N*-phenylbenzamide.

[1449]

(103.1)

3-Indazolone

3. Imidazole or its *N*-Oxide

Thermally induced interaction between an activated methylene and a neighbouring nitroso group gives a high yield of a reduced benzimidazole. An anion can convert a nitro group into a cyclic *N*-oxide as in this synthesis of a benzimidazole *N*-oxide (reviews [B-18, B-34]).

[478]

4-Benzimidazolone

R = H, Me, MeO, Cl

EtONa, EtOH, Δ
40–82%

Benzimidazole 3–oxide

[257]

4. Isoxazole

2-Nitrophenylacetic ester in a basic medium may be converted into its isonitroso (hydroxyimino) derivative which on heating further with a base cyclizes to 1,2-benzisoxazole. The nitro group has been displaced; such reactions have been reviewed [1398, 1399, 1432].

R = H, NO₂

$\dfrac{i}{66–78\%}$

$\begin{bmatrix} -\text{C}=\text{NOH} \\ | \\ \text{COOEt} \end{bmatrix}$

NaH, diglyme, Δ

—COOEt

1,2–Benzisoxazole
(review [1900])

R = H, NO₂
i, C₅H₁₁ONO, EtONa

[1486, 1867]

5. Thiazole N-Oxide

Dehydrative cyclization of a methylene-nitro compound requires heating with a base and when the methylene is attached to a sulphur atom, a fused thiazole N-oxide is obtained (review [B-34]).

R = COOMe, COOEt

EtOH, TEA, Δ
~48%

[274]

Thiazolo[5,4-d]pyrimidine-
5,7–dione 1–oxide

Another ring system synthesized similarly:

[1718]

Thiazolo[4,5-g]quinazoline-
5,7–dione 3–oxide

II.   FORMATION OF A SIX-MEMBERED RING

1.   Pyrimidine-2, 4-dione

When the benzoyl group of reactant (**103.1**) (Section I.2) is replaced by a cyano, the product, obtained in high yield, is a quinazolinedione.

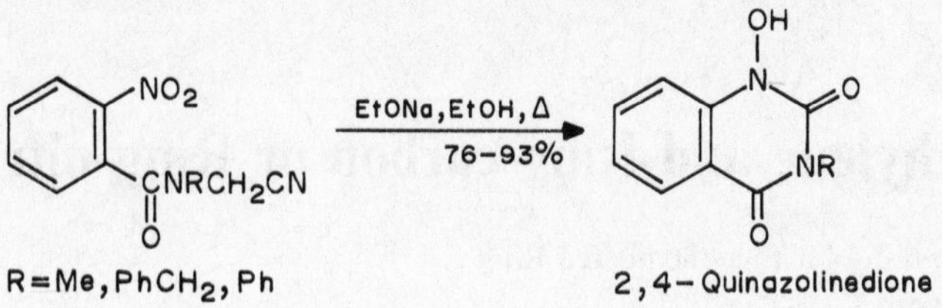

R=Me, PhCH₂, Ph   2,4-Quinazolinedione

[1449]

2.   Pyrazin-2-one or its 4-Oxide

Compounds which contain an acetoacetylamino side-chain *ortho* to a nitro group, yield a fused pyrazin-2-one 4-oxide on treatment with alkali but the formation of the *N*-oxide may be avoided by including a mild deoxygenator in the reaction sequence.

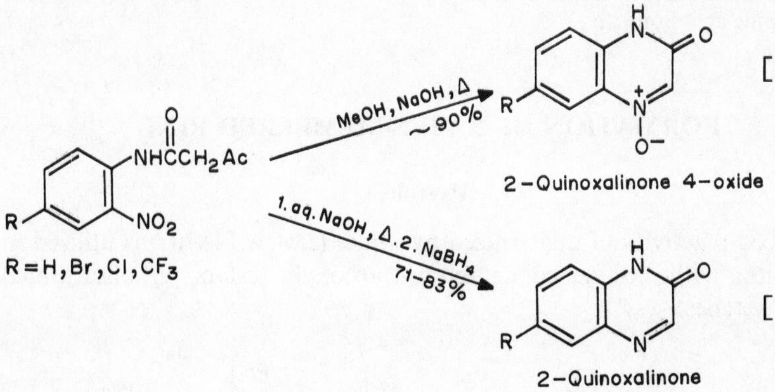

2-Quinoxalinone 4-oxide

[1526, 1850]

R=H, Br, Cl, CF₃

2-Quinoxalinone

[1526, 1850]

CHAPTER 104

Methylene and Ring-carbon or Ring-nitrogen

I. FORMATION OF A FIVE-MEMBERED RING

1. Pyrrole

The enhanced reactivity of quaternized pyridines (review [1870]) is utilized in their reaction with acetylenedi(or mono-)carboxylic esters, $\alpha\beta$-unsaturated ketones or ketenes.

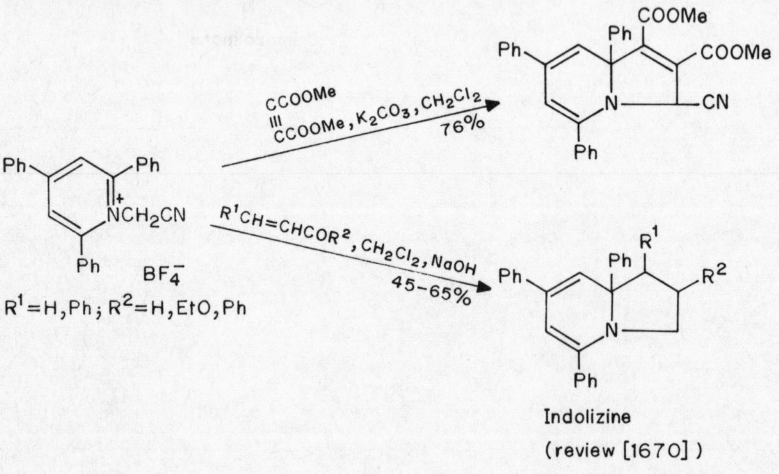

[713]

[713]

$R^1 = H, Ph; R^2 = H, EtO, Ph$

Indolizine
(review [1670])

In the absence of the 2- and 6-phenyl groups and using potassium carbonate-alumina [1286] or sodium hydride [1068] as base, the acetylenediester reaction gives a fully unsaturated product.

$R^1 = CN, COOMe$; $R^2 = H$, benzo ;
$R^3 = Ph, COOMe$

Indolizine

[1286]

Pyrrolo[2,1-a]isoquinoline

[283]

α-Methylazines yield anions with bases and reaction with halogenocarbonyl compounds results in ring closure. The same overall effect is sometimes achieved by heating the reactants on their own, in a solvent or with phosphorus oxychloride [1487].

Pyrrolo[1,2-c]quinazoline
(review [1542])

[391]

$R^1 = H, Ph$; $R^2 = H, COOEt$

Pyrrolo[1,2-a]-1,3,5-triazine-2,4-dione

[1835]

Other ring systems synthesized similarly:

Pyrrolo[1,2-c]quinazolin-5-one

[391]

Benzo[b]pyrido[2,3,4-gh]pyrrolizine

[769]

Pyrrolo[1,2-a]pyrazine
(review [1659])

[1657]

Pyrrolo[1,2-a]imidazole

[1323]

Pyrrolo[1,5-b:2,3-c']dipyridazine

[1487]

Pyrrolo[1,2-a]benzimidazole

[1323]

2. Pyrazole

Cyclization of the dithioketal (104.1) depends on the reactivity of both the N-substituted pyridinium ring and methylene group but its course may involve the initial formation of another ring.

(104.1)

R = H, Me

Pyrazolo[1,5-a]pyridine

[1290]

3. Imidazole

Nitrosoarenes and 6-methylaminouracils react in refluxing acetic anhydride to give a fused imidazole in good yield.

R = H, Me, Ph;
Ar = Ph, 4-ClC$_6$H$_4$

2,6-Purinedione

[992]

4. 1, 2, 3-Triazole

α-Methylpyridines when heated with phenylsulphonyl azides without a solvent react by annulation of the methyl and ring-nitrogen with loss of the phenylsulphonyl group.

[1746]

[1,2,3]Triazolo[1,5-a]-
quinoline

Another ring system synthesized similarly:

[1746]

[1,2,3]Triazolo[5,1-a]-
isoquinoline

5. Furan

Vilsmeier reagents supply a carbon atom and convert a reactive methyleneoxy group into a furan under mild conditions. This type of cyclization has been reviewed [1676].

[1933]

R = MeO, EtO, Et$_2$N   Benzofuran

II. FORMATION OF A SIX-MEMBERED RING

1. Pyridine or Pyridin-2-one

By taking advantage of the reactivity of the methylene in a cyanomethyl group, it and the neighbouring ring-nitrogen become part of a new pyridine ring fused at the CH$_2$—C=N— atoms of the substrate. The cyclization is usually promoted by a base but a thermally induced reaction is sometimes successful.

R = CN, PhCO, COOEt

Thiazolo[3,2-a]pyridin-3-one
(review [1865])

[1089]

[1150]

4-Quinolizinone
(reviews [1671,1692])

[538]

Pyrido[2,1-b]benzothiazol-1-one

Bridging a methyl and a ring-NH in different rings is achieved by reaction of butyllithium and a carboxylic ester.

R = PhCH₂, PhCHOH, 4-ClC₆H₄,
2,4-Cl₂C₆H₃

[661]

Imidazo[2,1-a]isoquinoline

2. Pyrimidin-4-one or Pyrimidine-4-thione

A pyrimidinone ring is formed by treatment of a reactive methylene in an α-acylamino side-chain with phosgene; the product retains a chlorine which itself may be displaced in a later reaction. When the substrate contains an α-N= CHOEt group, reaction with dichloroketene gives a new fused pyrimidinone.

R = H, Me

[310]

Pyrido[1,2-a]pyrimidin-4-one

R = H, Me

[1509]

Pyrido[1,2-a]pyrimidin-4-one

An activated methylene adjacent to a ring-nitrogen reacts with benzoyl isothiocyanate in pyridine with the formation of a new pyrimidine-2-thione in high yield.

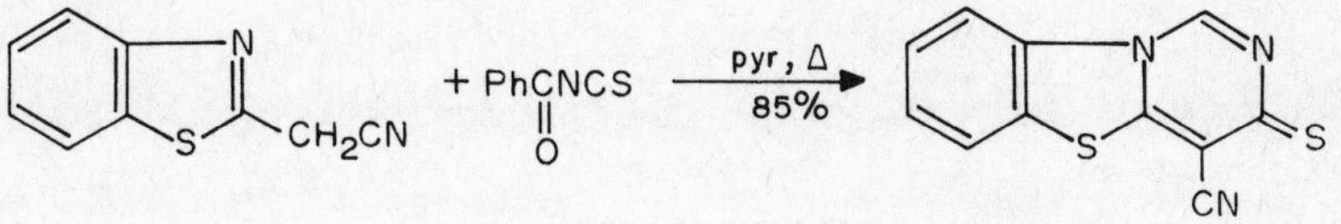

[538]

Pyrimido[6,1-*b*]benzothiazole-
3-thione

3. 1,2,4-Triazin-3-one

The methylene of a cyanomethyl group couples with diazonium salts to form a
hydrazone-type structure. This readily cyclizes on to the neighbouring ring
nitrogen to form a triazinone ring.

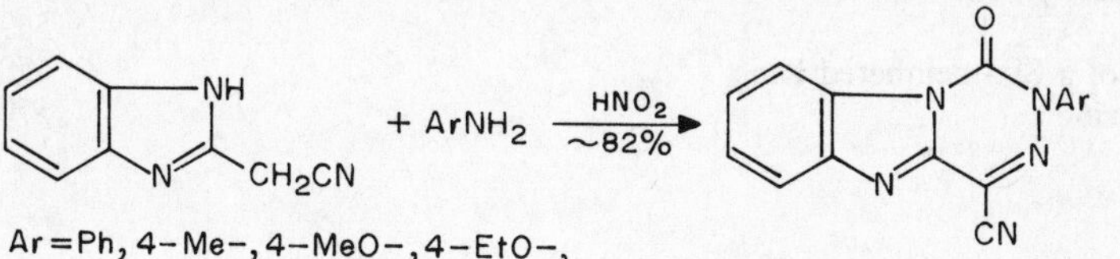

[1521, 2022]

Ar = Ph, 4-Me-, 4-MeO-, 4-EtO-,
4-Br-C$_6$H$_4$, 5-Me pyrazol-3-yl

[1,2,4]Triazino[4,5-*a*]-
benzimidazol-1-one

III. FORMATION OF A SEVEN-MEMBERED RING

1. Azepine-2,5-dione

A reactive methylene in an acylamino side-chain reacts with DMF and oxalyl
chloride (in a way reminiscent of the Vilsmeier reaction) and gives a fused and
highly substituted azepinedione ring in high yield.

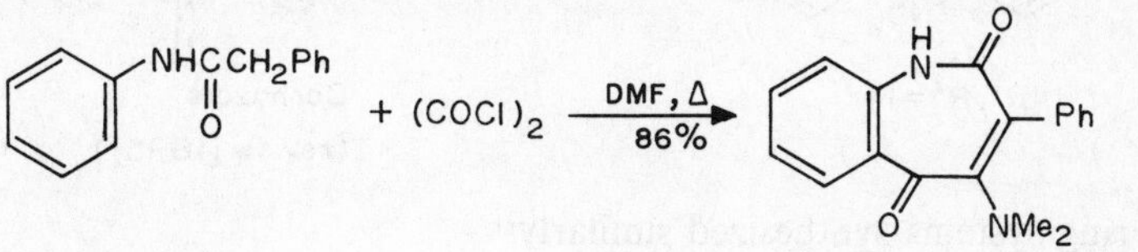

[1556]

1-Benazepine-2,5-dione

CHAPTER 105

Nitro or *N*-oxide and Ring-carbon or Ring-nitrogen

I. FORMATION OF A FIVE-MEMBERED RING

1. Pyrrole or Pyrrol-2-one

Triethyl phosphite deoxygenatively cyclizes nitro groups, probably forming a transient nitrene which attacks a neighbouring ring-carbon atom to form a pyrrole ring (review [1918]).

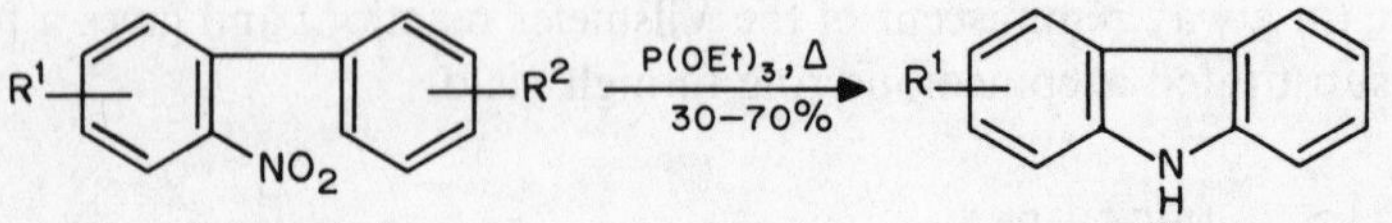

R¹, R² = Me

Carbazole
(review [1695])

[575]

Other ring systems synthesized similarly:

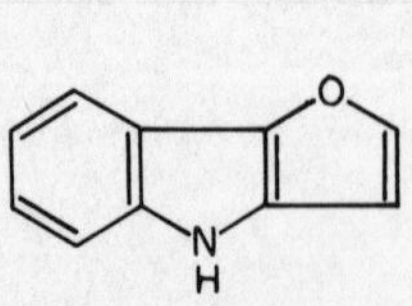

Furo[3,2-*b*]indole

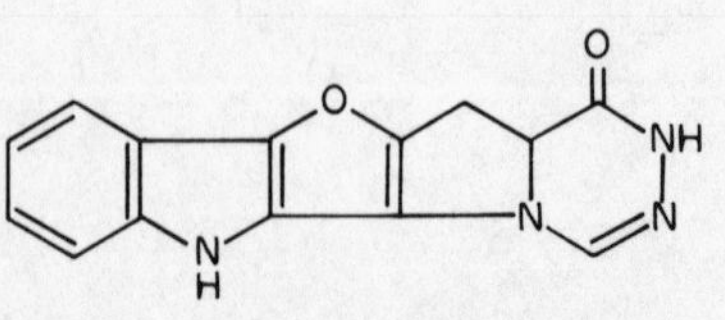

[1244]

[1,2,4]Triazino[4″,5″:1′,5′]pyrrolo-
[2′,3′:4,5]furo[3,2-*b*]indol-1-one

[1424]

2-Nitrostyrenes are cyclized by heating with an isonitrile to form 1-hydroxyindoles.

568

$R^1 = Me, Et$; $R^2 = H, MeO, NO_2$;
$R^3 = tBu, t\text{-}octyl, cyclohexyl$

Indole

(review [B-15a])

[1160]

A 2-indolone was obtained when a β-nitrostyrene was stirred at 0 °C with iron(III) chloride and acetyl chloride. Aluminium or titanium chloride was ineffective and the temperature had to be raised slightly for the nuclear nitro-derivative to cyclize. Among the byproducts are the chloro-ketone, ArCHCl·CO·NOHAc, which may be an intermediate in the cyclization.

R = H, Me, Cl, F, NO₂

2-Indolone

[543]

On heating a 2'-nitroflavanone briefly with ethanolic sodium hydroxide, ring-closure between nitro and C-3 of the flavanone gave an N-hydroxypyrrole ring.

[1]Benzopyrano[3,2-b]—
indol-11-one   [697]

Propiolic (propynoic) acid esters (two moles) add across the C–N bond of a heterocyclic N-oxide in refluxing benzene to form a fused pyrrole ring; this is accompanied by deoxygenation and deformylation of the intermediate.

R = H, MeO

Pyrrolo[2,1-a]isoquinoline

[1907]

2. Pyrazole

Deoxygenation of nitro groups by triethyl phosphite (see preceding section) may result in the formation of a N–N bond by attack at a conveniently positioned ring-nitrogen.

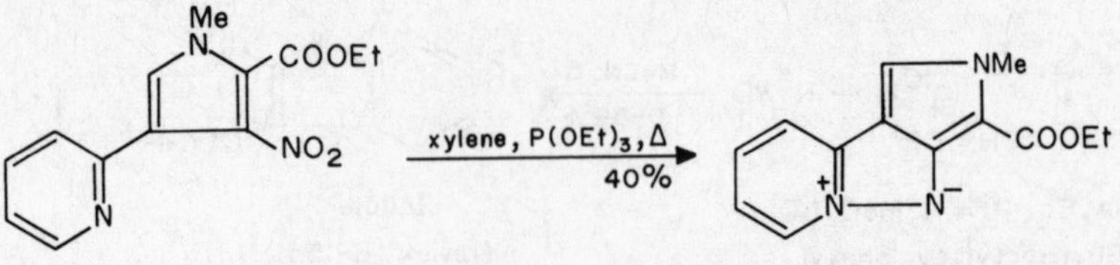

Pyrrolo[3′,4′:3,4]pyrazolo–
[1,5-*a*]pyridinium ylide

II. FORMATION OF A SIX-MEMBERED RING

1. 1,4-Thiazine

A thiazine ring is formed when a 2-nitrophenylsulphide is reduced with triethyl phosphite at about 150 °C.

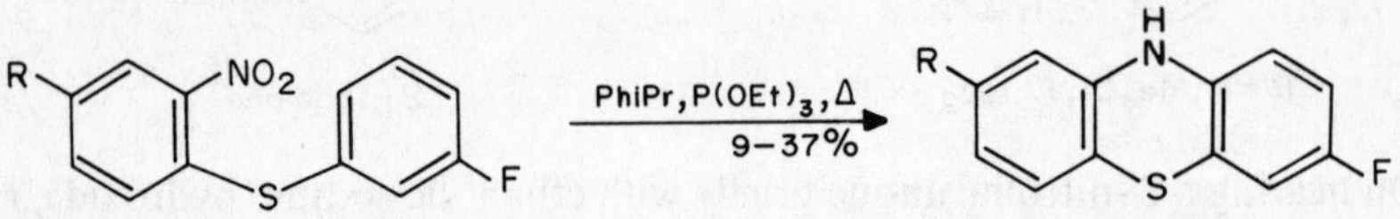

R=H,MeO,MeS,Cl,CF$_3$,CN,Ac

Phenothiazine

[1012]

CHAPTER 106

Ring-carbon and Ring-nitrogen

I. FORMATION OF A FIVE-MEMBERED RING

1. Pyrrole

A ring-carbon may be joined to a ring-nitrogen in another ring by reaction with either phenacyl bromide or butyllithium followed by a carboxylic ester.

Pyrano[2,3,4-*hi*]-
indolizine

R = tBu, 4-ClC$_6$H$_4$, 2-furyl

Imidazo[2,1-*a*]isoindole

2. 1, 2, 4-Triazole

In an unusual cyclization, a triazinone is heated in water with hydrazine and a carboxylic acid. A fused 1, 2, 4-triazole is produced in yields which vary according to the carboxylic acid used, the highest yield being from formic acid.

R=H,Me,Ph

1,2,4-Triazolo[3,4-f]-
[1,2,4]triazin-8-one   [1946]

Nitrilimines, generated from α-chlorohydrazones (review [1753]), add on to heterocyclic C=N bonds to provide a new fused 1,2,4-triazole.

R = Me, Ph

[1800]

1,2,4-Triazolo[4,3-a]-
benzimidazole

II.   FORMATION OF A SIX-MEMBERED RING

1.   Pyridine-4-one

Another addition to a C=N bond in a heterocyclic ring is that of the trimethylsilyloxybutadiene which results in the formation of a fused pyridinone ring but when the methyl group is replaced by a more electron-releasing substituent, the reaction fails.

[1580]

Pyrido[1,2-b][1,2]benzisothiazol-
2-one 6,6-dioxide

2.   1,3,5-Triazine-2,4-dione

4-Pyridinone reacts with two molar equivalents of methyl isocyanate to give a triazine-fused product in theoretical yield.

[1539]

Pyrido[1,2-a]-1,3,5-
triazine-2,4,8-trione

CHAPTER 107

Ring-nitrogen and Thiol or Lactam Thiocarbonyl

Two kinds of cyclizations are discussed in this chapter. The more common type is that in which both C=S and NH of a thiolactam react to form a ring (Scheme 107.1). An example is included of the parallel but of the less common annulation of a lactone carbonyl to its nitrogen. Some examples are also mentioned of compounds in which the thiocarbonyl and nitrogen functions are separated by one or more carbon atoms or are in different rings.

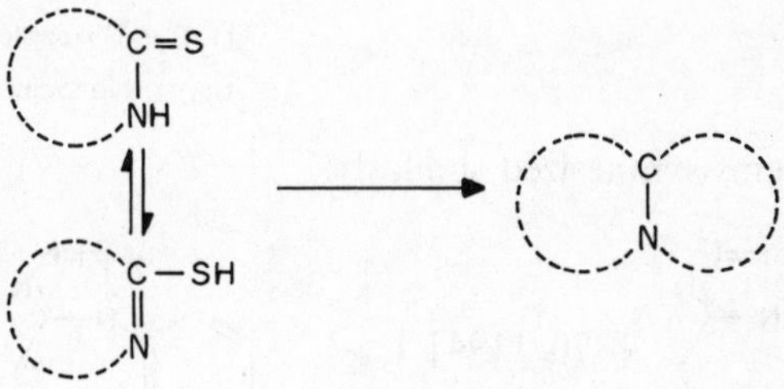

Scheme 107.1

I. FORMATION OF A FIVE-MEMBERED RING

1. Imidazole

Reaction of a thiolactam with a 2-alkynylamine leads to the formation of a fused imidazole ring in moderate yield. A lactam is annulated by reaction with an isonitrile in the presence of a base.

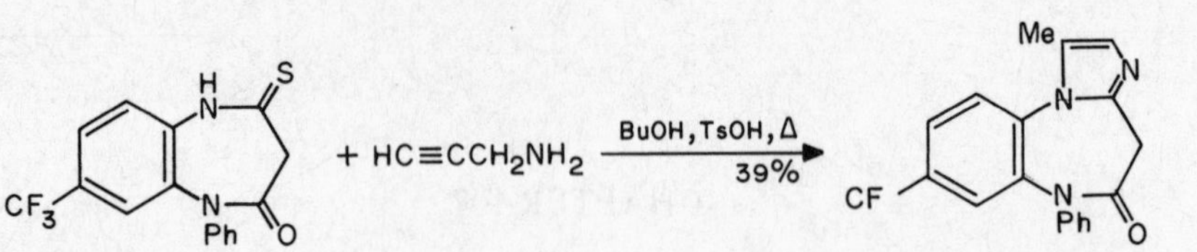

Imidazo[1,2-a][1,5]-
benzodiazepin-5-one

[1194]

[830]

R = H, Me

Imidazo[5,1-c][1,4]-
benzothiazine

2. 1, 2, 4-Triazole

A hydrazide condenses with a thiolactam group on heating with or without an
acidic catalyst—a reaction which has been widely used in medicinal chemistry.

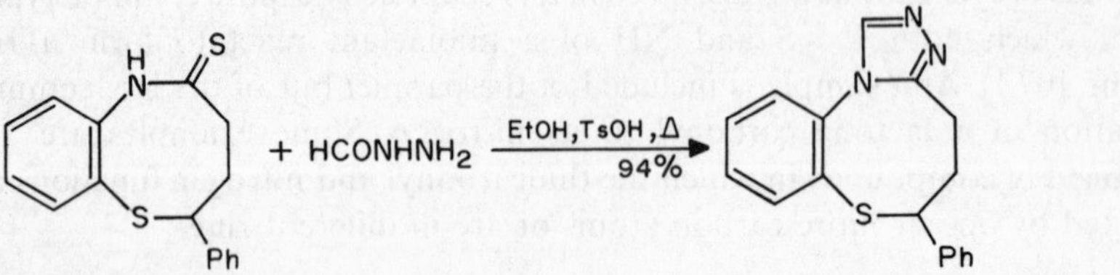

[829]

[1,2,4]Triazolo[3,4-e][1,6]-
benzothiazocine

Other ring systems synthesized similarly:

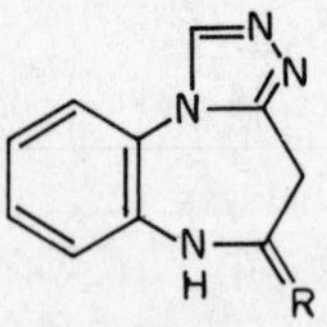

[770, 1194]

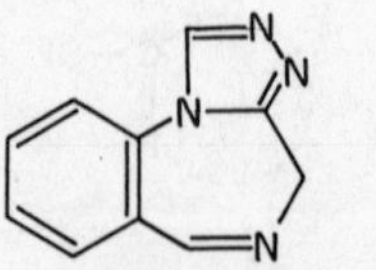

R = H₂
[1,2,4]Triazolo[4,3-a]-
[1,5]benzodiazepine
R = O
[1,2,4]Triazolo[4,3-a]-
[1,5]benzodiazepin-5-one

[1879]

[1,2,4]Triazolo[4,3-a]-
[1,4]benzodiazepine

3. Isothiazole

A thiol group may be linked by a bond to a nitrogen in another ring either by reaction with *N*-chlorosuccinimide or in an oxidative cyclization with bromine.

Pyrido[1,2-*b*][1,2]-
benzisothiazolium
[836]

X=CH,N
Benzimidazo[1,2-*b*][1,2]benzisothiazole
Pyrido[3',2':4,5]isothiazolo[2,3-*a*]-
benzimidazole
[1847]

4. Thiazole

For the purpose of classfication, compounds such as imidazole-2-thione and 1,2,4-triazole-3-thione (or their tautomers) are regarded as thiolactams. The sulphur function and its neighbouring nitrogen may be annulated by reaction with 2-halogenoketones in refluxing ethanol; the less reactive bromoacetalde-hyde dimethylacetal also gives a good yield in this type of cyclization [1860].

R=CH₂COOEt; Ar=Ph, 2-or
4-ClC₆H₄

Thiazolo[3,2-*b*][1,2,4]-
triazole
[662]

R=Ph, 4-MeO-, 4-Cl-C₆H₄

Pyrazolo[5,1-*b*]thiazole
[1860]

Another ring system synthesized similarly:

Imidazo[2,1-*b*]thiazole
[663]

5. Thiazol-4-one or Thiazole-4, 5-dione

The thiolactam group adds across the triple bond of acetylenedicarboxylic ester (the synthetic use of this compound has been reviewed [1725]) under very mild conditions and a thiazolone ring is formed. A rather lower yield is obtained in the cycloaddition of maleic anhydride to a thiolactam but oxalyl chloride-TEA gives a good yield of the fused thiazoledione.

R = H, Me, PhCH₂, Ph

Thiazolo[3,2-*b*][1,2,4]-triazine-3,7-dione

[1311]

Thiazolo[3,2-*a*]benzimidazol-1-one

[1475]

R = Ph, 4-MeO-, 4-Cl-C₆H₄

Pyrazolo[5,1-*b*]thiazole-2,3-dione

[1860]

II. FORMATION OF A SIX-MEMBERED RING

1. 1, 3-Oxazin-4-one

This ring is formed by heating 2-benzimidazolethiol with diketene but the yield is only moderate.

[1,3]Oxazino[3,2-*a*]-benzimidazol-4-one

[1114]

2. 1,3-Thiazin-4-one

The normal product of the reaction of thiolactams with diketene is the thiazinone ring; it is also formed when ethyl 2-butynoate, diethyl ethoxymethylenemalonate or acetylenedicarboxylic ester is used instead of diketne. The acetylene diester cyclization needs prolonged refluxing [912] compared with a similar cyclization on a different substrate (Section II.5). 1,3-Benzothiazin-4-ones have been reviewed [2010].

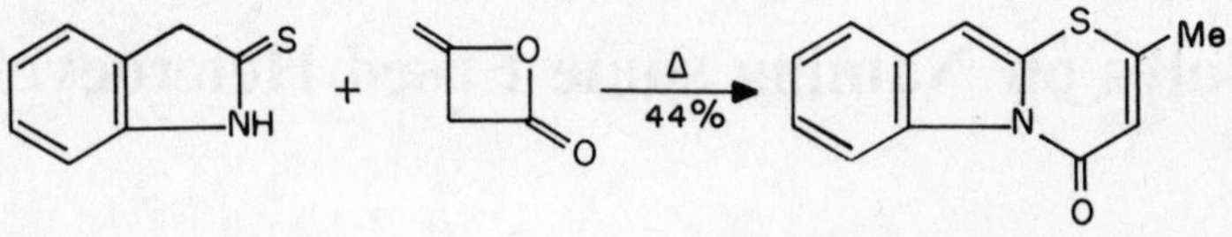

[1,3]Thiazino[3,2−a]−
indol−4−one

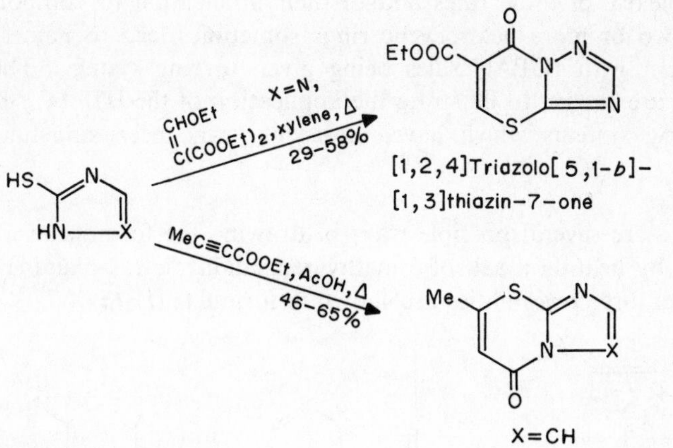

[1114]

[1,2,4]Triazolo[5,1−b]−
[1,3]thiazin−7−one

[739]

X=CH

[739]

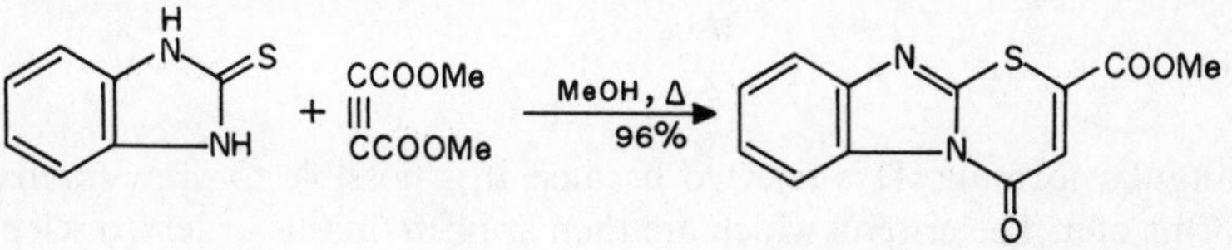

[1,3]Thiazino[3,2−a]−
benzimidazol−4−one

[912]

APPENDIX:

Notes on Naming some Fused Heterocycles

Basic rules for assigning a name to a fused heterocycle are given in the official IUPAC rules [B-1] and these have been amplified in a review [2014]. However, the complexity of these rules and of their application to compounds which contain two or more heterocyclic rings sometimes lead to names which are inconsistent with IUPAC rules being given to ring systems. The following examples are chosen to illustrate the application of the IUPAC rules to some of the ring systems which have, in recent years, been sometimes wrongly named.

A. There are several possible ways of drawing the formula of a compound prepared by heating a salt of 3-methylbenzimidazole-1,2-diamine with acetic anhydride; three possibilities are shown in formulae (1)–(3).

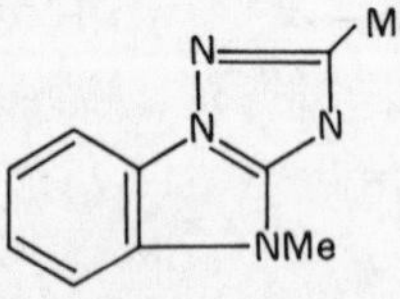

(1) 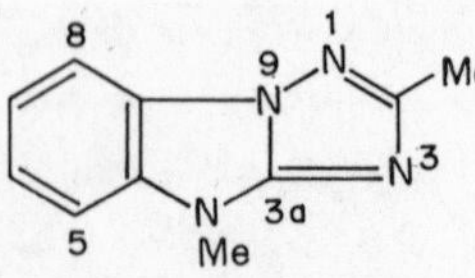   (2) 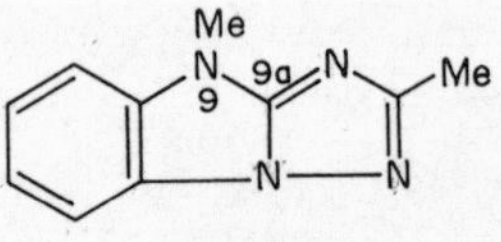   (3)

The angular formula (1) is rejected because it is possible to draw (preferred) linear forms of it. The criteria which are then applied (in this order) to select the correct formula from (2) and (3) are as follows:

(1) Hetero atoms should have the lowest possible numbers. In this example, both (2) and (3) lead to the hetero atoms being given the numbers 1, 3, 4, 9 and a choice cannot be made.

(2) Carbon atoms which are shared by two rings should follow the lowest possible numbers. In formula (2), the carbon atom which is common to both hetero-rings is numbered 3a while in (3), it is 9a. The correct formula is therefore (2) which also shows its correct alignment. It follows (from IUPAC rules) that peripheral numbering begins in the ring which is furthest to the

right and at the most anticlockwise (but not a bridgehead) atom.

(4)

(5)

Of the two ring systems [1, 2, 4-triazole (4) and benzimidazole (5)] which constitute compound (2), benzimidazole takes precedence as the base or last part of the name because it has more rings than the triazole. The name will therefore be of the type [1, 2, 4]triazolo[x, y − z]benzimidazole. The numerals 1, 2, 4 are placed in square brackets because the three triazole nitrogen atoms are not in positions 1, 2 and 4 of the peripheral numbering of (2). The three characters in the second pair of square brackets (the locant) identify the bond which is common to the two ring systems, x and y being numbers in the triazole ring (4) and z the bond as viewed from the (tautomeric) benzimidazole (5) ring. Where a choice of numbering is available, that which gives the lowest numerals and the alphabetically earliest letter is chosen; in this example, therefore, the locant is [1, 5-a] and (2) is 2, 4-dimethyl-4H-[1, 2, 4]triazolo[1, 5-a]benzimidazole.

(6)

(7)

(8)

B.   It is important to ensure that numerals such as 1, 2, 4 in the above example are those which are correct for that ring when it is isolated from (i.e. not fused to) the other ring but the position of substituents must be identified in terms of the peripheral numbering of the ring system. For example, to determine the IUPAC-approved name of compound (6), it is necessary to separate the two rings (omitting substituents at this stage) as in (7) and (8) and to name each one: pyrazole and 1, 2, 6-thiadiazine respectively. The latter is the base component by

(9)

(10)

(11)

virtue of the size of the ring; the locant [3, 4-c] follows from the details shown in (7) and (8). The letters a, b, c, . . . are given to bonds in the base component in the same order as for numbering ring (8); this begins with oxygen, or if this is absent with sulphur or, failing that, with nitrogen. The numbering (and alphabetizing of the bonds) then proceeds towards the nearest heteroatom and continues to any others present. A rudimentary name for the ring system of (6) is therefore pyrazolo[3, 4-c][1, 2, 6]thiadiazine and to this must be added the names of substituents to give 1, 7-dihydro-3-methyl-7-phenylpyrazolo[3, 4-c][1, 2, 6]-thiadiazin-4-one 2, 2-dioxide.

It is worth noting that replacement of the pyrazole ring of (6) by a benzene has an important effect on the numbers which appear in the name. Compound (9) is called 3-methyl-2, 1, 3-benzothiadiazin-4-one 2, 2-dioxide; the numbers 2, 1, 3 are derived from the peripheral numbering and so do not need square brackets. In benzoheterocycles, peripheral numbering begins at one of the atoms which are attached to the benzene ring irrespective of the numbering of the isolated heterocycle; for example, compound (10) is 2, 1-benzisothiazole. In such compounds, the numerals correspond to the peripheral numbers while in (6), the term [1, 2, 6] is derived from the numbering of the thiadiazine in isolation. Another example is shown in compound (11) which is called 7-methylthieno[3, 2-d][1, 6, 2]oxathiazepin-2-one.

List of Books and Monographs

B–1   Rigaudy, J. and Klesney, S. P. (eds), *IUPAC: Nomenclature of Organic Chemistry*, Pergamon Press, Oxford, 1979.

B–2   van der Plas, H. C. *Ring Transformations of Heterocycles*, Volumes 1 and 2, Academic Press, New York, 1973.

B–3   Meyers, A. I. *Heterocycles in Organic Synthesis*, Wiley, New York, 1974.

B–4   Katritzky, A. R. and Rees, C. W. (eds), *Comprehensive Heterocyclic Chemistry*, Volumes 1–8, Pergamon Press, Oxford, 1984.

(Books listed as B-5 to B-20 form part of the continuing series 'The Chemistry of Heterocyclic Compounds', edited by A. Weissberger and E. C. Taylor and published by Wiley, New York and Chichester.)

B–5   Armarego, W. L. F. *Quinazolines*, 1967.

B–6   Mustafa, A. *Furopyrans and Furopyrones*, 1967.

B–7   Lister, J. H. *Purines*, 1971.

B–8   Rosowsky, A. *Seven-membered Heterocyclic Compounds containing O and S*, 1972.

B–9   Castle, R. N. (ed.), *Condensed Pyridazines*, 1972.

B–10   Acheson, R. M. (ed.), *Acridines*, 2nd edn, 1973.

B–11   Mustafa, A. *Benzofurans*, 1974.

B–12   Ellis, G. P. (ed.), *Chromenes, Chromanones and Chromones*, 1977.

B–13   Jones, G. (ed.), *Quinoline and its Derivatives*, Parts 1 and 2, 1977 and 1982.

B–14   Neunhoffer, H. and Wiley, P. F. *Chemistry of 1,2,3-Triazines and 1,2,4-Triazines, Tetrazines and Pentazines*, 1978.

B–15   Cheeseman, G. W. H. and Cookson, R. F. *Condensed Pyrazines*, 1979.

B–15a   Houlihan, W. J. (ed.), *Indoles*, 1979.

B–16   Ellis, G. P. and Lockhart, I. M. (eds), *Chromans and Tocopherols*, 1980.

B–17   Grethe, G. (ed.), *Isoquinolines*, 1981.

B–18   Preston, P. N. (ed.), *Benzimidazoles and Congeneric Tricyclic Compounds*, Parts 1 and 2, 1980 and 1981.

B–19   Renfroe, B. Harrington, C. Proctor, G. and Rosowsky, A. *Azepines* (including Diazepines and Triazepines), Parts 1 and 2, 1984.

B–20   Newkome G. R. (ed.), *Pyridine and its Derivatives*, Part 5, 1985.

B–21   Rowlands, D. A. in *Synthetic Reagents*, J. S. Pizey (ed.), Ellis Horwood, Chichester, Volume 6, 1985.

B–22   Pizey, J. S. *Synthetic Reagents*, Ellis Horwood, Chichester, Volume 1, 1974.

B–23   Lwowski, W. '*Nitrenes*, Wiley, New York, 1970.

B–24   Schutz, H. *Benzodiazepines: A Handbook*, Springer-Verlag, Berlin, 1982.

B–25   Sandler, S. R. and Karo, W. *Organic Functional Group Preparations*, Academic Press, New York, Volume 2, 1971.

B–26 Sammes, P. G. (ed.), *Topics in Antibiotic Chemistry*, Ellis Horwood, Chichester, Volume 4, 1980.

B–27 Brown, A. G. and Roberts, S. M. (eds), *Recent Advances in the Chemistry of Beta-lactam Antibiotics*, Royal Society of Chemistry, London, 1985.

B–28 Saunders, K. H. and Allen, R. M. L. *Aromatic Diazo Compounds*, 3rd edn, E. Arnold, London, 1985.

B–29 Robinson, B. *The Fischer Indole Synthesis*, Wiley, Chichester, 1982.

B–30 Blomquist, A. T. (ed.), *The Chemistry of Indoles*, Academic Press, New York, 1970.

B–31 Trahanovsky, W. S. (ed.), *Oxidation in Organic Chemistry*, Academic Press, New York, Part C, 1978.

B–32 Harborne, J. B., Mabry, T. J. and Mabry, H. *The Flavonoids*, Chapman & Hall, London, 1975.

B–33 Harborne, J. B. and Mabry, T. J. (eds), *The Flavonoids: Advances in Research*, Chapman & Hall, London, 1982.

B–34 Katritzky, A. R. and Lagowski, J. M. *Chemistry of the Heterocyclic N-Oxides*, Academic Press, London, 1971.

B–35 Coffey, S. (ed.), *Rodd's Chemistry of Carbon Compounds*, 2nd edn and Supplements (M. F. Ansell, ed.), Elsevier, Amsterdam, Volume IV, 1973–1986.

B–36 Butler, R. N. in *Synthetic Reagents*, J. S. Pizey (ed.), Ellis Horwood, Chichester, Volume 3, 1977.

B–37 Butler, R. N. in *Synthetic Reagents*, J. S. Pizey (ed.), Ellis Horwood, Chichester, Volume 4, 1981.

B–38 Turner, A. B. in *Synthetic Reagents*, J. S. Pizey (ed.), Ellis Horwood, Chichester, Volume 3, 1977.

B–39 Murray, R. D. H. *Natural Coumarins*, Wiley, Chichester, 1982.

References

1. Svetlik, J. and Martvon, A., *Collect. Czech. Chem. Commun.*, 1981, **46**, 428.
2. Bell, S. C. and Wei, P. H. L., *J. Heterocycl. Chem.*, 1969, **6**, 599.
3. Kesler, E., *Monatsh. Chem.* 1982, **113**, 1217.
4. Deodkar, K. D., D'Sar, A. D., Pednekar, S. R. and Kanekar, D. S., *Synthesis*, 1982, 853.
5. Shiotani, S. and Morita, H., *J. Heterocycl. Chem.*, 1982, **19**, 1207.
6. Murahashi, S.-I., Yoshimura, N., Tsumiyama, T. and Kojima, T., *J. Am. Chem. Soc.*, 1983, **105**, 5002.
7. Vinot, N. and Maitte, P., *J. Heterocycl. Chem.*, 1982, **19**, 349.
8. Geyser, H. M., Martin, L. L., Crichlow, C. A., Dekow, F. W., Ellis, D. B., Kruse, H., Setescak, L. L. and Worm, M., *J. Med. Chem.*, 1982, **25**, 340.
9. Albert, A. and Ohta, K., *J. Chem. Soc. (C)*, 1971, 2357.
10. Harper, J. F. and Wibberley, D. G., *J. Chem. Soc. (C)*, 1971, 2985.
11. Harper, J. F. and Wibberley, D. G., *J. Chem. Soc. (C)*, 1971, 2991.
12. Merchant, J. R. and Venkatesh, M. S., *Chem. Ind. (London)*, 1979, 478.
13. Peet, N. P., Sunder, S. and Barbuch, R. J., *J. Heterocycl. Chem.*, 1980, **17**, 1513.
14. Dean, W. D. and Papadopoulos, E. P., *J. Heterocycl. Chem.*, 1982, **19**, 1117.
15. Meth-Cohn, O. and Narine, B., *Tetrahedron Lett.*, 1978, 2045.
16. Manning, W. B. and Horak, V., *Synthesis*, 1978, 363.
17. Cupps, T. L., Wise, D. S. and Townsend, L. B., *J. Org. Chem.*, 1983, **48**, 1060.
18. Taylor, E. C. and Shvo, Y., *J. Org. Chem.*, 1968, **33**, 1719.
19. Evans, D. J. and Eastwood, F. W., *Aust. J. Chem.*, 1974, **27**, 537.
20. Jones, C. D. and Shuarez, T., *J. Org. Chem.*, 1972, **37**, 3622.
21. Chantegrel, B., Nadi, A. I. and Gelin, S., *J. Heterocycl. Chem.*, 1985, **22**, 81.
22. Hymans, W. E. and Cruickshank, P. A., *J. Heterocycl. Chem.*, 1974, **11**, 231.
23. Monge Vega, A., Martinez, M. T., Palop, J. A., Mateo, J. M. and Fernandez-Alvarez, E., *J. Heterocycl. Chem.*, 1981, **18**, 889.
24. Coppola, G. M., *J. Heterocycl. Chem.*, 1981, **18**, 845.
25. Soth, S., Farnier, M. and Paulmier, C., *Can. J. Chem.*, 1978, **56**, 1429.
26. Czerney, P. and Hartmann, H., *J. Prakt. Chem.*, 1981, **323**, 691.
27. Wurm, G., *Arch. Pharm. (Weinheim)*, 1975, **308**, 259.
28. Mosti, L., Schenone, P. and Menozzi, G., *J. Heterocycl. Chem.*, 1979, **16**, 913.
29. Mosti, L., Schenone, P., Menozzi, G. and Cafaggi, S., *J. Heterocycl. Chem.*, 1982, **19**, 1031.
30. Gupta, R. K. and George, M. V., *Tetrahedron*, 1975, **31**, 1263.
31. Baumgarten, H. E., Barkley, R. P., Chin, S. H. L. and Thompson, R. D., *J. Heterocycl. Chem.*, 1981, **18**, 925.
32. Papadopoulos, E. P. and Torres, C. D., *Heterocycles*, 1982, **19**, 1039.
33. Sakaguchi, M., Miyata, Y., Ogura, H., Gonda, K., Koga, S. and Okamoto, T., *Chem. Pharm. Bull.*, 1979, **27**, 1094.

34. Boulton, A. J., *J. Chem. Soc., Chem. Commun.*, 1982, 1328.
35. Skoetsch, C. and Breitmaier, E., *Chem. Ztg*, 1978, **102**, 264.
36. Danishefsky, S., Bryan, T. A. and Puthenpurayil, J., *J. Org. Chem.*, 1975, **40**, 796.
37. Althius, T. R., Moore, P. F. and Hess, H.-J., *J. Med. Chem.*, 1979, **22**, 44.
38. Althius, T. R., Kadin, S. B., Czuba, L. J., Moore, P. F. and Hess, H.-J., *J. Med. Chem.*, 1980, **23**, 262.
39. Clark, R. D. and Repke, D. B., *J. Heterocycl. Chem.*, 1985, **22**, 121.
40. Tabakovic, K., Tabakovic, I., Trkovnik, M. and Trinajstic, N. *J. Heterocycl. Chem.*, 1980, **17**, 801.
41. Barton, D. H. R., Halpern, B., Porter, Q. N. and Collins, D. J., *J. Chem. Soc. (C)*, 1971, 2166.
42. Birchall, G. R., Galbraith, M. N., Gray, R. W., King, R. R. and Whalley, W. B., *J. Chem. Soc. (C)*, 1971, 3559.
43. Hall, J. H. and Dolan, F. W., *J. Org. Chem.*, 1978, **43**, 4608.
44. Osselaere, J. P. and Lapiere, C. L., *Eur. J. Med. Chem.*, 1974, **9**, 305.
45. Parish, H. A., Gilliom, R. D., Purcell, W. P., Browne, R. K., Spirk, R. F. and White, H. D., *J. Med. Chem.*, 1982, **25**, 98.
46. Althius, T. R. and Hess, H.-J., *J. Med. Chem.*, 1977, **20**, 146.
47. Hoffmann, R. W., Guhn, G., Preiss, M. and Dittrich, B., *J. Chem. Soc. (C)*, 1969, 769.
48. Koitz, G., Thierrichter, B. and Junek, H., *Heterocycles*, 1983, **20**, 2405.
49. Shealy, Y. F. and O'Dell, C. A., *J. Med. Chem.*, 1966, **9**, 733.
50. Manhas, M. S., Hoffman, W. A. and Bose, A. K., *J. Heterocycl. Chem.*, 1979, **16**, 711.
51. Sieper, H., *Chem. Ber.*, 1967, 100, 1646.
52. Kurihara, T., Tani, T., Maeyama, S. and Sakamoto, Y., *J. Heterocycl. Chem.*, 1980, 17, 945.
53. Montgomery, J. A., Laseter, A. G., Shortnacy, A. T., Clayton, S. J. and Thomas, H. J., *J. Med. Chem.*, 1975, **18**, 564.
54. Etson, S. R., Mattson, R. J. and Sowell, J. W., *J. Heterocycl. Chem.*, 1979, **16**, 929.
55. Taylor, E. C. and Garcia, E. E., *J. Org. Chem.*, 1964, **29**, 2121.
56. Taylor, E. C. and Hendess, R. W., *J. Am. Chem. Soc.*, 1965, **87**, 1995.
57. Taylor, E. C. and Ehrhart, W. A., *J. Am. Chem. Soc.*, 1960, **82**, 3138.
58. Taylor, E. C. and Hendess, R. W., *J. Am. Chem. Soc.*, 1965, **87**, 1981.
59. Tani, Y., Yamada, Y., Oine, T., Ochiai, T., Ishida, R. and Inoue, I., *J. Med. Chem.*, 1979, **22**, 95.
60. Parmar, S. S. and Singh, S. P., *J. Heterocycl. Chem.*, 1979, **16**, 449.
61. Bosch, J., Domingo, A. and Linares, A., *J. Org. Chem.*, 1983, **48**, 1075.
62. Coppola, G. M. and Shapiro, M. J., *J. Heterocycl. Chem.*, 1980, **17**, 1163.
63. Dave, K. G., Shishoo, C. J., Devani, M. B., Kalyanaraman, R., Ananthan, S., Ullas, G. V. and Bhadti, V. S. *J. Heterocycl. Chem.*, 1980, **17**, 1497.
64. Shawali, A. S., Sami, M., Sherif, S. M. and Parkanyi, C., *J. Heterocycl. Chem.*, 1980, **17**, 877.
65. Kim, D. H., *J. Heterocycl. Chem.*, 1981, **18**, 801.
66. Wamhoff, H., *Chem. Ber.*, 1968, **101**, 3377.
67. Cohen, E. and Klarberg, B., *J. Am. Chem. Soc.*, 1962, **84**, 1994.
68. Stanovnik, B., Tisler, M., Golob, V., Hvala, I. and Nikolic, O. *J. Heterocycl. Chem.*, 1980, **17**, 733.
69. Robba, M., Touzot, P. and El-Kashef, H., *J. Heterocycl. Chem.*, 1980, **17**, 923.
70. Albright, J. D., Moran, D. B., Wright, W. B., Collins, J. B., Beer, B., Lippa, A. S. and Greenblatt, E. N., *J. Med. Chem.*, 1981, **24**, 592.
71. Okada, J., Nakano, K. and Miyake, H., *Chem. Pharm. Bull.*, 1981, **29**, 667.
72. McCarty, J. E. *J. Org. Chem.*, 1962, **27**, 2672.
73. Papadopoulos, E. P. and Torres, C. D., *J. Heterocycl. Chem.*, 1982, **19**, 269.

74. Rizkalla, B. H., Broom, A. D., Stout, M. G. and Robins, R. K., *J. Org. Chem.*, 1972, **37**, 3975.
75. Keyser, G. E. and Leonard, N. J., *J. Org. Chem.*, 1976, **41**, 3529.
76. Bailey, A. S., Scott, P. W. and Vandrevala, M. H., *J. Chem. Soc., Perkin Trans., 1*, 1980, 97.
77. Stunic, Z., Trkovnik, M., Lacan, M. and Jankovic, R., *J. Heterocycl. Chem.*, 1981, **18**, 511.
78. Adegoke, E. A., Alo, B. I. and Ogunsulire, F. O., *J. Heterocycl. Chem.*, 1982, **19**, 1169.
79. Dattolo, G., Cirrincione, G., Almerico, A. M., D'Asidia, I. and Aiello, E., *J. Heterocycl. Chem.*, 1982, **19**, 1237.
80. Schwan, T. J. and Miles, N. J., *J. Heterocycl. Chem.*, 1982, **19**, 1257.
81. Hirota, K., Maruhashi, K., Asao, T., Kitamura, N., Maki, Y. and Senda, S., *Chem. Pharm. Bull.*, 1983, **31**, 3959.
82. Ishikawa, M., Azuma H., Eguchi, Y., Sugimoto, A., Ito, S., Takashima, Y., Ebisawa, H., Moriguchi, S., Kotoku, I. and Suzuki, H., *Chem. Pharm. Bull.*, 1982, **30**, 744.
83. Le Mahieu, R. A., Carson, M., Welton, A. F., Baruth, H. W. and Yaremko, B., *J. Med. Chem.*, 1983, **26**, 107.
84. Le Mahieu, R. A., Carson, M., Nason, W. C., Parish, D. R., Welton, A. F., Baruth, H. W. and Yaremko, B., *J. Med. Chem.*, 1983, **26**, 420.
85. Massa, S., Demartino, G. and Corelli, F., *J. Heterocycl. Chem.*, 1982, **19**, 1497.
86. Leymarie-Beljean, M. and Pays, M., *J. Heterocycl. Chem.*, 1980, **17**, 1175.
87. Dattolo, G., Cirrincione, G. and Aiello, E., *J. Heterocycl. Chem.*, 1980, **17**, 701.
88. Lin, A. J. and Kasina, S., *J. Heterocycl. Chem.*, 1981, **18**, 759.
89. Lumma, W. C., Randall, W. C., Cresson, E. L., Huff, J. R., Hartman, R. D. and Lyon, T. F., *J. Med. Chem.*, 1983, **26**, 357.
90. Paparao, C., Rao, K. V. and Sundaramurthy, V., *Synthesis*, 1981, 234.
91. Mogilaiah, K. and Sreenivasulu, B., *Indian J. Chem.*, 1982, **21B**, 582.
92. Godard, A. and Queguiner, G., *J. Heterocycl. Chem.*, 1982, **19**, 1289.
93. Imai, Y., Sato, S., Takasawa, R. and Ueda, M., *Synthesis*, 1981, 35.
94. Merour, J. Y., *J. Heterocycl. Chem.*, 1982, **19**, 1425.
95. Abdelrazek, F. M., Erian, A. W. and Hilmy, K. M. H., *Synthesis*, 1986, 74.
96. Osman, A. M., Metwally, S. A. M. and Youssef, M. S. K., *Can. J. Chem.*, 1976, **54**, 37.
97. Ito, Y., Kobayashi, K. and Saegusa, T., *J. Org. Chem.*, 1979, **44**, 2030.
98. Haugwitz, R. D., Angel, R. G., Jacobs, G. A., Maurer, B. V., Narayanan, V. L., Cruthers, L. R. and Szanto, J., *J. Med. Chem.*, 1982, **25**, 969.
99. Webb, R. L. and Labaw, C. S. *J. Heterocycl. Chem.*, 1982, **19**, 1205.
100. Kappe, T. and Ziegler, E., *Angew, Chem., Int. Ed. Engl.*, 1974, **13**, 491.
101. Hromatka, O., Binder, D. and Veit, W., *Monatsh. Chem.*, 1973, **104**, 979.
102. Radhakrishna, A. S. and Berlin, K. D., *Org. Prep. Proced. Int.*, 1978, **10**, 39.
103. Da Settimo, A., Primofiore, G., Livi, O., Ferrarini, P. L. and Spinelli, S., *J. Heterocycl. Chem.*, 1979, **16**, 169.
104. Nachman, R. J., *J. Heterocycl. Chem.*, 1982, **19**, 1545.
105. Buggy, T. and Ellis, G. P., *J. Chem. Res. (S)*, 1980, 317.
106. Kwok, R., *J. Heterocycl. Chem.*, 1978, **15**, 877.
107. DeGraw, J. I. and Tagawa, H., *J. Heterocycl. Chem.*, 1982, **19**, 1461.
108. Werbel, L. M., Newton, L. and Elslager, E. F., *J. Heterocycl. Chem.*, 1980, **17**, 497.
109. Meyer, R. B. and Skibo, E. B., *J. Med. Chem.*, 1979, **22**, 944.
110 Ashton, W. T. and Hynes, J. B., *J. Med. Chem.*, 1973, **16**, 1233.
111. Ferris, J. P., and Orgel, L. E., *J. Am. Chem. Soc.*, 1966, **88**, 3829.
112. Foster, S. A., Leyshon, L. J. and Saunders, G. D., *J. Chem. Soc., Chem. Commun.*, 1973, 29.
113. Haufel, J. and Breitmaier, E., *Angew. Chem., Int. Ed. Engl.*, 1974, **13**, 604.

114. Lespagnol, A., Lespagnol, C. and Bernier, J.-L., *Ann. Pharm. Fr.*, 1974, **32**, 125.

115. Brown, D. J. and Ienaga, K., *J. Chem. Soc., Perkin Trans., 1*, 1975, 2182.

116. Kato, T., *Acc. Chem. Res.*, 1974, **7**, 265.

117. Huang, B.-S., Chello, P. L., Yip, L. and Parham, J. C., *J. Med. Chem.*, 1980, **23**, 575.

118. Migliara, O., Petruso, S. and Sprio, V., *J. Heterocycl. Chem.*, 1979, **16**, 835.

119. Girard, Y., Atkinson, J. G. and Rokach, J., *J. Chem. Soc., Perkin Trans., 1*, 1979, 1043.

120. Vinot, N. and Maitte, P., *J. Heterocycl. Chem.*, 1985, **22**, 33.

121. Jaenecke, G., Meister, L., Richter, R., Voigt, H. and Voigt, D., *Z. Chem.*, 1984, **24**, 404.

122. Honkanen, E., Pippuri, A., Kaircsalo, P., Thaler, H., Koivisto, M. and Tuomi, S., *J. Heterocycl. Chem.*, 1980, **17**, 797.

123. Antonini, I., Cristalli, G., Franchetti, P., Grifantini, M. and Martelli, S., *J. Heterocycl. Chem.*, 1980, **17**, 155.

124. Papadopoulos, E. P., *J. Heterocycl. Chem.*, 1981, **18**, 515.

125. Werbel, L. M., Johnson, J., Elslager, E. F. and Worth, D. F., *J. Med. Chem.*, 1978, **21**, 337.

126. Daboun, H. A., Abdou, S. E. and Khader, M. M., *Heterocycles*, 1982, **19**, 1925.

127. Elslager, E. F., Jacob, P., Johnson, J. and Werbel, L. M., *J. Heterocycl. Chem.*, 1980, **17**, 129.

128. Elslager, E. F., Jacob, P., Johnson, J., Werbel, L. M., Worth, D. F. and Rane, L., *J. Med. Chem.*, 1978, **21**, 1059.

129. Rosowsky, A., Chen, K. K. N. and Lin, M., *J. Med. Chem.*, 1973, **16**, 191.

130. Rosowsky, A., Chaykovsky, M., Chen, K. K. N., Lin, M. and Modest, E. J., *J. Med. Chem.*, 1973, **16**, 185.

131. Hirano, H., Takamatsu, M., Sugiyama, K. and Kurihara, T., *Chem. Pharm. Bull.*, 1979, **27**, 374.

132. Castillon, S., Melendez, E. and Vilarrasa, J., *J. Heterocycl. Chem.*, 1982, **19**, 61.

133. Novinson, T., Okabe, T., Robins, R. K. and Dea, P., *J. Heterocycl. Chem.*, 1975, **12**, 1187.

134. Novinson, T., Springer, R. H., O'Brien, D. E., Scholten, M. B., Miller, J. P. and Robins, R. K., *J. Med. Chem.*, 1982, **25**, 420.

135. Connor, D. T., Sorenson, R. J., Tinney, F. J., Cetenko, W. A. and Kerbleski, J. J., *J. Heterocycl. Chem.*, 1982, **19**, 1185.

136. Elliott, A. J., Gruzik, H. and Soler, J. R., *J. Heterocycl. Chem.*, 1982, **19**, 1437.

137. Legraverend, M., Bisagni, E. and Lhoste, J.-M. *J. Heterocycl. Chem.*, 1981, **18**, 893.

138. Mizutani, M. and Sanemitsu, Y., *J. Heterocycl. Chem.*, 1982, **19**, 1577.

139. Sutter, P. and Weis, C. D., *J. Heterocycl. Chem.*, 1982, **19**, 997.

140. Khan, M. A. and Freitas, A. C. C., *J. Heterocycl. Chem.*, 1980, **17**, 1603.

141. Kraus, G. A. and Yue, S., *J. Org. Chem.*, 1983, **48**, 2936.

142. Bradsher, C. K. and Wallis, T. G., *J. Org. Chem.*, 1978, **43**, 3817.

143. Shah, K. R. and Blanton, C. D., *J. Org. Chem.*, 1982, **47**, 502.

144. Manhas, M. S. and Amin, S. G., *J. Med. Chem.*, 1977, **14**, 161.

145. El-Kashef, H. S., Sadek, K. U. and Elnagdi, M. H., *J. Chem. Eng. Data*, 1982, **27**, 103.

146. Kreutzberger, A. and Leger, M., *Arch. Pharm. (Weinheim)*, 1982, **315**, 438.

147. Lalezari, I. and Nabahi, S., *J. Heterocycl. Chem.*, 1980, **17**, 1121.

148. Gupta, R. R., Ojha, K. G. and Kumar, M., *J. Heterocycl. Chem.*, 1980, **17**, 1325.

149. Kuroki, M., Terashi, Y. and Tsunashima, Y., *J. Heterocycl. Chem.*, 1981, **18**, 873.

150. Liso, G., Trapani, G., Latrofa, A. and Marchini, P., *J. Heterocycl. Chem.*, 1981, **18**, 279.

151. Florio, S., Leng, J. L. and Stirling, C. J. M., *J. Heterocycl. Chem.*, 1982, **19**, 237.

152. Kakehi, A., Ito, S., Uchiyama, K., Konno, Y. and Kendo, K., *J. Org. Chem.*, 1977, **42**, 443.

153. Nodiff, E. A. and Taunk, P. C. *J. Heterocycl. Chem.*, 1982, **19**, 1313.
154. Baggaley, K. H., Jennings, L. J. A. and Tyrrell, A. W. R., *J. Heterocycl. Chem.*, 1982, **19**, 1393.
155. Armarego, W. L. F. and Reece, P. A., *Aust. J. Chem.*, 1981, **34**, 1561.
156. Narasimhan, N. S. and Bhagwat, S. P., *Synthesis*, 1979, 903.
157. Narasimhan, N. S., Mali, R. S. and Barve, M. V., *Synthesis*, 1979, 906.
158. Thyagarajan, B. S. and Rajagopalan, K., *Chem. Ind. (London)*, 1965, 1931.
159. Kim, D. H., *J. Heterocycl. Chem.*, 1981, **18**, 855.
160. Gronowska, J. and Mokhtar, H. M., *Tetrahedron*, 1982, **38**, 1657.
161. Brown, D. W., Dyke, S. F., Sainsbury, M. and Hardy, G., *J. Chem. Soc. (C)*, 1971, 3219.
162. Migliara, O. and Sprio, V., *J. Heterocycl. Chem.*, 1981, **18**, 271.
163. Castedo, L., Estevez, R. J., Saa, J. M. and Suau, R., *J. Heterocycl. Chem.*, 1982, **19**, 1319.
164. Miyano, S., Fujii, S., Yamashita, O., Toraishi, N. and Sumoto, K., *J. Heterocycl. Chem.*, 1982, **19**, 1465.
165. Kim, D. H., Guinoso, C. J., Buzby, G. C., Herbst, D. R. and McCaully, R. J., *J. Med. Chem.*, 1983, **26**, 394.
166. DeFusco, A. A. and Strauss, M. J., *J. Heterocycl. Chem.*, 1981, **18**, 351.
167. Khan, M. A. and Pedrotti, F., *Monatsh. Chem.*, 1982, **113**, 123.
168. Coppola, G. M. and Hardtmann, G. E., *J. Heterocycl. Chem.*, 1981, **18**, 917.
169. Suess, R., *Helv. Chim. Acta*, 1979, **62**, 1103.
170. Kametani, T., Kigasawa, K., Hiiragi, M., Ishimaru, H., Haga, S. and Shirayama, K., *J. Heterocycl. Chem.*, 1978, **15**, 369.
171. Pifferi, G., Gaviraghi, G., Pinza, M. and Ventura, P., *J. Heterocycl. Chem.*, 1977, **14**, 1257.
172. Mills, F. D., *J. Heterocycl. Chem.*, 1980, **17**, 1597.
173. Vora, M. M. and Blanton, C. D., *J. Heterocycl. Chem.*, 1981, **18**, 507.
174. Schafer, H., Gewald, K. and Seifert, M., *J. Prakt. Chem.*, 1976, **318**, 39.
175. Srivastava, P. C. and Robins, R. K., *J. Heterocycl. Chem.*, 1979, **16**, 1063.
176. Nohara, A., Umetani, T. and Sanno, Y., *Tetrahedron*, 1974, **30**, 3553.
177. Tamura, Y., Fujita, M., Chen, L. C., Ueno, K. and Kita, Y., *J. Heterocycl. Chem.*, 1982, **19**, 289.
178. Agui, H., Mitani, T., Nakashita, M. and Nakagome, T., *J. Heterocycl. Chem.*, 1971, **8**, 357.
179. Lamartina, L., Migliara, O. and Sprio, V., *J. Heterocycl. Chem.*, 1982, **19**, 1381.
180. Lombardino, J. G., *J. Org. Chem.*, 1971, **36**, 1843.
181. Gerhardt, G. A. and Castle, R. N., *J. Heterocycl. Chem.*, 1964, **1**, 247.
182. Nakanishi, S. and Massett, S. S., *Org. Prep. Proced. Int.*, 1980, **12**, 219.
183. Novello, F. C., Bell, S. C., Abrams, E. A. L., Ziegler, C. and Sprague, J. M., *J. Org. Chem.*, 1960, **25**, 970.
184. Yoneda, F., Nakagawa, K., Koshiro, A., Fujita, T. and Harima, Y., *Chem. Pharm. Bull.*, 1982, **30**, 172.
185. Eiden, F., Muller, H. and Bachman, G., *Arch. Pharm. (Weinheim)*, 1972, **305**, 2.
186. Omar, A. M. M. E., Habib, N. S. and Aboulwafa, O. M., *Synthesis*, 1977, 864.
187. Warren, J. D., Lee, V. J. and Angier, R. B., *J. Heterocycl. Chem.*, 1979, **16**, 1617.
188. Taylor, E. C., Maryanoff, C. A. and Skotnicki, J. S., *J. Org. Chem.*, 1980, **45**, 2513.
189. Vinot, N. and Maitte, P., *J. Heterocycl. Chem.*, 1980, **17**, 855.
190. Nasielski-Hinkens, R., Benedek-Vamos, M. and Maetens, D., *J. Heterocycl. Chem.*, 1980, **17**, 873.
191. Tominaga, Y., Fujito, H., Matsuda, Y. and Kobayashi, G., *Heterocycles*, 1979, **12**, 401.

192. Tavs, P., Sieper, H. and Beecken, H., *Liebigs Ann. Chem.*, 1967, **704**, 150.
193. Komin, A. P. and Carmack, M., *J. Heterocycl. Chem.*, 1975, **12**, 829.
194. Nicolaides, D. N., Tsakalidou, E. C. and Hatziantoniou, C. T., *J. Heterocycl. Chem.*, 1982, **19**, 1243.
195. Tong, Y. C., *J. Heterocycl. Chem.*, 1981, **18**, 751.
196. Hetzheim, A. and Schneider, D., *Z. Chem.*, 1982, **22**, 219.
197. Shealy, Y. F., Clayton, J. D. and Montgomery, J. A., *J. Org. Chem.*, 1962, **27**, 2154.
198. Abdel-Fattah, A. M., Hassain, S. M. and El-Reedy, A. M., *J. Heterocycl. Chem.*, 1982, **19**, 1341.
199. Mataka, S., Takahashi, K., Imura, T. and Tashiro, M., *J. Heterocycl. Chem.*, 1982, **19**, 1481.
200. Reid, W., Oremek, G., Guryn, R. and Erle, H.-E., *Chem. Ber.*, 1980, **113**, 2818.
201. Lalezari, I. and Levy, Y., *J. Heterocycl. Chem.*, 1974, **11**, 327.
202. Press, J. B., Hofmann, C. M. and Wiegand, G. E., *J. Heterocycl. Chem.*, 1982, **19**, 391.
203. Aiello, E., Dattolo, G., Cirrincione, G., Plescia, S. and Davidone, G., *J. Heterocycl. Chem.*, 1979, **16**, 209.
204. Beljean-Leymarie, M., Pays, M. and Richer, J. C., *Can. J. Chem.*, 1983, **61**, 2563.
205. Ishikawa, F. and Watanabe, Y., *Chem. Pharm. Bull.*, 1980, **28**, 1307.
206. Tashiro, M., Fukuda, Y. and Yamato, T., *Heterocycles*, 1981, **16**, 771.
207. Matsuo, T., Tsukamoto, Y., Takagi, T. and Yaginuma, H., *Chem. Pharm. Bull.*, 1982, **30**, 1030.
208. Bonsignore, L., Cabiddu, S., Loy, G. and Secci, M., *J. Heterocycl. Chem.*, 1982, **19**, 1241.
209. Cassis, R., Tapia, R. and Valderrama, J., *J. Heterocycl. Chem.*, 1982, **19**, 381.
210. Repic, O., Mattner, P. G. and Shapiro, M. J., *J. Heterocycl. Chem.*, 1982, **19**, 1201.
211. Nixon, W. J., Garland, J. T. and Blanton, C. D., *Synthesis*, 1980, 56.
212. Kawahara, N., Nakajima, T. and Ogura, H., *Heterocycles*, 1981, **16**, 729.
213. Buckle, D. R. and Rockell, C. J. M., *J. Chem. Soc., Perkin Trans.*, 1, 1985, 2443.
214. Coppola, G. M., Hardtmann, G. and Pfister, O. R., *J. Heterocycl. Chem.*, 1982, **19**, 717.
215. Albert, A. and Trotter, A. M., *J. Chem. Soc., Perkin Trans.*, 1, 1979, 922.
216. Albert, A., *J. Chem. Soc., Perkin Trans.*, 1, 1979, 1574.
217. Alkhader, M. A., Perera, R. C., Sinha, R. P. and Smalley, R. K., *J. Chem. Soc., Perkin Trans.*, 1, 1979, 1056.
218. Connor, D. T., Young, P. A. and Von Strandtmann, M., *J. Heterocycl. Chem.*, 1981, **18**, 697.
219. Abou-Teim, O., Hacker, N. P., Jansen, R. B., McOmie, J. F. W. and Perry, D. H., *J. Chem. Soc., Perkin Trans.*, 1, 1981, 988.
220. Sundberg, R. J. and Ellis, J. E., *J. Heterocycl. Chem.*, 1982, **19**, 585.
221. Albert A., *J. Chem. Soc., Perkin Trans.*, 1, 1976, 291.
222. Ege, G. and Gilbert, K., *J. Heterocycl. Chem.*, 1981, **18**, 675.
223. Grunert, C. and Wiechert, K., *Z. Chem.*, 1970, **10**, 396.
224. Womack, C. H., Martin, L. M., Martin, G. E. and Smith, K., *J. Heterocycl. Chem.*, 1982, **19**, 1447.
225. Gelin, S., Chantegrel, B. and Deshayes, C., *J. Heterocycl. Chem.*, 1982, **19**, 989.
226. Ibrahim, Y. A., Badawy, M. A. and El-Bahaie, S., *J. Heterocycl. Chem.*, 1982, **19**, 699.
227. Murphy, B. P., *J. Org. Chem.*, 1985, **50**, 5873.
228. Lowe, W. and Berthold, G., *Arch. Pharm. (Weinheim)*, 1982, **315**, 892.
229. Cecchetti, V., Fravolini, A. and Schiaffella, F., *J. Heterocycl. Chem.*, 1982, **19**, 1045.
230. Lemke, T. L. and Sawhney, K. N., *J. Heterocycl. Chem.*, 1982, **19**, 1335.
231. Matsuo, T., Tsukamoto, Y., Takagi, T. and Sato, M., *Chem. Pharm. Bull.*, 1982, **30**, 832.
232. Yamaguchi, H. and Ishikawa, F., *Chem. Pharm. Bull.*, 1982, **30**, 28.

233. Sugasawa, T., Adachi, M., Toyoda, T. and Sasakura, K., *J. Heterocycl. Chem.*, 1979, **16**, 445.
234. Ashby, J. and Ramage, E. M., *J. Heterocycl. Chem.*, 1979, **16**, 189.
235. Shafiee, A. and Sattari, S., *J. Heterocycl. Chem.*, 1982, **19**, 227.
236. Coppola, G. M. and Shapiro, M. J., *J. Heterocycl. Chem.*, 1981, **18**, 495.
237. Ishii, H., Murakami, Y., Watanabe, T., Iwazeki, A., Suzuki, H., Masaka, T. and Mizuma, Y., *J. Chem. Res (S)*, 1984, 326.
238. Caldwell, S. R., Martin, G. E., Simonsen, S. H., Inners, R. R. and Willcott, M. R., *J. Heterocycl. Chem.*, 1981, **18**, 479.
239. Kurihara, T., Tani, T., Maeyama, S. and Sakamoto, Y., *J. Heterocycl. Chem.*, 1980, **17**, 947.
240. Plescia, S., Agozzino, P. and Fabra, I., *J. Heterocycl. Chem.*, 1977, **14**, 1431.
241. Uno, H. and Kurokawa, M., *Chem. Pharm. Bull.*, 1982, **30**, 333.
242. Okumura, K., Adachi, T., Tomie, M., Kondo, K. and Inoue, I., *J. Chem. Soc., Perkin Trans., 1*, 1972, 173.
243. Pfister, J. R., *J. Heterocycl. Chem.*, 1982, **19**, 1255.
244. Kocevar, M., Stanovnik, B. and Tisler, M., *J. Heterocycl. Chem.*, 1982, **19**, 1397.
245. Eiden, F. and Rademacher, G., *Arch. Pharm. (Weinheim)*, 1983, **316**, 34.
246. Clark-Lewis, J. W. and Thompson, M. J., *J. Chem. Soc.*, 1957, 430.
247. Puig-Torres, S., Martin, G. E., Ford, J. J., Willcott, M. R. and Smith, K., *J. Heterocycl. Chem.*, 1982, **19**, 1441.
248. Puig-Torres, S., Womack, C. H., Martin, G. E. and Smith, K. *J. Heterocycl. Chem.*, 1982, **19**, 1561.
249. Naito, T., Miyata, O., Ninomiya, I. and Pakrashi, S. C., *Heterocycles*, 1981, **16**, 725.
250. Wadodkar, K. N. and Doifode, K. B., *Indian J. Chem.*, 1979, **18B**, 458.
251. Ozaki, S., *Chem. Rev.*, 1972, **72**, 457.
252. Hirota, K., Yamada, Y., Asao, T. and Senda, S., *J. Chem. Soc., Perkin Trans., 1*, 1982, 277.
253. Daisley, R. W., Elagbar, Z. A. and Walker, J., *J. Heterocycl. Chem.*, 1982, **19**, 1013.
254. Sarkis, G. Y. and Faisal, E. D., *J. Heterocycl. Chem.*, 1985, **22**, 725.
255. Lofthouse, G. J., Suschitzky, H., Wakefield, B. J., Whittaker, R. and Tuck, B., *J. Chem. Soc., Perkin Trans., 1*, 1979, 1634.
256. Freter, K. and Fuchs, V., *J. Heterocycl. Chem.*, 1982, **19**, 377.
257. Machin, J. and Smith, D. M., *J. Chem. Soc., Perkin Trans., 1*, 1979, 1371.
258. Henn, L., Hickey, D. M. B., Moody, C. J. and Rees, C. W., *J. Chem. Soc., Perkin Trans., 1*, 1984, 2189.
259. Kauffman, J. M. and Taraporewala, I. B., *J. Heterocycl. Chem.*, 1982, **19**, 1557.
260. Singh, B. and Lesher, G. Y., *J. Heterocycl. Chem.*, 1982, **19**, 1581.
261. Abe, Y., Ohsawa, A. and Igeta, H., *Chem. Pharm. Bull.*, 1982, **30**, 881.
262. Bondinell, W. E., Chapin, F. W., Girard, G. R., Kaiser, C., Krog, A. J., Pavloff, A. M., Schwartz, M. S., Silvestri, J. S. and Vaidya, P. D., *J. Med. Chem.*, 1980, **23**, 506.
263. Yamada, K., Ikezaki, M., Umino, N., Ohtsuka, H., Itoh, N., Ikezawa, K., Kiyomoto, A. and Iwakuma, T., *Chem. Pharm. Bull.*, 1981, **29**, 744.
264. Lenz, G. R. and Woo, C. M., *J. Heterocycl. Chem.*, 1981, **18**, 691.
265. Laduree, D. and Robba, M., *Chem. Pharm. Bull.*, 1982, **30**, 789.
266. Glennon, R. A. and Liebowitz, S. M., *J. Med. Chem.*, 1982, **25**, 393.
267. Beck, J. R. and Suhr, R. G., *J. Heterocycl. Chem.*, 1974, **11**, 227.
268. Harris, T. W., Smith, H. E., Mobley, P. L., Manier, D. H. and Sulser, F., *J. Med. Chem.*, 1982, **25**, 855.
269. Lantos, I., Oh, H., Razgaitis, C. and Loev, B., *J. Org. Chem.*, 1978, **43**, 4841.
270. Shafiee, A. and Kiaeay, G., *J. Heterocycl. Chem.*, 1981, **18**, 899.
271. Scheiner, P., Arwin, S., Eliacin, M. and Tu, J., *J. Heterocycl. Chem.*, 1985, **22**, 1435.
272. Carroll, F. I., Berrang, B. D. and Linn, C. P., *J. Heterocycl. Chem.*, 1981, **18**, 941.

273. Ishii, H., Murakami, Y., Hosoya, K., Takeda, H., Suzuki, Y. and Ikeda, N., *Chem. Pharm. Bull.*, 1973, **21**, 1481.
274. Senga, K., Ichiba, M., Kanazawa, H. and Nishigaki, S., *J. Heterocycl. Chem.*, 1982, **19**, 77.
275. Sarkis, G. Y. and Al-Badriz, H. T., *J. Heterocycl. Chem.*, 1980, **17**, 813.
276. Lancelot, J.-C., Maume, D. and Robba, M., *J. Heterocycl. Chem.*, 1980, **17**, 631.
277. Maeba, I. and Castle, R. N., *J. Heterocycl. Chem.*, 1980, **17**, 407.
278. Mitchell, J. A. and Reid, D. H., *J. Chem. Soc., Perkin Trans., 1*, 1982, 499.
279. Cue, B. W., Czuba, L. J. and Dirlam, J. P., *J. Org. Chem.*, 1978, **43**, 4125.
280. Yevich, J. P., Temple, D. L., Covington, R. R., Owens, D. A., Seidehamel, R. J. and Dungan, K. W., *J. Med. Chem.*, 1982, **25**, 864.
281. Ferrarini, P. L., Mori, C., Livi, O., Biagi, G. and Marini, A. M., *J. Heterocycl. Chem.*, 1983, **20**, 1063.
282. Kraus, G. A. and Pezzanite, J. O., *J. Org. Chem.*, 1979, **44**, 2480.
283. Sato, M., Kanuma, N. and Kato, T., *Chem. Pharm. Bull.*, 1982, **30**, 4359.
284. Beck, J. R., *J. Org. Chem.*, 1972, **37**, 3224.
285. Glazer, E. A. and Chappel, L. R., *J. Med. Chem.*, 1982, **25**, 766.
286. Papadopoulos, E. P. *J. Heterocycl. Chem.*, 1980, **17**, 1553.
287. Lacroix, A. and Fleury, J. P., *Tetrahedron Lett.*, 1978, 3469.
288. Houlihan, W. J., Cooke, G., Denzer, M. and Nicoletti, J., *J. Heterocycl. Chem.*, 1982, **19**, 1453.
289. Puig-Torres, S., Martin, G. E., Smith, K., Cacioli, P. and Reiss, J. A., *J. Heterocycl. Chem.*, 1982, **19**, 879.
290. Rivalle, C., Ducrocq, C. and Bisagni, E., *J. Chem. Soc., Perkin Trans., 1*, 1979, 138.
291. Merchan, F. L., Garin, J. and Tejero, T., *Synthesis*, 1982, 984.
292. Kiffer, D., *Bull. Soc. Chim. Fr.*, 1970, 2377.
293. Ladd, D. L., *J. Heterocycl. Chem.*, 1982, **19**, 917.
294. Lal, B., Dohadwalla, A. N., Dadkar, N. K., D'Sa, A. and De Souza, N. J., *J. Med. Chem.*, 1984, **27**, 1470.
295. Lee, C.-S., Hashimoto, Y., Shudo, K. and Okamoto, T., *Chem. Pharm. Bull.*, 1982, **30**, 1857.
296. Tsuchiya, T., Sashida, H. and Konoshita, A., *Chem. Pharm. Bull.*, 1983, **31**, 4568.
297. Hudlicky, T., Kutchan, T. M., Shen, G., Sutliff, V. E. and Coscia, C. J., *J. Org. Chem.*, 1981, **46**, 1738.
298. Khan, M. A., Cosenza, A. G. and Ellis, G. P., *J. Heterocycl. Chem.*, 1982, **19**, 1077.
299. Philipp, A., Jirkovsky, I. and Martel, R. R., *J. Med. Chem.*, 1980, **23**, 1372.
300. Pratap, R., Castle, R. N. and Lee, M. L., *J. Heterocycl. Chem.*, 1982, **19**, 439.
301. Katsuura, K., Ohta, M. and Mitsuhashi, K. *Chem. Pharm. Bull.*, 1982, **30**, 4379.
302. Nagarajan, K. and Shah, R. K., *Indian J. Chem.*, 1976, **14B**, 1.
303. Hajpal, I. and Berenyi, E., *J. Heterocycl. Chem.*, 1982, **19**, 313.
304. Merchan, F. L., *Synthesis*, 1981, 965.
305. Neidlein, R. and Krull, H., *Arch. Pharm. (Weinheim)*, 1971, **304**, 763.
306. Sakamoto, T., Kondo, Y. and Yamanaka, H., *Chem. Pharm. Bull.*, 1982, **30**, 2410.
307. Clarke, R. W., Garside, S. C., Lunts, L. H. C., Hartley, D., Hornby, R. and Oxford, A. W., *J. Chem. Soc., Perkin Trans., 1*, 1979, 1120.
308. Willette, R. E., *Adv. Heterocycl. Chem.*, 1968, **9**, 27.
309. Lamke, T. L., Sawhney, K. N. and Lemke, B. K., *J. Heterocycl. Chem.*, 1982, **19**, 363.
310. Yale, H. L. and Spitzmiller, E. R., *J. Heterocycl. Chem.*, 1977, **14**, 241.
311. Kunitomo, J., Oshikata, M., Nakayama, K., Suwa, K. and Murakami, Y., *Chem. Pharm. Bull.*, 1982, **30**, 4283.
312. Tanaka, K., Shimazaki, M. and Murakami, Y., *Chem. Pharm. Bull.*, 1982, **30**, 2714.
313. Tamura, Y., Chen, L. C., Fujita, M. and Kita, Y., *Chem. Pharm. Bull.*, 1982, **30**, 1257.
314. Kuroda, K., Nagamatus, T., Sakuma, Y. and Yoneda, F., *J. Heterocycl. Chem.*, 1982, **19**, 929.

315. Yoneda, F. and Sakuma, Y., *J. Heterocycl. Chem.*, 1973, **10**, 993.
316. Yoneda, F., Hirayama, R. and Yamashita, M., *J. Heterocycl. Chem.*, 1982, **19**, 301.
317. Sakamoto, T., Kondo, Y. and Yamanaka, H., *Chem. Pharm. Bull.*, 1982, **30**, 2417.
318. Eweiss, N. P., *Chem. Pharm. Bull.*, 1983, **19**, 273.
319. Snyder, C. A., Thorn, M. A., Klijanowicz, J. E. and Southwick, P. L., *J. Heterocycl. Chem.*, 1982, **19**, 603.
320. Ibrahim, Y. A., Abdel-Hady, S. A. L., Badawy, M. A. and Ghazala, M. A. H., *J. Heterocycl. Chem.*, 1982, **19**, 913.
321. Colburn, V. M., Iddon, B., Suschitzky, H. and Gallagher, P. T., *J. Chem. Soc., Perkin Trans.*, *1*, 1979, 1337.
322. Ishiguro, T., Ukawa, K., Sugihara, H. and Nohara, A., *Heterocycles*, 1981, **16**, 733.
323. Yoneda, F., Koga, R., Nishigaki, S. and Fukazawa, S., *J. Heterocycl. Chem.*, 1982, **19**, 949.
324. Szabo, W. A., *Aldrichim. Acta*, 1977, **10**, 23.
325. Brogden, P. J. and Hepworth, *J. Chem. Soc., Perkin Trans. 1*, 1983, 827.
326. Bartsch, H. and Schwarz, O., *J. Heterocycl. Chem.*, 1982, **19**, 1189.
327. Ivanov, Kh. and Chorbadzhiev, S., *Izv. Khim.*, 1982, **15**, 242.
328. Mason, J. C. and Tennant, G., *J. Chem. Soc. (B)*, 1970, 911.
329. Gorvin, J. H. and Whalley, D. P., *J. Chem. Soc., Perkin Trans.*, *1*, 1979, 1364.
330. Shimizu, M., Ishikawa, M., Komoda, Y., Matsubara, Y. and Nakajima, T., *Chem. Pharm. Bull.*, 1982, **30**, 4529.
331. Repke, D. B. and Ferguson, W. J., *J. Heterocycl. Chem.*, 1982, **19**, 845.
332. Lancelot, J.-C., Gazengel, J.-M., Rault, S. and Robba, M., *Chem. Pharm. Bull.*, 1982, **30**, 1674.
333. Ishi, H., Koyama, K., Chen, I. S. and Ishikawa, T., *Chem. Pharm. Bull.*, 1982, **30**, 1992.
334. Hisano, T., Ichikawa, M., Tsumoto, K. and Tasaki, M., *Chem. Pharm. Bull.*, 1982, **30**, 2996.
335. Hirota, K., Maruhashi, K., Asao, T. and Senda, S., *Chem. Pharm. Bull.*, 1982, **30**, 3377.
336. Kokosi, J., Hermecz, I., Szasz, G., Meszaros, Z., Toth, G. and Csakvari-Pongor, M., *J. Heterocycl. Chem.*, 1982, **19**, 909.
337. Nair, M. D., Sudarsanam, V. and Desai, J. A., *Indian J. Chem.*, 1982, **21B**, 1030.
338. Saari, W. S., Halczenko, W., Freedman, M. B. and Arison, W., *J. Heterocycl. Chem.*, 1982, **19**, 837.
339. Smith, J. R. L., Norman, R. O. C., Rose, M. E. and Curran, A. W. C., *J. Chem. Soc., Perkin Trans.*, *1*, 1979, 1185.
340. Effland, R. E. and Helsley, G. C., *J. Heterocycl. Chem.*, 1982, **19**, 537.
341. Ghosh, C. K. and Mukhopadhyay, K. K., *Synthesis*, 1978, 779.
342. Bosch, J., Bonjoch, J. and Serret, I., *J. Heterocycl. Chem.*, 1982, **19**, 489.
343. Ohnuma, T., Kasuya, H., Kimura, Y. and Ban, Y., *Heterocycles*, 1982, **17**, 377.
344. Kanao, M., Hashizuma, T., Ichikawa, Y., Irie, K., Satoh, Y. and Isoda, S., *J. Heterocycl. Chem.*, 1982, **19**, 180.
345. Shridhar, D. R., Sarma, C. R., Krishna, R. R., Prasad, R. S. and Sachdeva, Y. P., *Org. Prep. Proced. Int.*, 1978, **10**, 163.
346. Daidone, G. and Plescia, S., *J. Heterocycl. Chem.*, 1982, **19**, 689.
347. Fajgelj, S., Stanovnik, B. and Tisler, M., *Heterocycles*, 1986, **24**, 379.
348. Acheson, R. M., Prince, R. J. and Proctor, G., *J. Chem. Soc., Perkin Trans.*, *1*, 1979, 595.
349. Hirose, T., Mishio, S., Matsumoto, J. and Minami, S., *Chem. Pharm. Bull.*, 1982, **30**, 2399.
350. Coburn, R. A. and Taylor, M. D., *J. Heterocycl. Chem.*, 1982, **19**, 567.
351. Chantegrel, B., Nadi, A. I. and Gelin, S., *Synthesis*, 1983, 214.

352. Coburn, R. A. and Gala, D., *J. Heterocycl. Chem.*, 1982, **19**, 757.
353. Clausen, K. and Lawesson, S. O., *Bull. Soc. Chim. Belg.*, 1979, **88**, 305.
354. Barker, G. and Ellis, G. P., *J. Chem. Soc. (C)*, 1971, 1482.
355. Casy, A. F., Needle, R. J. and Upton, C., *J. Chem. Res. (S)*, 1986, 4.
356. Morris, L. R. and Collins, L. R., *J. Heterocycl. Chem.*, 1975, **12**, 309.
357. Devaux, G., Renandie, C., Boineau, F., Mesnard, P. and Demarquez, N., *Eur. J. Med. Chem.*, 1974, **9**, 44.
358. Ellis, G. P. and Jones, R. T., *J. Chem. Soc., Perkin Trans.*, *1*, 1974, 903.
359. Mosti, L., Schenone, P., Menozzi, G., Romussi, G. and Baccichetti, F., *J. Heterocycl. Chem.*, 1982, **19**, 1227.
360. Barker, G., Ellis, G. P. and Wilson, D. A., *J. Chem. Soc. (C)*, 1971, 2079.
361. Elliott, A. J. and Gibson, M. S., *J. Org. Chem.*, 1980, **45**, 3677.
362. Taylor, E. C. and Martin, S. F., *J. Org. Chem.*, 1972, **37**, 3958.
363. Khan, M. A., Pagotto, M. C. and Ellis, G. P., *Heterocycles*, 1977, **6**, 983.
364. Becket, G. J. P., Ellis, G. P. and Trindade, M. I. U., *J. Chem. Res. (S)*, 1978, 47.
365. Wobig, D., *Liebigs Ann. Chem.*, 1984, 1994.
366. Rene, L., Risse, S., Demerseman, P. and Royer, R., *Eur. J. Med. Chem.*, 1979, **14**, 281.
367. Ryabukhin, Y. I., Karpenko, V. D., Mezheritskii, V. V. and Dorofeenko, G. N., *Chem. Heterocycl. Compds*, 1975, **11**, 1029.
368. Geffken, D., *Liebigs Ann. Chem.*, 1981, 1513.
369. Butler, D. E. and Alexander, S. M., *J. Heterocycl. Chem.*, 1982, **19**, 1173.
370. Marxer, A. and Siegrist, M., *Helv. Chim. Acta*, 1979, **62**, 1753.
371. Kurihara, T., Nasu, K., Ishimori, F. and Tani, T., *J. Heterocycl. Chem.*, 1981, **18**, 163.
372. Kurihara, T., Nasu, K., Byakuno, J. and Tani, T., *Chem. Pharm. Bull.*, 1982, **30**, 1289.
373. Yoneda, F., Motokura, M., Kamishimoto, M., Nagamatsu, T., Otagiri, M., Uekama, K. and Takamoto, M., *Chem. Pharm. Bull.*, 1982, **30**, 3187.
374. Schulze, J., Tanneberg, H. and Matschiner, H., *Z. Chem.*, 1980, **20**, 436.
375. Manhas, M. S. and Amin, S. G., *J. Heterocycl. Chem.*, 1977, **14**, 161.
376. Petersen, U. and Heitzer, H., *Liebigs Ann. Chem.*, 1976, 1659.
377. Ardakani, M. A. and Smalley, R. K., *Tetrahedron Lett.*, 1979, 4765.
378. Ardakani, M. A. and Smalley, R. K., *Tetrahedron Lett.*, 1979, 4769.
379. Hickey, D. M. B., MacKenzie, A. R., Moody, C. J. and Rees, C. W., *J. Chem. Soc., Chem. Commun.*, 1984, 776.
380. Paul, R. and Menschik, J., *J. Heterocycl. Chem.*, 1979, **16**, 277.
381. Joshi, K. C., Pathak, V. N. and Garg, U., *J. Heterocycl. Chem.*, 1980, **17**, 789.
382. Maeba, I., Mori, K. and Castle, R. N., *J. Heterocycl. Chem.*, 1979, **16**, 1559.
383. Taylor, E. C. and Loux, H. M., *J. Am. Chem. Soc.*, 1959, **81**, 2474.
384. Nasielski-Hinkens, R., Maetens, D. and Vandendijk, D., *Bull. Soc. Chim. Belg.*, 1979, **88**, 169.
385. Trust, R. I., Albright, J. D., Lovell, F. M. and Parkinson, N. A., *J. Heterocycl. Chem.*, 1979, **16**, 1393.
386. Friary, R. and Sunday, B. R., *J. Heterocycl. Chem.*, 1979, **16**, 1277.
387. Lancelot, J.-C., Maume, D. and Robba, M., *J. Heterocycl. Chem.*, 1981, **18**, 1319.
388. Looker, J. H., McMechan, J. H. and Mader, J. W., *J. Org. Chem.*, 1978, **43**, 2344.
389. Ogura, H., Mineo, S. and Nakagawa, K., *Chem. Pharm. Bull.*, 1981, **29**, 1518.
390. Eloy, F. and Deryckere, A. *Chim. Ther.*, 1971, **6**, 48.
391. Bandurco, V. T., Wong, E. M., Levine, S. D. and Hajos, Z. G., *J. Med. Chem.*, 1981, **24**, 1455.
392. Schafer, H., Gewald, K. and Hartmann, H., *J. Prakt. Chem.*, 1974, **316**, 169.
393. Berti, C., Ettling, B. V., Greci, L. and Marchetti, L., *J. Heterocycl. Chem.*, 1979, **16**, 17.
394. Lalezari, I. and Sadeghi-Milani, S., *J. Heterocycl. Chem.*, 1979, **16**, 707.
395. Liu, Y.-L., Thom, E. and Liebman, A. A., *J. Heterocycl. Chem.*, 1979, **16**, 799.
396. Lalezari, I., *J. Heterocycl. Chem.*, 1979, **16**, 603.

397. Galun, A., Markus, A. and Kampf, A., *J. Heterocycl. Chem.*, 1979, **16**, 221.
398. Nagahara, K., Tagaki, K. and Ueda, T., *Chem. Pharm. Bull.*, 1976, **24**, 1310.
399. Ozaki, K. Y., Yamada, Y. and Oine, T., *J. Org. Chem.*, 1981, **46**, 1571.
400. Unangst, P. C., Brown, R. E., Fabian, A. and Fontsere, F., *J. Heterocycl. Chem.*, 1979, **16**, 661.
401. Maguire, J. H. and McKee, R. L., *J. Heterocycl. Chem.*, 1979, **16**, 133.
402. Elnagdi, M. H., Fahmy, S. M., Elmoghayar, M. R. H. and Kandeel, E. M., *J. Heterocycl. Chem.*, 1979, **16**, 61.
403. Cohen, V. I., *J. Heterocycl. Chem.*, 1979, **16**, 13.
404. Taylor, E. C. and Dumas, D. J., *J. Org. Chem.*, 1981, **46**, 1394.
405. Yan, S.-J., Weinstock, L. T. and Cheng, C.-C., *J. Heterocycl. Chem.*, 1979, **16**, 541.
406. Deady, L. W. and Stanborough, M. S., *J. Heterocycl. Chem.*, 1979, **16**, 187.
407. Cheeseman, G. W. H. and Greenberg, S. G., *J. Heterocycl. Chem.*, 1979, **16**, 241.
408. Bernier, J.-L. and Henichart, J.-P., *J. Heterocycl. Chem.*, 1979, **16**, 717.
409. Bernath, G., Fulop, F., Hermecz, I., Meszaros, Z. and Toth, G., *J. Heterocycl. Chem.*, 1979, **16**, 137.
410. Fulop, F., Hermecz, I., Meszaros, Z., Dombi, G. and Bernath, G., *J. Heterocycl. Chem.*, 1979, **16**, 457.
411. Le Corre, M., Hercouet, A. and Le Baron, H., *J. Chem. Soc., Chem. Commun.*, 1981, 14.
412. Medvedeva, M. M., Pozharskii, A. F., Kuz'menko, V. V., Bessonov, V. V. and Tertov, B. A., *Chem. Heterocycl. Compds*, 1979, **15**, 166.
413. Tanaka, A., Yakushijin, K. and Yoshina, S., *J. Heterocycl. Chem.*, 1979, **16**, 785.
414. Iwaso, M. and Kuraishi, T., *J. Heterocycl. Chem.*, 1979, **16**, 689.
415. Sharanin, Yu. A., Lopatinskaya, K. Y., Sharanina, L. G. and Baranov, S. H., *Chem. Heterocycl. Compds*, 1974, **10**, 1125.
416. Maeba, I., Ando, M., Yoshina, S. and Castle, R. N., *J. Heterocycl. Chem.*, 1979, **16**, 245.
417. Fravolini, A., Schiaffella, F. and Strappaghetti G., *J. Heterocycl. Chem.*, 1979, **16**, 29.
418. Maeba, I. and Castle, R. N., *J. Heterocycl. Chem.*, 1979, **16**, 249.
419. Migliara, O. and Petruso, S., *J. Heterocycl. Chem.*, 1979, **16**, 203.
420. Mokrushina, G. A., Kotovskaya, S. K. and Postovskii, I. Ya., *Chem. Heterocycl. Compds*, 1979, **15**, 118.
421. Chen, S.-F. and Panzica, R. P., *J. Org. Chem.*, 1981, **46**, 2467.
422. Obase, H., Takai, H., Teranishi, M. and Nakamizo, N., *J. Heterocycl. Chem.*, 1983, **20**, 565.
423. Wright, G. E. and Gambino, J., *J. Heterocycl. Chem.*, 1979, **16**, 401.
424. Bradsher, C. K. and Reames, D. C., *J. Org. Chem.*, 1981, **46**, 1384.
425. Nasr, M. and Burckhalter, J. H., *J. Heterocycl. Chem.*, 1979, **16**, 497.
426. Yamazaki, T., Takahata, H., Hama, Y., Takano, Y., Nagata, M. and Castle, R. N., *J. Heterocycl. Chem.*, 1979, **16**, 525.
427. Sunder, S. and Peet, N. P., *J. Heterocycl. Chem.*, 1979, **16**, 33.
428. Hollins, R. A. and Martins, P. R. C., *J. Heterocycl. Chem.*, 1979, **16**, 681.
429. Yamazaki, T., Matoba, K., Imai, T. and Castle, R. N., *J. Heterocycl. Chem.*, 1979, **16**, 517.
430. Oliver, J. E., *J. Heterocycl. Chem.*, 1985, **22**, 1165.
431. Kawase, Y., Yamaguchi, S., Maeda, O., Hayashi, A., Hayashi, I., Tabata, K. and Kondo, M., *J. Heterocycl. Chem.*, 1979, **16**, 487.
432. Miyano, S., Fuji, S., Yamashita, O., Toraishi, N. and Sumoto, K., *J. Org. Chem.*, 1981, **46**, 1737.
433. Lempert-Sreter, M., Lempert, K. and Moller, J., *J. Chem. Soc., Perkin Trans.*, 1, 1983, 2011.
434. Yamaguta, K., Tomioka, Y., Yamazaki, M. and Noda, K., *Chem. Pharm. Bull.*, 1983, **31**, 401.

435. Kuhlmann, K. F. and Mosher, C. W., *J. Med. Chem.*, 1981, **24**, 1333.
436. Dunn, J. P., Muchowski, J. M. and Nelson, P. H., *J. Med. Chem.*, 1981, **24**, 1097.
437. Bennett, L. R., Blankley, C. J., Fleming, R. W., Smith, R. D. and Tessman, D. K., *J. Med. Chem.*, 1981, **24**, 382.
438. Gandasegui, M. T. and Alvarez-Builla, J., *J. Chem. Res. (S)*, 1986, 74.
439. Ghosh, C. K., Bandyopadhyay, C. and Morin, C., *J. Chem. Soc., Perkin Trans., 1*, 1983, 1989.
440. Orlov, V. D., Roberman, A. I. and Lavrushin, V. F., *Chem. Heterocycl. Compds*, 1979, **15**, 117.
441. Prokopov, A. A. and Yakhontov, L. N., *Chem. Heterocycl. Compds*, 1979, **15**, 76.
442. Bennett, G. B., Babington, R. G., Deacon, M. A., Eden, P. L., Kerestan, S. P., Leslie, G. H., Ryan, E. A., Mason, R. B. and Minor, H. E., *J. Med. Chem.*, 1981, **24**, 490.
443. Matoba, K., Miyata, Y. and Yamazaki, T., *Chem. Pharm. Bull.*, 1983, **31**, 476.
444. Tinney, F. J., Cetenko, W. A., Kerbleski, J. J., Connor, D. T., Sorenson, R. J. and Herzig, D. J., *J. Med. Chem.*, 1981, **24**, 878.
445. DeWald, H. A., Lobbestael, S. and Poschel, B. P. H., *J. Med. Chem.*, 1981, **24**, 982.
446. Senga, K., Novinson, T., Wilson, H. R. and Robins, R. K., *J. Med. Chem.*, 1981, **24**, 610.
447. Bochis, R. J., Olen, L. E., Waksmunski, F. S., Moczik, H., Eskola, P., Kulsa, P., Wilks, G., Taylor, J. E., Egerton, J. R., Ostlind, D. A. and Olson G., *J. Med. Chem.*, 1981, **24**, 1518.
448. Lombardino, J. G. and Otterman, J. G., *J. Med. Chem.*, 1982, **24**, 830.
449. Zenker, N., Talaty, C. N., Galbry, P. S., Wright, J. and Hubbard, L. J., *J. Heterocycl. Chem.*, 1983, **20**, 435.
450. Bartsch, H. and Schwarz, O., *J. Heterocycl. Chem.*, 1983, **20**, 45.
451. Bartholomew, D. and Kay, I. T., *Tetrahedron Lett.*, 1979, 2827.
452. Rault, S., Effi, Y., De Sevricourt, M. C., Lancelot, J.-C. and Robba, M., *J. Heterocycl. Chem.*, 1983, **20**, 17.
453. Turck, A., Brument, J.-F. and Queguiner, G., *J. Heterocycl. Chem.*, 1983, **20**, 101.
454. Foster, R. H. and Leonard, N. J., *J. Org. Chem.*, 1979, **44**, 4609.
455. Paulmier, C. and Outurquin, F. *J. Heterocycl. Chem.*, 1983, **20**, 113.
456. Ponticello, G. S. and Baldwin, J. J., *J. Org. Chem.*, 1979, **44**, 4003.
457. Ishikawa, T., Sano, M., Isagawa, K. and Fushizaki, Y., *Bull. Chem. Soc. Jpn*, 1970, **43**, 135.
458. Delia, T. J., Kirt, D. D. and Sani, S. M., *J. Heterocycl. Chem.*, 1983, **20**, 145.
459. Huang, B.-S. and Parham, J. C., *J. Org. Chem.*, 1979, **44**, 4046.
460. Traxler, J. T., *J. Org. Chem.*, 1979, **44**, 4971.
461. Trybulski, E. J., Benjamin, L., Vitone, S., Walser, A. and Fryer, R. I., *J. Med. Chem.*, 1983, **26**, 367.
462. Yoneda, F., Higuchi, M. and Nitta, Y., *J. Heterocycl. Chem.*, 1980, **17**, 869.
463. Ohtsuka, Y., Tohma, E., Kojima, S. and Tomita, N., *J. Org. Chem.*, 1979, **44**, 4871.
464. Lap, B. V., Boux, L. J., Cheung, H. T. A. and Holder, G. M., *J. Heterocycl. Chem.*, 1983, **20**, 281.
465. Perillo, I. A., Schapira, C. B. and Lamdan, S., *J. Heterocycl. Chem.*, 1983, **20**, 155.
466. Tam, S. Y.-K., Klein, R. S., Wampen, I. and Fox, J. J., *J. Org. Chem.*, 1979, **44**, 4547.
467. Acheson, R. M., Bite, M. G. and Kemp, J. E. G., *J. Med. Chem.*, 1981, **24**, 1300.
468. Draper, R. E. and Castle, R. N., *J. Heterocycl. Chem.*, 1983, **20**, 193.
469. Heine, H. W., Ludovici, D. W., Pardoen, J. A., Weber, R. C., Bonsall, E. and Osterhout, K. R., *J. Org. Chem.*, 1979, **44**, 3843.
470. Tong, Y. C. and Kerlinger, H. O., *J. Heterocycl. Chem.*, 1983, **20**, 365.
471. Cignarella, G., Sanna, P., Miele, E., Anania, V. and Desole, M. S., *J. Med. Chem.*, 1981, **24**, 1003.
472. Heyes, G., Holt, G. and Lewis, A., *J. Chem. Soc., Perkin Trans., 1*, 1972, 2351.
473. Nore, P. and Honkanen, E., *J. Heterocycl. Chem.*, 1980, **17**, 985.

474. Madhav, R., Snyder, C. A. and Southwick, P. L., *J. Heterocycl. Chem.*, 1980, **17**, 1231.
475. Yoneda, F. and Higuchi, M., *J. Heterocycl. Chem.*, 1980, **17**, 1365.
476. Rabilloud, G. and Sillion, B., *J. Heterocycl. Chem.*, 1980, **17**, 1065.
477. Pilgram, K., *J. Heterocycl. Chem.*, 1980, **17**, 1413.
478. Veronese, A. C., Cavicchioni, G., Servadio, G. and Vecchiati, G., *J. Heterocycl. Chem.*, 1980, **17**, 1723.
479. Heindel, N. D. and Reid, J. R., *J. Heterocycl. Chem.*, 1980, **17**, 1087.
480. Chorvat, R. J. and Desai, B. N., *J. Heterocycl. Chem.*, 1980, **17**, 1313.
481. Kim, D. H., *J. Heterocycl. Chem.*, 1980, **17**, 1647.
482. Kano, S., Shibuya, S. and Yuasa, Y., *J. Heterocycl. Chem.*, 1980, **17**, 1559.
483. Middleton, R. W. and Wibberley, D. G., *J. Heterocycl. Chem.*, 1980, **17**, 1757.
484. Joshi, K. C. and Chand, P., *J. Heterocycl. Chem.*, 1980, **17**, 1783.
485. Bobowski, G., *J. Heterocycl. Chem.*, 1973, **20**, 267.
486. Batu, G. and Stevenson, R., *J. Org. Chem.*, 1979, **44**, 3948.
487. Scannell, R. T. and Stevenson, R., *J. Heterocycl. Chem.*, 1980, **17**, 1727.
488. Ghera, E., Ben-David, Y. and Rapoport, H., *J. Org. Chem.*, 1983, **48**, 774.
489. Walker, K. A., Boots, M. R., Stubbings, J. F., Rogers, M. E. and Davis, C. W., *J. Med. Chem.*, 1983, **26**, 174.
490. Schmidt, D. G., Seemuth, P. D. and Zimmer, H., *J. Org. Chem.*, 1983, **48**, 1914.
491. Temple, C., Wheeler, G. P., Elliott, R. D., Rose, J. D., Comber, R. N. and Montgomery, J. A., *J. Med. Chem.*, 1983, **26**, 91.
492. Benjamin, L. E., Fryer, R. I., Gilman, N. W. and Trybulski, E. J., *J. Med. Chem.*, 1983, **26**, 100.
493. Press, J. B., Hofmann, C. M. and Safir, S. R., *J. Heterocycl. Chem.*, 1980, **17**, 1361.
494. Al-Sammerrai, D. A.-J., Ralph, J. T. and West, D. E., *J. Heterocycl. Chem.*, 1980, **17**, 1705.
495. Chavdarian, C. G., Seeman, J. I. and Wooton, J. B., *J. Org. Chem.*, 1983, **48**, 492.
496. Zvilichovsky, G. and David, M., *J. Org. Chem.*, 1983, **48**, 575.
497. Buckle, D. R., Outred, D. J., Rockell, C. J. M., Smith, H. and Spicer, B. A., *J. Med. Chem.*, 1983, **26**, 251.
498. Le Bris, M.-T., *J. Heterocycl. Chem.*, 1985, **22**, 1275.
499. Coppola, G. M. and Damon, R. E., *J. Heterocycl. Chem.*, 1980, **17**, 1729.
500. Khan, M. A., Rolim, A. M. C. and Guarconi, A. E., *J. Heterocycl. Chem.*, 1983, **20**, 475.
501. Fravolini, A., Schiaffella, F., Brunelli, C. and Cecchetti, V., *J. Heterocycl. Chem.*, 1980, **17**, 125.
502. Moron, J., Nguyen, C. H. and Bisagni, E., *J. Chem. Soc., Perkin Trans.*, **1**, 1983, 225.
503. Taylor, E. C. and Wachsen, E., *J. Org. Chem.*, 1978, **43**, 4154.
504. Sugiyama, Y., Sasaki, T. and Nagato, N., *J. Org. Chem.*, 1978, **43**, 4485.
505. Von Angerer, E. and Prekajac, J., *J. Med. Chem.*, 1983, **26**, 113.
506. Okafor, C. O. and Castle, R. N., *J. Heterocycl. Chem.*, 1983, **20**, 199.
507. Zvilichovsky, G. and David, M., *J. Chem. Soc., Perkin Trans.*, 1, 1983, 11.
508. Prisbe, E. J., Verheyden, J. P. H. and Moffatt, J. G., *J. Org. Chem.*, 1978, **43**, 4784.
509. Fryer, R. I., Blount, J., Reeder, E., Trybulski, E. J. and Walser, A., *J. Org. Chem.*, 1978, **43**, 4480.
510. Kokel, B., Guillaumel, J. and Royer R., *J. Heterocycl. Chem.*, 1983, **20**, 575.
511. Unangst, P. C., *J. Heterocycl. Chem.*, 1983, **20**, 495.
512. Cheeseman, G. W. H. and Hawi, A. A., *J. Heterocycl. Chem.*, 1983, **20**, 585.
513. Lancelot, J.-C., Maume, D. and Robba, M., *J. Heterocycl. Chem.*, 1982, **19**, 817.
514. Cheeseman, G. W. H. and Hawi, A. A., *J. Heterocycl. Chem.*, 1983, **20**, 591.
515. Molock, F. F. and Boykin, D. W., *J. Heterocycl. Chem.*, 1983, **20**, 681.
516. Merchant, J. R., Martyres, G. and Koshti, N. M., *J. Heterocycl. Chem.*, 1983, **20**, 775.

517. Al-Jobour, N. H. and Shandala, M. Y., *J. Heterocycl. Chem.*, 1980, **17**, 941.
518. Matoba, K., Fukushima, A., Takahata, H., Hirai, J. and Yamazaki, T., *Chem. Pharm. Bull.*, 1982, **30**, 1300.
519. De Settimo, A., Biagi, G., Primofiore, G., Ferrarini, P. L., Livi, O. and Marini, A. M., *J. Heterocycl. Chem.*, 1980, **17**, 1225.
520. Akiba, M., Kosugi, Y. and Takada, T., *J. Org. Chem.*, 1978, **43**, 4472.
521. Menozzi, G., Mosti, L., Schenone, P. and Cafaggi, S., *J. Heterocycl. Chem.*, 1982, **19**, 937.
522. Baldwin, J. J., Mensler, K. and Poticello, G. S., *J. Org. Chem.*, 1978, **43**, 4878.
523. Simay, A. and Takacs, K., *J. Heterocycl. Chem.*, 1982, **19**, 809.
524. Mosti, L., Schenone, P., Menozzi, G. and Romussi, G., *J. Heterocycl. Chem.*, 1982, **19**, 1057.
525. Taylor, E. C. and McKillop, A., *Adv. Org. Chem.*, 1970, **7**, 226.
526. Markovac, A., Wu, G. S., LaMontagne, M. P., Blumbergs, P. and Ao, M. S., *J. Heterocycl. Chem.*, 1982, **19**, 829.
527. Abdelhamid, A. O., Hassaneen, H. M., Shawali, A. S. and Parkanyi, C., *J. Heterocycl. Chem.*, 1983, **20**, 639.
528. Elfahham, H. A. E., Abdel-Galil, F. M., Ibraheim, Y. R. and Elnagdi, M. H., *J. Heterocycl. Chem.*, 1983, **20**, 667.
529. Okabe, T., Bhooshan, B., Novinson, T., Hillyard, I. W., Garner, G. E. and Robins, R. K., *J. Heterocycl. Chem.*, 1983, **20**, 735.
530. Kunstlinger, M. and Breitmaier, E., *Synthesis*, 1983, 161.
531. Plattner, J. J. and Parks, J. A., *J. Heterocycl. Chem.*, 1983, **20**, 1059.
532. Bisagni, E., Ducrocq, C. and Civier, A., *Tetrahedron*, 1976, **32**, 1383.
533. Padwa, A. and Nahm, S., *J. Org. Chem.*, 1981, **46**, 1402.
534. Sainsbury, M., Weerasinghe, D. and Dolman, D., *J. Chem. Soc., Perkin Trans., 1*, 1982, 587.
535. Patton, J. R. and Dudley, K. H., *J. Heterocycl. Chem.*, 1979, **16**, 257.
536. Borisova, L. N. and Kartashova, T. A., *Chem. Heterocycl. Compds*, 1979, **15**, 162.
537. Ohtsuka Y., *Bull. Chem. Soc. Jpn*, 1970, **43**, 187.
538. Elgemeie, G. E. H. and Aal, F. A. E. M. E., *Heterocycles*, 1986, **24**, 349.
539. Hara, T., Kayama, Y. and Sunami, T., *J. Org. Chem.*, 1978, **43**, 4865.
540. Sinhababu, A. K. and Borchardt, R. T., *J. Org. Chem.*, 1983, **48**, 3347.
541. Laduree, D., Florentin, D. and Robba, M., *J. Heterocycl. Chem.*, 1980, **17**, 1189.
542. Mitsuhashi, K., Itho, E., Kawahara, T. and Tanaka, K., *J. Heterocycl. Chem.*, 1983, **20**, 1103.
543. Guillaumel, J., Demerseman, P., Clavel, J.-C. and Royer, R., *J. Heterocycl. Chem.*, 1980, **17**, 1531.
544. Pankina, Z. A. and Shchukina, M. N., *Chem.-Pharm. J.*, 1972, **6**, 633.
545. Rosowsky, A. and Chen, K. K. N., *J. Org. Chem.*, 1973, **38**, 2071.
546. Bourah, R. C., Sandhu, J. S. and Thyagarajan, G., *J. Heterocycl. Chem.*, 1981, **18**, 1081.
547. Girard, Y., Atkinson, J. G., Belanger, P. C., Fuentes, J. J., Rokach, J., Rooney, C. S., Remy, D. C. and Hant, C. A., *J. Org. Chem.*, 1983, **48**, 3220.
548. Belanger, P. C., Atkinson, J. G., Rooney, C. S., Britcher, S. F. and Remy, D. C., *J. Org. Chem.*, 1983, **48**, 3234.
549. Charonnat, J. A., Muchowski, J. M. and Nelson, P. H., *J. Heterocycl. Chem.*, 1983, **20**, 1085.
550. Augustine, R. L., Gustavsen, A. J., Wanat, S. F., Pattison, I. C., Houghton, K. S. and Koletar, G., *J. Org. Chem.*, 1973, **38**, 3004.
551. Van Der Plas, H. C., *Acc. Chem. Res.*, 1978, **11**, 462.
552. Galons, H., Girardeau, J.-F., Farnoux, C. C. and Miocque, M., *J. Heterocycl. Chem.*, 1981, **18**, 561.
553. Abushanab, E., Bindra, A. P. and Goodman, L., *J. Heterocycl. Chem.*, 1975, **12**, 207.
554. Chiodini, L., Di Ciommo, M. and Merlini, L., *J. Heterocycl. Chem.*, 1981, **18**, 23.

555. Ruiz, V. M., Tapia, R., Valderrama, J. and Vega, J. C., *J. Heterocycl. Chem.*, 1981, **18**, 1161.
556. Meyers, A. I., Nolen, R. L., Collington, E. W., Narwid, T. A. and Strickland, R. C., *J. Org. Chem.*, 1973, **38**, 1974.
557. Shishoo, C. J., Devani, M. B., Ullas, G. V., Ananthan, S. and Bhadti, V. S., *J. Heterocycl. Chem.*, 1981, **18**, 43.
558. Beck, J. R. and Yahner, J. A., *J. Org. Chem.*, 1973, **38**, 2450.
559. Schneller, S. W. and Christ, W. J., *J. Heterocycl. Chem.*, 1981, **18**, 539.
560. Ram, V. J., Pandey, H. K. and Vlietinck, A. J., *J. Heterocycl. Chem.*, 1981, **18**, 1277.
561. Cecchi, L. and Filacchioni, G., *J. Heterocycl. Chem.*, 1983, **20**, 871.
562. Daisley, R. W. and Hanbali, J. R., *J. Heterocycl. Chem.*, 1983, **20**, 999.
563. Walser, A., Flynn, T. and Fryer, R. I., *J. Heterocycl. Chem.*, 1983, **20**, 791.
564. Jourdan, G. P. and Dreikorn, B. A., *J. Org. Chem.*, 1982, **47**, 5255.
565. Frydman, B., Buldain, G. and Repetto, J. C., *J. Org. Chem.*, 1973, **38**, 1824.
566. Krapcho, A. P. and Shaw, K. J., *J. Org. Chem.*, 1983, **48**, 3341.
567. Essassi, E. M., Zniber, R., Bernardini, A. and Viallefont, P., *J. Heterocycl. Chem.*, 1983, **20**, 1015.
568. Cava, M. P., Noguchi, I. and Buck, K. T., *J. Org. Chem.*, 1973, **38**, 2394.
569. Taylor, E. C. and Kobayashi, T., *J. Org. Chem.*, 1973, **38**, 2817.
570. White R. L., Schwan, T. J. and Alaimo, R. J., *J. Heterocycl. Chem.*, 1980, **17**, 817.
571. Somanathan, R. and Smith, K. M., *J. Heterocycl. Chem.*, 1981, **18**, 1077.
572. Yoneda, F., Nagamatsu, T. and Takamoto, M., *Chem. Pharm. Bull.*, 1983, **31**, 344.
573. Anderson, P. L., Hasak, J. P., Kahle, A. D., Paolella, N. A. and Shapiro, M. J., *J. Heterocycl. Chem.*, 1981, **18**, 1149.
574. Kelley, J. L. and McLean, E. W., *J. Heterocycl. Chem.*, 1981, **18**, 671.
575. Kuroki, M. and Tsunashima, Y., *J. Heterocycl. Chem.*, 1981, **18**, 709.
576. Peterson, L. H., Douglas, A. W. and Tolman, R. L., *J. Heterocycl. Chem.*, 1981, **18**, 659.
577. Bevan, P. S. and Ellis, G. P., *J. Chem. Soc., Perkin Trans.,* 1, 1983, 1705.
578. Garcia, F. and Galvez, C., *Synthesis*, 1985, 143.
579. Wade, J. J., Toso, C. B., Matson, C. J. and Stelzer, V. L., *J. Med. Chem.*, 1983, **26**, 608.
580. Hough, T. L., *J. Heterocycl. Chem.*, 1983, **20**, 1003.
581. Gupta, C. L. and Mital, R. L., *J. Heterocycl. Chem.*, 1983, **20**, 803.
582. Okafor, C. O., Castle, R. N. and Wise, D. S., *J. Heterocycl. Chem.*, 1983, **20**, 1047.
583. Ueno, Y., Takeuchi, Y., Koshitani, J. and Yoshida, T., *J. Heterocycl. Chem.*, 1981, **18**, 645.
584. Takada, K., Woon, T. K. and Boulton, A. J., *J. Org. Chem.*, 1982, **47**, 4323.
585. Rault, S., De Sevricourt, M. C., Dung, N.-H. and Robba, M., *J. Heterocycl. Chem.*, 1981, **18**, 739.
586. Senga, K., Fukami, K., Kanazawa, H. and Nishigaki, S., *J. Heterocycl. Chem.*, 1982, **19**, 805.
587. Sanna, P., Savelli, F. and Cignarella, G., *J. Heterocycl. Chem.*, 1981, **18**, 475.
588. Jackson, A. H., Stewart, G. W., Charnock, G. A. and Martin, J. A., *J. Chem. Soc., Perkin Trans.,* 1, 1974, 1911.
589. Lancelot, J.-C., Maume, D. and Robba, M., *J. Heterocycl. Chem.*, 1981, **18**, 743.
590. Tseng, C. K., Simone, R. A. and Walker, F. H., *J. Org. Chem.*, 1973, **38**, 1746.
591. Chorvat, R. J., Desai, B. N., Radak, S. E., Bloss, J., Hirsch, J. and Tenen, S., *J. Med. Chem.*, 1983, **26**, 845.
592. Ishizumi, K., Inaba, S. and Yamamoto, H., *J. Org. Chem.*, 1973, **38**, 2617.
593. Hayashi, K., Ozaki, Y., Nunami, K., Uchida, T., Kato, J., Kinashi, K. and Yoneda, N., *Chem. Pharm. Bull.*, 1983, **31**, 570.
594. Peet, N. P. and Sunder, S., *J. Heterocycl. Chem.*, 1981, **18**, 1123.
595. Connor, D. T. and Sorenson, R. J., *J. Heterocycl. Chem.*, 1981, **18**, 587.

596. Suschitzky, H., Wakefield, B. J., Walocha, K., Hughes, N. and Nelson, A. J., *J. Chem. Soc., Perkin Trans., 1*, 1983, 637.
597. Turley, J. C., Martin, G. E. and Inners, R. R., *J. Heterocycl. Chem.*, 1981, **18**, 1169.
598. Okafor, C. O., *J. Heterocycl. Chem.*, 1981, **18**, 405.
599. Elliott, A. J., *J. Heterocycl. Chem.*, 1981, **18**, 799.
600. Press, J. B. and Eudy, N. H., *J. Heterocycl. Chem.*, 1981, **18**, 1261.
601. TenBrink, R. E. and McCall, J. M., *J. Heterocycl. Chem.*, 1981, **18**, 821.
602. Massa, S., Corelli, F. and Stefancich, G., *J. Heterocycl. Chem.*, 1981, **18**, 829.
603. Mataka, S., Takahashi, K. and Tashiro, M., *J. Heterocycl. Chem.*, 1981, **18**, 1073.
604. Acton, E. M. and Tong, G. L., *J. Heterocycl. Chem.*, 1981, **18**, 1141.
605. Unangst, P. C., *J. Heterocycl. Chem.*, 1981, **18**, 1257.
606. Kano, S. and Yuasa, Y., *J. Heterocycl. Chem.*, 1981, **18**, 769.
607. Tzeng, C. C. and Panzica, R. P., *J. Heterocycl. Chem.*, 1983, **20**, 1123.
608. Schneller, S. W. and Christ, W. J., *J. Heterocycl. Chem.*, 1981, **18**, 653.
609. Ogura, H., Sakaguchi, M., Okamoto, T., Gonda, K. and Koga, S. *Heterocycles*, 1979, **12**, 359.
610. Potts, K. T. and Elliott, A. J., *J. Org. Chem.*, 1973, **38**, 1769.
611. Bargagna, A., Bignardi, G., Schenone, P. and Longobardi, M., *J. Heterocycl. Chem.*, 1983, **20**, 839.
612. Chen, W.-Y. and Gilman, N. W., *J. Heterocycl. Chem.*, 1983, **20**, 663.
613. Taylor, E. C. and Turchi, I. J., *Chem. Rev.*, 1979, **79**, 181.
614. Yamamori, T., Hiramatsu, Y. and Adachi, I., *J. Heterocycl. Chem.*, 1981, **18**, 347.
615. Veeraraghavan, S. and Popp, F. D., *J. Heterocycl. Chem.*, 1981, **18**, 71.
616. Pinto, A. C., Schirch, P. and Hollins, R. A., *J. Heterocycl. Chem.*, 1983, **20**, 467.
617. Merchan, F. L., Garin, J., Melendez, E. and Tejero, T., *Synthesis*, 1983, 154.
618. Bourguignon, J., Becue, C. and Queguiner, G., *J. Heterocycl. Chem.*, 1981, **18**, 425.
619. Ram, V. J., Pandey, H. K. and Vlietinck, A. J., *J. Heterocycl. Chem.*, 1980, **17**, 1305.
620. Morrison, R. W., Mallory, W. R. and Styles, V. L., *J. Org. Chem.*, 1978, **43**, 4844.
621. Lovelette, C. A., *J. Heterocycl. Chem.*, 1979, **16**, 555.
622. Campaigne, E. and McLaughlin, A. R., *J. Heterocycl. Chem.*, 1983, **20**, 781.
623. Omar, A. M. M. E., Kasem, M. G., Laabota, I. M. and Bourdais, J., *J. Heterocycl. Chem.*, 1981, **18**, 499.
624. Bates, D. K. and Jones, M. C., *J. Org. Chem.*, 1978, **43**, 3856.
625. Elsohly, M. A., Boeren, E. G. and Turner, C. E., *J. Heterocycl. Chem.*, 1978, **15**, 699.
626. Bongini, A., Cardillo, G., Orena, M., Porzi, G. and Sandri, S., *Tetrahedron Lett.*, 1979, 2545.
627. Hercouet, A. and Le Corre, M., *Tetrahedron Lett.*, 1979, 2995.
628. Begasse, B. and Le Corre, M., *Tetrahedron*, 1980, **36**, 3409.
629. Bohlmann, F. and Stohr, F.-M., *Liebigs Ann. Chem.*, 1980, 185.
630. Krohn, K., Bruckner, G. and Tietjen, H.-P., *Chem. Ber.*, 1978, **111**, 1284.
631. Antus, S., Gottsegen, A., Nogradi, M. and Gergely, A., *Chem. Ber.*, 1979, **112**, 3879.
632. Chan, T. Y. and Sammes, M. P., *J. Chem. Res. (S)*, 1986, 92.
633. Jain, A. C., Khazanchi, R. and Kumar, A., *Bull. Chem. Soc. Jpn*, 1979, **52**, 1203.
634. Shimizu, T., Hayashi, Y., Yamada, Y., Nishio, T. and Teramura, K., *Bull. Chem. Soc. Jpn*, 1981, **54**, 217.
635. Bouvier, P., Andrieux, J. and Molho, D., *Tetrahedron Lett.*, 1974, 1033.
636. Camps, F., Coll, J., Messeguer, A. and Pericas, M. A., *J. Heterocycl. Chem.*, 1980, **17**, 1377.
637. Maffrand, J.-P., Boigegrain, R., Courregelongue, J., Ferrand, G. and Frehel, D., *J. Heterocycl. Chem.*, 1981, **18**, 727.
638. Jacob, J. N., Nichols, D. E., Kohli, J. D. and Glock, D., *J. Med. Chem.*, 1981, **24**, 1013.
639. Anastasis, P. and Brown, P. E., *J. Chem. Soc., Perkin Trans., 1*, 1983, 197.

640. Campaigne, E., Smith, H. A., Sandhu, J. S. and Kim, C. S., *J. Heterocycl. Chem.*, 1983, **20**, 55.
641. Lancelot, J.-C., Gazengel, J.-M. and Robba, M., *J. Heterocycl. Chem.*, 1981, **18**, 1281.
642. Eberle, M. K. and Kahle, G. G., *J. Heterocycl. Chem.*, 1981, **18**, 525.
643. Bantick, J. R. and Suschitzky, J. L., *J. Heterocycl. Chem.*, 1981, **18**, 679.
644. Kaufman, K. D., Erb, D. J., Block, T. M., Carlson, R. W., Knoechel, D. J., McBride, L. and Zeitlow, T., *J. Heterocycl. Chem.*, 1982, **19**, 1051.
645. Ahluwalia, V. K., Jolly, R. S. and Tehim, A. K., *J. Chem. Soc., Perkin Trans., 1*, 1983, 1229.
646. Chenault, J. and Dupin, J.-F. E., *Heterocycles*, 1983, **20**, 437.
647. Neumeyer, J. L. and Boyce, C. B., *J. Org. Chem.*, 1973, **38**, 2291.
648. Coates, R. M. and McManus, P. A., *J. Org. Chem.*, 1982, **47**, 4822.
649. Gall, M. and Kamdar, B. V., *J. Org. Chem.*, 1981, **46**, 1575.
650. Park, D. J., Fulmer, T. D. and Beam, C. F., *J. Heterocycl. Chem.*, 1981, **18**, 649.
651. Nair, M. G., Salter, O. C., Kisliuk, R. L., Gaumont, Y. and North, G., *J. Med. Chem.*, 1983, **26**, 1164.
652. Weber, K.-H. and Daniel, H., *Liebigs Ann. Chem.*, 1979, 328.
653. Rault, S., Lancelot, J.-C., Effi, Y. and Robba, M., *Heterocycles*, 1983, **20**, 477.
654. Okuda, H., Tominaga, Y., Matsuda, Y. and Kobayashi, G., *Heterocycles*, 1979, **12**, 485.
655. Yurugi, S., Hieda, M., Fushimi, T., Kawamatsu, Y., Sugihara, H. and Tomimoto, M., *Chem. Pharm. Bull.*, 1972, **20**, 1513.
656. Wright, G. C., Gray, J. E. and Yu, C. N., *J. Med. Chem.*, 1974, **17**, 244.
657. Koyama, T., Hirota, T., Shinoara, Y., Fukuoka, S., Yamato, M. and Ohmori, S., *Chem. Pharm. Bull.*, 1975, **23**, 494.
658. Rajappa, S., Sreenivasan, R. and Khalwadekar, A., *J. Chem. Res. (S)*, 1986, 160.
659. Gilchrist, T. L. and Roberts, T. G., *J. Chem. Soc., Perkin Trans., 1*, 1983, 1283.
660. De La Torre, M., Garcia, F. and Cruz, R., *J. Heterocycl. Chem.*, 1981, **18**, 1251.
661. Houlihan, W. J. and Parrino, V. A., *J. Org. Chem.*, 1982, **47**, 5177.
662. Moskowitz, H., Mignot, A. and Miocque, M., *J. Heterocycl. Chem.*, 1980, **17**, 1321.
663. Powers, L. J., Fogt, S. W., Ariyan, Z. S., Rippin, D. J., Heilman, R. D. and Matthews, R. J., *J. Med. Chem.*, 1981, **24**, 604.
664. Bourdais, J., Rajniakova, O. and Povazanec, F., *J. Heterocycl. Chem.*, 1980, **17**, 1351.
665. Ochi, H. and Miyasaka, T., *Chem. Pharm. Bull.*, 1983, **31**, 1228.
666. Wynberg, H. and Cabell, M., *J. Org. Chem.*, 1973, **38**, 2814.
667. Chen, S.-F. and Panzica, R. P., *J. Heterocycl. Chem.*, 1981, **18**, 303.
668. Alberola, A., Ortega, A. G., Pedrosa, R., Bragado, J. L. P. and Amo, J. F. R., *J. Heterocycl. Chem.*, 1983, **20**, 715.
669. Barnard, I. F. and Elvidge, J. A., *J. Chem. Soc., Perkin Trans., 1*, 1983, 1137.
670. Sartori, G., Casiraghi, G., Bolzoni, L. and Casnati, G., *J. Org. Chem.*, 1979, **44**, 803.
671. Anastasis, P. and Brown, P. E., *J. Chem. Soc., Perkin Trans., 1*, 1983, 1431.
672. Furukawa, Y. and Shima, S., *Chem. Pharm. Bull.*, 1976, **24**, 979.
673. Tominaga, Y., Okuda, H., Mitsutomi, Y., Matsuda, Y., Kobayashi, G. and Sakemi, K., *Heterocycles*, 1979, **12**, 503.
674. Imafuku, K., Sumio, M. and Matsumura, H., *J. Heterocycl. Chem.*, 1980, **17**, 1057.
675. El-Kashef, H., Rault, S., De Sevricourt, M. C., Touzot, P. and Robba, M., *J. Heterocycl. Chem.*, 1980, **17**, 1399.
676. Westerlund, C., *J. Chem. Soc., Perkin Trans., 1*, 1974, 534.
677. Honkanen, E., Pippuri, A., Kaircsalo, P., Nore, P., Karppanen, H. and Paakkari, I., *J. Med. Chem.*, 1983, **26**, 1433.
678. Robba, M., Lancelot, J.-C., Maume, D. and Rabaron, A., *J. Heterocycl. Chem.*, 1978, **15**, 1159.
679. Brooke, G. M., *J. Chem. Soc., Perkin Trans., 1*, 1983, 821.

680. Begley, M. J., Mohamed, S. E., Whiting, D. A., D'Souza, F. and Hatam, N. A. R., *J. Chem. Soc., Perkin Trans., 1*, 1983, 883.

681. Sunder, S. and Peet, N. P., *J. Heterocycl. Chem.*, 1980, **17**, 1527.

682. Walser, A. and Flynn, T., *J. Heterocycl. Chem.*, 1980, **17**, 1697.

683. Neunhoeffer, H. and Hammann, H., *Tetrahedron Lett.*, 1983, **24**, 1767.

684. Garin, J., Melendez, E., Merchan, F. L., Tejel, C. and Tejero, T., *Synthesis*, 1983, 375.

685. Kreutzberger, A. and Leger, M., *Arch. Pharm. (Weinheim)*, 1983, **316**, 582.

686. Palfreyman, M. N. and Wooldridge, K. R. H., *J. Chem. Soc., Perkin Trans., 1*, 1974, 57.

687. Badger, R. J., Brown, D. J. and Lister, J. H., *J. Chem. Soc., Perkin Trans., 1*, 1974, 152.

688. Anderson, R. K. and Cheeseman, G. W. H., *J. Chem. Soc., Perkin Trans., 1*, 1974, 129.

689. Kemp, M. S., Burden, R. S. and Loeffler, R. S. T., *J. Chem. Soc., Perkin Trans., 1*, 1983, 2267.

690. Grigg, R., Gunaratne, H. Q. N. and McNaughton, E., *J. Chem. Soc., Perkin Trans., 1*, 1983, 185.

691. Stevenson, T. M., Kazmierczak, F. and Leonard, N. J., *J. Org. Chem.*, 1986, **51**, 616.

692. Robev, S. K., *Tetrahedron Lett.*, 1983, **24**, 4351.

693. Ingram, A. S., Reid, D. H. and Symon, J. D., *J. Chem. Soc., Perkin Trans., 1*, 1974, 242.

694. Furukawa, M., Kawanabe, K., Yoshimi, A., Okawara, T. and Noguchi, Y., *Chem. Pharm. Bull.*, 1983, **31**, 2473.

695. Katagiri, N., Koshihara, A., Atsuumi, S. and Kato, T., *Chem. Pharm. Bull.*, 1983, **31**, 2288.

696. Dennis, N., Katritzky, A. R. and Parton, S. K., *J. Chem. Soc. (C)*, 1974, 750.

697. Dean, F. M., Patampongse, C. and Podimuang, V., *J. Chem. Soc., Perkin Trans., 1*, 1974, 583.

698. Brown, D. J. and Lynn, R. K., *J. Chem. Soc., Perkin Trans., 1*, 1974, 349.

699. Stokker, G. E., *J. Org. Chem.*, 1983, **48**, 2613.

700. Dinh, T. H., Kolb, A., Barnathan, G. and Igolen, J., *J. Chem. Soc., Chem. Commun.*, 1973, 680.

701. Le Count, D. J. and Greer, A. T., *J. Chem. Soc., Perkin Trans., 1*, 1974, 297.

702. Katagiri, N., Atsuumi, S. and Kato, T., *Chem. Pharm. Bull.*, 1983, **31**, 2540.

703. Adembri, G., Chimichi, S., De Sio, F., Nesi, R. and Scotton, M., *J. Chem. Soc., Perkin Trans., 1*, 1974, 1022.

704. Siddiqui, M. S. S. and Stevens, M. F. G., *J. Chem. Soc. (C)*, 1974, 609.

705. Garcia-Lopez, M. T., De Las Heras, F. G. and Stud, M., *J. Chem. Soc., Perkin Trans., 1*, 1978, 483.

706. Van Heerden, F. R., Brandt, E. V. and Roux, D. G., *J. Chem. Soc., Perkin Trans., 1*, 1978, 137.

707. Davies, R. V., Iddon, B., Suschitzky, H. and Gitlos, M. W., *J. Chem. Soc., Perkin Trans., 1*, 1978, 180.

708. Kametani, T., Ohsawa, T., Ihara, M. and Fukumoto, K., *J. Chem. Soc., Perkin Trans., 1*, 1978, 460.

709. Albert, A., *J. Chem. Soc., Perkin Trans., 1*, 1978, 513.

710. Ames, D. E., Chandrasekhar, S. and Hansen, K. J., *J. Chem. Soc., Perkin Trans., 1*, 1978, 539.

711. Lancelot, J.-C., Rault, S., Dung, N.-H. and Robba, M., *Chem. Pharm. Bull.*, 1983, **31**, 3160.

712. Witiak, D. T., Stratford, E. S., Nazareth, R., Wagner, G. and Feller, D. R., *J. Med. Chem.*, 1971, **14**, 758.

713. Katritzky, A. R., Yeung, W. K., Patel, R. C. and Burgess, K., *Heterocycles*, 1983, **20**, 623.

714. Laidlaw, G. M., Collins, J. C., Archer, S., Rossi, D. and Schulenberg, J. W., *J. Org. Chem.*, 1973, **38**, 1743.
715. Gewald, K., Bottcher, H. and Schinke, E., *Chem. Ber.*, 1966, **99**, 94.
716. Clarke, D. G., Crombie, L. and Whiting, D. A., *J. Chem. Soc., Perkin Trans., 1*, 1974, 1007.
717. Dupas, G., Duflos, J. and Queguiner, G., *J. Heterocycl. Chem.*, 1983, **20**, 967.
718. Tashiro, M., Fukuda, Y. and Fukata, G., *Heterocycles*, 1983, **20**, 633.
719. Chapleo, C. B., Myers, P. L., Butler, R. C. M., Doxey, J. C., Roach, A. G. and Smith, C. F. C., *J. Med. Chem.*, 1983, **26**, 823.
720. Novak, J. and Salemink, C. A., *J. Chem. Soc., Perkin Trans., 1*, 1983, 2867.
721. Ishikawa, F., Kosasayama, A., Yamaguchi, H., Watanabe, Y., Saegusa, J., Shiba-mura, S., Ashida, S. and Abiko, Y., *J. Med. Chem.*, 1981, **24**, 376.
722. Shiraiwa, M., Sakamoto, T. and Yamanaka, H., *Chem. Pharm. Bull.*, 1983, **21**, 2275.
723. Kurihara, T., Nasu, K. and Adachi, Y., *J. Heterocycl. Chem.*, 1983, **20**, 81.
724. Lopatin, W., Sheppard, C. and Owen, T. C., *J. Org. Chem.*, 1978, **43**, 4678.
725. Stevens, H. N. E. and Stevens, M. F. G., *J. Chem. Soc. (C)*, 1970, 765.
726. Rahman, L. K. A. and Scrowston, R. M., *J. Chem. Soc., Perkin Trans., 1*, 1983, 2973.
727. Appel, R. and Siegmund, G., *Z. Anorg. Allg. Chem.*, 1968, **363**, 183.
728. Ardakani, M. A., Smalley, R. K. and Smith, R. H., *J. Chem. Soc., Perkin Trans., 1*, 1983, 2501.
729. Ardakani, M. A. and Smalley, R. K., *Synthesis*, 1979, 308.
730. Zanirato, P., Spagnolo, P. and Zanardi, G., *J. Chem. Soc., Perkin Trans., 1*, 1983, 2551.
731. Garanti, L. and Zecchi, G., *J. Chem. Soc., Perkin Trans., 1*, 1980, 116.
732. Asherson, J. L., Bilgic, O. and Young, D. W., *J. Chem. Soc., Perkin Trans., 1*, 1980, 522.
733. Howes, P. D. and Pianka, M., *J. Chem. Soc., Perkin Trans., 1*, 1980, 762.
734. Dorgan, R. J. J., Parrick, J. and Hardy, C. R., *J. Chem. Soc., Perkin Trans., 1*, 1980, 939.
735. Huppatz, J. L. and Moore, R. M. J., *Aust. J. Chem.*, 1971, **24**, 405.
736. Hallberg, A. and Martin, A. R., *J. Heterocycl. Chem.*, 1981, **18**, 1255.
737. Clarke, K., Fox, W. R. and Scrowston, R. M., *J. Chem. Soc., Perkin Trans., 1*, 1980, 1029.
738. Yoneda, F., Mori, K., Sakuma, Y. and Yamaguchi, H., *J. Chem. Soc., Perkin Trans., 1*, 1980, 978.
739. Clayton, J. P., O'Hanlon, P. J. and King, T. J., *J. Chem. Soc., Perkin Trans., 1*, 1980, 1352.
740. Tromein, A., Demerseman, P. and Royer, R., *Synthesis*, 1985, 1074.
741. Charles, I., Latham, D. W. S., Hartley, D., Oxford, A. W. and Scopes, D. I. C., *J. Chem. Soc., Perkin Trans., 1*, 1980, 1139.
742. McKittrick, B. A. and Stevenson, R., *J. Chem. Soc., Perkin Trans., 1*, 1983, 2423.
743. Nomura, T., Fukai, T., Hano, Y. and Tsukamoto, K., *Heterocycles*, 1983, **20**, 661.
744. Ames, D. E. and Brohi, M. I., *J. Chem. Soc., Perkin Trans., 1*, 1980, 1384.
745. Potts, K. T., Cipullo, M. J., Ralli, P. and Theodoridis, G., *J. Org. Chem.*, 1983, **48**, 4841.
746. Taylor, E. C., Palmer, D. C., George, T. J., Fletcher, S. R., Tseng, C. P., Harrington, P. J. and Beardsley G. P., *J. Org. Chem.*, 1983, **48**, 4852.
747. Melhado, L. L. and Leonard, N. J., *J. Org. Chem.*, 1983, **48**, 5130.
748. Daunis, J. and Follet, M., *Bull. Soc. Chim. Fr.*, 1976, 1178.
749. Battesti, P., Battesti, O. and Selim, M., *Bull. Soc. Chim. Fr.*, 1976, 1549.
750. Chartier, O., Lhommet, G. and Maitte, P., *Bull. Soc. Chim. Fr.*, 1976, 1916.
751. Barker, J. M., Huddleston, P. R., Jones, A. W. and Edwards, M., *J. Chem. Res. (S)*, 1980, 4.

752. Barker, J. M., Huddleston, P. R., Chadwick, N. and Keenan, G. J., *J. Chem. Res. (S)*, 1980, 6.
753. Senga, K., Robins, R. K. and O'Brien, D. E., *J. Heterocycl. Chem.*, 1975, **12**, 899.
754. Sharanin, Yu. A. and Klokol, G. V., *Chem. Heterocycl. Compds*, 1983, **19**, 232.
755. Serafin, B. and Konopski, L., *Pol. J. Chem.*, 1978, **52**, 51.
756. Buggle, K., Ghogain, U. N., Nangle, M. and MacManus, P. A., *J. Chem. Soc., Perkin Trans.*, 1, 1983, 1427.
757. Ogura, H., Mineo, S. and Nakagawa, K., *Heterocycles*, 1980, **14**, 1125.
758. Romanchick, W. A. and Joullie, M. M., *Heterocycles*, 1980, **14**, 1139.
759. Elliott, A. J. and Guzik, H., *J. Heterocycl. Chem.*, 1981, **18**, 861.
760. Kido, K. and Watanabe, Y., *Heterocycles*, 1980, **14**, 1151.
761. Gupta, R. R., Ojha, K. G., Kalwania, G. S. and Kumar, M., *Heterocycles*, 1980, **14**, 1145.
762. Pelter, A. and Foot, S., *Synthesis*, 1976, 326.
763. Roudier, J. F. and Foucaud, A., *Synthesis*, 1984, 159.
764. Talapatra, S. K., Chaudhuri, P. and Talapatra, B., *Heterocycles*, 1980, **14**, 1279.
765. Tsukayama, M., Horie, T., Yamashita, Y., Masumura, M. and Nakayama, M., *Heterocycles*, 1980, **14**, 1283.
766. Kametani, T., Higuchi, M., Noguchi, M., Hashiguchi, Y. and Yoneda, F., *Heterocycles*, 1980, **14**, 1295.
767. Rufer, C., Bahlmann, F. and Kapp, J. F., *Eur. J. Med. Chem.*, 1978, **13**, 193.
768. Coyle, J. D., Addison, P. L., Farmer, J. L., Haws, E. J. and Small, P. W., *Synthesis*, 1980, 403.
769. Atta-ur-Rahman and Ghazala, M. A. H., *Synthesis*, 1980, 372.
770. Szarvazi, E., Grand, M., Depin, J.-C. and Betbeder-Matibet, A., *Eur. J. Med. Chem.*, 1978, **13**, 113.
771. Kurihara, T., Uno, T. and Sakamoto, Y., *J. Heterocycl. Chem.*, 1980, **17**, 231.
772. Khan, M. A. and Gemal, A. L., *J. Heterocycl. Chem.*, 1977, **14**, 1009.
773. Lancelot, J.-C., Gazengel, J.-M. and Robba, M., *Chem. Pharm. Bull.*, 1983, **31**, 2652.
774. Sargent, M. V., *J. Chem. Soc., Perkin Trans.*, 1, 1982, 403.
775. Afzal, M. and Al-Hassan, J. M., *Heterocycles*, 1980, **14**, 1173.
776. Wade, L. G., Acker, K. J., Earl, R. A. and Osteryoung, R. A., *J. Org. Chem.*, 1979, **44**, 3724.
777. Graham, R. and Lewis, J. R., *J. Chem. Soc., Perkin Trans.*, 1, 1978, 876.
778. Jorgensen, A., *Heterocycles*, 1986, **24**, 997.
779. Kurihara, T., Imai, H. and Nasu, K., *Chem. Pharm. Bull.*, 1980, **28**, 2972.
780. Ege, G., Gilbert, K. and Franz, H., *Synthesis*, 1977, 556.
781. Ege, G. and Gilbert, K., *Tetrahedron Lett.*, 1979, 4253.
782. Paulmier, C., *Bull. Soc. Chim. Fr.*, 1980, **II**, 151.
783. Rusinov, V. L., Petrov, A. Y. and Postovskii, I. Ya., *Chem. Heterocycl. Compds*, 1980, **16**, 974.
784. Jagodzinski, T., Kost, A. N. and Sagitullin, R. S., *Chem. Heterocycl. Compds*, 1979, **15**, 179.
785. Seha, Z. and Weis, C. D., *Helv. Chim. Acta*, 1980, **63**, 413.
786. Attia, A. El-H. and Michael, M., *Gazz. Chim. Ital.*, 1982, **112**, 387.
787. Alaka, B. V., Patnaik, D. and Rout, M. K., *J. Indian Chem. Soc.*, 1982, **59**, 1168.
788. Kost, A. A., *Chem. Heterocycl. Compds*, 1980, **16**, 903.
789. Mosti, L., Schenone, P. and Menozzi, G., *J. Heterocycl. Chem.*, 1980, **17**, 61.
790. Slouka, J. and Bekarek, V., *Collect. Czech. Chem. Commun.*, 1980, **45**, 1379.
791. Miller, T. K. and Sharp, J. T., *J. Chem. Soc., Perkin Trans.*, 1, 1984, 223.
792. Knutsen, L. J. S., Judkins, B. D., Mitchell, W. L., Newton, R. F. and Scopes, D. I. C., *J. Chem. Soc., Perkin Trans.*, 1, 1984, 229.
793. Rivalle, C. and Bisagni, E., *J. Heterocycl. Chem.*, 1980, **17**, 245.

794. Monge Vega, A., Palop, J. A., Martinez, M. T. and Fernandez-Alvarez, E., *J. Heterocycl. Chem.*, 1980, **17**, 249.
795. Bourguignon, J., Lemachand, M. and Queguiner, G., *J. Heterocycl. Chem.*, 1980, **17**, 257.
796. Pene, C. and Hubert-Habart, M., *J. Heterocycl. Chem.*, 1980, **17**, 331.
797. Glennon, R. A., Rogers, M. E. and El-Said, M. K., *J. Heterocycl. Chem.*, 1980, **17**, 337.
798. Moffett, R. B., *J. Heterocycl. Chem.*, 1980, **17**, 341.
799. Monge Vega, A., Aldama, I., Rabbani, M. M. and Fernandez-Alvarez, E., *J. Heterocycl. Chem.*, 1980, **17**, 77.
800. Brunelli, C., Fravolini, A., Grandolini, G. and Tiralti, M. C., *J. Heterocycl. Chem.*, 1980, **17**, 121.
801. Knierzinger, A. and Wolfbeis, O. S., *J. Heterocycl. Chem.*, 1980, **17**, 225.
802. Brunel, S., Montginoul, C., Torreilles, E. and Giral, L., *J. Heterocycl. Chem.*, 1980, **17**, 235.
803. Abdel-Hady, S. A. L., Badawy, M. A., Ibrahim, Y. A. and Pfleiderer, W., *Chem. Ber.*, 1984, **117**, 1077.
804. Stanova, E., Haimova, M. and Ognyanov, V., *Liebigs Ann. Chem.*, 1984, 389.
805. Outurquin, F. and Paulmier, C. *Bull. Soc. Chim. Fr.*, 1983, **II**, 153, 159.
806. Liso, G., Trapani, G., Berardi, V. and Marchini, P., *J. Heterocycl. Chem.*, 1980, **17**, 377.
807. Orito, K., Kaga, H., Itoh, M., De Silva, S. O., Manske, R. H. and Rodrigo, R., *J. Heterocycl. Chem.*, 1980, **17**, 417.
808. Singerman, G. M., *J. Heterocycl. Chem.*, 1975, **12**, 877.
809. Danylec, B. and Davis, M., *J. Heterocycl. Chem.*, 1980, **17**, 533.
810. Danylec, B. and Davis, M., *J. Heterocycl. Chem.*, 1980, **17**, 537.
811. Bourdais, J. and Omar, A. M. M. E., *J. Heterocycl. Chem.*, 1980, **17**, 555.
812. Malesani, G., Galiano, F., Ferlin, M. G. and Masiero, S., *J. Heterocycl. Chem.*, 1980, **17**, 563.
813. Hester, J. B., *J. Heterocycl. Chem.*, 1980, **17**, 575.
814. Charlton, J. L., Lypka, G. and Sayced, V., *J. Heterocycl. Chem.*, 1980, **17**, 593.
815. Gilis, P. M., Haemers, A. and Bollaert, W., *J. Heterocycl. Chem.*, 1980, **17**, 717.
816. Kokel, B., Menichi, G. and Hubert-Habart, M., *Tetrahedron Lett.*, 1984, **25**, 1557.
817. Cook, C. E., Corley, R. C. and Wall, M. E., *J. Org. Chem.*, 1965, **30**, 411.
818. Neidlein, R. and Jeromin, G., *J. Chem. Res. (S)*, 1980, 233.
819. Merchan, F. L., Garin, J., Martinez, V. and Melendez, E., *Synthesis*, 1982, 482.
820. Molina, P., Arques, A. and Hernandez, H., *Synthesis*, 1983, 1021.
821. Niwas, S., Kumar, S. and Bhaduri, A. P., *Synthesis*, 1983, 1027.
822. Evans, J. M., Fake, C. S., Hamilton, T. C., Poyser, R. H. and Watts, E. A., *J. Med. Chem.*, 1983, **26**, 1582.
823. Trybulski, E. J., Benjamin, L. E., Earley, J. V., Fryer, R. I., Gilman, N. W., Reeder, E., Walser, A., Davidson, A. B., Horst, W. D., Sepinwall, J., O'Brien, R. A. and Dairman, W., *J. Med. Chem.*, 1983, **26**, 1589.
824. Temple, C., Wheeler, G. P., Comber, R. N., Elliott, R. D. and Montgomery, J. A., *J. Med. Chem.*, 1983, **26**, 1614.
825. Almirante, N. and Forti, L., *J. Heterocycl. Chem.*, 1983, **20**, 1523.
826. Kornet, M. J., Varia, T. and Beaven, W., *J. Heterocycl. Chem.*, 1983, **20**, 1553.
827. Tomasik, D., Tomasik, P. and Abramovitch, R. A., *J. Heterocycl. Chem.*, 1983, **20**, 1539.
828. Ong-Lee, A., Sylvester, L. and Wasley, J. W. F., *J. Heterocycl. Chem.*, 1983, **20**, 1565.
829. Press, J. B. and Eudy, N. H., *J. Heterocycl. Chem.*, 1983, **20**, 1593.
830. Fryer, R. I., Lauer, R. F., Trybulski, E. J., Vitone, S., Walser, A. and Zenchoff, G., *J. Heterocycl. Chem.*, 1983, **20**, 1605.
831. Sliwa, H., Blondeau, D. and Rydzkowski, R., *J. Heterocycl. Chem.*, 1983, **20**, 1613.

832. Sanemitsu, Y., Nakayama, Y. and Shiroshita, M., *J. Heterocycl. Chem.*, 1983, **20**, 1671.
833. Campaigne, E. and Kim, C. S., *J. Heterocycl. Chem.*, 1983, **20**, 1701.
834. Ueno, Y., Maeda, K., Koshitani, J. and Yoshida, T. *J. Heterocycl. Chem.*, 1982, **19**, 189.
835. Chen, S.-F., Panzica, R. P., Dexter, D. L., Chu, M.-Y. W. and Calabresi, P., *J. Heterocycl. Chem.*, 1982, **19**, 285.
836. Abramovitch, R. A., Inbasekaran, M. N., Miller, A. L. and Hanna, J. M., *J. Heterocycl. Chem.*, 1982, **19**, 509.
837. Sundberg, R. J. and Ellis, J. E., *J. Heterocycl. Chem.*, 1982, **19**, 573.
838. Zvilichovsky, G., Garbi, H. and Nemes, E., *J. Heterocycl. Chem.*, 1982, **19**, 205.
839. Rene, L., Faulques, M. and Royer, R., *J. Heterocycl. Chem.*, 1982, **19**, 691.
840. Senda, S., Hirota, K., Asao, T. and Maruhashi, K., *J. Am. Chem. Soc.*, 1978, **100**, 7661.
841. Chandler, C. J., Deady, L. W., Reiss, J. A. and Tzimos, V., *J. Heterocycl. Chem.*, 1982, **19**, 1017.
842. Descours, D. and Festal, D., *Synthesis*, 1983, 1033.
843. Clark, B. J. and Grayshan, R., *J. Chem. Res. (S)*, 1981, 324.
844. Al-Hassan, S. S., Sterling, I. and Wood, H. C. S., *J. Chem. Res. (S)*, 1980, 278.
845. Dorn, H. and Zubek, A., *Chem. Ber.*, 1968, **101**, 3265.
846. Turchi, I. J. and Maryanoff, C. A., *Synthesis*, 1983, 837.
847. Murthy, A. K., Sailaja, S., Rajanarendar, E. and Rao, C. J., *Synthesis*, 1983, 839.
848. Saito, K., Kambe, S., Nakano, Y., Sakurai, A. and Midorikawa, H., *Synthesis*, 1983, 210.
849. Krutosikova, A., Kovac, J., Dandarova, M., Lesko, J. and Ferik, S., *Collect. Czech. Chem. Commun.*, 1981, **46**, 2564.
850. Ram, S., Wise, D. S. and Townsend, L. B., *Heterocycles*, 1984, **22**, 1789.
851. Marchalin, M., Svetlik, J. and Martvon, A., *Collect. Czech. Chem. Commun.*, 1981, **46**, 2557.
852. Marchalin, M., Svetlik, J. and Martvon, A., *Collect. Czech. Chem. Commun.*, 1981, **46**, 2428.
853. Ciattini, P. G., Morera, E. and Ortar, G., *Synthesis*, 1983, 311.
854. Talley, J. J., *Synthesis*, 1983, 845.
855. Kovac, T., Oklobdzija, M., Comisso, G., Dacorte, E., Fajdiga, T., Moimas, F., Angeli, C., Zonno, F., Toso, R. and Sunjic, V., *J. Heterocycl. Chem.*, 1983, **20**, 1339.
856. Gatta, F. and Settimj, G., *J. Heterocycl. Chem.*, 1983, **20**, 1251.
857. Clark, R. D., *J. Heterocycl. Chem.*, 1983, **20**, 1393.
858. Oklobdzija, M., Comisso, G., Dacorte, E., Fajdiga, T., Gratton, G., Moimas, F., Toso, R. and Sunjic, V., *J. Heterocycl. Chem.*, 1983, **20**, 1329.
859. Nagano, M., Matsui, T., Tobitsuka, J. and Oyamada, K., *Chem. Pharm. Bull.*, 1972, **20**, 2626.
860. George, B. and Papadopoulos, E. P., *J. Heterocycl. Chem.*, 1983, **20**, 1127.
861. Della Vecchia, L., Dellureficio, J., Kisis, B. and Vlattas, I., *J. Heterocycl. Chem.*, 1983, **20**, 1287.
862. Schwan, T. J., Freedman, R. and Pollack, J. R., *J. Heterocycl. Chem.*, 1983, **20**, 1351.
863. Harrington, P. J. and Hegedus, L. S., *J. Org. Chem.*, 1984, **49**, 2657.
864. Petridou-Fischer, J. and Papadopoulos, E. P., *J. Heterocycl. Chem.*, 1983, **20**, 1159.
865. Dean, W. D. and Papadopoulos, E. P., *J. Heterocycl. Chem.*, 1982, **19**, 171.
866. Duflos, J., Dupas, G. and Queguiner, G., *J. Heterocycl. Chem.*, 1983, **20**, 1191.
867. D'Amico, J. J., Stults, B. R., Ruminski, P. G. and Wood, K. V., *J. Heterocycl. Chem.*, 1983, **20**, 1283.
868. Eid, M. M., Badawy, M. A. and Ibrahim, Y. A., *J. Heterocycl. Chem.*, 1983, **20**, 1255.
869. Nachman, R. J., *J. Heterocycl. Chem.*, 1983, **20**, 1423.

870. Mochalov, S. S., Fedotov, A. N. and Shabarov, Yu. S., *J. Org. Chem. USSR*, 1979, **15**, 847.
871. Earley, J. V., Fryer, R. I. and Gilman, N. W., *J. Heterocycl. Chem.*, 1983, **20**, 1195.
872. Koren, B., Kovac, F., Petric, A., Stanovnik, B. and Tisler, M., *Tetrahedron*, 1976, **32**, 493.
873. Shutske, G. M., *J. Org. Chem.*, 1984, **49**, 180.
874. Albert, A., *Adv. Heterocycl. Chem.*, 1982, **32**, 1.
875. Kamal, A. and Sattur, P. B., *Synthesis*, 1981, 272.
876. Falsone, G., Spur, B. and Wingen, P. H., *Arch. Pharm. (Weinheim)*, 1983, **316**, 763.
877. Duflos, J. and Queguiner, G., *J. Heterocycl. Chem.*, 1984, **21**, 49.
878. Gammill, R. B., *Synthesis*, 1979, 901.
879. Bartsch, H., Eiden, F. and Buchborn, H., *Arch. Pharm. (Weinheim)*, 1982, **315**, 481.
880. Maquestiau, A., Van Haverbeke, Y. and Vanden Eynde, J.-J., *Bull. Soc. Chim. Belg.*, 1979, **88**, 665.
881. Abdou, S. E., Fahmy, S. M., Khader, M. M. and Elnagdi, M. H., *Monatsh. Chem.*, 1982, **113**, 985.
882. Sauter, F. and Deinhammer, W., *Monatsh. Chem.*, 1973, **104**, 1586.
883. Sauter, F. and Deinhammer, W., *Monatsh. Chem.*, 1973, **104**, 1593.
884. Remp, W. and Junek, H., *Monatsh. Chem.*, 1973, **104**, 1101.
885. Hromatka, O., Binder, D. and Eichinger, K., *Monatsh. Chem.*, 1973, **104**, 1513.
886. Hromatka, O., Binder, D. and Eichinger, K., *Monatsh. Chem.*, 1973, **104**, 1599.
887. Davidson, J. S., *Chem. Ind. (London)*, 1982, 660.
888. Kocevar, M., Vercek, B., Stanovnik, B. and Tisler, M., *Monatsh. Chem.*, 1982, **113**, 731.
889. Soliman, F. S. G. and Kappe, T., *Monatsh. Chem.*, 1982, **113**, 475.
890. Stevens, M. F. G., Hickman, J. A., Stone, R., Gibson, N. W., Baig, G. U., Lunt, E. and Newton, C. G., *J. Med. Chem.*, 1984, **27**, 196.
891. Buckle, D. R., Rockell, C. J. M., Smith, H. and Spicer, B. A., *J. Med. Chem.*, 1984, **27**, 223.
892. Sundberg, R. J. and Laurino, J. P., *J. Org. Chem.*, 1984, **49**, 249.
893. Bakthavachalam, V., D'Alarcao, M. and Leonard, N. J., *J. Org. Chem.*, 1984, **49**, 289.
894. Coffen, D. L., Schaer, B. W., Bizzaro, F. T. and Cheung, J. B., *J. Org. Chem.*, 1984, **49**, 296.
895. Gallastegui, J., Lago, J. M. and Palomo, C., *J. Chem. Res. (S)*, 1984, 170.
896. Rahman, L. K. A. and Scrowston, R. M., *J. Chem. Soc., Perkin Trans.*, 1, 1984, 385.
897. Clancy, M. G., Hesabi, M. M. and Meth-Cohn, O., *J. Chem. Soc., Perkin Trans.*, 1, 1984, 429.
898. Yamauchi, M., Katayama, S., Nakashita, Y. and Watanabe, T., *J. Chem. Soc., Perkin Trans.*, 1, 1984, 503.
899. Martin, L. L., Setescak, L. L., Spaulding, T. C. and Helsley, G. C., *J. Med. Chem.*, 1984, **27**, 372.
900. Rees, C. W. and Storr, R. C., *J. Chem. Soc. (C)*, 1969, 756.
901. Wade, J. J., Hegel, R. F. and Toso, C. B., *J. Org. Chem.*, 1979, **44**, 1811.
902. Yoneda, F. and Higuchi, M., *J. Chem. Soc., Perkin Trans.*, 1, 1977, 1336.
903. Akermark, B., Eberson, L., Jonsson, E. and Pettersson, E., *J. Org. Chem.*, 1975, **40**, 1365.
904. Ichiba, M., Nishigaki, S. and Senga, K., *J. Org. Chem.*, 1978, **43**, 469.
905. Bailey, J., *J. Chem. Soc., Perkin Trans.*, 1, 1977, 2047.
906. Uff, B. C., Budhram, R. S., Consterdine, M. F., Hicks, J. K., Slingsby, B. P. and Pemblington, J. A., *J. Chem. Soc., Perkin Trans.*, 1, 1977, 2018.
907. Somei, M., Karasawa, Y., Shoda, T. and Kaneko, C., *Chem. Pharm. Bull.*, 1981, **29**, 249.
908. Yoneda, F., Nakagawa, K., Noguchi, M. and Higuchi, M., *Chem. Pharm. Bull.*, 1981, **29**, 379.

909. Kametani, T. and Takahashi, K., *Heterocycles*, 1978, **9**, 293.
910. Daboun, H. A. F., Abdou, S. E., Hussein, M. M. and Elnagdi, M. H., *Synthesis*, 1982, 502.
911. El-Shafei, A. K., Ghattas, A.-B. A. C., Sultan, A., El-Kashef, H. S. and Vernin, G., *Gazz. Chim. Ital.*, 1982, **112**, 345.
912. Wade, J. J., *J. Org. Chem.*, 1979, **44**, 1816.
913. DeCroix, B., Strauss, M. J., DeFusco, A. and Palmer, D. C., *J. Org. Chem.*, 1979, **44**, 1700.
914. Witte, J. and Boekelheide, V., *J. Org. Chem.*, 1972, **37**, 2849.
915. Zupan, M., Stanovnik, B. and Tisler, M., *J. Org. Chem.*, 1972, **37**, 2960.
916. McCabe, R. W., Young, D. W. and Davies, G. M., *J. Chem. Soc., Chem. Commun.*, 1981, 395.
917. Dash, B., Dora, E. K. and Panda, C. S., *Indian J. Chem.*, 1981, **208**, 369.
918. Broughton, B. J., Chaplen, P., Knowles, P., Lunt, E., Marshall, S. M., Pain, D. L. and Wooldridge, K. R. H., *J. Med. Chem.*, 1975, **18**, 1117.
919. Junek, H., Schmidt, H.-W. and Gfrerer, G., *Synthesis*, 1982, 791.
920. Schmidt, D. M. and Bonvicino, G. E., *J. Org. Chem.*, 1984, **49**, 1664.
921. Yakhontov, L. N., *Chem. Heterocycl. Compds*, 1982, **18**, 873.
922. Novak, J. and Salemink, C. A., *J. Chem. Soc., Perkin Trans., 1*, 1984, 729.
923. Majeed, A. J., Sainsbury, M. and Hall, S. A., *J. Chem. Soc., Perkin Trans., 1*, 1984, 833.
924. Ichikawa, M., Nabeya, S., Muraoka, K. and Hisano, T., *Chem. Pharm. Bull.*, 1979, **27**, 1255.
925. Asakawa, H. and Matano, M., *Chem. Pharm. Bull.*, 1979, **27**, 1287.
926. Kato, T., Chiba, T. and Okada, T., *Chem. Pharm. Bull.*, 1979, **27**, 1186.
927. Naka, T. and Furukawa, Y., *Chem. Pharm. Bull.*, 1979, **27**, 1328.
928. Glamkowski, E. J. and Fortunato, J. M., *J. Heterocycl. Chem.*, 1979, **16**, 865.
929. Albrecht, W. L., Corona, J. A. and Edwards, M. L., *J. Heterocycl. Chem.*, 1979, **16**, 1349.
930. Robert, J. F. and Panouse, J. J., *J. Heterocycl. Chem.*, 1979, **16**, 1201.
931. Saint-Ruf, G. and Silou, T., *J. Heterocycl. Chem.*, 1979, **16**, 1535.
932. Stefancich, G., Artico, M., Massa, S. and Vomero, S., *J. Heterocycl. Chem.*, 1979, **16**, 1443.
933. Moskowitz, H., Mignot, A. and Miocque, M., *J. Heterocycl. Chem.*, 1979, **16**, 1077.
934. Fuhrer, H., Sutter, P. and Weis, C. D., *J. Heterocycl. Chem.*, 1979, **16**, 1121.
935. Robba, M., De Sevricourt, M. C. and Godard, A.-M., *J. Heterocycl. Chem.*, 1979, **16**, 1175.
936. Omar, A. M. M. E., Ashour, F. A. and Bourdais, J., *J. Heterocycl. Chem.*, 1979, **16**, 1435.
937. Yoneda, F., Shinozuka, K., Tsukuda, K. and Koshiro, A., *J. Heterocycl. Chem.*, 1979, **16**, 1365.
938. Benchidmi, M., Bouchet, P. and Lazaro, R., *J. Heterocycl. Chem.*, 1979, **16**, 1599.
939. Shafiee, A., Mazloumi, A. and Cohen, V. I., *J. Heterocycl. Chem.*, 1979, **16**, 1563.
940. Maume, D., Lancelot, J.-C. and Robba, M., *J. Heterocycl. Chem.*, 1979, **16**, 1217.
941. Migliara, O., Petruso, S. and Sprio, V., *J. Heterocycl. Chem.*, 1979, **16**, 833.
942. Catsoulacos, P. and Camoutsis, C., *J. Heterocycl. Chem.*, 1979, **16**, 1503.
943. Amer, A. and Zimmer, H., *J. Heterocycl. Chem.*, 1983, **20**, 1231.
944. Khan, M. A. and Morley, M. L. DeM., *J. Heterocycl. Chem.*, 1979, **16**, 997.
945. Johnson, J., Elslager, E. F. and Werbel, L. M., *J. Heterocycl. Chem.*, 1979, **16**, 1101.
946. Tanaka, C., Nasu, K., Yamamoto, N. and Shibata, M., *Chem. Pharm. Bull.*, 1982, **30**, 4195.
947. Ogura, H., Kawano, M. and Itoh, T., *Chem. Pharm. Bull.*, 1973, **21**, 2019.
948. Upadhyaya, V. P. and Srinivasan, V. R., *Indian J. Chem.*, 1981, **20B**, 161.
949. Hickson, C. L. and McNab, H., *Synthesis*, 1981, 464.

950. March, L. C. and Joullie, M. M., *J. Heterocycl. Chem.*, 1970, **7**, 39.
951. Klemm, L. H., Johnson, W. O. and White, D. V., *J. Heterocycl. Chem.*, 1970, **7**, 463.
952. Catsoulacos, P., *J. Heterocycl. Chem.*, 1970, **7**, 409.
953. Davis, J. C., Ballard, H. H. and Jones, J. W., *J. Heterocycl. Chem.*, 1970, **7**, 405.
954. Khan, M. A. and Lynch, B. M., *J. Heterocycl. Chem.*, 1970, **7**, 247.
955. Oh, C. S. and Greco, C. V., *J. Heterocycl. Chem.*, 1970, **7**, 261.
956. Niess, R. and Robins, R. K., *J. Heterocycl. Chem.*, 1970, **7**, 243.
957. Rynbrandt, R. H., *J. Heterocycl. Chem.*, 1970, **7**, 191.
958. Gasco, A., Rua, G., Menziani, E., Nano, G. M. and Tappi, G., *J. Heterocycl. Chem.*, 1970, **7**, 131.
959. Tweit, R. C., *J. Heterocycl. Chem.*, 1970, **7**, 687.
960. Van Allan, J. A., Reynolds, G. A., Petropoulos, C. C. and Maier, D. P., *J. Heterocycl. Chem.*, 1970, **7**, 495.
961. Prota, G., Petrillo, O., Santacroce, G. and Sica, D., *J. Heterocycl. Chem.*, 1970, **7**, 555.
962. Pilgram, K., Zupan, M. and Skiles, R. D., *J. Heterocycl. Chem.*, 1970, **7**, 629.
963. Portnoy, S., *J. Heterocycl. Chem.*, 1970, **7**, 703.
964. Crohare, R., Merkuza, V. M., Gonzalez, H. A. and Ruveda, E. A., *J. Heterocycl. Chem.*, 1970, **7**, 729.
965. Cajipe, G. J. B., Landen, G., Semler, B. and Moore, A. W., *J. Org. Chem.*, 1975, **40**, 3874.
966. Mosher, W. A. and Bechara, I. S., *J. Heterocycl. Chem.*, 1970, **7**, 843.
967. Long, R. A., Gerster, J. F. and Townsend, L. B., *J. Heterocycl. Chem.*, 1970, **7**, 863.
968. Gray, E. J., Stevens, H. N. E. and Stevens, M. F. G., *J. Chem. Soc., Perkin Trans., 1*, 1978, 885.
969. Baig, G. U. and Stevens, M. F. G., *J. Chem. Soc., Perkin Trans., 1*, 1981, 1424.
970. Smirnov-Zamkov, I. V., Zborovskii, Y. L. and Staninets, V. I., *J. Org. Chem. USSR*, 1979, **15**, 1602.
971. Fedenko, V.S., Avramenko, V. I., Khmel, M. P. and Solomko, Z. F., *J. Org. Chem. USSR*, 1979, **15**, 1495.
972. Potts, K. T. and Surapaneni, C. R., *J. Heterocycl. Chem.*, 1970, **7**, 1019.
973. Heindel, N. D. and Ko, C. C. H., *J. Heterocycl. Chem.*, 1970, **7**, 1007.
974. Mosher, W. A. and Innes, J. E., *J. Heterocycl. Chem.*, 1970, **7**, 1083.
975. Garcia, E. E., Arfaei, A. and Fryer, R. I., *J. Heterocycl. Chem.*, 1970, **7**, 1161.
976. Blazevic, N. and Kajfez, F., *J. Heterocycl. Chem.*, 1970, **7**, 1173.
977. Temple, C., Laseter, A. G., Rose, J. D. and Montgomery, J. A., *J. Heterocycl. Chem.*, 1970, **7**, 1195.
978. Jacobs, R. L., *J. Heterocycl. Chem.*, 1970, **7**, 1337.
979. Maeba, I. and Castle, R. N., *J. Heterocycl. Chem.*, 1979, **16**, 1213.
980. Chiodini, L., Garanti, L. and Zecchi, G., *Synthesis*, 1978, 603.
981. Wurm, G. and Loth, H., *Arch. Pharm. (Weinheim)*, 1970, **303**, 413.
982. Coppola, G. M. and Dodsworth, R. W., *Synthesis*, 1981, 523.
983. Hirai, K., Sugimoto, H. and Ishiba, T., *J. Org. Chem.*, 1980, **45**, 253.
984. Press, J. B., Eudy, N. H. and Safir, S. R., *J. Org. Chem.*, 1980, **45**, 497.
985. Langdon, S. P., Simmonds, R. J. and Stevens, M. F. G., *J. Chem. Soc., Perkin Trans., 1*, 1984, 993.
986. Heine, H. W., Williard, P. G. and Hoye, T. R., *J. Org. Chem.*, 1972, **37**, 2980.
987. Hershenson, F. M., *J. Org. Chem.*, 1972, **37**, 3111.
988. Mosher, W. A. and Lipp, D. W., *J. Org. Chem.*, 1972, **37**, 3190.
989. Jones, C. D., *J. Org. Chem.*, 1972, **37**, 3624.
990. Walker, G. N., Engle, A. R. and Kempton, R. J., *J. Org. Chem.*, 1972, **37**, 3755.
991. Rizkalla, B. H. and Broom, A. D., *J. Org. Chem.*, 1972, **37**, 3981.
992. Taylor, E. C. and Yoneda, F., *J. Org. Chem.*, 1972, **37**, 4464.
993. Rossy, P. A., Hoffmann, W. and Muller, N., *J. Org. Chem.*, 1980, **45**, 617.

994. Lalezari, I. and Stein, C. A., *J. Heterocycl. Chem.*, 1984, **21**, 5.
995. Chantegrel, B., Nadi, A. I. and Gelin, S., *J. Heterocycl. Chem.*, 1984, **21**, 13.
996. Godard, A. and Queguiner, G., *J. Heterocycl. Chem.*, 1984, **21**, 27.
997. Eras, J., Galvez, C. and Garcia, F., *J. Heterocycl. Chem.*, 1984, **21**, 215.
998. Nakazumi, H., Ueyama, T. and Kitao, T., *J. Heterocycl. Chem.*, 1984, **21**, 193.
999. Robba, M. and Lancelot, J.-C., *J. Heterocycl. Chem.*, 1984, **21**, 91.
1000. Slouka, J., Pec, P. and Urbanova, J., *Acta Univ. Palackianae Olomucensis Fac. Rerum Nat.*, 1972, **37**, 481.
1001. Schmidt, H.-W. and Junek, H., *Monatsh. Chem.*, 1978, **109**, 1075.
1002. Sauter, F., Stanetty, P., Schrom, E. and Sengstschmid, G., *Monatsh. Chem.*, 1978, **109**, 53.
1003. Schafer, H. and Seifert, M., *Monatsh. Chem.*, 1978, **109**, 527.
1004. Lamm, B. and Aurell, C.-J., *Acta Chem. Scand.*, 1982, **36B**, 435.
1005. Iddon, B., Suschitzky, H. and Taylor, D. S., *J. Chem. Soc., Perkin Trans.*, 1, 1974, 579.
1006. Lancelot, J.-C., Rault, S. and Robba, M., *Chem. Pharm. Bull.*, 1984, **32**, 452.
1007. Baig, G. U., Stevens, M. F. G. and Vaughan, K., *J. Chem. Soc., Perkin Trans.*, 1, 1984, 999.
1008. Moody, C. J., Rees, C. W. and Tsoi, S. C., *J. Chem. Soc., Perkin Trans.*, 1, 1984, 915.
1009. Mornet, R., Leonard, N. J., Theiler, J. B. and Doree, M., *J. Chem. Soc., Perkin Trans.*, 1, 1984, 879.
1010. Rao, U. and Balasubramanian, K. K., *Tetrahedron Lett.*, 1983, **24**, 5023.
1011. Moody, C. J., *J. Chem. Soc., Perkin Trans.*, 1, 1984, 1333.
1012. Schneider, R. and Buege, A., *Pharmazie*, 1984, 22.
1013. Osuka, A., Uno, Y., Horiuchi, H, and Suzuki, H. *Synthesis*, 1984, 145.
1014. Coltman, S. C. W., Eyley, S. C. and Raphael, R. A., *Synthesis*, 1984, 150.
1015. Hollins, R. A. and Pinto, A. C., *J. Heterocycl. Chem.*, 1978, **15**, 711.
1016. Deshayes, C., Chabannet, M. and Gelin, S., *J. Heterocycl. Chem.*, 1984, **21**, 301.
1017. Senga, K., Sato, J., Kanamori, Y., Ichiba, M. and Nishigaki, S., *J. Heterocycl. Chem.*, 1978, **15**, 781.
1018. Caluwe, P., *Tetrahedron*, 1980, **36**, 2359.
1019. Lepage, L. and Lepage, Y., *J. Heterocycl. Chem.*, 1978, **15**, 793.
1020. Gelin, S. and Hartmann, D., *J. Heterocycl. Chem.*, 1978, **15**, 813.
1021. Bazile, Y., De Cointet, P. and Pigerol, C., *J. Heterocycl. Chem.*, 1978, **15**, 859.
1022. Da Rocha, J. F. *J. Heterocycl. Chem.*, 1978, **15**, 913.
1023. Madronero, R. and Vega, S., *J. Heterocycl. Chem.*, 1978, **15**, 1127.
1024. Barker, J. M., Huddleston, P. R. and Holmes, D., *J. Chem. Res. (S)*, 1985, 214.
1025. De Sevricourt, M. C. and Robba, M. *J. Heterocycl. Chem.*, 1978, **15**, 977.
1026. Maury, G., Paugam, J.-P. and Paugam, R., *J. Heterocycl. Chem.*, 1978, **15**, 1041.
1027. Giardina, D., Ballini, R., Ferappi, M. and Casini, G., *J. Heterocycl. Chem.*, 1978, **15**, 993.
1028. Corral, C., Lissavetzky, J. and Quintanilla, G., *J. Heterocycl. Chem.*, 1978, **15**, 1137.
1029. Kuhla, D. and Watson, H. A. *J. Heterocycl. Chem.*, 1978, **15**, 1149.
1030. Kocevar, M., Stanovnik, B. and Tisler, M., *J. Heterocycl. Chem.*, 1978, **15**, 1175.
1031. Daniel, J. K. and Peet, N. P. *J. Heterocycl. Chem.*, 1978, **15**, 1309.
1032. Robba, M., De Sevricourt, M. C. and Godard, A.-M., *J. Heterocycl. Chem.*, 1978, **15**, 1387.
1033. Khan, M. A. and Morley, M. L. DeM., *J. Heterocycl. Chem.*, 1978, **15**, 1399.
1034. Schneller, S. W. and Hosmane, R. S., *J. Heterocycl. Chem.*, 1978, **15**, 1505.
1035. Iijima, I. and Rice, K. C., *J. Heterocycl. Chem.*, 1978, **15**, 1527.
1036. Shafiee, A. and Behnam, E., *J. Heterocycl. Chem.*, 1978, **15**, 1459.
1037. Brewster, K., Harrison, J. M. and Inch, T. D., *J. Heterocycl. Chem.*, 1978, **15**, 1497.
1038. Shafiee, A. and Mazloumi, A., *J. Heterocycl. Chem.*, 1978, **15**, 1455.

1039. Cohen, V. I., *J. Heterocycl. Chem.*, 1978, **15**, 1415.
1040. Ellis, G. P. and Romney-Alexander, T. M., *J. Chem. Res. (S)*, 1984, 350.
1041. Kokel, B., Menichi, G. and Hubert-Habart, M., *Tetrahedron Lett.*, 1984, **25**, 3837.
1042. Antonini, I., Cristalli, G., Franchetti, P., Grifantini, M., Martelli, S., Lapidi, G. and Riva, F., *J. Med. Chem.*, 1984, **27**, 274.
1043. Setescak, L. L., Dekow, F. W., Kitzen, J. M. and Martin, L. L., *J. Med. Chem.*, 1984, **27**, 401.
1044. Dillard, R. D., Pavey, D. E. and Bensley, D. N., *J. Med. Chem.*, 1973, **16**, 251.
1045. Bonsignore, L., Loy, G., Secci, M. and Cabiddu, S., *Synthesis*, 1984, 266.
1046. Kametani, T., Sota, K. and Shio, M., *J. Heterocycl. Chem.*, 1970, **7**, 815.
1047. Henrie, R. N., Lazarus, R. A. and Benkovic, S. J., *J. Med. Chem.*, 1983, **26**, 559.
1048. Gupta, R. R., Gautam, R. K. and Kumar, R., *Heterocycles*, 1984, **22**, 1143.
1049. Ito, K. and Maruyama, J., *Heterocycles*, 1984, **22**, 1057.
1050. Gupta, R. R. and Kumar, R., *Heterocycles*, 1984, **22**, 1169.
1051. Murakami, Y., Yokoyama, Y., Miura, T., Hirasawa, H., Kamimura, Y. and Izaki, M., *Heterocycles*, 1984, **22**, 1211.
1052. Legrand, L. and Lozac'h, N., *Bull. Soc. Chim. Fr.*, 1983, **II**, 226.
1053. Still, I. W. J. and Sayeed, V. A., *Synth. Commun.*, 1983, **13**, 1181.
1054. Khan, M.S. and LaMontagne, M. P. *J. Med. Chem.*, 1979, **22**, 1105.
1055. Itoh, T., Imini, T., Ogura, H., Kawahara, N., Nakajima, T. and Watanabe, K. A., *Heterocycles*, 1983, **20**, 2177.
1056. Jensen, E. M. and Papadopoulos, E. P., *Heterocycles*, 1983, **20**, 2233.
1057. Dominguez, E. and Lete, E., *Heterocycles*, 1983, **20**, 1247.
1058. Mataka, S., Takahashi, K. and Tashiro, M., *Heterocycles*, 1983, **20**, 1285.
1059. Youssef, M. S. K., *Heterocycles*, 1983, **20**, 1335.
1060. Press, J. B., Hofmann, C. M., Eudy, N. H., Fanshaw, W. J., Day, I. P., Greenblatt, E. N. and Safir, S. R., *J. Med. Chem.*, 1979, **22**, 725.
1061. Chantegrel, B., Nadi, A. I. and Gelin, S., *Heterocycles*, 1983, **20**, 1801.
1062. Svetlik, J., *Heterocycles*, 1983, **20**, 1495.
1063. Matsumoto, K., Uchida, T., Sugi, T. and Kobayashi, T., *Heterocycles*, 1983, **20**, 1525.
1064. MacKensie N. E., Thomson, R. H. and Greenhalgh C. W., *J. Chem. Soc., Perkin Trans., 1*, 1980, 2923.
1065. Shawali, A. S., *Heterocycles*, 1983, **20**, 2239.
1066. Krbavcic, A., Pouse, L. and Stanovnik, B., *Heterocycles*, 1983, **20**, 2347.
1067. Ito, Y., Kobayashi, K., Seko, N. and Saegusa, T., *Bull. Chem. Soc. Jpn.*, 1984, **57**, 73.
1068. Stetinova, J., Kovac, J., Povazanec, F., Dandarova, M. and Pajchortova, A., *Collect. Czech. Chem. Commun.*, 1984, **49**, 533.
1069. Chakraborty, D. P., Mandal, A. K. and Roy, S. K., *Synthesis*, 1981, 977.
1070. Barlin, G. B., *Aust. J. Chem.*, 1983, **36**, 983.
1071. Barlin, G. B., Brown, D. J., Kadunc, Z., Petric, A., Stanovnik, B. and Tisler, M., *Aust. J. Chem.*, 1983, **36**, 1215.
1072. Barlin, G. B., *Aust. J. Chem.*, 1982, **35**, 2299.
1073. Schwender, C. F., Sunday, B. R. and Herzig, D. J., *J. Med. Chem.*, 1979, **22**, 114.
1074. Temple, D. L., Yevich, J. P., Covington, R. R., Hanning, C. A., Seidehamel, R. J., Mackey, H. K. and Bartek, M. J., *J. Med. Chem.*, 1979, **22**, 505.
1075. Kurasawa, Y., Ichikawa, M. and Takada, A., *Heterocycles*, 1983, **20**, 269.
1076. Dattolo, G., Cirrincione, G., Almerico, A. M. and Aiello, E., *Heterocycles*, 1983, **20**, 255.
1077. Lin, S.-C., Holmes, G. P., Dunn, D. L. and Skinner, C. G., *J. Med. Chem.*, 1979, **22**, 741.
1078. Sliwa, W. and Thomas, A., *Heterocycles*, 1983, **20**, 71.
1079. Unangst, P. C. and Brown, R. E., *J. Heterocycl. Chem.*, 1984, **21**, 283.
1080. Shishoo, C. J., Devani, M. B., Pathak, U. S., Ananthan, S., Bhadti, V. S., Ullas,

G. V., Jain, K. S., Rathod, I. S., Talati, D. S. and Doshi, N. H., *J. Heterocycl. Chem.*, 1984, **21**, 375.

1081. Monge, A., Palop, J. A., Goni, T., Martinez, A. and Fernandez-Alvarez, E., *J. Heterocycl. Chem.*, 1984, **21**, 381.
1082. Leysen, D. C., Haemers, A. and Bollaert, W., *J. Heterocycl. Chem.*, 1984, **21**, 401.
1083. Katagiri, N., Kato, T. and Niwa, R., *J. Heterocycl. Chem.*, 1984, **21**, 407.
1084. Badr, M. Z. A., El-Sherief, H. A., El-Naggar, G. M. and Mahgoub, S. A., *J. Heterocycl. Chem.*, 1984, **21**, 471.
1085. Parkanyi, C., Abdelhamid, A. O. and Shawali, A. S., *J. Heterocycl. Chem.*, 1984, **21**, 521.
1086. Guerrera, F., Siracusa, M. A., Tornetta, B., Bousquet, E., Agozzino, P. and Lamartina, L., *J. Heterocycl. Chem.*, 1984, **21**, 587.
1087. Stokker, G. E., *J. Heterocycl. Chem.*, 1984, **21**, 609.
1088. Kappe, T. and Mayer, C., *Synthesis*, 1981, 524.
1089. Elmoghayar, M. R. H., Ibraheim, M. K. A., Elghandour, A. H. H. and Elnagdi, M. H., *Synthesis*, 1981, 635.
1090. Nicolaides, D. N. and Gallos, J. K., *Synthesis*, 1981, 638.
1091. Baraldi, P. G., Guerneri, M., Moroder, F., Simoni, D. and Benetti, S., *Synthesis*, 1981, 727.
1092. Rao, P. P. and Srimannarayana, G., *Synthesis*, 1981, 887.
1093. Cheng, C.-C. and Yan, S.-J., *Org. React.*, 1982, **28**, 37.
1094. Gewald, K., Schafer, H. and Sattler, K., *Monatsh. Chem.*, 1979, **110**, 1189.
1095. Heinisch, G. and Kirchner, I., *Monatsh. Chem.*, 1979, **110**, 365.
1096. Elnagdi, M. H., Elfahham, H. A. E. and Elgemeie, G. E. H., *Heterocycles*, 1983, **20**, 519.
1097. Katagiri, N., Niwa, R. and Kato, T., *Heterocycles*, 1983, **20**, 597.
1098. Allah, S. O. A., Ead, H. A., Kassab, N. A. and Metwali, N. H., *Heterocycles*, 1983, **20**, 637.
1099. Capuano, L., Urhahn, G. and Willmes, A., *Chem. Ber.*, 1979, **112**, 1012.
1100. Ackroyd, J. and Scheinmann, F., *J. Chem. Soc., Chem. Commun.*, 1981, 339.
1101. Baccolini, G. and Todesco, P. E., *J. Chem. Soc., Chem. Commun.*, 1981, 563.
1102. Uchida, Y. and Kozuka, S., *Bull. Chem. Soc. Jpn*, 1982, **55**, 1183.
1103. Soni, R. P., *J. Prakt. Chem.*, 1981, **323**, 516.
1104. Martin, D., Graubaum, H., Kempter, G. and Ehrlichmann, W., *J. Prakt. Chem.*, 1981, **323**, 303.
1105. Soni, R. P., *J. Prakt. Chem.*, 1981, **323**, 853.
1106. Sato, R., Senzaki, T., Goto, T. and Saito, M., *Chem. Lett.*, 1984, 1599.
1107. Arduini, A., Bigi, F., Casiraghi, G., Casnati, G. and Sartori, G., *Synthesis*, 1981, 975.
1108. Hegedus, L. S., Allen, G. F., Bozell, J. J. and Waterman, E. L., *J. Am. Chem. Soc.*, 1978, **100**, 5800.
1109. Takada, S., and Makisumi, Y., *Chem. Pharm. Bull.*, 1984, **32**, 872.
1110. Lancelot, J.-C., Landelle, H. and Robba, M., *Chem. Pharm. Bull.*, 1984, **32**, 902.
1111. Takahashi, H., Nimura, N. and Orgura, H., *Chem. Pharm. Bull.*, 1979, **27**, 1147.
1112. Yogo, M., Hirota, K., Maki, Y. and Senda, S., *Chem. Pharm. Bull.*, 1984, **32**, 1761.
1113. Ozaki, K. Y., Yamada, Y. and Oine, T., *Chem. Pharm. Bull.*, 1984, **32**, 2160.
1114. Sakamoto, M., Akimoto, T., Fukutomi, K. and Ishii, K., *Chem. Pharm. Bull.*, 1984, **32**, 2516.
1115. Dorn, H. and Ozegowski, R., *J. Prakt. Chem.*, 1979, **321**, 881.
1116. Latif, N., Mishriky, N. and Assad, F. M., *Aust. J. Chem.*, 1982, **35**, 1037.
1117. James, F. C. and Krebs, H. D., *Aust. J. Chem.*, 1982, **35**, 385.
1118. James, F. C. and Krebs, H. D., *Aust. J. Chem.*, 1982, **35**, 393.
1119. Buchanan, J. G., Stobie, A. and Wightman, R. H., *J. Chem. Soc., Perkin Trans., 1*, 1981, 2374.
1120. Balli, H. and Felder, L., *Helv. Chim. Acta*, 1978, **61**, 108.

1121. Mataka, S. Takahashi, K., Tsuda, Y. and Tashiro, M., *Heterocycles*, 1980, **14**, 789.
1122. Yoneda, F., Higuchi, M. and Matsumoto, S., *J. Chem. Soc., Perkin Trans.*, *1*, 1977, 1754.
1123. Elnagdi, M. H., Abdel-Galil, F. M., Riad, B. Y. and Elgemeie, G. E. H., *Heterocycles*, 1983, **20**, 2437.
1124. Cerri, R., Boido, A. and Speratore, F., *J. Heterocycl. Chem.*, 1979, **16**, 1005.
1125. Watanabe, Y., Yamamoto, M., Shim, S. C., Mitsudo, T. and Takegawi, Y., *Chem. Lett.*, 1979, 1025.
1126. Tong, Y. C., *J. Heterocycl. Chem.*, 1975, **12**, 1127.
1127. Drewes, S. E. and Upfold, U. J., *J. Chem. Soc., Perkin Trans.*, *1*, 1977, 1901.
1128. Mataka, S., Takahashi, K., Tashiro, M. and Tsuda, Y., *Synthesis*, 1980, 842.
1129. Schafer, H., Sattler, K. and Gewald, K., *J. Prakt. Chem.*, 1979, **321**, 695.
1130. Brennan, J., Cadogan, J. I. G. and Sharp, J. T., *J. Chem. Soc., Perkin Trans.*, *1*, 1977, 1844.
1131. Cook, P. D., Rousseau, R. J., Mian, A. M., Dea, P., Meyer, R. B. and Robins, R. K., *J. Am. Chem. Soc.*, 1976, **98**, 1492.
1132. Ito, Y., Kobayashi, K. and Saegusa, T., *J. Am. Chem. Soc.*, 1977, **99**, 3532.
1133. Khan, M. A. and Gemal, A. L., *J. Heterocycl. Chem.*, 1978, **15**, 159.
1134. Connor, D. T. and Von Strandtmann, M., *J. Heterocycl. Chem.*, 1978, **15**, 113.
1135. Connor, D. T., Young, P. A. and Von Strandtmann, M., *J. Heterocycl. Chem.*, 1978, **15**, 115.
1136. Kluge, A. F., *J. Heterocycl. Chem.*, 1978, **15**, 119.
1137. Friary, R., *J. Heterocycl. Chem.*, 1978, **15**, 77.
1138. Schneller, S. W. and Moore, D. R., *J. Heterocycl. Chem.*, 1978, **15**, 319.
1139. Lin, Y., Fields, T. L. and Lang, S. A., *J. Heterocycl. Chem.*, 1978, **15**, 311.
1140. Ito, K., Yakushijin, K., Yoshima, S., Tanaka, A. and Yamamoto, K., *J. Heterocycl. Chem.*, 1978, **15**, 301.
1141. Senga, K., Kanamori, Y., Kanazawa, H. and Nishigaki, S., *J. Heterocycl. Chem.*, 1978, **15**, 359.
1142. Grehn, L., *J. Heterocycl. Chem.*, 1978, **15**, 81.
1143. Uchida, T., *J. Heterocycl. Chem.*, 1978, **15**, 241.
1144. Ochoa, C. and Stud, M., *J. Heterocycl. Chem.*, 1978, **15**, 221.
1145. Netchitailo, P., DeCroix, B., Morel, J. and Pastour, P., *J. Heterocycl. Chem.*, 1978, **15**, 337.
1146. Carvalho, C. F. and Sargent, M. V., *J. Chem. Soc., Perkin Trans.*, *1*, 1984, 1613.
1147. Manson, J. W. and Hodgkins, T. G., *J. Heterocycl. Chem.*, 1978, **15**, 545.
1148. Schneller, S. W. and Bartholomew, D. G., *J. Heterocycl. Chem.*, 1978, **15**, 439.
1149. Messmer, A. and Gelleri, A., *Angew. Chem., Int. Ed. Engl.*, 1967, **6**, 261.
1150. Hermecz, I., Meszaros, Z., Simon, K., Saszlo, L. and Pal, Z., *J. Chem. Soc., Perkin Trans.*, *1*, 1984, 1795.
1151. Beck, J. R., *J. Heterocycl. Chem.*, 1978, **15**, 513.
1152. Cairns, H. and Payne, A. R., *J. Heterocycl. Chem.*, 1978, **15**, 551.
1153. Clark, J. and Hitiris, G., *J. Chem. Soc., Perkin Trans.*, *1*, 1984, 2005.
1154. Clark, J. and Varvounis, G., *J. Chem. Soc., Perkin Trans.*, *1*, 1984, 1475.
1155. Vasvari-Debreczy, L., Hermecz, I. and Meszaros. Z., *J. Chem. Soc., Perkin Trans.*, *1*, 1984, 1799.
1156. Babudri, F., Florio, S., Vitrani, A. M. and DiNunno, L., *J. Chem. Soc., Perkin Trans.*, *1*, 1984, 1899.
1157. Atkinson, R. S., Malpass, J. R., Skinner, K. L. and Woodthorpe, K. L., *J. Chem. Soc., Perkin Trans.*, *1*, 1984, 1905.
1158. Clarke, K., Goulding, J. and Scrowston, R. M., *J. Chem. Soc., Perkin Trans.*, *1*, 1984, 1501.
1159. Huang, X. and Chan, C.-C., *Synthesis*, 1984, 851.
1160. Person, H., Pardo, M. D. A. and Foucaud, A., *Tetrahedron Lett.*, 1980, **21**, 281.

1161. Keir, W. F. and Wood, H. C. S., *J. Chem. Soc., Perkin Trans., 1*, 1976, 1847.
1162. Muller, J.-C., Ramuz, H., Daly, J. and Schonholzer, P., *Helv. Chim. Acta*, 1982, **65**, 1454.
1163. Hajos, G. and Messmer, A., *J. Heterocycl. Chem.*, 1978, **15**, 463.
1164. Cuadrado, F. J., Perez, M. A. and Soto, J. L., *J. Chem. Soc., Perkin Trans., 1*, 1984, 2447.
1165. Skaric, V., Skaric, D. and Cizmek, A., *J. Chem. Soc., Perkin Trans., 1*, 1984, 2221.
1166. Leiby, R. W. and Heindel, N. D., *J. Pharm. Sci.*, 1977, **66**, 605.
1167. Yamato, M., Takeuchi, Y., Hashigaki, K., Hattori, K., Muroga, E. and Hirota, T., *Chem. Pharm. Bull.*, 1983, **31**, 1733.
1168. Plattner, J. J., Fung, A. K. L., Parks, J. A., Pariza, R. J., Crowley, S. R., Pernet, A. G., Bunnell, R. P. and Dodge, P. W., *J. Med. Chem.*, 1984, **27**, 1016.
1169. Griffith, R. C., Gentile, R. J., Robichaud, R. C. and Frankenheim, J., *J. Med. Chem.*, 1984, **27**, 995.
1170. Appriou, P., Trebaul, C., Brelivet, J., Garnier, F. and Guglielmetti, R., *Bull. Soc. Chim. Fr.*, 1976, 2039.
1171. Ellis, G. P. and Wathey, W. B., *J. Chem. Res. (S)*, 1984, 384.
1172. Wood, S. G., Dalley, N. K., George, R. D., Robins, R. K. and Revankar, G. R., *J. Org. Chem.*, 1984, **49**, 3534.
1173. Sato, N. and Adachi, J., *J. Org. Chem.*, 1978, **43**, 341.
1174. Dainter, R. S., Julia, L., Suschitzky, H. and Wakefield, B. J., *J. Chem. Soc., Perkin Trans., 1*, 1982, 2897.
1175. Rosowsky, A., Huang, P. C. and Modest, E. J., *J. Heterocycl. Chem.*, 1970, **7**, 197.
1176. Bacon, R. G. R. and Hamilton, S. D., *J. Chem. Soc., Perkin Trans., 1*, 1974, 1975.
1177. Bryce, M. R., *J. Chem. Soc., Perkin Trans., 1*, 1984, 2591.
1178. Kornis, G., Marks, P. J. and Chidester, C. G., *J. Org. Chem.*, 1980, **45**, 4860.
1179. Kane, J. M., *J. Org. Chem.*, 1980, **45**, 5396.
1180. Kakehi, A., Ito, S., Watanabe, K. A., Kitagawa, M., Takeuchi, S. and Hashimoto, T., *J. Org. Chem.*, 1980, **45**, 5100.
1181. Kaufman, K. D. and Hewitt, L. E., *J. Org. Chem.*, 1980, **45**, 738.
1182. Kabbe, H. J., *Synthesis*, 1978, 886.
1183. Devlin, J. P., Bauen, A., Possanza. G. J. and Stewart, P. B., *J. Med. Chem.*, 1978, **21**, 480.
1184. Taylor, E. C., Beardsley, G. P. and Maki, Y., *J. Org. Chem.*, 1971, **36**, 3211.
1185. Komin, A. P. and Carmack, M., *J. Heterocycl. Chem.*, 1976, **13**, 13.
1186. Eichler, E., Rooney, C. S. and Williams, H. W. R., *J. Heterocycl. Chem.*, 1976, **13**, 41.
1187. Eichler, E., Rooney, C. S. and Williams, H. W. R., *J. Heterocycl. Chem.*, 1976, **13**, 43.
1188. Rivalle, C., Bisagni, E. and Lloste, J. M., *J. Heterocycl. Chem.*, 1976, **13**, 89.
1189. Da Settimo, A., Primofiore, G., Biagi, G. and Santerini, V., *J. Heterocycl. Chem.*, 1976, **13**, 97.
1190. Antonini, I., Franchetti, P., Grifantini, M. and Martelli, S., *J. Heterocycl. Chem.*, 1976, **13**, 111.
1191. Shafiee, A., Lalezari, I. and Mirrashed, M., *J. Heterocycl. Chem.*, 1976, **13**, 117.
1192. Weis, C. D., *J. Heterocycl. Chem.*, 1976, **13**, 145.
1193. Crenshaw, R. R., Luke, G. M. and Whitehead, D. F., *J. Heterocycl. Chem.*, 1976, **13**, 155.
1194. Chow, A. W., Gyurik, R. J. and Parish, R. C., *J. Heterocycl. Chem.*, 1976, **13**, 163.
1195. Lauer, R. F. and Zenchoff, G., *J. Heterocycl. Chem.*, 1976, **13**, 291.
1196. Shafiee, A., *J. Heterocycl. Chem.*, 1976, **13**, 301.
1197. Kappe, T. and Soliman, F. S. G., *J. Heterocycl. Chem.*, 1976, **13**, 377.
1198. Decormeille, A., Gugnant, F., Queguiner, G. and Pastour, P., *J. Heterocycl. Chem.*, 1976, **13**, 387.
1199. Garcia-Munoz, G., Madronero, R., Ochoa, C., Stud, M. and Pfleiderer, W., *J. Heterocycl. Chem.*, 1976, **13**, 793.

1200. Plescia, S., Aiello, E., Daidone, G. and Sprio, V., *J. Heterocycl. Chem.*, 1976, **13**, 395.
1201. Beam, C. F., Heindel, N. D., Chun, M. and Stefanski, A., *J. Heterocycl. Chem.*, 1976, **14**, 421.
1202. Bourdais, J., Abenhaim, D., Sabourault, B. and Lorre, A., *J. Heterocycl. Chem.*, 1976, **13**, 491.
1203. Wright, G. E., *J. Heterocycl. Chem.*, 1976, **13**, 539.
1204. Klein, R. S., De Las Heras, F. G., Tam, S. Y.-K., Wempen, I. and Fox, J. J., *J. Heterocycl. Chem.*, 1976, **13**, 589.
1205. Wright, G. C., *J. Heterocycl. Chem.*, 1976, **13**, 601.
1206. Manhas, M. S., Amin, S. G. and Dayal, B., *J. Heterocycl. Chem.*, 1976, **13**, 633.
1207. Aiello, E., Dattolo, G. and Plescia, S., *J. Heterocycl. Chem.*, 1976, **13**, 645.
1208. Haddadin, M. J., Bitar, H. E. and Issidorides, C. H., *J. Heterocycl. Chem.*, 1976, **13**, 323.
1209. Ogura, H., Sakaguchi, M., Okamoto, T., Gonda, K. and Koga, S., *J. Heterocycl. Chem.*, 1976, **13**, 359.
1210. Schuda, P. F. and Phillips, J. L., *J. Heterocycl. Chem.*, 1984, **21**, 669.
1211. Chioccara, F., Della Gala, F., Novellino, E. and Prota, G., *J. Heterocycl. Chem.*, 1977, **14**, 773.
1212. Molina, P., Arques, A. and Hernandez, H., *J. Heterocycl. Chem.*, 1984, **21**, 685.
1213. Essassi, E. M., Lavergne, J. P. and Villafont, P., *J. Heterocycl. Chem.*, 1976, **13**, 885.
1214. Mitchell, W. L., Hill, M. L., Newton, R. F., Ravenscroft, P. and Scopes, D. I. C., *J. Heterocycl. Chem.*, 1984, **21**, 697.
1215. Schneller, S. W., Ibay, A. C. and Christ, W. J., *J. Heterocycl. Chem.*, 1984, **21**, 791.
1216. Saxena, M. P. and Ahmed, S. R., *J. Heterocycl. Chem.*, 1977, **14**, 595.
1217. Catsoulacos, P. and Camoutsis, C., *J. Heterocycl. Chem.*, 1976, **13**, 1309.
1218. Melani, F., Cecchi, L. and Filacchioni, G., *J. Heterocycl. Chem.*, 1984, **21**, 813.
1219. Silwa, H. and Cordonnier, G., *J. Heterocycl. Chem.*, 1977, **14**, 169.
1220. Garcia-Munoz, G., Ochoa, C., Stud, M. and Pfleiderer, W., *J. Heterocycl. Chem.*, 1977, **14**, 427.
1221. McCord, T. J., DuBose, C. E., Shafer, P. L. and Davis, A. L., *J. Heterocycl. Chem.*, 1984, **21**, 643.
1222. Camparini, A., Ponticelli, F. and Tedeschi, P., *J. Heterocycl. Chem.*, 1977, **14**, 435.
1223. Manhas, M. S. and Amin, S. G., *J. Heterocycl. Chem.*, 1976, **13**, 903.
1224. Hymans, W. E., *J. Heterocycl. Chem.*, 1976, **13**, 1141.
1225. Kukla, M. J., *J. Heterocycl. Chem.*, 1977, **14**, 933.
1226. Agarwal, N. L., Bohnstengel, H. and Schafer, W., *J. Heterocycl. Chem.*, 1984, **21**, 825.
1227. Merchand, G., DeCroix, B. and Morel, J., *J. Heterocycl. Chem.*, 1984, **21**, 877.
1228. Atland, H. W. and Molander, G. A., *J. Heterocycl. Chem.*, 1977, **14**, 129.
1229. Aiello, E., Dattolo, G., Cirrincione, G. and Almerico, A. M., *J. Heterocycl. Chem.*, 1984, **21**, 7721.
1230. Peet, N. P. and Sunder, S., *J. Heterocycl. Chem.*, 1976, **13**, 967.
1231. Cohen, V. I. and Pourabass, S., *J. Heterocycl. Chem.*, 1977, **14**, 1321.
1232. Rao, K. V., *J. Heterocycl. Chem.*, 1977, **14**, 653.
1233. Gilman, N. W. and Fryer, R. I., *J. Heterocycl. Chem.*, 1977, **14**, 1171.
1234. Imafuku, K. and Shimazu, A., *J. Heterocycl. Chem.*, 1984, **21**, 653.
1235. Germain, C. and Bourdais, J., *J. Heterocycl. Chem.*, 1976, **13**, 1209.
1236. Hadjimihalakis, P. M., *J. Heterocycl. Chem.*, 1976, **13**, 1327.
1237. Prime, J., Stanovnik, B. and Tisler, M., *J. Heterocycl. Chem.*, 1976, **13**, 899.
1238. Boyer, S. K., Fitchett, G., Wasley, J. W. F. and Zaunius, G., *J. Heterocycl. Chem.*, 1984, **21**, 833.
1239. Lalezari, I., *J. Heterocycl. Chem.*, 1976, **13**, 1249.
1240. Tam, S. Y.-K., Hwang, J.-S., De Las Heras, F. G., Klein, R. S. and Fox, J. J., *J. Heterocycl. Chem.*, 1976, **13**, 1305.
1241. Yale, H. L., *J. Heterocycl. Chem.*, 1977, **14**, 207.

1242. Peet, N. P. and Sunder, S., *J. Heterocycl. Chem.*, 1977, **14**, 561.

1243. Fryer, R. I. and Earley, J. V., *J. Heterocycl. Chem.*, 1977, **14**, 1435.

1244. Tanaka, A., Yakushijin, K. and Yoshina, S., *J. Heterocycl. Chem.*, 1977, **14**, 975.

1245. Stavropoulos, G. and Theodoropoulos, D., *J. Heterocycl. Chem.*, 1977, **14**, 1139.

1246. Schneller, S. W. and Hosmane, R. S., *J. Heterocycl. Chem.*, 1977, **14**, 1291.

1247. De Sevricourt, M. C. and Robba, M., *J. Heterocycl. Chem.*, 1977, **14**, 777.

1248. Trimarco, P. and Lastrucci, C., *J. Heterocycl. Chem.*, 1976, **13**, 913.

1249. Singh, S. P., Parmar, S. S. and Pandey, B. R., *J. Heterocycl. Chem.*, 1977, **14**, 1093.

1250. McPherson, M. L. and Ponder, B. W., *J. Heterocycl. Chem.*, 1976, **13**, 909.

1251. Harrison, D. R., Kennewell, P. D. and Taylor, J. B., *J. Heterocycl. Chem.*, 1977, **14**, 1191.

1252. Khan, M. A. and Guarconi, A. E., *J. Heterocycl. Chem.*, 1977, **14**, 807.

1253. Kametani, T., Kigasawa, K., Hirugi, M., Wakisaka, K., Kusama, O., Sugi, H. and Kawasaki, K., *J. Heterocycl. Chem.*, 1977, **14**, 1175.

1254. Matsumoto, J., Miyomoto, T., Minamida, A., Nishimura, Y., Egawa, H. and Nishimura, H., *J. Heterocycl. Chem.*, 1984, **21**, 673.

1255. Moffett, R. B., Evenson, G. N. and Von Voigtlander, P. F., *J. Heterocycl. Chem.*, 1977, **14**, 1231.

1256. Anderson, G. L., *J. Heterocycl. Chem.*, 1985, **22**, 1469.

1257. Pilgram, K. and Pollard, G. E., *J. Heterocycl. Chem.*, 1976, **13**, 1225.

1258. Vincent, M., Remond, G. and Volland, J.-P., *J. Heterocycl. Chem.*, 1977, **14**, 493.

1259. Denzel, T. and Hohn, H., *J. Heterocycl. Chem.*, 1977, **14**, 813.

1260. Goya, P., Paez, J. A. and Pfleiderer, W., *J. Heterocycl. Chem.*, 1984, **21**, 861.

1261. Unangst, P. C. and Carethers, M. E., *J. Heterocycl. Chem.*, 1984, **21**, 709.

1262. Ogawa, K., Terada, T. and Honna, T., *Chem. Pharm. Bull.*, 1984, **32**, 930.

1263. Monson, R. S., *J. Heterocycl. Chem.*, 1976, **13**, 893.

1264. Yamaguchi, S., Yoshimoto, Y., Murai, R., Masuda, F., Yamada, M. and Kawase, Y., *J. Heterocycl. Chem.*, 1984, **21**, 737.

1265. Robba, M., Maume, D. and Lancelot, J.-C., *J. Heterocycl. Chem.*, 1977, **14**, 1365.

1266. Santilli, A. A. and Scotese, A. C., *J. Heterocycl. Chem.*, 1977, **14**, 361.

1267. Adembri, G., De Sio, F., Nesi, R. and Scotton, M., *J. Heterocycl. Chem.*, 1976, **13**, 1155.

1268. Peet, N. P. and Sunder, S., *J. Heterocycl. Chem.*, 1977, **14**, 1147.

1269. Cabiddu, S., Ciuccatosta, F., Cocco, M. T., Loi, G. and Secci, M., *J. Heterocycl. Chem.*, 1977, **14**, 123.

1270. Iwao, M. and Kuraishi, T., *J. Heterocycl. Chem.*, 1977, **14**, 993.

1271. Chu, C. K. and Bardos, T. J., *J. Heterocycl. Chem.*, 1977, **14**, 1053.

1272. Grunhaus, H., Pailer, M. and Stof, S., *J. Heterocycl. Chem.*, 1976, **13**, 1161.

1273. Walser, A. and Zenchoff, G., *J. Heterocycl. Chem.*, 1976, **13**, 907.

1274. Shaikh, Y. A., *J. Heterocycl. Chem.*, 1977, **14**, 1049.

1275. Yale, H. L. and Spitzmiller, E. R., *J. Heterocycl. Chem.*, 1977, **14**, 637.

1276. Lamm, B. and Aurell, C.-J., *Acta Chem. Scand.*, 1981, **35B**, 197.

1277. Kirk, K. L., *J. Heterocycl. Chem.*, 1976, **13**, 1253.

1278. Munshi, K. L., Kohl, H. and De Souza, N. J., *J. Heterocycl. Chem.*, 1977, **14**, 1145.

1279. Chapman, D. and Hurst, J., *J. Chem. Soc., Perkin Trans.*, 1, 1980, 2398.

1280. Dennis, N., Katritzky, A. R. and Parton, S. K., *J. Chem. Soc., Perkin Trans.*, 1, 1980, 750.

1281. Foster, H. E. and Hurst, J., *J. Chem. Soc., Perkin Trans.*, 1, 1976, 507.

1282. Molina, P., Alajarin, M. and Saez, J. R., *Synthesis*, 1984, 983.

1283. Banks, M. R., Barker, J. M. and Huddleston, P. R., *J. Chem. Res. (S)*, 1984, 27.

1284. Ali, S. M., Iqbal, J. and Ilyas, M., *J. Chem. Res. (S)*, 1984, 236.

1285. Shibata, T., Sugimura, Y., Sato, S. and Kawazoe, K., *Heterocycles*, 1985, **23**, 3069.

1286. Alvarez-Builla, J., Quintanilla, M. G., Abril, C. and Gandasegui, M. T., *J. Chem. Res. (S)*, 1984, 202.

1287. Hendriksen, J. B. and Rodriguez, C., *J. Org. Chem.*, 1983, **48**, 3344.
1288. Arya, V. P., Dave, K. G., Shenoy, S. J., Khadse, V. G. and Nayak, R. H., *Indian J. Chem.*, 1973, **11**, 744.
1289. Kawahara, N., Nakajima, T., Itoh, T. and Ogura, H., *Heterocycles*, 1984, **22**, 2217.
1290. Kakehi, A., Ito, S., Ito, M. and Yotsuya, T., *Heterocycles*, 1984, **22**, 2237.
1291. Dattolo, G., Cirrincione, G., Almerico, A. M., Presti, G. and Aiello, E., *Heterocycles*, 1984, **22**, 2269.
1292. Herdeis, C. and Dimmerling, A., *Heterocycles*, 1984, **22**, 2277.
1293. Gray, E. J., Stevens, M. F. G., Tennant, G. and Vevers, R. J. S., *J. Chem. Soc., Perkin Trans., 1*, 1976, 1496.
1294. Gray, E. J. and Stevens, M. F. G., *J. Chem. Soc., Perkin Trans., 1*, 1976, 1492.
1295. Hermecz, I., Kokosi, J., Meszaros, Z., Szasz, G., Toth, G. and Almasy, A., *Heterocycles*, 1984, **22**, 2285.
1296. Isomura, Y., Ito, N., Homma, H., Abe, T. and Kubo, K., *Chem. Pharm. Bull.*, 1983, **31**, 3168.
1297. Iddon, B., Suschitzky, H., Taylor, D. S. and Pickering, M. W., *J. Chem. Soc., Perkin Trans., 1*, 1974, 575.
1298. Cadogan, J. I. G., Done, J. N., Lunn, G. and Lim, P. K. K., *J. Chem. Soc., Perkin Trans., 1*, 1976, 1749.
1299. Flowers, W. T., Holt, G., Poulos, C. P. and Poulos, K., *J. Chem. Soc., Perkin Trans., 1*, 1976, 1757.
1300. Isomura, Y., Ito, N., Sakamoto, S., Homma, H., Abe, T. and Kubo, K., *Chem. Pharm. Bull.*, 1983, **31**, 3179.
1301. Baig, G. U. and Stevens, M. F. G., *J. Chem. Soc., Perkin Trans., 1*, 1984, 2765.
1302. Anastasis, P., Brown, P. E. and Marcus, W. Y., *J. Chem. Soc., Perkin Trans., 1*, 1984, 2815.
1303. Moody, C. J. and Ward, J. G., *J. Chem. Soc., Perkin Trans., 1*, 1984, 2895.
1304. Moody, C. J. and Ward, J. G., *J. Chem. Soc., Perkin Trans., 1*, 1984, 2903.
1305. Hammouda, H. A., El-Barbary, A. A. and Sharaf, M. A. F., *J. Heterocycl. Chem.*, 1984, **21**, 945.
1306. Obase, H., Nakamizo, N., Takai, H., Teranishi, M., Kubo, K., Shuto, K., Kasuya, Y., Shigenobu, K. and Hashikami, M., *Chem. Pharm. Bull.*, 1983, **31**, 3186.
1307. Bellemin, R. and Festal, D., *J. Heterocycl. Chem.*, 1984, **21**, 1017.
1308. Parkanyi, C., Abdelhamid, A. O., Cheng, J. C. S. and Shawali, A. S., *J. Heterocycl. Chem.*, 1984, **21**, 1029.
1309. Chan, E., Putt, S. R., Showalter, H. D. H. and Baker, D. C., *J. Org. Chem.*, 1982, **47**, 3457.
1310. Kanazawa, H., Neshigaki, S. and Senga, K., *J. Heterocycl. Chem.*, 1984, **21**, 969.
1311. Giannda, L. I., Giammona, G., Palazzo, S. and Lamartina, L., *J. Chem. Soc., Perkin Trans., 1*, 1984, 2707.
1312. Ozaki, K. Y., Yamada, Y. and Oine, T., *Chem. Pharm. Bull.*, 1983, **31**, 2234.
1313. Revankar, G. R., Gupta, P. K., Adams, A. D., Dalley, N. K., McKernan, P. A., Cook, P. D., Canonico, P. G. and Robins, R. K., *J. Med. Chem.*, 1984, **27**, 1389.
1314. Abdelhamid, A. O., Parkanyi, C. and Shawali, A. S., *J. Heterocycl. Chem.*, 1984, **21**, 1049.
1315. Bartsch, R. A. and Yang, I.-W., *J. Heterocycl. Chem.*, 1984, **21**, 1063.
1316. Lutz, W. B., McNamara, C. R., Olinger, M. R., Schmidt, D. F., Doster, D. E. and Fiedler, M. D., *J. Heterocycl. Chem.*, 1984, **21**, 1183.
1317. Chern, J.-W., Wise, D. S. and Townsend, L. B., *J. Heterocycl. Chem.*, 1984, **21**, 1245.
1318. Belgodere, E., Bossio, R., Cencioni, R., Marcaccini, S. and Pepino, R., *J. Heterocycl. Chem.*, 1984, **21**, 1241.
1319. Sunder, S. and Peet, N. P., *J. Org. Chem.*, 1977, **42**, 2551.
1320. Peet, N. P. and Sunder, S., *J. Org. Chem.*, 1975, **40**, 1909.

1321. Simonov, A. M., Koshchienko, Y. V., Suvorova, G. M., Tertov, B. A. and Malysheva, E. N., *Chem. Heterocycl. Compds*, 1976, **12**, 1151.

1322. Kuz'menko, T. A.., Simonov, A. M. and Kuz'menko, V. V., *Chem. Heterocycl. Compds*, 1976, **12**, 1370.

1323. Kochergin, P. M., Druzhinina, A. A. and Palei, R. M., *Chem. Heterocycl. Compds*, 1976, **12**, 1274.

1324. Kametani, T., Takagi, N., Kanaya, N. and Honda, T., *Heterocycles*, 1982, **19**, 535.

1325. Joshua, C. P. and Thomas, S. K., *Heterocycles*, 1982, **19**, 531.

1326. Elnagdi, M. H., Zayed, E. M. and Abdou, S. E., *Heterocycles*, 1982, **19**, 559.

1327. Abushanab, E., Bindra, A. P., Lee, D.-Y. and Goodman, L., *J. Heterocycl. Chem.*, 1975, **12**, 211.

1328. Trepanier, D. L. and Sunder, S., *J. Heterocycl. Chem.*, 1975, **12**, 321.

1329. Tamura, Y., Lee, D.-Y., Miki, Y., Hayashi, H. and Ikeda, M., *J. Heterocycl. Chem.*, 1975, **12**, 481.

1330. Walser, A., Silverman, G., Flynn, T. and Fryer, R. I., *J. Heterocycl. Chem.*, 1975, **12**, 351.

1331. Hardtmann, G. E., Koletar, G. and Pfister, O. R., *J. Heterocycl. Chem.*, 1975, **12**, 565.

1332. Ratajczyk, J. D. and Swett, L. R., *J. Heterocycl. Chem.*, 1975, **12**, 517.

1333. Bell, S. C., Gochman, C. and Wei, P. H. L., *J. Heterocycl. Chem.*, 1975, **12**, 1207.

1334. Solomko, Z. F., Fedenko, V. S., Avramenko, V. I. and Bozhanova, N. Y., *J. Org. Chem. USSR*, 1979, **15**, 120.

1335. Paterson, T. McC., Smalley, R. K. and Suschitzky, H., *Tetrahedron Lett.*, 1977, 3973.

1336. Santilli, A. A., Wanser, S. V., Kim, D. H. and Scotese, A. C., *J. Heterocycl. Chem.*, 1975, **12**, 311.

1337. Kocak, A. and Bekaroglu, O., *Helv. Chim. Acta*, 1984, **67**, 1503.

1338. Tamura, Y., Kim, J.-H. and Ikeda, M., *J. Heterocycl. Chem.*, 1975, **12**, 107.

1339. Tong, Y. C., *J. Heterocycl. Chem.*, 1975, **12**, 451.

1340. Adembri, G., De Sio, F., Nesi, R. and Scotton, M., *J. Heterocycl. Chem.*, 1975, **12**, 95.

1341. Campaigne, E. and Mais, D. E., *J. Heterocycl. Chem.*, 1975, **12**, 267.

1342. Bisagni, E., Civier, A. and Marquet, J.-P., *J. Heterocycl. Chem.*, 1975, **12**, 461.

1343. Robba, M., De Sevricourt, M. C. and Lecomte, J. M., *J. Heterocycl. Chem.*, 1975, **12**, 525.

1344. La Noce, T. and Guiliani, A. M., *J. Heterocycl. Chem.*, 1975, **12**, 551.

1345. Petyunin, P. A., Petyunin, G. P., Bondarchuk, I. I. and Kolesnikova, T. A., *Chem. Heterocycl. Compds*, 1976, **12**, 873.

1346. Steiner, G., *Liebigs Ann. Chem.*, 1978, 643.

1347. Bliesener, J.-U., *Liebigs Ann. Chem.*, 1978, 259.

1348. Walser, A., Flynn, T. and Fryer, R. I., *J. Heterocycl. Chem.*, 1975, **12**, 717.

1349. Yurugi, S., Hieda, M., Fushimi, T. and Tomimoto, M., *Chem. Pharm. Bull.*, 1971, **19**, 2354.

1350. Beck, J. R., *J. Heterocycl. Chem.*, 1975, **12**, 1037.

1351. Senga, K., Robins, R. K. and O'Brien, D. E., *J. Heterocycl. Chem.*, 1975, **12**, 1043.

1352. Plescia, S., Aiello, E. and Sprio, V., *J. Heterocycl. Chem.*, 1975, **12**, 199.

1353. DePompei, M. F. and Paudler, W. W., *J. Heterocycl. Chem.*, 1975, **12**, 861.

1354. Snyder, H. R., *J. Heterocycl. Chem.*, 1975, **12**, 1301.

1355. Anderson, R. C. and Hsiao, Y. Y., *J. Heterocycl. Chem.*, 1975, **12**, 883.

1356. Walser, A., Flynn, T. and Fryer, R. I., *J. Heterocycl. Chem.*, 1975, **12**, 737.

1357. Gal, M., Tihanyi, E. and Dvortsak, P., *Heterocycles*, 1984, **22**, 1985.

1358. Essassi, E. M., Lavergne, J. P., Viallefont, P. and Daunis, J., *J. Heterocycl. Chem.*, 1975, **12**, 661.

1359. Lavergne, J. P., Viallefont, P. and Daunis, J., *J. Heterocycl. Chem.*, 1975, **12**, 1095.

1360. Schneller, S. W. and Clough, F. W., *J. Heterocycl. Chem.*, 1975, **12**, 513.

1361. Schram, K. H., Manning, S. J. and Townsend, L. B., *J. Heterocycl. Chem.*, 1975, **12**, 1021.
1362. Crawley, L. S. and Safir, S. R., *J. Heterocycl. Chem.*, 1975, **12**, 1075.
1363. Billings, B. K., Wagner, J. A., Cook, P. D. and Castle, R. N., *J. Heterocycl. Chem.*, 1975, **12**, 1221.
1364. Schweizer, E. E., Berninger, C. J., Crouse, D. M., Davis, R. A. and Logothetis, R. S., *J. Org. Chem.*, 1969, **34**, 207.
1365. Schweizer, E. E. and Anderson, S. E., *J. Org. Chem.*, 1974, **39**, 3038.
1366. Pilgram, K. and Skiles, R. D., *J. Org. Chem.*, 1973, **38**, 1578.
1367. Daunis, J., Lopez, H. and Maury, G., *J. Org. Chem.*, 1977, **42**, 1018.
1368. Gelin, S., Chantegrel, B. and Nadi, A. I., *J. Org. Chem.*, 1983, **48**, 4078.
1369. Styles, V. L. and Morrison, R. W., *J. Org. Chem.*, 1982, **47**, 585.
1370. Polanc, S., Vercek, B., Sek, B., Stanovnik, B. and Tisler, M., *J. Org. Chem.*, 1974, **39**, 2143.
1371. Beck, J. R. and Yahner, J. A., *J. Org. Chem.*, 1974, **39**, 3440.
1372. Madding, G. D., *J. Org. Chem.*, 1972, **37**, 1853.
1373. Nardi, D., Tajana, A. and Pennini, R., *J. Heterocycl. Chem.*, 1975, **12**, 139.
1374. Gauthier, J. and Duceppe, J. S., *J. Heterocycl. Chem.*, 1984, **21**, 1081.
1375. Elliott, R. D., Temple, C. and Montgomery, J. A., *J. Org. Chem.*, 1970, **35**, 1676.
1376. Taylor, E. C. and Reiter, L. A., *J. Org. Chem.*, 1982, **47**, 528.
1377. Leadbetter, G., Fost, D. L., Ekwuribe, N. N. and Remers, W. A., *J. Org. Chem.*, 1974, **39**, 3580.
1378. Elliott, R. D., Temple, C., Frye, J. L. and Montgomery, J. A., *J. Org. Chem.*, 1971, **36**, 2818.
1379. Peet, N. P., *Synthesis*, 1984, 1065.
1380. Sandison, A. A. and Tennant, G., *J. Chem. Soc., Chem. Commun.*, 1974, 752.
1381. Potts, K. T., Datta, S. K. and Marshall, J. L., *J. Org. Chem.*, 1979, **44**, 622.
1382. Schweizer, E. E., Kim, C. S., Labaw, C. S. and Murray, W. P., *J. Chem. Soc., Chem. Commun.*, 1973, 7.
1383. Yale, H. L. and Sheehan, J. T., *J. Heterocycl. Chem.*, 1973, **10**, 143.
1384. Meth-Cohn, O., *Synthesis*, 1986, 76.
1385. Beck, J. R., *J. Org. Chem.*, 1973, **38**, 4086.
1386. Watson, A. A. and Brown, G. B., *J. Org. Chem.*, 1972, **37**, 1867.
1387. LaMattina, J. L. and Taylor, R. L., *J. Org. Chem.*, 1981, **46**, 4179.
1388. Burmistov, S. I. and Sannikova, V. M., *Chem. Heterocycl. Compds*, 1981, **17**, 606.
1389. Castellino, A. J. and Rapoport, H., *J. Org. Chem.*, 1984, **49**, 4399.
1390. Gribble, G. W., Saulmir, M. G., Sibi, M. P. and Obaza-Nutaitis, J. A., *J. Org. Chem.*, 1984, **49**, 4518.
1391. Temple, C., Elliott, R. D. and Montgomery, J. A., *J. Org. Chem.*, 1982, **47**, 761.
1392. Hercouet, A., Le Corre, M. and Le Floc'h, Y., *Synthesis*, 1982, 597.
1393. Muhmel, G., Hanke, R. and Breitmaier, E., *Synthesis*, 1982, 673.
1394. Taylor, R. T. and Cassell, R. A., *Synthesis*, 1982, 672.
1395. Chow, A. W., Jakas, D. R., Trotter, B. P., Hall, N. M. and Hoover, J. R. E., *J. Heterocycl. Chem.*, 1973, **10**, 71.
1396. Kada, R., Kovac, J., Jurasek, A. and David, L., *Collect. Czech. Chem. Commun.*, 1973, **38**, 1700.
1397. Flitsch, W. and Jones, G., *Adv. Heterocycl. Chem.*, 1984, **37**, 1.
1398. Migachev, G. I. and Danilenko, V. A., *Chem. Heterocycl. Compds*, 1982, **18**, 649.
1399. Beck, J. R., *Tetrahedron*, 1978, **34**, 2057.
1400. Wagner, K., Heitzer, H. and Oehlmann, L., *Chem. Ber.*, 1973, **106**, 640.
1401. Bestmann, H. J. and Schade, G., *Chem. Lett.*, 1983, 997.
1402. Kofman, T. P., Vasil'eva, I. V. and Pevzner, M. S., *Chem. Heterocycl. Compds*, 1977, **13**, 1129.

1403. Roubinek, F., Bydzovsky, V. and Budesinsky, Z., *Collect. Czech. Chem. Commun.*, 1984, **49**, 285.
1404. Kadin, S. B. and Lamphere, C. H., *J. Org. Chem.*, 1984, **49**, 4999.
1405. Hibino, S., Kano, S., Mochizuki, N. and Sugino, E., *J. Org. Chem.*, 1984, **49**, 5006.
1406. Watanabe, T., Nakashita, Y., Katayama, S. and Yamauchi, M., *J. Chem. Soc., Chem. Commun.*, 1977, 493.
1407. Petersen, U. and Heitzer, H., *Liebigs Ann. Chem.*, 1976, 1663.
1408. Cortes, M. J., Haddad, G. R. and Valderrama, J. A., *Heterocycles*, 1984, **22**, 1951.
1409. Roberge, G. and Brassard, P., *Synth. Commun.*, 1979, **9**, 129.
1410. Ermili, A., Balbi, A., Di Braccio, M. and Roma, G., *Farmaco, Sci. Ed.*, 1977, **32**, 713.
1411. Valenti, P., Zanelli, P. and Da Re, P., *Arch. Pharm. (Weinheim)*, 1976, **309**, 1006.
1412. Kato, T., Takeda, A. and Ueda, T., *Chem. Pharm. Bull.*, 1976, **24**, 431.
1413. Lhommet, G., Sliwa, H. and Maitte, P., *Bull. Soc. Chim. Fr.*, 1972, 1442.
1414. Bisagni, E., Marquet, J.-P. and Andre-Louisfert, J., *Bull. Soc. Chim. Fr.*, 1972, 1483.
1415. Eloy, F. and Deryckere, A., *J. Heterocycl. Chem.*, 1971, **8**, 57.
1416. Showell, G. A., *Synth. Commun.*, 1980, **10**, 241.
1417. McKillop, A. and Sayer, T. S. B., *J. Org. Chem.*, 1976, **41**, 1079.
1418. McKillop, A. and Ford, M, E., *Synthesis*, 1977, 760.
1419. McKillop, A., Henderson, A., Ray, P. S., Avendano, C. and Molinero, E., *Tetrahedron Lett.*, 1982, **23**, 3357.
1420. Bell, M. R. and Zalay, A. W., *J. Heterocycl. Chem.*, 1975, **12**, 1001.
1421. Ibrahim, N. S., Abed, N. A. and Kandal, Z. E., *Heterocycles*, 1984, **22**, 1677.
1422. Beylin, V. G. and Townsend, L. B., *Heterocycles*, 1984, **22**, 1693.
1423. Hayakawa, I. and Tanaka, Y., *Heterocycles*, 1984, **22**, 1697.
1424. Korenova, A., Krutosikova, A., Dandarova, M. and Kovac, J., *Collect. Czech. Chem. Commun.*, 1984, **49**, 1529.
1425. Nixon, N. S., Scheinmann, F. and Suschitzky, J. L., *J. Chem. Res. (S)*, 1984, 380.
1426. Wagner, H., Farkas, L., Flores, G. and Strelisky, J., *Chem. Ber.*, 1974, **107**, 1049.
1427. Bormann, D., *Liebigs Ann. Chem.*, 1974, 1391.
1428. Manecke, G. and Zerpner, D., *Chem. Ber.*, 1972, **105**, 1943.
1429. Rastogi, R. and Sharma, S., *Synthesis*, 1983, 861.
1430. Simoneau, B. and Brassard, P., *J. Chem. Soc., Perkin Trans., 1*, 1984, 1507.
1431. Hurst, J. and Wibberley, D. G., *J. Chem. Soc., (C)*, 1968, 1487.
1432. Preston, P. N. and Tennant, G., *Chem. Rev.*, 1972, **72**, 627.
1433. Maki, Y., Sako, M. and Taylor, E. C., *Tetrahedron Lett.*, 1971, 4271.
1434. Krohnke, F. and Renschling, D. B., *Chem. Ber.*, 1971, **104**, 2103.
1435. Propipcak, J. M. and Forte, P. A., *Can. J. Chem.*, 1970, **48**, 3059.
1436. Tamura, Y., Hayashi, H., Kim, J.-H. and Ikeda, M., *Chem. Pharm. Bull.*, 1979, **27**, 2521.
1437. Hafez, E. A. A., Abed, N. M., Elmoghayar, M. R. H. and El-Agamey, A. G. A., *Heterocycles*, 1984, **22**, 1821.
1438. Gupta, R. R. and Kumar, R., *Heterocycles*, 1984, **22**, 87.
1439. Clark, R. D. and Repke, D. B., *Heterocycles*, 1984, **22**, 195.
1440. Kozikowski, A. P., Ishida, H. and Chen, Y. Y., *J. Org. Chem.*, 1980, **45**, 3350.
1441. Somei, M., Inoue, S., Tokutake, S., Yamada, F. and Kaneko, C., *Chem. Pharm. Bull.*, 1981, **29**, 726.
1442. Yamamoto, M., *J. Chem. Soc., Perkin Trans., 1*, 1979, 3161.
1443. Kruse, L. I., *Heterocycles*, 1981, **16**, 1119.
1444. Kurasawa, Y., Ichikawa, M., Sakakura, A. and Takada, A., *Chem. Pharm. Bull.*, 1983, **32**, 4140.
1445. Niwa, R., Katagiri, N. and Kato, T., *Chem. Pharm. Bull.*, 1984, **32**, 4149.
1446. Somei, M. and Shoda, T., *Heterocycles*, 1982, **17**, 417.
1447. Weiss, S., Michaud, H., Prietzel, H. and Krommer, H., *Angew. Chem., Int. Ed. Engl.*, 1973, **12**, 841.

1448. Albert, A. H., Robins, R. K. and O'Brien, D. E., *J. Heterocycl. Chem.*, 1973, **10**, 885.
1449. Spence, T. W. M. and Tennant, G., *J. Chem. Soc., Perkin Trans.*, 1, 1972, 97.
1450. Kaji, K., Nagashima, H., Ohta, Y., Tabashi, K. and Oda, H., *J. Heterocycl. Chem.*, 1984, **21**, 1249.
1451. Cadogan, J. I. G., Marshall, R., Smith, D. M. and Todd, M. J., *J. Chem. Soc. (C)*, 1970, 2441.
1452. Meth-Cohn, O. and Taljaard, H. C., *Tetrahedron Lett.*, 1983, **24**, 4607.
1453. Lloyd, D. H. and Nichols, D. E., *Tetrahedron Lett.*, 1983, **24**, 4561.
1454. Wagner, H., Maurer, G., Horhammer, L. and Farkas, L., *Chem. Ber.*, 1971, **104**, 3357.
1455. Sano, T., Toda, J., Kashiwaba, N., Tsuda, Y. and Iitaka, Y., *Heterocycles*, 1981, **16**, 1151.
1456. Okabayashi, I. and Fujiwara, H., *J. Heterocycl. Chem.*, 1984, **21**, 1401.
1457. Hlavka, J. J., Bitha, P., Lin, Y. and Strohmeyer, T., *J. Heterocycl. Chem.*, 1984, **21**, 1537.
1458. Tsukamoto, G., Yoshino, K., Kohno, T., Oktaka, T., Kagaya, H. and Ito, K. *J. Med. Chem.*, 1980, **23**, 734.
1459. Klioze, S. S., Ehrgott, F. J. and Glamkowski, E. J., *J. Heterocycl. Chem.*, 1984, **21**, 1257.
1460. Monge Vega, A., Gil, M. J. and Fernandez-Alvarez, E., *J. Heterocycl. Chem.*, 1984, **21**, 1271.
1461. Kametani, T., Higy, T., Loc, C. V., Ihara, M., Koizumi, M. and Fukumoto, R., *J. Am. Chem. Soc.*, 1976, **98**, 6186.
1462. Clemence, F., Le Martet, O. and Collard, J., *J. Heterocycl. Chem.*, 1984, **21**, 1345.
1463. New, J. S. and Yevich, J. P., *J. Heterocycl. Chem.*, 1984, **21**, 1355.
1464. Leysen, D. C., Haemers, A. and Bollaert, W., *J. Heterocycl. Chem.*, 1984, **21**, 1361.
1465. Peet, N. P., *J. Heterocycl. Chem.*, 1984, **21**, 1389.
1466. Papadopoulos, E. P., *J. Heterocycl. Chem.*, 1984, **21**, 1411.
1467. Belleney, J., Vebrel, J. and Cerutti, E., *J. Heterocycl. Chem.*, 1984, **21**, 1431.
1468. Menozzi, G., Mosti, L. and Schenone, P., *J. Heterocycl. Chem.*, 1984, **21**, 1437.
1469. Menozzi, G., Mosti, L. and Schenone, P., *J. Heterocycl. Chem.*, 1984, **21**, 1441.
1470. Neelima, X., Bhat, B. and Bhaduri, A. P., *J. Heterocycl. Chem.*, 1984, **21**, 1469.
1471. Colbry, N. L., Elslager, E. F. and Werbel, L. M., *J. Heterocycl. Chem.*, 1984, **21**, 1521.
1472. Kornet, M. J., Varia, T. and Beaven, W., *J. Heterocycl. Chem.*, 1984, **21**, 1533.
1473. Su, T.-L. and Watanabe, K. A., *J. Heterocycl. Chem.*, 1984, **21**, 1543.
1474. Youssef, M. S. K., Hassan, K. M., Alta, F. M. and Abbady, M. S., *J. Heterocycl. Chem.*, 1984, **21**, 1565.
1475. McKillop, A., Bellinger, G. C. A., Preston, P. N., Davidson, A. B. and King, T. J., *Tetrahedron Lett.*, 1978, 2621.
1476. Aiello, E., Dattolo, G. and Cirrincione, G., *J. Chem. Soc., Perkin Trans.*, 1, 1981, 1.
1477. Meth-Cohn, O., Rhonati, S., Tarnowski, B. and Robinson, A., *J. Chem. Soc., Perkin Trans.*, 1, 1981, 1537.
1478. Meth-Cohn, O., Narine, B. and Tarnowski, B., *J. Chem. Soc., Perkin Trans.*, 1, 1981, 1531.
1479. Katritzky, A. R., Ballesteros, P. and Tomas, A. T., *J. Chem. Soc., Perkin Trans.*, 1, 1981, 1495.
1480. Dauzonne, D. and Royer, R., *Synthesis*, 1984, 348.
1481. Cushman, M. and Abbaspour, A., *J. Org. Chem.*, 1984, **49**, 1280.
1482. Gorlitzer, K. and Englet, E., *Arch. Pharm. (Weinheim)*, 1980, **313**, 385.
1483. Girgis, N. S., Jorgensen, A. and Pedersen, E. B., *Synthesis*, 1985, 101.
1484. Gunzenhauser, S. and Balli, H., *Helv. Chim. Acta*, 1985, **68**, 56.
1485. Aizpurua, J. M. and Palomo, C., *Bull. Soc. Chim. Fr.*, 1984, **II-142**.
1486. Kemp, D. S. and Paul, K. G., *J. Am. Chem. Soc.*, 1975, **97**, 7305.

1487. Lund, H. and Gruhn, S., *Acta Chem. Scand.*, 1966, **20**, 2637.
1488. Godard, A., Queguiner, G. and Pastour, P., *Bull. Soc. Chim. Fr.*, 1972, 1588.
1489. Lhommet, G., Sliwa, H. and Maitte, P., *Bull. Soc. Chim. Fr.*, 1972, 1435.
1490. Ghosh, C. K., *J. Heterocycl. Chem.*, 1983, **20**, 1437.
1491. Perera, R. C., Smalley, R. K. and Rogerson, L. G., *J. Chem. Soc. (C)*, 1971, 1348.
1492. Forster, D. L., Gilchrist, T. L. and Rees, C. W., *J. Chem. Soc. (C)*, 1971, 993.
1493. McKittrick, B. A., Scannell, R. T. and Stevenson, R., *J. Chem. Soc., Perkin Trans., 1*, 1982, 3017.
1494. Gallagher, P. T., Iddon, B. and Suschitzky, H., *J. Chem. Soc., Perkin Trans., 1*, 1980, 2358.
1495. Bailey, A. S., Heaton, M. W. and Murphy, J. I., *J. Chem. Soc. (C)*, 1971, 1211.
1496. Adamson, J., Forster, D. L., Gilchrist, T. L. and Rees, C. W., *J. Chem. Soc. (C)*, 1971, 981.
1497. Harnisch, H., *Liebigs Ann. Chem.*, 1972, **765**, 8.
1498. Gupta, A. K., Singh, V. K. and Pant, U. C., *Indian J. Chem.*, 1983, **22**, 1057.
1499. Couture, A. and Grandclaudon, P., *Heterocycles*, 1984, **22**, 1383.
1500. Larock, R. C. and Harrison, L. W., *J. Am. Chem. Soc.*, 1984, **106**, 4218.
1501. Campbell, C. D. and Rees, C. W., *J. Chem. Soc. (C)*, 1969, 742.
1502. Gilmore, W. F. and Clark, R. N., *J. Heterocycl. Chem.*, 1969, **6**, 809.
1503. Eger, K., *Arch. Pharm. (Weinheim)*, 1981, **314**, 176.
1504. Phadke, C. P., Kelkar, S. L. and Wadia, M. S., *Synth. Commun.*, 1984, **14**, 407.
1505. Grigg, R. and Gunaratne, H. Q. N., *J. Chem. Soc., Chem. Commun.*, 1984, 661.
1506. Taylor, E. C. and Evans, B. E., *J. Chem. Soc., Chem. Commun.*, 1971, 189.
1507. Gewald, K., Calderon, O., Schafer, H. and Hain, V., *Liebigs Ann. Chem.*, 1984, 1390.
1508. Slouka, J., Bekarek, V. and Sternberk, V., *Collect. Czech. Chem. Commun.*, 1978, **43**, 960.
1509. Katagiri, N., Niwa, R. and Kato, T., *Chem. Pharm. Bull.*, 1983, **31**, 2899.
1510. Sakamoto, T., Kondo, Y. and Yamanaka, H., *Heterocycles*, 1984, **22**, 1347.
1511. Sharanin, Yu. A. and Klokol, G. V., *J. Org. Chem. USSR*, 1983, **19**, 1582.
1512. Slouka, J., *Collect. Czech. Chem. Commun.*, 1979, **44**, 2438.
1513. Sato, R., Senzaki, T., Shikagaki, Y., Goto, T. and Saito, M., *Chem. Lett.*, 1984, 1423.
1514. Tritschler, W., Sturm, E., Kiesele, H. and Deltrozzo, E., *Chem. Ber.*, 1984, **117**, 2703.
1515. Martin, P., *Helv. Chim. Acta*, 1984, **67**, 1647.
1516. Wrubel, J. and Mayer, R., *Z. Chem.*, 1984, **24**, 254.
1517. Nalepa, K. and Slouka, J., *Pharmazie*, 1984, **39**, 504.
1518. Butler, R. N. and Johnston, S. M., *J. Chem. Soc., Perkin Trans., 1*, 1984, 2109.
1519. Moriarty, R. M., Prakash, O., Prakash, I. and Musallam, H. A., *J. Chem. Soc., Chem. Commun.*, 1984, 1342.
1520. Babin, P. and Dunogues, J., *Tetrahedron Lett.*, 1984, **25**, 4389.
1521. Slouka, J. and Bekarek, V., *Collect. Czech. Chem. Commun.*, 1984, **49**, 275.
1522. Troxler, F. and Weber, H. P., *Helv. Chim. Acta*, 1974, **57**, 2356.
1523. Solomko, Z. F., Tkachenko, V. S., Kost, A. N., Budylin, V. A. and Pikalov, V. L., *Chem. Heterocycl. Compds*, 1975, **11**, 470.
1524. Stanovnik, B., Boznar, B., Koren, B., Petricek, S. and Tisler, M., *Heterocycles*, 1985, **23**, 1.
1525. Varma, R. S. and Kabalka, G. W., *Heterocycles*, 1985, **23**, 139.
1526. Sakata, G., Makino, K. and Morimoto, K., *Heterocycles*, 1985, **23**, 143.
1527. Ames, D. E. and Opalko, A., *Synthesis*, 1983, 234.
1528. Glushkov, R. G. and Stezhko, T. V., *Chem. Heterocycl. Compds*, 1980, **16**, 853.
1529. Ames, D. E., Leung, O. T. and Singh, A. G., *Synthesis*, 1983, 52.
1530. Yoneda, F., Koga, M. and Nagamatsu, T., *Synthesis*, 1983, 75.
1531. Takahata, H., Nakano, M. and Yamazaki, T., *Synthesis*, 1983, 225.
1532. Dmowski, W., *Synthesis*, 1983, 396.
1533. Kurasawa, Y., Ichikawa, M., Sakakura, A. and Takada, A., *Synthesis*, 1983, 399.

1534. Molina, P. and Tarraga, A., *Synthesis*, 1983, 411.
1535. Molina, P., Alajarin, M. and Vilaplana, M. J., *Synthesis*, 1983, 415.
1536. Granoth, I., Segall, Y. and Kalir, A., *J. Chem. Soc., Perkin Trans., 1*, 1973, 1972.
1537. Gelin, S. and Deshayes, C., *Synthesis*, 1983, 566.
1538. Sharma, K. S., Singh, R. P. and Kumari, S., *Synthesis*, 1983, 581.
1539. Sawada, M., Furukawa, Y., Takai, Y. and Hanafusa, T., *Synthesis*, 1983, 595.
1540. Molina, P., Alajarin, M. and Benzal, R., *Synthesis*, 1983, 759.
1541. Dauzonne, D. and Royer, R., *Synthesis*, 1983, 836.
1542. Hermecz, I. and Vasvary-Debrezy, *Adv. Heterocycl. Chem.*, 1986, **39**, 281.
1543. Imai, Y., Mochizuki, A. and Kakimoto, M., *Synthesis*, 1983, 851.
1544. Kutschy, P., Imrich, J. and Bernat, J., *Synthesis*, 1983, 929.
1545. Daneshtalab, M. and Tehrani, M. H. H., *Heterocycles*, 1984, **22**, 1095.
1546. Dean, F. M. and Varma, R. S., *J. Chem. Soc., Perkin Trans., 1*, 1982, 1193.
1547. Iqbal, J., Fatma, W., Shaida, W. A. and Rahman, W., *J. Chem. Res. (S)*, 1982, 92.
1548. Muchowski, J. M. and Venuti, M. C., *J. Org. Chem.*, 1980, **45**, 4798.
1549. Narasimhan, N. S. and Mali, R. S., *Synthesis*, 1983, 957.
1550. Gschwend, H. W. and Rodriguez, H. R., *Org. React.*, 1979, **26**, 1.
1551. Kato, T. and Masuda, S., *Chem. Pharm. Bull.*, 1974, **22**, 1542.
1552. Kato, T. and Masuda, S., *Chem. Pharm. Bull.*, 1975, **23**, 2251.
1553. Somei, M., Matsubara, M. and Natsume, M., *Chem. Pharm. Bull.*, 1975, **23**, 2891.
1554. Ho, R. I. and Day, A. R., *J. Org. Chem.*, 1973, **38**, 3084.
1555. Sakamoto, T., Kondo, Y. and Yawanaka, H., *Heterocycles*, 1986, **24**, 31.
1556. Le Count, D. J., Pearce, R. J. and Marson, A. P., *Synthesis*, 1984, 349.
1557. Flitsch, W. and Lubisch, W., *Chem. Ber.*, 1984, **117**, 1424.
1558. Nardi, D., Tajana, A. and Rossi, S., *J. Heterocycl. Chem.*, 1973, **10**, 815.
1559. Yamamoto, K. and Watanabe, H., *Chem. Lett.*, 1982, 1225.
1560. Fitton, A. O., Frost, J. R., Houghton, P. G. and Suschitzky, H., *J. Chem. Soc., Perkin Trans., 1*, 1977, 1450.
1561. Slouka, J., *Collect. Czech. Chem. Commun.*, 1977, **42**, 894.
1562. Peet, N. P. and Sunder, S., *J. Heterocycl. Chem.*, 1984, **21**, 1807.
1563. Scheiner, P., Frank, L., Giusti, I., Arwin, S., Pearson, S. A., Excellent, F. and Harper, A. P., *J. Heterocycl. Chem.*, 1984, **21**, 1817.
1564. Leiby, R. W., *J. Heterocycl. Chem.*, 1984, **21**, 1825.
1565. Suzuki, H., Thiruvikraman, S. V. and Osuka, A., *Synthesis*, 1984, 616.
1566. Boger, D. L. and Panek, J. S., *Tetrahedron Lett.*, 1984, **23**, 3175.
1567. Smolders, R. R., Hanuise, J. and Voglet, N., *Bull. Soc. Chim. Belg.*, 1984, **93**, 239.
1568. Iinuma, M., Matsuura, S. and Tanaka, T., *Chem. Pharm. Bull.*, 1984, **32**, 1472.
1569. Elmoghayar, M. R. H., El-Agamey, A. G. A., Nasr, M. Y. A. S. and Sallam, M. M. M., *J. Heterocycl. Chem.*, 1984, **21**, 1885.
1570. Carabateas, P. M., Brundage, R. P., Gelotte, K. O., Gruett, M. D., Lorenz, R. R., Opalka, C. J., Singh, B., Thielking, W. H., Williams, G. L. and Lesher, G. Y., *J. Heterocycl. Chem.*, 1984, **21**, 1857.
1571. Duceppe, J. S. and Gauthier, J., *J. Heterocycl. Chem.*, 1984, **21**, 1685.
1572. Battistini, C., Scarafile, C., Foglio, M. and Franceschi, G., *Tetrahedron Lett.*, 1984, **25**, 2395.
1573. Govindachari, T. R., Chinnasamy, P., Rajeswari, S., Chandrasekaran, S., Premila, M. S., Nagarajan, K., Nagarajan, K. and Pai, B. R., *Heterocycles*, 1984, **22**, 585.
1574. Bremner, J. B., Browne, E. J. and Gunawardana, I. W. K., *Aust. J. Chem.*, 1984, **37**, 129.
1575. Ford, N. F., Browne, L. J., Campbell, T., Gemenden, C., Goldstein, R., Gude, C. and Wasley, J. W. F., *J. Med. Chem.*, 1985, **28**, 164.
1576. Sawada, M., Furukawa, Y., Takai, Y. and Hanafusa, T., *Heterocycles*, 1984, **22**, 501.
1577. Kaji, K., Nagashima, H., Ohta, Y., Nagao, S., Hirose, Y. and Oda, H., *Heterocycles*, 1984, **22**, 479.

1578. Matyus, P., Sohar, P. and Wamhoff, H., *Heterocycles*, 1984, **22**, 513.

1579. Mackay, C. and Waigh, R. D., *Heterocycles*, 1984, **22**, 687.

1580. Abramovitch, R. A. and Stowers, J. R., *Heterocycles*, 1984, **22**, 671.

1581. Kaji, K., Nagashima, H. and Oda, H., *Heterocycles*, 1984, **22**, 675.

1582. Tamura, Y., Miki, Y., Sumida, Y. and Ikeda, M., *J. Chem. Soc., Perkin Trans., 1*, 1973, 2580.

1583. Reid, A. A., Sharp, J. T., Snood, H. R. and Thorogood, P. B., *J. Chem. Soc., Perkin Trans., 1*, 1973, 2543.

1584. Loev, B., Musser, J. H., Brown, R. E., Jones, H., Kahen, R., Huang, F.-C., Khandwala, A., Sonnino-Goldman, P. and Leibowitz, M. J., *J. Med. Chem.*, 1985, **28**, 363.

1585. Miki, Y., Nishikubo, N., Nomura, Y., Kinoshita, H., Takamura, S. and Ikeda, M., *Heterocycles*, 1984, **22**, 2467.

1586. Aziz, S. J., Abd-Allah, S. O. and Ibrahim, N. S., *Heterocycles*, 1984, **22**, 2523.

1587. Flitsch, W. and Russkamp, P., *Heterocycles*, 1984, **22**, 541.

1588. Sword, I. P., *J. Chem. Soc. (C)*, 1970, 1916.

1589. Atkins, P. R., Glue, S. E. J. and Kay, I. T., *J. Chem. Soc., Perkin Trans., 1*, 1973, 2644.

1590. Bossio, R., Marcaccini, S., Parrini, V. and Pepino, R., *Heterocycles*, 1985, **23**, 391.

1591. Albert, A., *J. Chem. Soc., Perkin Trans., 1*, 1973, 2659.

1592. Volovenko, Y. M., Litenko, V. A., Deshkovskaya, E. V. and Babichev, F. S., *Chem. Heterocycl. Compds*, 1982, **18**, 796.

1593. Belgodere, E., Bossio, R., Marcaccini, S., Parti, S. and Pepino, R., *Heterocycles*, 1985, **23**, 349.

1594. Puar, M. S. and Vogt, B. R. *Tetrahedron*, 1978, **34**, 2887.

1595. Dyall, L. K., *Aust. J. Chem.*, 1979, **32**, 643.

1596. Bradbury, S., Rees, C. W. and Storr, R. C., *J. Chem. Soc., Perkin Trans., 1*, 1972, 72.

1597. Jaques, B. and Wallace, R. G., *J. Chem. Soc., Chem. Commun.*, 1972, 397.

1598. Hull, R., *J. Chem. Soc., Perkin Trans., 1*, 1973, 2911.

1599. Foster, H. E. and Hurst, J., *J. Chem. Soc., Perkin Trans., 1*, 1973, 2901.

1600. Hardy, C. R., *Adv. Heterocycl. Chem.*, 1984, **36**, 343.

1601. Suster, D. C., Feyns, L. V., Ciustea, G., Botez, G., Dobre, V., Bick, R. and Niculescu-Duvaz, I., *J. Med. Chem.*, 1974, **17**, 758.

1602. Ushirogochi, A., Tominaga, Y., Matsuda, Y. and Kobayashi, G., *Heterocycles*, 1980, **14**, 7.

1603. Munro, D. P. and Sharp, J. T., *J. Chem. Soc., Perkin Trans.*, 1, 1980, 1718.

1604. Munro, D. P. and Sharp, J. T., *J. Chem. Soc., Perkin Trans., 1*, 1984, 849.

1605. Beugelmans, R. and Roussi, G., *J. Chem. Soc., Chem. Commun.*, 1979, 950.

1606. Barton, J. W., Lapham, D. J. and Rowe, D. J., *J. Chem. Soc., Perkin Trans., 1*, 1985, 131.

1607. Johnson, M. P. and Moody, C. J., *J. Chem. Soc., Perkin Trans., 1*, 1985, 71.

1608. Taylor, E. C., Skotnicki, J. S. and Fletcher, S. R., *J. Org. Chem.*, 1985, **50**, 1005.

1609. Ghosh, P. B. and Everitt, B. J., *J. Med. Chem.*, 1974, **17**, 203.

1610. Kim, Y.-H. and Mautner, H. G., *J. Med. Chem.*, 1974, **17**, 369.

1611. Alaimo, R. J., Pelosi, S. S., Hatton, C. J. and Gray, J. E., *J. Med. Chem.*, 1974, **17**, 775.

1612. Parmar, S. S., Pandey, B. R., Dwivedi, C. and Ali, B., *J. Med. Chem.*, 1974, **17**, 1031.

1613. Tinney, F. J., Sanchez, J. P. and Nogas, J. A., *J. Med. Chem.*, 1974, **17**, 624.

1614. Styles, V. L. and Morrison, R. W., *J. Org. Chem.*, 1985, **50**, 346.

1615. Bonnett, R. and North, S. A., *Adv. Heterocycl. Chem.*, 1981, **29**, 341.

1616. Keene, B. R. T. and Tissington, P., *Adv. Heterocycl. Chem.*, 1971, **13**, 315.

1617. Davis, M., *Adv. Heterocycl. Chem.*, 1972, **14**, 43.

1618. Meth-Cohn, O. and Suschitzky, H., *Adv. Heterocycl. Chem.*, 1972, **14**, 211.

1619. Lloyd, D. H. and Cleghorn, H. P., *Adv. Heterocycl. Chem.*, 1974, **17**, 27.

1620. Kasparek, S., *Adv. Heterocycl. Chem.*, 1974, **17**, 45.

1621. Merlini, L., *Adv. Heterocycl. Chem.*, 1975, **18**, 159.
1622. Cagniant, P. and Cagniant, D., *Adv. Heterocycl. Chem.*, 1975, **18**, 337.
1623. Kobylecki, R. J. and McKillop, A., *Adv. Heterocycl. Chem.*, 1976, **19**, 215.
1624. Wentrup, C., *Helv. Chim. Acta*, 1978, **61**, 1755.
1625. Nielsen, F. E., Pedersen, E. B. and Begtrup, M., *Liebigs Ann. Chem.*, 1984, 1848.
1626. Gaston, J. L., Greer, R. J. and Grundon, M. F., *J. Chem. Res. (S)*, 1985, 135.
1627. Andresen, O. R. and Pedersen, E. B., *Liebigs Ann. Chem.*, 1982, 1012.
1628. Nielsen, K. E. and Pedersen, E. B., *Acta Chem. Scand.*, 1980, **34B**, 637.
1629. Matyus, P., Sohar, P. and Wamhoff, H., *Liebigs Ann. Chem.*, 1984, 1653.
1630. De Diego, C., Gomez, E. and Avendano, C., *Heterocycles*, 1985, **23**, 649.
1631. Molina, P., Alajarin, M. and Ciencias, M.-J., *Heterocycles*, 1985, **23**, 641.
1632. Potts, K. T. and McKeough, D., *J. Am. Chem. Soc.*, 1974, **96**, 4268.
1633. Potts, K. T. and McKeough, D., *J. Am. Chem. Soc.*, 1974, **96**, 4276.
1634. Kato, T., Chiba, T. and Tanaka, S., *Chem. Pharm. Bull.*, 1974, **22**, 744.
1635. Mackie, R. K., McKenzie, S., Reid, D. H. and Webster, R. G., *J. Chem. Soc., Perkin Trans., 1*, 1973, 657.
1636. Lapage, L. and Lapage, Y., *J. Heterocycl. Chem.*, 1978, **15**, 1185.
1637. Cotterill, W. D., Frame, C. J., Livingstone, R. and Atkinson, J. R., *J. Chem. Soc., Perkin Trans., 1*, 1972, 817.
1638. Zhmurenko, L. A., Glozman, O. M. and Zagorevskii, V. A., *Chem. Heterocycl. Compds*, 1978, **14**, 141.
1639. Velezheva, V. S., Yarosh, A. V., Kozik, T. A. and Suvorov, N. N., *Chem. Heterocycl. Compds*, 1978, **14**, 1217.
1640. Goszczynski, S. and Kucherenko, A. I., *J. Org. Chem. USSR*, 1972, **8**, 2635.
1641. Seybold, G. and Eilingsfeld, H., *Liebigs Ann. Chem.*, 1979, 1271.
1642. Gotthardt, H. and Nieberl, S., *Liebigs Ann. Chem.*, 1980, 867.
1643. Ruchardt, C. and Hassmann, V., *Liebigs Ann. Chem.*, 1980, 908.
1644. Hall, J. H., Behr, F. E. and Reed, R. L., *J. Am. Chem. Soc.*, 1972, **94**, 4952.
1645. Chauhan, M. S. and Still, I. W. J., *Can. J. Chem.*, 1975, **53**, 2880.
1646. Lynch, B. M., Khan, M. A., Sharma, S. C. and Teo, H. C., *Can. J. Chem.*, 1975, **53**, 119.
1647. Irwin, W. J. and Wibberley, D. G., *Adv. Heterocycl. Chem.*, 1969, **10**, 149.
1648. Paudler, W. W. and Sheets, R. M., *Adv. Heterocycl. Chem.*, 1983, **33**, 147.
1649. Scrowston, R. M., *Adv. Heterocycl. Chem.*, 1981, **29**, 171.
1650. Iddon, B., *Adv. Heterocycl. Chem.*, 1972, **14**, 331.
1651. Schneller, S. W., *Adv. Heterocycl. Chem.*, 1975, **18**, 59.
1652. Markert, J. and Hagen, H., *Liebigs Ann. Chem.*, 1980, 768.
1653. Bruche, L., Garanti, L. and Zecchi, G., *J. Chem. Soc., Perkin Trans., 1*, 1981, 2245.
1654. Tsuchiya, T., Kato, M. and Sashida, H., *Chem. Pharm. Bull.*, 1984, **32**, 4666.
1655. Cabello, J. A., Campelo, J. M., Garcia, A., Luna, D. and Marinas, J. M., *J. Org. Chem.*, 1984, **49**, 5195.
1656. Anderson, A. G. and Ko, R. P., *J. Org. Chem.*, 1984, **49**, 4769.
1657. Buchan, R., Fraser, M. and Lin, P. V. S. K. T., *J. Org. Chem.*, 1985, **50**, 1324.
1658. Abramenko, P. I., Ponomareva, T. K. and Priklonskikh, G. I., *Chem. Heterocycl. Compds*, 1979, **15**, 387.
1659. Kuhla, D. E. and Lombardino, J. G., *Adv. Heterocycl. Chem.*, 1977, **21**, 1.
1660. Schramm, N. I. and Konshin, M. E., *Khim. Geterotsikl. Soedin*, 1985, 114.
1661. Barker, J. M., *Adv. Heterocycl. Chem.*, 1977, **21**, 65.
1662. Hiremath, S. P. and Hooper, M., *Adv. Heterocycl. Chem.*, 1978, **22**, 123.
1663. Elguero, J., Claramunt, R. M. and Summers, A. J. H., *Adv. Heterocycl. Chem.*, 1978, **22**, 183.
1664. Crombie, L., Redshaw, S. D., Slack, D. A. and Whiting, D. A., *J. Chem. Soc., Perkin Trans., 1*, 1983, 1411.
1665. Crombie, L., Jones, R. C. F. and Palmer, C. J., *Tetrahedron Lett.*, 1985, **26**, 2933.

1666. Pelter, A., Ward, R. S. and Ashdown, D. H. J., *Synthesis*, 1978, 843.
1667. Zamet, J.-J., Haddadin, M. J. and Issidorides, C. H., *J. Chem. Soc., Perkin Trans., 1*, 1974, 1687.
1668. Tarzia, G. and Panzone, G., *Gazz. Chim. Ital.*, 1978, **108**, 591.
1669. Cheeseman, G. W. H. and Werstiuk, E. S. G., *Adv. Heterocycl. Chem.*, 1978, **22**, 367.
1670. Swinbourne, F. J., Hunt, J. H. and Klinkert, G., *Adv. Heterocycl. Chem.*, 1978, **23**, 103.
1671. Armarego, W. L. F., *Adv. Heterocycl. Chem.*, 1979, **24**, 1.
1672. Meth-Cohn, O. and Tarnowski, B., *Adv. Heterocycl. Chem.*, 1980, **26**, 115.
1673. Kappe, T. and Stadlbauer, W., *Adv. Heterocycl. Chem.*, 1981, **28**, 127.
1674. Smalley, R. K., *Adv. Heterocycl. Chem.*, 1981, **29**, 1.
1675. Lojka, J., Novacek, L. and Vacek, J., *Chem. Prum.*, 1983, **33**, 435.
1676. Meth-Cohn, O. and Tarnowski, B., *Adv. Heterocycl. Chem.*, 1982, **31**, 207.
1677. Di Nunno, L., Florio, S. and Todesco, P. E., *J. Chem. Soc., Perkin Trans., 1*, 1973, 1954.
1678. Gotthardt, H. and Hoffmann, N., *Liebigs Ann. Chem.*, 1985, 529.
1679. Karrick, G. L. and Peet, N. P., *J. Heterocycl. Chem.*, 1986, **23**, 1055.
1680. Baronnet, R., Callendret, R., Blanchard, L., Foussard-Blanpin, O. and Bretaudon, J., *Eur. J. Med. Chem.*, 1983, **18**, 241.
1681. Madhav, R., *Org. Prep. Proced. Int.*, 1982, **14**, 403.
1682. Sahu, J. K., Dehuri, S. N., Naik, S. and Nayak, A., *Indian J. Chem.*, 1984, **23B**, 117.
1683. Stanovnik, B. and Tisler, M., *Synthesis*, 1974, 120.
1684. Troschutz, R. and Klein, H. J., *Arch. Pharm. (Weinheim)*, 1985, **318**, 87.
1685. Lemke, T. L. and Parker, G. R., *J. Heterocycl. Chem.*, 1980, **17**, 1519.
1686. Molina, P., Fresneda, P. M. and Lajara, M. C., *J. Heterocycl. Chem.*, 1985, **22**, 113.
1687. Buckle, D. R., *J. Heterocycl. Chem.*, 1985, **22**, 77.
1688. Ghosh, C. K. and Tewari, N., *J. Org. Chem.*, 1980, **45**, 1964.
1689. Ghosh, C. K., Sinharoy, D. K. and Mukhopadhyay, K. K., *J. Chem. Soc., Perkin Trans., 1*, 1979, 1964.
1690. Boulton, A. J. and Ghosh, P. B., *Adv. Heterocycl. Chem.*, 1969, **10**, 1.
1691. Friedrichsen, W., *Adv. Heterocycl. Chem.*, 1980, **26**, 135.
1692. Jones, G., *Adv. Heterocycl. Chem.*, 1982, **31**, 1.
1693. Hermecz, I. and Meszaros, Z., *Adv. Heterocycl. Chem.*, 1983, **33**, 241.
1694. Jones, G. and Sliskovic, D. R., *Adv. Heterocycl. Chem.*, 1983, **34**, 79.
1695. Joule, J. A., *Adv. Heterocycl. Chem.*, 1984, **35**, 83.
1696. Faull, A. W., Griffiths, D., Hull, R. and Seden, T. P., *J. Chem. Soc., Perkin Trans., 1*, 1980, 2587.
1697. Adger, B. M., Bradbury, S., Keating, M., Rees, C. W., Storr, R. C. and Williams, M. T., *J. Chem. Soc., Perkin Trans., 1*, 1975, 31.
1698. Rüfenacht, K., Kristinson, H. and Mattern, G., *Helv. Chim. Acta*, 1976, **59**, 1593.
1699. Kreysig, D., Kempter, G. and Stroh, H.-H., *Z. Chem.*, 1969, **9**, 230.
1700. McKittrick, B. A. and Stevenson, R., *J. Chem. Soc., Perkin Trans., 1*, 1984, 709.
1701. Niess, R. and Eilingsfeld, H., *Liebigs Ann. Chem.*, 1974, 2019.
1702. Pfleiderer, W., *Liebigs Ann. Chem.*, 1974, 2030.
1703. Grohe, K. and Heitzer, H., *Liebigs Ann. Chem.*, 1974, 2066.
1704. Deady, L. W. and Stanborough, M. S., *Aust. J. Chem.*, 1981, **34**, 1295.
1705. Takeno, H. and Hashimoto, M., *J. Chem. Soc., Chem. Commun.*, 1981, 474.
1706. Booth, B. L. and Proenca, M. F., *J. Chem. Soc., Chem. Commun.*, 1981, 788.
1707. Couture, A. and Grandclaudon, P., *Synthesis*, 1985, 533.
1708. Sakamoto, T., Kondo, Y. and Yamanaka, H., *Chem. Pharm. Bull.*, 1985, **33**, 626.
1709. Hashiyama, T., Watanabe, A., Inoue, H., Kondo, M., Takeda, M., Murata, S. and Nagao, T., *Chem. Pharm. Bull.*, 1985, **33**, 634.
1710. Eiden, F. and Dobinsky, H., *Liebigs Ann. Chem.*, 1974, 1981.
1711. Klutchko, S., Shavel, J. and Strandtmann, M. Von, *J. Org. Chem.*, 1974, **39**, 2436.

1712. Narwid, T. A. and Marks, A., *J. Org. Chem.*, 1974, **39**, 2572.
1713. El-Sheikh, M. I., Marks, A. and Biehl, E. R., *J. Org. Chem.*, 1981, **46**, 3256.
1714. El-Telbany, F. and Hutchins, R. O., *J. Heterocycl. Chem.*, 1985, **22**, 401.
1715. Sard, H., Meltzer, P. C. and Razdan, R. K., *J. Heterocycl. Chem.*, 1985, **22**, 257.
1716. Byers, M., Forrester, A. R., John, I. L. and Thomson, R. H., *J. Chem. Soc., Perkin Trans., 1*, 1981, 1092.
1717. Nacci, V., Garofalo, A. and Fiorini, I., *J. Heterocycl. Chem.*, 1985, **22**, 259.
1718. Kanazawa, H., Senga, K. and Tamura, Z., *Chem. Pharm. Bull.*, 1985, **33**, 618.
1719. Cheeseman, G. W. H., Hawi, A. A. and Varvounis, G., *J. Heterocycl. Chem.*, 1985, **22**, 423.
1720. Oliver, J. E. and Brown, R. T., *J. Org. Chem.*, 1974, **39**, 2228.
1721. Tominaga, Y., Sakai, S., Kohra, S., Tsuka, J., Matsuda, Y. and Kobayashi, G., *Chem. Pharm. Bull.*, 1985, **33**, 962.
1722. Kaji, K., Nagashima, H., Hirose, Y. and Oda, H., *Chem. Pharm. Bull.*, 1985, **33**, 982.
1723. Fahmy, S. M. and Mohareb, R. M., *Liebigs Ann. Chem.*, 1985, 1492.
1724. Zbiral, E., *Synthesis*, 1974, 775.
1725. Acheson, R. M. and Elmore, N. F., *Adv. Heterocycl. Chem.*, 1978, **23**, 263.
1726. Gasco, A., *Adv. Heterocycl. Chem.*, 1981, **29**, 251.
1727. Barnish, I. T. and Hauser, C. R., *J. Org. Chem.*, 1968, **33**, 1372.
1728. Gammill, R. B., *J. Org. Chem.*, 1981, **46**, 3340.
1729. Agasimundin, Y. S., Oakes, F. T. and Leonard, N. J., *J. Org. Chem.*, 1985, **50**, 2474.
1730. Robertson, D. W., Beedle, E. E., Krushinski, J. H., Pollock, G. D., Wilson, H. R., Wyss, V. L. and Hayes, J. S., *J. Med. Chem.*, 1985, **28**, 717.
1731. Victory, P., Nomen, R., Colomina, O., Garriga, M. and Crespo, A., *Heterocycles*, 1985, **23**, 1135.
1732. Rodios, N. A., *J. Chem. Soc., Perkin Trans., 1*, 1985, 1167.
1733. Huang, J. J., *J. Org. Chem*, 1985, **50**, 2293.
1734. Carrington, D. E. L., Clarke, K. and Scrowston, R. M., *J. Chem. Soc., Perkin Trans., 1*, 1971, 3903.
1735. Korbonits, D., Kiss, P., Simon, K. and Kolonits, P., *Chem. Ber.*, 1984, **117**, 3183.
1736. Chantegrel, B. and Gelin, S., *Synthesis*, 1985, 548.
1737. Sharanin, Yu. A., Shcherbina, L. N., Sharanina, L. G. and Puzanova, V. V., *J. Org. Chem. USSR*, 1983, **19**, 150.
1738. Lamotte, G., Demerseman, P. and Royer, R., *Synthesis*, 1984, 1068.
1739. Seela, F. and Richter, R., *Chem. Ber.*, 1978, **111**, 2925.
1740. Flament, I., Sonnay, P. and Ohloff, G., *Helv. Chim. Acta*, 1977, **60**, 1872.
1741. Kazimierczuk, Z. and Pfleiderer, W., *Liebigs Ann. Chem.*, 1982, 754.
1742. Reddy, K. V., Mogilaiah, K. and Sreenivasulu, B., *Indian J. Chem.*, 1984, **23B**, 866.
1743. Goerdeler, J., Hohage, H. and Zeid, I., *Chem. Ber.*, 1976, **109**, 3108.
1744. Gogoi, M., Bhuyan, P., Sandhu, J. S. and Baruah, J. N., *J. Chem. Soc., Chem. Commun.*, 1984, 1549.
1745. Alaimo, R. J., *J. Heterocycl. Chem.*, 1973, **10**, 769.
1746. Abramovitch, R. A. and Takaya, T., *J. Org. Chem.*, 1972, **37**, 2022.
1747. Grohe, K., *Synthesis*, 1975, 645.
1748. Sianesi, E., Redcelli, R., Magistretti, M. J. and Massarani, E., *J. Med. Chem.*, 1973, **16**, 1133.
1749. Krapcho, J. and Turk, C. F., *J. Med. Chem.*, 1973, **16**, 776.
1750. Taylor, E. C., Harrington, P. J., Fletcher, S. R., Beardsley, G. P. and Moran, R. G., *J. Med. Chem.*, 1985, **28**, 914.
1751. Wamhoff, H. and Ertas, M., *Synthesis*, 1985, 190.
1752. Semenov, A. A. and Tolstikhina, V. V., *Chem. Heterocycl. Compds*, 1984, **20**, 345.
1753. Shawali, A. S. and Parkanyi, C., *J. Heterocycl. Chem.*, 1980, **17**, 833.
1754. Ried, W. and Erle, H.-E., *Chem. Ber.*, 1982, **115**, 475.
1755. Babin, P., Dunogues, J. and Petraud, M., *Tetrahedron*, 1981, **37**, 1131.

1756. Allen, G. R., *Org. React.*, 1973, **20**, 337.
1757. Liepa, A. J., *Aust. J. Chem.*, 1984, **37**, 2545.
1758. Ohishi, Y., Doi, Y. and Nakanishi, T., *Chem. Pharm. Bull.*, 1984, **32**, 4260.
1759. Thummel, R. P. and Kohli, D. K., *J. Heterocycl. Chem.*, 1977, **14**, 685.
1760. Boyd, G. V., Hewson, D. and Newberry, R. A., *J. Chem. Soc. (C)*, 1969, 935.
1761. Parr, R. W. and Reiss, J. A., *Aust. J. Chem.*, 1984, **37**, 1263.
1762. Kobe, J., Robins, R. K. and O'Brien, D. E., *J. Heterocycl. Chem.*, 1974, **11**, 199.
1763. Itaya, T. and Ogawa, K., *Chem. Pharm. Bull.*, 1985, **33**, 1906.
1764. Petrie, C. R., Cotham, H. B., McKernan, P. A., Robins, R. K. and Revankar, G. R., *J. Med. Chem.*, 1985, **28**, 1010.
1765. Abdelhamid, A. O., Hassaneen, H. M. and Shawali, A. S., *J. Heterocycl. Chem.*, 1985, **22**, 453.
1766. Allah, S. O. A., Ead, H. A., Kassab, N. A. and Zayed, M. M., *J. Heterocycl. Chem.*, 1983, **20**, 189.
1767. Peet, N. P., Sunder, S. and Barbuch, R. J., *J. Heterocycl. Chem.*, 1983, **20**, 511.
1768. Ranganathan, D., Bamezai, S. and Ramachandran, P. V., *Heterocycles*, 1985, **23**, 623.
1769. Gorlitzer, K., *Arch. Pharm. (Weinheim)*, 1974, **307**, 523.
1770. Schweizer, E. E. and Light, K. K., *J. Org. Chem.*, 1966, **31**, 870.
1771. Bernier, J.-L. and Henichart, J.-P., *J. Org. Chem.*, 1981, **46**, 4197.
1772. Brown, C. and Davidson, R. M., *Adv. Heterocycl. Chem.*, 1985, **38**, 135.
1773. Lombardino, J. G. and Kuhla, D. E., *Adv. Heterocycl. Chem.*, 1981, **28**, 73.
1774. Moody, C. J., *Adv. Heterocycl. Chem.*, 1982, **30**, 1.
1775. Davis, M., *Adv. Heterocycl. Chem.*, 1985, **38**, 106.
1776. Wamhoff, H., *Adv. Heterocycl. Chem.*, 1985, **38**, 299.
1777. Murphy, B. P. and Schultz, T. M., *J. Org. Chem.*, 1985, **50**, 2790.
1778. Taylor, E. C., Skotnicki, J. S. and Dumes, D. J., *Heterocycles*, 1985, **23**, 1703.
1779. Bevan, P. S., Ellis, G. P., Hudson, H. V., Romney-Alexander, T. M. and Williams, J. M., *J. Chem. Soc., Perkin Trans., 1*, 1986, 1643.
1780. Groziak, M. P., Chern, J. W. and Townsend, L. B., *J. Org. Chem.*, 1986, **51**, 1065.
1781. Lancelot, J.-C., Rault, S., Laduree, D. and Robba, M., *Chem. Pharm. Bull.*, 1985, **33**, 2798.
1782. Nagamatsu, T., Ukai, M., Yoneda, F. and Brown, D. J., *Chem. Pharm. Bull.*, 1985, **33**, 3113.
1783. Lancelot, J.-C., Laduree, D. and Robba, M., *Chem. Pharm. Bull.*, 1985, **33**, 3122.
1784. Oakes, F. T. and Leonard, N. J., *J. Org. Chem.*, 1985, **50**, 4986.
1785. Robins, R. K., Revankar, G. R., O'Brien, D. E., Springer, R. H., Novinson, T., Albert, A., Senga, K., Miller, J. P. and Streeter, D. G., *J. Heterocycl. Chem.*, 1985, **22**, 601.
1786. Torres, T., Eswaran, S. V. and Schafer, W., *J. Heterocycl. Chem.*, 1985, **22**, 697.
1787. Bayomi, S. M., Price, K. E. and Sowell, J. W., *J. Heterocycl. Chem.*, 1985, **22**, 729.
1788. Cheeseman, G. W. H., Eccleshall, S. A. and Thornton, T. *J. Heterocycl. Chem.*, 1985, **22**, 809.
1789. Shishoo, C. J., Devani, M. B., Ullas, G. V., Ananthan, S. and Bhadti, V. S., *J. Heterocycl. Chem.*, 1985, **22**, 825.
1790. Al-Hassan, S. S., Cameron, R., Nicholson, S. H., Robinson, D. H., Suckling, C. J. and Wood, H. C. S., *J. Chem. Soc., Perkin Trans., 1*, 1985, 2145.
1791. Cameron, R., Nicholson, S. H., Robinson, D. H., Suckling, C. J. and Wood, H. C. S., *J. Chem. Soc., Perkin Trans., 1*, 1985, 2133.
1792. Ross, J. R. and Sowell, J. W., *J. Heterocycl. Chem.*, 1985, **22**, 817.
1793. Hirota, T., Fukumota, M., Sasaki, K., Namba, T. and Hayakawa, S., *Heterocycles*, 1986, **24**, 143.
1794. El-Bayouki, K. A. M., Nielsen, F. E. and Pedersen, E. B., *J. Heterocycl. Chem.*, 1985, **22**, 853.

1795. Chioccara, F. and Novellino, E., *J. Heterocycl. Chem.*, 1985, **22**, 1021.
1796. Kornet, M. J., Beaven, W. and Varia, T., *J. Heterocycl. Chem.*, 1985, **22**, 1089.
1797. Martin, L. L., Agnew, M. N. and Setescak, L. L., *J. Heterocycl. Chem.*, 1985, **22**, 1105.
1798. Walia, J. S., Walia, A. S., Lankin, D. C., Patterson, R. C. and Singh, J., *J. Heterocycl. Chem.*, 1985, **22**, 1117.
1799. Schafer, H., Bartho, B. and Gewald, K., *Z. Chem.*, 1973, **13**, 294.
1800. Sayanna, E., Venkataratnam, R. V. and Thyagarajan, G., *Heterocycles*, 1985, **23**, 2183.
1801. Tominaga, Y. and Matsuda, Y., *J. Heterocycl. Chem.*, 1985, **22**, 937.
1802. Bridson, P. K., Davis, R. A. and Renner, L. S., *J. Heterocycl. Chem.*, 1985, **22**, 753.
1803. Bellassoued-Fergeau, M.-C., Graffe, B., Sacquet, M.-C. and Maitte, P., *J. Heterocycl. Chem.*, 1985, **22**, 713.
1804. Ellames, G. J., Newington, I. M. and Stobie, A., *J. Chem. Soc., Perkin Trans., 1*, 1985, 2087.
1805. Link, P. A. J., Van Der Plas, H. C. and Muller, V., *J. Heterocycl. Chem.*, 1985, **22**, 873.
1806. Rodriguez, J. G. and Canoira, L., *J. Heterocycl. Chem.*, 1985, **22**, 883.
1807. Cascaval, A., *Synthesis*, 1985, 428.
1808. Ram, S., Wise, D. S. and Townsend, L. B., *Org. Prep. Proced. Int.*, 1985, **17**, 215.
1809. Ahluwalia, V. K., Singh, D. and Singh, R. P., *Monatsh. Chem.*, 1985, **116**, 869.
1810. Gomez-Parra, V., Sanchez, F. and Torres, T., *Monatsh. Chem.*, 1985, **116**, 639.
1811. Gakkar, H. K. and Gill, J. K., *Monatsh. Chem.*, 1985, **116**, 633.
1812. Khan, M. A. and Rolim, A. M. C., *Monatsh. Chem.*, 1983, **114**, 1079.
1813. Sokolova, E. A., Maretina, I. A. and Petrov, A. A., *J. Org. Chem. USSR*, 1984, **20**, 1641.
1814. Jacobson, K. A., Kirk, K. L., Padgett, W. L. and Daly, J. W., *J. Med. Chem.*, 1985, **28**, 1334.
1815. Sircar, I., *J. Heterocycl. Chem.*, 1985, **22**, 1045.
1816. Effland, R. C. and Davis, L., *J. Heterocycl. Chem.*, 1985, **22**, 1071.
1817. Zborovskii, Y. L., Smirnov-Zamkov, I. V. and Staninets, V. I., *J. Org. Chem. USSR*, 1984, **20**, 1617.
1818. Traven, N. I. and Safonova, T. S., *Chem. Heterocycl. Compds*, 1985, **21**, 176.
1819. Abed, N. M., Hafez, E. A. A., Ibrahim, N. S. and Mustafa, M. E., *Monatsh. Chem.*, 1985, **116**, 223.
1820. Wilde, H., Hauptmann, S., Kanitz, A., Franzheld, M. and Mann, G., *J. Prakt. Chem.*, 1985, **327**, 297.
1821. Martin, D., Tittelbach, F. and Wenzel, A., *J. Prakt. Chem.*, 1984, **326**, 159.
1822. Eiden, F. and Rademacher, G., *Arch. Pharm. (Weinheim)*, 1985, **318**, 926.
1823. Stegman, H. B., Klotz, D. and Weiss, J. E., *Chem. Ber.*, 1985, **118**, 4632.
1824. Westling, M. and Livinghouse, T., *Tetrahedron Lett.*, 1985, **26**, 5389.
1825. Takahashi, K., Suenobu, K., Ogura, K. and Iida, H., *Chem. Lett.*, 1985, 1487.
1826. Cherkasov, R. A., Kutyrev, G. A. and Pudovik, A. N., *Tetrahedron*, 1985, **41**, 2567.
1827. El-Bayouki, K. A. M., Nielsen, F. E. and Pedersen, E. B., *Synthesis*, 1985, 104.
1828. Capuano, L., Hell, W. and Wamprecht, C., *Liebigs Ann. Chem.*, 1986, 132.
1829. Al-Hassan, S. S., Cameron, R., Curran, A. W. C., Lyall, W. J. S., Nicholson, S. H., Robinson, D. H., Stuart, A., Suckling, C. J., Stirling, I. and Wood, H. C. S., *J. Chem. Soc., Perkin Trans., 1*, 1985, 1645.
1830. Jones, G., Monat, D. J. and Tonkinson, D. J., *J. Chem. Soc., Perkin Trans., 1*, 1985, 2719.
1831. Caldirola, P., Di Amici, M., De Micheli, C. and Pevarello, P., *Heterocycles*, 1985, **23**, 2479.
1832. Seela, F. and Steker, H., *Heterocycles*, 1985, **23**, 2521.

1833. Tabakovic, I., Tabakovic, K., Grujic, R., Trinajstic, N. and Meid, Z., *Heterocycles*, 1985, **23**, 2539.

1834. Duffy, T. D. and Wibberley, D. G., *J. Chem. Soc., Perkin Trans., 1*, 1974, 1921.

1835. Traynor, J. R. and Wibberley, D. G., *J. Chem. Soc., Perkin Trans., 1*, 1974, 1781.

1836. Traynor, J. R. and Wibberley, D. G., *J. Chem. Soc., Perkin Trans., 1*, 1974, 1786.

1837. Wood, H. C. S., Wigglesworth, R., Yeowell, D. A., Gurney, F. W. and Hurlbert, B. S., *J. Chem. Soc., Perkin Trans., 1*, 1974, 1225.

1838. Gait, S. F., Peck, M. E., Rees, C. W. and Storr, R. C., *J. Chem. Soc., Perkin Trans., 1*, 1974, 1248.

1839. Gilchrist, T. L., Nunn, E. E. and Rees, C. W., *J. Chem. Soc., Perkin Trans., 1*, 1974, 1262.

1840. Jain, P. K., Makrandi, J. K. and Grover, S. K., *Synthesis*, 1982, 221.

1841. Clark, J. and Southon, I. W., *J. Chem. Soc., Perkin Trans., 1*, 1974, 1814.

1842. Bancroft, K. C. C., Ward, T. J. and Brown, K., *J. Chem. Soc., Perkin Trans., 1*, 1974, 1852.

1843. Alexandrou, N. E. and Nicolaides, D. N., *J. Chem. Soc. (C)*, 1969, 2319.

1844. Molina, P., Alajarin, M. and De Vega, M.-J. P., *Heterocycles*, 1985, **23**, 2613.

1845. Cirrincione, G., Dattolo, G., Almerico, A. M. and Aiello, E., *Heterocycles*, 1985, **23**, 2635.

1846. Dal Paiz, V., Ciciani, G. and Chimichi, S., *Heterocycles*, 1985, **23**, 2639.

1847. Bossio, R., Marcaccini, S., Parrini, V. and Pepino, R., *Heterocycles*, 1985, **23**, 2705.

1848. Ali, M. I., Abou-state, M. A. and Ibrahim, A. F., *J. Prakt. Chem.*, 1974, **316**, 147.

1849. El-Meligy, M. S. A. and Mohamed, S. A., *J. Prakt. Chem.*, 1974, **316**, 154.

1850. Makino, K., Sakata, G. and Morimoto, K., *Heterocycles*, 1985, **23**, 2069.

1851. Makino, R., Sakata, G., Morimoto, K. and Ochiai, Y., *Heterocycles*, 1985, **23**, 2025.

1852. Balicki, R., Hosmane, R. S. and Leonard, N. J., *J. Org. Chem.*, 1983, **48**, 3.

1853. Loriga, M. and Paglietti, G., *J. Chem. Res. (S)*, 1986, 16.

1854. Brettle, R. and Mosedale, A. J., *J. Chem. Res. (S)*, 1986, 36.

1855. Dunn, A. D., *J. Heterocycl. Chem.*, 1984, **21**, 961.

1856. Bowie, R. A., Edwards, P. N., Nicholson, S., Taylor, P. J. and Thomson, D. A., *J. Chem. Soc., Perkin Trans., 2*, 1979, 1708.

1857. Dattolo, G., Cirrincione, G. and Almerico, A. M., *Heterocycles*, 1982, **19**, 681.

1858. Daboun, H. A., Aziz, M. A. A. and Aal, F. A. A., *Heterocycles*, 1982, **19**, 677.

1859. Itaya, T. and Harada, T., *Heterocycles*, 1982, **19**, 687.

1860. Pescke, K., Vogel, C., Blaesche, J. and Kollhof, K.-H., *J. Prakt. Chem.*, 1982, **324**, 639.

1861. Molina, P., Arques, A., Cartagena, I. and Valcarel, M. V., *Heterocycles*, 1980, **14**, 1139.

1862. Matyus, P., Lowinger, L., and Wamhoff, H., *Heterocycles*, 1985, **23**, 2057.

1863. Hammad, M. A., Nawwar, G. A. M., Elgemeie, G. E. H. and Elnagdi, M. H., *Heterocycles*, 1985, **23**, 2177.

1864. Chu, D. T. W., *J. Heterocycl. Chem.*, 1985, **22**, 1033.

1865. Okafor, C. O., *Int. J. Sulfur Chem. Part B*, 1972, **7**, 121.

1866. Kreysig, D., Kempter, G. and Stroh, H.-H., *Z. Chem.*, 1969, **9**, 930.

1867. Kemp, D. S., Cox, D. D. and Paul, K. G., *J. Am. Chem. Soc.*, 1975, **97**, 7312.

1868. Furlan, B., Stanovnik, B. and Tisler, M., *Synthesis*, 1986, 78.

1869. Yoneda, F., Sakuma, Y., Ichiba, M. and Shinomura, K., *J. Am. Chem. Soc.*, 1976, **98**, 830.

1870. Sliwa, W., *Heterocycles*, 1986, **24**, 181.

1871. Basil, B., Coffee, E. C. J., Gell, D. L., Maxwell, D. R., Sheffield, D. J. and Wooldridge, K. R. H., *J. Med. Chem.*, 1970, **13**, 403.

1872. Bremner, J. B., Browne, E. J. and Gunawardana, I. W. K., *Heterocycles*, 1982, **19**, 709.

1873. Stoss, P., *Arch. Pharm. (Weinheim)*, 1977, **310**, 509.

1874. Sliskovic, D. R., Siegel, M. and Lin, Y., *Synthesis*, 1986, 71.

1875. Clerin, D., Lacroix, A. and Fleury, J.-P., *Tetrahedron Lett.*, 1976, 2899.

1876. Fleury, J.-P., *Heterocycles*, 1980, **14**, 1581.

1877. Barnish, I. T., Cross, P. E., Dickinson, R. P., Gadsby, B., Perry, M. J., Randall, M. J. and Sinclair, I. W., *J. Med. Chem.*, 1980, **23**, 117.

1878. Pfister, J. R., Wymann, W. E., Schuler, M. E. and Roszkowski, A. R., *J. Med. Chem.*, 1980, **23**, 335.

1879. Hester, J. B., Rudzik, A. D. and Von Voigtlander, P. F., *J. Med. Chem.*, 1980, **33**, 392.

1880. Van Angerer, E., Taneja, A. K., Ringshandl, R. and Schonenberger, H., *Liebigs Ann. Chem.*, 1980, 409.

1881. Lovgren, K. L., Hedberg, A. and Nilsson, J. L. G., *J. Med. Chem.*, 1980, **23**, 624.

1882. Kano, S., Ozaki, T. and Hibino, S., *Heterocycles*, 1979, **12**, 489.

1883. Truce, W. E., Kreider, E. M. and Brand, W. W., *Org. React.*, 1970, **18**, 99.

1884. Ito, Y., Kobayashi, K. and Saegusa, T., *Tetrahedron Lett.*, 1978, 2087.

1885. Lancelot, J.-C., Laduree, D. and Robba, M., *Chem. Pharm. Bull.*, 1985, **33**, 4242.

1886. Takahata, H., Nakajima, T., Nakano, M., Taniguchi, A. and Yamazaki, T., *Chem. Pharm. Bull.*, 1985, **33**, 4299.

1887. Haider, N. and Heinisch, G., *J. Chem. Soc., Perkin Trans.*, 1, 1986, 169.

1888. Sliwa, H. and Zamarlik, H., *Heterocycles*, 1979, **12**, 529.

1889. Sakamoto, T., Kondo, Y. and Yamanaka, H., *Chem. Pharm. Bull.*, 1985, **33**, 4764.

1890. Kawahara, N., Nakajima, T., Itoh, T. and Ogura, H., *Chem. Pharm. Bull.*, 1985, **33**, 4740.

1891. Lancelot, J.-C., Laduree, D., Rault, S. and Robba, M., *Chem. Pharm. Bull.*, 1985, **33**, 4769.

1892. Pangon, G., *Bull. Soc. Chim. Fr.*, 1970, 1993.

1893. Kawase, H., Yamaguchi, S., Horita, H., Takeno, J. and Kameyama, H., *Bull. Chem. Soc. Jpn*, 1982, **55**, 1153.

1894. Nakagumi, H., Asada, A. and Kitao, T., *Bull. Chem. Soc. Jpn*, 1980, **53**, 2046.

1895. Hamor, T. A. and Martin, I. L., *Prog. Med. Chem.*, 1983, **20**, 157.

1896. Murray, M., Ryan, A. J. and Little, P. J., *J. Med. Chem.*, 1982, **25**, 887.

1897. Moody, C. J., *J. Chem. Soc., Perkin Trans.*, 1, 1985, 2505.

1898. Panetta, J. A. and Rapoport, H., *J. Org. Chem.*, 1982, **47**, 2626.

1899. Molina, P., Alajarin, M. and De Vega, M.-J. P., *Synthesis*, 1986, 342.

1900. Wunsch, K. H. and Boulton, A. J., *Adv. Heterocycl. Chem.*, 1967, **8**, 277.

1901. Rousseau, R. J., May, J. E., Robins, R. K. and Townsend, L. B., *J. Heterocycl. Chem.*, 1974, **11**, 233.

1902. Tilley, J. W., Le Mahieu, R. A., Carson, M., Kierstead, R. W., Baruth, H. W. and Yaremko, B., *J. Med. Chem.*, 1980, **23**, 92.

1903. Baldwin, J. J., Engelhardt, E. L., Hirschmann, R., Ponticello, G. S., Atkinson, J. G., Wasson, B. K., Sweet, C. S. and Scriabine, A., *J. Med. Chem.*, 1980, **23**, 45.

1904. Victory, P. and Garriga, M., *Heterocycles*, 1985, **23**, 2853.

1905. Newton, R. F., Sainsbury, M. and Stanley, P. L. R., *Heterocycles*, 1982, **19**, 2037.

1906. Weber, R. W. and Cook, J. M. *Heterocycles*, 1982, **19**, 2089.

1907. Kano, S., Yokomatsu, T., Yuasa, Y. and Shibuya, S., *Heterocycles*, 1982, **19**, 2143.

1908. Suschitzky, H., *Lect. Heterocycl. Chem.*, 1980, **5**, S1.

1909. Farnum, D. G. and Yates, P., *J. Am. Chem. Soc.*, 1962, **84**, 1399.

1910. Brown, A. D., Shim, J. L. and Anderson, G. L., *J. Org. Chem.*, 1976, **41**, 1095.

1911. Jiang, J. B. and Urbanski, M. J., *Tetrahedron Lett.*, 1985, **26**, 259.

1912. Fleming, I., Loreto, M. A., Wallace, I. H. M. and Michael, J. P., *J. Chem. Soc., Perkin Trans.*, 1, 1986, 349.

1913. McKillop, A. and Young, D. W., *Synthesis*, 1979, 401.

1914. Pillai, R. K. M., Naiksatam, P., Johnson, F., Rajagopalan, R., Watts, P. C., Cricchio, R. and Borras, S., *J. Org. Chem.*, 1986, **51**, 717.

1915. Rigo, B. and Kolocouris, N., *J. Heterocycl. Chem.*, 1983, **20**, 893.

1916. Black, D. St. C., Kumar, N. and Wang, L. C. H., *Aust. J. Chem.*, 1986, **39**, 15.
1917. Popp, F. D. and Noble, A. C., *Adv. Heterocycl. Chem.*, 1967, **8**, 21.
1918. Iddon, B., Meth-Cohn, O., Scriven, E. F. V., Suschitzky, H. and Gallagher, P. T., *Angew. Chem., Int. Ed. Engl.*, 1979, **18**, 900.
1919. Meier, H. and Zeller, K.-P., *Angew. Chem., Int. Ed. Engl.*, 1975, **14**, 32.
1920. Summers, L. A., *Adv. Heterocycl. Chem.*, 1978, **22**, 1.
1921. Burton, D. E., Lambie, A. J., Law, D. W. J., Newbold, G. T. and Percival, A., *J. Chem. Soc.*, 1968. 1268.
1922. Schafer, J. P. and Bloomfield, J. J., *Org. React.*, 1967, **15**, 1.
1923. Ionescu, M. and Mantsch, H., *Adv. Heterocycl. Chem.*, 1967, **8**, 83.
1924. Cama, L. D. and Christensen, B. G., *J. Am. Chem. Soc.*, 1974, **96**, 7582.
1925. Bargagna, A., Schenone, P. and Longobardi, M., *J. Heterocycl. Chem.*, 1985, **22**, 1471.
1926. Canoira, L. and Rodriguez, J. G., *J. Heterocycl. Chem.*, 1985, **22**, 1511.
1927. Armand, J., Armand, Y., Boulares, L., Bellec, C. and Pinson, J., *J. Heterocycl. Chem.*, 1985, **22**, 1519.
1928. Kosik, D., Kristian, P., Gonda, J. and Dandarova, E., *Collect. Czech. Chem. Commun.*, 1983, **48**, 3315.
1929. Wolfe, S., Ducep, J.-B., Tin, K.-C. and Lee, S.-L., *Can. J. Chem.*, 1974, **52**, 3996.
1930. Sowmithran, D. and Prasad, K. J. R., *Heterocycles*, 1986, **24**, 711.
1931. Gramatica, P., Gianotti, M. P., Speranza, G. and Manitto, P., *Heterocycles*, 1986, **24**, 743.
1932. Todoriki, R., Ono, M. and Tamura, S., *Heterocycles*, 1985, **24**, 755.
1933. Hirota, T., Fujita, H., Sasaki, K., Namba, T. and Hayakawa, S., *Heterocycles*, 1986, **24**, 771.
1934. Langlois, N., Bourrel, P. and Andriamialisa, R. Z., *Heterocycles*, 1986, **24**, 777.
1935. Popp, F. D., *Adv. Heterocycl. Chem.*, 1968, **9**, 1.
1936. Tozuka, Z., Yazawa, H., Murata, M. and Takaya, T., *J. Antibiot.*, 1983, **36**, 1699.
1937. Lindley, J. M., McRobbie, I. M., Meth-Cohn, O. and Suschitzky, H., *J. Chem. Soc., Perkin Trans., 1*, 1977, 2194.
1938. Nakashima, Y., Kawashima, Y., Sato, M., Okuyama, S., Amanuma, F. and Kameyama, T., *Chem. Pharm. Bull.*, 1985, **33**, 5250.
1939. Camparini, A., Ponticelli, F. and Tedeschi, P., *J. Heterocycl. Chem.*, 1985, **22**, 1561.
1940. Barili, P. L., Biagi, G., Livi, O. and Scartoni, V., *J. Heterocycl. Chem.*, 1985, **22**, 1607.
1941. Nielsen, F. E. and Pedersen, E. B., *J. Heterocycl. Chem.*, 1985, **22**, 1693.
1942. Anderson, G. L. and Richardson, S. G., *J. Heterocycl. Chem.*, 1985, **22**, 1735.
1943. Crenshaw, M. D. and Zimmer, H., *J. Heterocycl. Chem.*, 1984, **21**, 623.
1944. Flitsch, W., Russkamp, P. and Langer, W., *Liebigs Ann. Chem.*, 1985, 1413.
1945. Akiba, K., Negishi, Y. and Inamoto, N., *Bull. Chem. Soc. Jpn*, 1984, **57**, 2188.
1946. Heinisch, L., *J. Prakt. Chem.*, 1974, **316**, 667.
1947. McKillop, A. and Young, D. W., *Synthesis*, 1979, 481.
1948. Mase, T., Arima, A., Tomioka, K., Yamada, T. and Murase, K., *J. Med. Chem.*, 1986, **29**, 386.
1949. Itoh, T., Fuji, I., Tomii, Y., Nishimura, H., Ogura, H., Mizuno, Y., Kawahara, N. and Shimamori, T., *Heterocycles*, 1986, **24**, 927.
1950. Wamhoff, H., *Chem. Ber.*, 1972, **105**, 743.
1951. Barker, J. M., Huddleston, P. R. and Holmes, D., *J. Chem. Res. (S)*, 1986, 155.
1952. Takagi, K., *Chem. Lett.*, 1986, 265.
1953. Heine, H. W., Empfield, J. R., Golobish, T. D., Williams, E. A. and Garbauskas, M. F., *J. Org. Chem.*, 1986, **51**, 829.
1954. Cohnen, E. and Mahuke, J., *Chem. Ber.*, 1972, **105**, 757.
1955. Johnson, J. L. and Pattison, I., *J. Heterocycl. Chem.*, 1986, **23**, 249.
1956. Taylor, E. C., Katz, A. H., Alvarado, S. I. and McKillop, A., *J. Org. Chem.*, 1986, **51**, 1607.

1957. Rajappa, S., Sreenivasan, R. and Khalwadekar, A., *J. Chem. Res. (S)*, 1986, 158.
1958. Peet, N. P., *J. Heterocycl. Chem.*, 1986, **23**, 193.
1959. Benjamin, L. E., Earley, J. V. and Gilman, N. W., *J. Heterocycl. Chem.*, 1986, **23**, 119.
1960. Ogura, H. and Sakaguchi, M., *Chem. Pharm. Bull.*, 1973, **21**, 2014.
1961. Reimlinger, H., *Chem. Ber.*, 1971, **104**, 2232.
1962. Nugent, R. A. and Murphy, M., *J. Heterocycl. Chem.*, 1986, **23**, 245.
1963. Flitsch, W. and Neumann, U., *Chem. Ber.*, 1971, **104**, 2170.
1964. Lim, M.-I., Klein, R. S. and Fox, J. J., *J. Org. Chem.*, 1979, **44**, 3826.
1965. Akimoto, H., Kawai, A. and Nomura, H., *Bull. Chem. Soc. Jpn*, 1985, **58**, 123.
1966. Kreher, R. P. and Pfister, J. R., *Chem. Ztg*, 1984, **108**, 275.
1967. Pfleiderer, W., Mengel, R. and Hemmerich, P., *Chem. Ber.*, 1971, **104**, 2273.
1968. Kurasawa, Y., Yamazaki, K., Tajima, S., Okamoto, Y. and Takeda, A., *J. Heterocycl. Chem.*, 1986, **23**, 281.
1969. Forti, L., Gelmi, M. L., Pocar, D. and Varallo, M., *Heterocycles*, 1986, **24**, 1401.
1970. Reimlinger, H. and Peiren, M. A., *Chem. Ber.*, 1971, **104**, 2237.
1971. Bhattacharya, B. K., *J. Heterocycl. Chem.*, 1986, **23**, 113.
1972. Balbi, A., Roma, G., Di Braccio, M. and Ermili, A., *Farmaco, Sci. Ed.*, 1978, **33**, 807.
1973. Mazzei, M., Roma, G. and Ermili, A., *Farmaco, Sci. Ed.*, 1979, **34**, 52.
1974. Sargent, M. V. and Stransky, P. O., *Adv. Heterocycl. Chem.*, 1984, **35**, 1.
1975. Reed, G. W. B., Cheng, P. T. W. and McLean, S., *Can. J. Chem.*, 1982, **60**, 419.
1976. Kuz'menko, V. V., Komissarov, V. N. and Simonov, A. M., *Chem. Heterocycl. Compds*, 1981, **17**, 1090.
1977. Ivashchenko, A. V. and Garicheva, O. N., *Chem. Heterocycl. Compds*, 1982, **18**, 429.
1978. Davis, M., Pettett, M., Scanlon, D. B. and Ferrito, V., *Aust. J. Chem.*, 1978, **31**, 1053.
1979. Davis, M., Pettett, M., Scanlon, D. B. and Ferrito, V., *Aust. J. Chem.*, 1977, **30**, 2289.
1980. Bass, R. J., *J. Chem. Soc., Chem. Commun.*, 1970, 322.
1981. Campbell, S. F., Hardstone, J. D. and Palmer, M. J., *Tetrahedron Lett.*, 1984, **25**, 4813.
1982. Akimoto, H., Kawai, A. and Nomura, H., *Bull. Chem. Soc. Jpn*, 1984, **58**, 123.
1983. Balogh, M., Hermecz, I., Naray-Szabo, G., Simon, K. and Meszaros, Z., *J. Chem. Soc., Perkin Trans., 1*, 1986, 753.
1984. Okamoto, Y. and Ueda, T., *Chem. Pharm. Bull.*, 1975, **23**, 1391.
1985. Golgolab, H., Lalezari, I. and Hosseini-Gohari, L., *J. Heterocycl. Chem.*, 1973, **10**, 387.
1986. Fryer, R. I., Coffen, D. L., Earley, J. V. and Walser, A., *J. Heterocycl. Chem.*, 1973, **10**, 473.
1987. Cook, P. D. and Castle, R. N., *J. Heterocycl. Chem.*, 1973, **10**, 807.
1988. Werbel, L. M., Elslager, E. F. and Chu, V. P., *J. Heterocycl. Chem.*, 1973, **10**, 631.
1989. Suzuki, N., *Chem. Pharm. Bull.*, 1980, **28**, 761.
1990. Israel, M. and Jones, L. C., *J. Heterocycl. Chem.*, 1973, **10**, 201.
1991. Elguero, J., Jacquier, R. and Mignonac-Mondon, S., *J. Heterocycl. Chem.*, 1973, **10**, 411.
1992. DeMilo, A. B., Oliver, J. E. and Gilardi, R. D., *J. Heterocycl. Chem.*, 1973, **10**, 231.
1993. Lombardino, J. G. and Wiseman, E. H., *J. Med. Chem.*, 1971, **14**, 973.
1994. Suzuki, N., Tanaka, Y. and Dohmori, R., *Chem. Pharm. Bull.*, 1980, **28**, 235.
1995. Abushanab, E., Lee, D.-Y. and Goodman, L., *J. Heterocycl. Chem.*, 1973, **10**, 181.
1996. Werbel, L. M., Curry, A., Elslager, E. F. and Hess, C., *J. Heterocycl. Chem.*, 1973, **10**, 363.
1997. Unangst, P. C. and Southwick, P. L., *J. Heterocycl. Chem.*, 1973, **10**, 399.
1998. Coburn, M. D., *J. Heterocycl. Chem.*, 1973, **10**, 743.
1999. Williams, R. L. and Shalaby, S. W., *J. Heterocycl. Chem.*, 1973, **10**, 891.
2000. Cushman, M., Wong, W. C. and Bacher, A., *J. Chem. Soc., Perkin Trans., 1*, 1986, 1043.

2001. Nishigaki, S., Sato, J., Shimizu, K., Furukawa, K., Senga, K. and Yoneda, F., *Chem. Pharm. Bull.*, 1980, **28**, 142.

2002. Stapler, J. T. and Bornstein, J., *J. Heterocycl. Chem.*, 1973, **10**, 983.

2003. Knabe, J., *Adv. Heterocycl. Chem.*, 1986, **40**, 105.

2004. Ames, D. E. and Ribeiro, O., *J. Chem. Soc., Perkin Trans.*, 1, 1975, 1390.

2005. Gilchrist, T. L., Gymer, G. E. and Rees, C. W., *J. Chem. Soc., Perkin Trans.*, 1, 1975, 1747.

2006. Bussy, T. and Ellis, G. P., *J. Chem. Res. (S)*, 1980, 317.

2007. Smith, A. B., Visnick, M., Haseltine, J. N. and Sprengler, P.R., *Tetrahedron*, 1986, **42**, 2957.

2008. Kozikowski, A. P., *Heterocycles*, 1981, **16**, 267.

2009. Wozniak, M. and Van Der Plas, H. C., *Heterocycles*, 1982, **19**, 363.

2010. Szabo, J., *Chem. Heterocycl. Compds*, 1979, **15**, 239.

2011. George, M. V., Khetan, S. K. and Gupta, R. K., *Adv. Heterocycl. Chem.*, 1976, **19**, 279.

2012. Somei, M., Ohuishi, H. and Shoken, Y., *Chem. Pharm. Bull.*, 1986, **34**, 677.

2013. Eiden, F. and Schunemann, J., *Arch. Pharm. (Weinheim)*, 1985, **318**, 1096.

2014. McNaught, A. D., *Adv. Heterocycl. Chem.*, 1976, **20**, 175.

2015. Varma, R. S., Kadkhayan, M. and Kabalka, G. W., *Synthesis*, 1986, 486.

2016. Slouka, J., Bekarek, V., Nalepa, K. and Lycka, A., *Collect. Czech. Chem. Commun.*, 1984, **49**, 2628.

2017. Yokoyama, M. and Imamoto, T., *Synthesis*, 1984, 797.

2018. Clemens, R. J., *Chem. Rev.*, 1986, **86**, 241.

2019. Trybulski, E. J., Fryer, R. I., Reeder, E., Vitone, S. and Todaro, L., *J. Org. Chem.*, 1986, **51**, 2191.

2020. Barker, G. and Ellis, G. P., *J. Chem. Soc. (C)*, 1970, 2230.

2021. Nagasawa, K., Kanbara, H., Matsushita, K. and Ito, K., *Tetrahedron Lett.*, 1985, **26**, 6477.

2022. Slouka, J., *Monatsh. Chem.*, 1969, **100**, 91.

2023. Albert, A., *Adv. Heterocycl. Chem.*, 1986, **39**, 117.

2024. Glazer, E. A. and Presslitz, J. E., *J. Med. Chem.*, 1982, **25**, 868.

2025. Glazer, E. A. and Presslitz, J. E., *J. Med. Chem.*, 1982, **25**, 868.

2026. Groziak, M. P., Chern, J.-W. and Townsend, L. B., *J. Org. Chem.*, 1986, **51**, 1065.

2027. Albert, A., *Adv. Heterocycl. Chem.*, 1986, **40**, 129.

General Index

Pairs of functional groups which are not in the contents list but are mentioned in the text are listed below.

633

Index of Ring Systems